# Maths**Practice**

## Edexcel GCSE
## **Mathematics** 1MA1

Foundation

Step-by-step guidance and practice

**Published by**
PG Online Limited
The Old Coach House
35 Main Road
Tolpuddle
Dorset
DT2 7EW
United Kingdom

sales@pgonline.co.uk
www.clearrevise.com
www.pgonline.co.uk
**2021**

# PREFACE

A crisp, clear and accessible guide to every maths topic, these pages are everything you need to ace the three exams in this course and beam with pride. Each topic is clearly presented in a format that is approachable and as concise and simple as possible.

Each section of the specification is indicated to help you cross-reference your work or revision. The checklist on the contents pages will help you keep track of what you have already worked through and what's left before the big day.

We have included worked example questions with answers for each objective. This is followed by a series of related questions that gently increase in their level of challenge. You can check your answers against those given at the end of the book.

# ACKNOWLEDGMENTS

Every effort has been made to trace and acknowledge ownership of copyright. The publishers will be happy to make any future amendments with copyright owners that it has not been possible to contact. The publisher would like to thank the following companies and individuals who granted permission for the use of their images in this textbook.

Design and artwork: Michael Bloys and Jessica Webb / PG Online Ltd
Photographic images: © Shutterstock
Toblerone image: © emka74 / Shutterstock.com

First edition 2021. 10 9 8 7 6 5 4 3 2 1
A catalogue entry for this book is available from the British Library
ISBN: 978-1-910523-16-2

Printed on FSC certified paper by Bell and Bain Ltd, Glasgow, UK.

# THE SCIENCE OF LEARNING AND REVISION

### Variation theory

Procedural variation is the gradual change from one question to the next, in order to subtly increase difficulty. It focuses on not only what changes, but on what does not change. This enables misconceptions and misunderstandings to be identified more easily when only one variable is altered. Seeing these differences, rather than seeing sameness, better enables pupils to reason and make connections.[1]

### Retrieval of information

Retrieval practice encourages students to come up with answers to questions.[2] The closer the question is to one you might see in a real examination, the better. Also, the closer the environment in which a student revises is to the 'examination environment', the better. Students who had a test 2–7 days away did 30% better using retrieval practice than students who simply read, or repeatedly reread material. Students who were expected to teach the content to someone else after their revision period did better still.[3] What was found to be most interesting in other studies is that students using retrieval methods and testing for revision were also more resilient to the introduction of stress.[4]

### Ebbinghaus' forgetting curve and spaced learning

Ebbinghaus' 140-year-old study examined the rate in which we forget things over time. The findings still hold true. However, the act of forgetting things and relearning them is what cements things into the brain.[5] Spacing out learning or revision is more effective than cramming – we know that, but students should also know that the space between revisiting material should vary depending on how far away the examination is. A cyclical approach is required. An examination 12 months away necessitates revisiting covered material about once a month. A test in 30 days should have topics revisited every 3 days – intervals of roughly a tenth of the time available.[6]

### Summary

Students: the more tests and past questions you do, in an environment as close to examination conditions as possible, the better you are likely to perform on the day. If you prefer to listen to music while you revise, tunes without lyrics will be far less detrimental to your memory and retention. Silence is most effective.[5] If you choose to study with friends, choose carefully – effort is contagious.[7]


1. Kullberg, Angelika, Ulla Runesson Kempe, and Ference Marton. What is made possible to learn when using the variation theory of learning in teaching mathematics?. *ZDM* (2017), 1-11.
2. Roediger III, H. L., & Karpicke, J.D. (2006). Test-enhanced learning: Taking memory tests improves long-term retention. *Psychological Science*, 17(3), 249-255.
3. Nestojko, J., Bui, D., Kornell, N. & Bjork, E. (2014). Expecting to teach enhances learning and organisation of knowledge in free recall of text passages. *Memory and Cognition*, 42(7), 1038-1048.
4. Smith, A. M., Floerke, V. A., & Thomas, A. K. (2016) Retrieval practice protects memory against acute stress. *Science*, 354(6315), 1046-1048.
5. Perham, N., & Currie, H. (2014). Does listening to preferred music improve comprehension performance? *Applied Cognitive Psychology*, 28(2), 279-284.
6. Cepeda, N. J., Vul, E., Rohrer, D., Wixted, J. T. & Pashler, H. (2008). Spacing effects in learning a temporal ridgeline of optimal retention. *Psychological Science*, 19(11), 1095-1102.
7. Busch, B. & Watson, E. (2019), *The Science of Learning*, 1st ed. Routledge.

# CONTENTS AND CHECKLIST

## Section 1 Integers



## Section 2 Primes, factors and multiples



## Section 3 Algebraic expressions



## Section 4 Decimals



## Section 5 Measures

## Section 6 Fractions



## Section 7 Straight line graphs



## Section 8 Fractions, decimals, and percentages



## Section 9 Probability



## Section 10 Ratio

## Section 11 Shapes and transformations



## Section 12 Sequences



## Section 13 Proportion



## Section 14 Data



## Section 15 Properties of shapes

## Section 16 Applications of number



## Section 17 Further graphs



## Section 18 Geometry



## Section 19 Equations and identities



## Section 20 Trigonometry

## Section 21 Statistics



## Section 22 Probability diagrams



## Section 23 Mensuration



## Section 24 Applications of Ratio



## Section 25 Further equations

# FOUNDATION TIER
## Mathematics (1MA1)

## This qualification is assessed over three examination papers.

**Specification coverage**

The content for assessment in each paper will be drawn from each of the six areas of mathematics:

1. Number
2. Algebra
3. Ratio, proportion and rates of change
4. Geometry and measures
5. Probability
6. Statistics

**Assessment**

**Three written exams: 1 hour 30 minutes**

**Each 80 marks**

All questions are mandatory

Each paper is 33.33% of the qualification grade

Calculators are permitted in Papers 2 and 3 only.

**Assessment overview**

Each paper will consist of a range of question types.

The total mark across the three equally weighted Foundation tier papers is used to form a combined GCSE grade from 1 to 5.

SECTION 1

# INTEGERS

## 1.1 INTEGERS AND PLACE VALUE

N1 N2

### Integers

**Integers** are whole numbers. They can be positive, zero or negative.

e.g. 6, −10, 0, 107 **are** integers

2.3, $\frac{1}{2}$, −2300.9 are **not** integers

### Objectives

**Understand and use place value**

The value of a digit depends on its position or **place value**.

e.g. The number 4603 can be shown as:

| thousands 1000 | hundreds 100 | tens 10 | units (ones) 1 |
|---|---|---|---|
| 4 | 6 | 0 | 3 |

**Place integers on a number line**

**Number line**

**Numbers** can be placed on a number line. Negative numbers are numbers below zero.

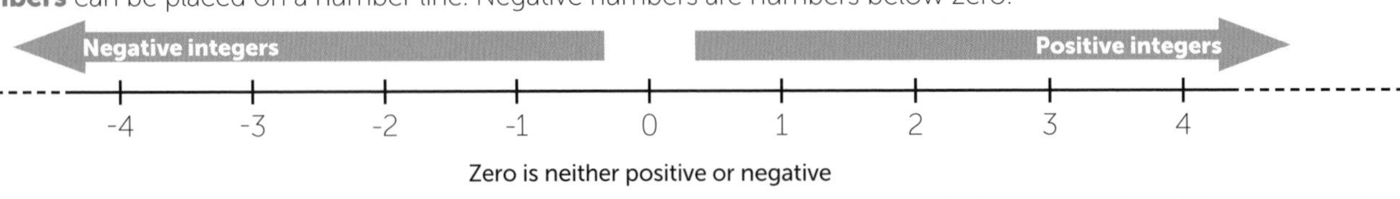


Zero is neither positive or negative

**Sort integers into ascending or descending order**

**Ascending order**

Numbers sorted in ascending order up go up in size.

e.g. −11, −3, 15, 34, 101 are sorted in ascending order.

**Descending order**

Numbers sorted in descending order go down in size.

e.g. 9, 3, 0, −5 are sorted in descending order.

### Practice questions

1. a) In the number 16 289 what is the value of the 6?
   b) Write the number 30 209 in words.
   c) Write the number two million, one thousand and thirty six as a number.

2. This is a list of numbers: 8, 3, −9, 0, −2, 1, −4, −6, 2
   a) Draw a number line with a scale from −10 to 10. Mark the position of each number on the number line.
   b) Write these numbers in descending order.
   c) All the numbers are doubled. What effect will this have on the descending order?

3. Tom has five numbered cards.

a) Write down the largest number that he can make using all the cards once only.

b) Write down the largest possible number less than 50 000, that he can make using all the cards once only.

c) Write down the smallest possible number that is greater than 20, 000 that he can make using all the cards once only.

4. Jodie has tried to write these numbers in ascending order.

She has made mistakes. Spot the mistakes and correct them.

a) 4443, 30 400, 44 104, 44 033, 400 300

b) 280 880, 28 888, 208 088, 8882, 888

5. Here is a map of the British Isles showing the temperatures on 1st December.

List the temperatures in ascending order.

6. A three-digit number has a tens digit which is two times as big as its hundreds digit. It has a zero units (ones) digit.

Write down the possible numbers it could be.

7. A three-digit number has one repeated digit. Its digits are all prime numbers. It is an even number greater than 700.

Write down the possible numbers it could be.

8. Use these clues to find the missing number.

- There is a 2 in the ten thousand and tens place.
- The units digit is 3 times the digit in the ten thousand place.
- The digit in the thousands place is 2 less than 5.
- The digit in the hundred thousand place is 2 more than half of the units (ones) digit.
- The digit in the tens place is 2 more than the digit in the hundreds place.

9. Numbers are written on four cards as shown below. Two of the cards are blank.

| 280 350 | 247 350 | 238 350 | 258 350 | | |
|---|---|---|---|---|---|

a) Choose numbers to write on the blank cards so that the six cards can be used to complete a number pattern.

b) Write the six numbers in ascending order.

10. A three-digit number has digits which are ascending consecutive numbers.

Another three-digit number has digits which are descending consecutive numbers.

The difference between the two numbers is 24.

What are the two numbers?

# 1.2 NEGATIVE INTEGERS

## Objectives

### Add and subtract positive and negative integers

**Adding and subtracting integers**

A number line can help when adding and subtracting integers.

Move up the line when adding and down the line when subtracting a positive number.

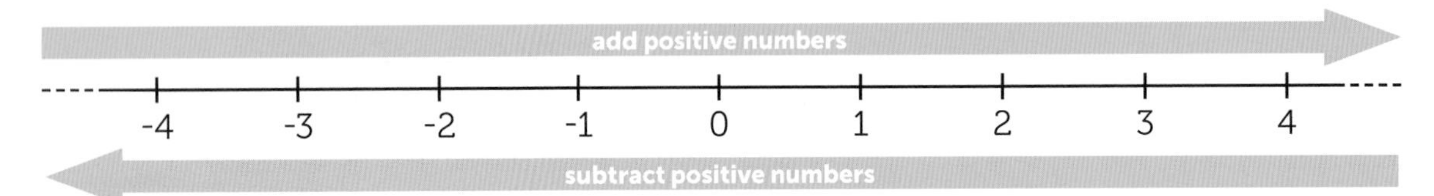


**Adding** a negative number is the same as subtracting a positive number. e.g. $-5 + (-8) = -5 - 8 = -13$

**Subtracting** a negative number is the same as adding a positive number. e.g. $-5 - (-8) = -5 + 8 = 3$

### Use a calculator to evaluate arithmetic operations involving positive and negative integers

**Calculators**

A calculator can be used to check an answer is correct. You do not have to put brackets around negative numbers.

e.g. $-124 - -213 = 89$

The table shows the temperatures in two different cities on a Monday.

Work out the difference in temperature between Athens and Oslo on that Monday.

| Athens | 7°C |
|---|---|
| Oslo | −14°C |

Difference in temperature is

$7 - (-14)$

$= 7 + 14$

$= 21°C$

## Practice questions

1. Work out the answers to these without a calculator.

   a) $13 - 20$   b) $-10 + 18$   c) $-10 - 2$   d) $8 - 20$

   e) $-6 - 4$   f) $-4 + 1$   g) $-7 + 9$   h) $-21 + 10$

2. Work out the answers to these without a calculator.

   a) $13 + (-7)$   b) $-10 + (-1)$   c) $-15 + (-2)$   d) $10 + (-20)$

   e) $-12 - (-3)$   f) $4 - 1$   g) $9 - (-7)$   h) $-31 - (-8)$

3. Work out the answers to these using a calculator.

   a) $137 + -198$   b) $-509 - -237$   c) $-2489 - -379$   d) $-3478 - +9691$

4. Place two or three of the numbers 7, −11 and −3 into each of the following, to make the calculation correct.

   a) _____ + _____ = 4   b) _____ − _____ = 18

   c) _____ − _____ = − 8   d) _____ − _____ − _____ = 1

5. Tom says that $-10 - 24 = 34$. By referring to a number line, give a reason why Tom is wrong.

   Give the correct answer to the calculation.

6. At 6pm the temperature is 2°C. By midnight, the temperature drops to −8°C.
   By how many degrees did the temperature fall?

7. Work out the missing numbers in each calculation.
   Use a calculator to check your answers.
   a) 297 + ____ = 129 b) 1032 + ______ = 1006
   c) ____+ −358 = 1008 d) ____ + − 762 = −2101

8. The table shows the surface temperatures of different planets.

| | **Minimum surface temperature (°C)** | **Maximum surface temperature (°C)** |
|---|---|---|
| Mercury | −170 | 449 |
| Mars | −125 | 20 |
| Earth | −89 | 58 |

   a) What is the difference between the minimum temperatures on Mercury and on Earth?
   b) What is the temperature difference on Mars?
   c) Which planet has the greatest temperature difference? What is it?

9. In a magic square all the rows, columns and diagonals add up to the same total. Complete the magic square.

| | | |
|---|---|---|
| 1 | | |
| −16 | −7 | |
| | | −15 |

10. In an addition pyramid each number is the sum of the two numbers beneath it.
    Copy and complete these addition pyramids.

a)

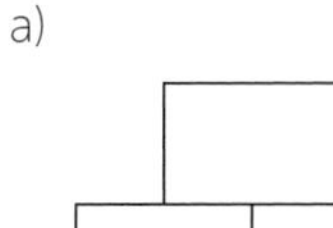

b)

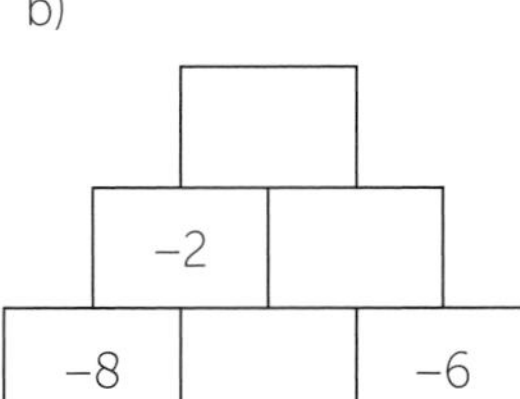


c)

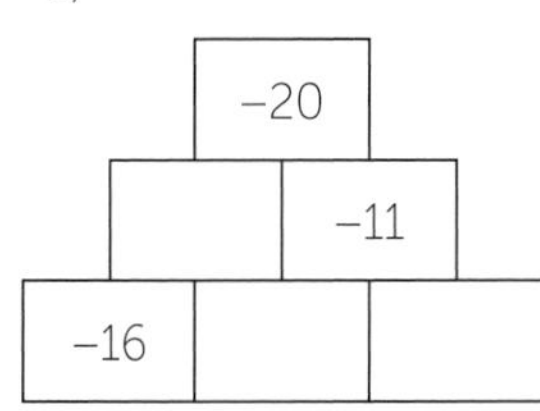

# 1.3 CALCULATING WITH NEGATIVE INTEGERS

## Objectives

### Multiply and divide positive and negative integers

You can use these rules when multiplying and dividing positive and negative integers.

| positive × positive = positive |
|---|
| positive × negative = negative |
| negative × positive = negative |
| negative × negative = positive |

| positive ÷ positive = positive |
|---|
| positive ÷ negative = negative |
| negative ÷ positive = negative |
| negative ÷ negative = positive |

**e.g.** $3 \times -6 = -18$  $-5 \times 7 = -35$

$-24 \div 6 = -4$  $-55 \div -5 = 11$

### Use the symbols =, ≠, <, >, ≤, ≥

Symbols such as: =, ≠, <, >, ≤, ≥, can be used to make number sentences true.

| < less than |
|---|
| > greater than |
| = equal to |
| ≤ less than or equal to |
| ≥ greater than or equal to |
| ≠ not equal to |

**e.g.** $-6 > -10$  $5 - 7 < 7 + -7$

$15 - 20 \neq 15 \div -5$  $4 \times -3 = -4 \times 3$

### Use a calculator to evaluate arithmetic operations involving positive and negative integers

## Practice questions

Do not use a calculator with these questions unless instructed to do so.

1. Work out these calculations.
   a) $3 \times -6$  b) $-32 \div 4$  c) $-9 \times -6$  d) $72 \div -6$
   e) $-2 \times 6$  f) $24 \div -8$  g) $-8 \times -6$  h) $-100 \div -5$

2. Decide if each of these calculations is true or false. If the statement is false, correct it.
   a) $9 \times -6 = 54$  b) $-48 \div -4 = 12$  c) $16 \div -4 = 4$  d) $-7 \times -4 = -28$

3. Use a calculator to work out each of these calculations.
   a) $59 \times -38$  b) $-238 \times -45$  c) $2960 \div -80$  d) $-1416 \div -12$

4. The temperature of a metal changes by −3°C every hour.
   What will the total change in temperature be after 6 hours?

5. Match each calculation with its answer. (12, −12, 48, −48)
   a) $6 \times -8$  b) $36 \div -3$  c) $-24 \div 2$  d) $-12 \times -4$  e) $-12 \times -1$

6. Work out these calculations and write the answers in ascending order.
   a) $12 \times -11$  b) $-36 \div 6$  c) $125 \div -5$  d) $0 \times -10$  e) $-200 \div 2$

7. Here are some number cards.

| −6 | 4 | −7 | 1 | 9 |
|---|---|---|---|---|

   a) Which two cards when multiplied give the largest possible answer?
   b) Which two cards when multiplied give the smallest possible answer?

## 1.4 MULTIPLICATION

### Objectives

**Understand and use the terms sum, difference, product, quotient**

**Sum, difference, product, quotient**

| | |
|---|---|
| The sum of two numbers is obtained by adding them. | e.g. the sum of 11 and 8 is 19 |
| The difference of two numbers is obtained by subtracting them. | e.g. the difference of 15 and 8 is 7 |
| The product of two or more numbers is obtained by multiplying the numbers together. | e.g. the product of 2 and −3 is −6 |
| The quotient of two numbers is obtained by dividing one number by the other. | e.g. the quotient of 12 and −3 is −4 |

**Use non-calculator methods to complete multiplications**

**Non-calculator methods for multiplication**

**The grid method** is one example of a non-calculator method for multiplication, e.g. work out 34 × 26

| × | 30 | 4 | | |
|---|---|---|---|---|
| 20 | 600 | 80 | | |
| 6 | 180 | 24 | | |
| total | 780 | 104 | **884** | So **34 × 26 = 884** |

Traditional **long multiplication** method is another example, e.g. work out 34 × 27

```
  34
× 27
 238
   2
 680
 1
 918
```

So **34 × 27 = 918**

### Practice questions

1 Work out the sum of these numbers:

a) 11 and 23 b) -1 and 1 c) 105 and 24 d) −6 and −2

2. Work out the difference between these numbers:

a) 12 and 5 b) 3 and 9 c) −5 and 7 d) 6 and −6

3. Here is a list of numbers: −60, −25, −9, −5, 10, 18

Choose two numbers from the list so that:

a) their product is −250 b) their quotient is 5 c) their quotient is the largest positive integer

4. Use a non-calculator method to work out the following calculations:

Show your working.

a) 34 × 19 b) 25 × 14 c) 18 × 103 d) 124 × 22

5. One tin of paint costs £17.
   Work out the cost of 25 tins of paint.

6. Potatoes cost 64p per kilogram. Sally buys 12 kg of potatoes.
   Work out the cost of the potatoes. Give your answer in pounds and pence.

7. A rectangular swimming pool measures 18 m long by 12 m wide.
   Work out the surface area of the pool.

8. a) Copy and complete this grid.
   b) Use the grid to work out the value of $25 \times 47$

| × | | 7 |
|---|---|---|
| 20 | 800 | |
| | | 35 |
| total | | |

9. Issy knows that $50 \times 20 = 1000$.
   She wants to find the answer to $50 \times 19$ without a calculator.
   a) Explain one method to do this.
   b) Write down the answer to $50 \times 19$

10. James has worked out $62 \times 37$ using the traditional long multiplication method.
    He has made a mistake.
    Here are his workings.

```
   6 2
×  3 7
 4 3 4
 1 8 6
 6 2 0
```

Find James' mistake and correct it.

# 1.5 DIVISION

**Division** is the inverse of multiplication. It can be written in a variety of ways.

e.g. 175 divided by 7 can be written as $175 \div 7$ or $7\overline{)175}$ or $\frac{175}{7}$

## Objectives

### Use long division for written methods

**Long division** is used to divide one number by the other, particularly for larger numbers.

e.g. Work out $544 \div 16$

```
     34
16 ) 544
     48
     64
    −64
      0
```

```
     3 4
16 ) 5⁵4⁶4
```

So **544 ÷ 16 = 34**

### Work out the remainder when numbers do not divide exactly

Sometimes when dividing there is a **remainder** left over. The remainder has not been divided.

e.g. Work out $350 \div 13$

```
     26
13 ) 350
     26
     90
    −78
     12
```

or

```
     2 6
13 ) 35⁹0¹²
```

So **350 ÷ 13 = 26 remainder 12**

## Practice questions

1. Copy and complete these statements:
   a) 45 lots of 19 is 855, so 855 divided by ______ equals 45
   b) The product of 36 and 42 is 1512, so 1512 divided by 42 equals _______
   c) 58 multiplied by 24 equals 1392, so 1392 divided by 24 equals _______

2. Use a non-calculator method to work out:
   a) $280 \div 8$ b) $9\overline{)387}$ c) $\frac{444}{12}$ d) $315 \div 15$

3. Use a non-calculator method to work out these calculations.
   a) $475 \div 19$ b) $950 \div 25$ c) $870 \div 29$ d) $1352 \div 26$

4. Use a non-calculator method to work out these divisions with remainders.
   a) $278 \div 13$ b) $300 \div 21$ c) $437 \div 20$ d) $703 \div 25$

5. Eamon says that $1289 \div 28 = 45$ remainder 29. Without doing any calculations how can you tell that Eamon is wrong?

6. There are 16 biscuits in each packet. Joanne needs 520 biscuits for the party. How many packets does she need to buy?

7. There are 432 people in the school theatre split into 24 equal groups. Work out how many people there are in each group.

8. Simon needs to divide 12 m of ribbon into pieces that are 55 cm long.
   a) How many pieces can he make? b) How long is the piece of ribbon left over?

9. 381 people need to travel on coaches. One coach holds 32 people. Work out the number of coaches needed.

10. A number divided by 54, goes 23 times with a remainder of 16. What is the number?

11. A number divided by 32 goes exactly 17 times. What is the remainder if the same number is divided by 16?

12. James collects model cars. He has less than 400 cars. If the number of cars is divided by 13 there is no remainder. If the number of cars is divided by 17 there is no remainder. How many cars does James have?

# 1.6 PRIORITY OF OPERATIONS

## Objectives

**Understand and use priority of operations, including brackets, powers and indices**

**Order of operations BIDMAS**

Mathematical operations are carried out in a certain order.

The order is: **B**rackets, **I**ndices, **D**ivision/**M**ultiplication, **A**ddition/**S**ubtraction

Do brackets first, then indices, then do any division or multiplication and finally do any addition or subtraction.

The acronym **BIDMAS** is a reminder of the order of operations.

e.g. Work out $9 \div 3 + 5^2 \times 2$

First indices: $= 9 \div 3 + 25 \times 2$

Then divide and multiply: $= 3 + 50$

Then add: $= 53$

For division or multiplication and addition or subtraction, if both operations are in the calculation then work from left to right.

e.g. Work out $5 + 9 - 8 + 11$

Work from left to right: $= 14 - 8 + 11$

$= 6 + 11$

$= 17$

**Use relationships between operations, including inverse operations**

**Inverse operations**

Multiplication and division are inverse operations.

Addition and subtraction are inverse operations.

e.g. $314 + 253 = 567$ so

$567 - 314 = 253$

$56 \times 18 = 1008$ so

$1008 \div 56 = 18$

Inverse operations can be used to check calculation answers.

## Practice questions

1. Work out:

   a) $2 \times 3 + 9$ b) $12 \div 6 + 6$ c) $6 + 7 - 9 + 3$ d) $4 \times 6 \div 3$

2. Work out:

   a) $(30 - 9) \times 2$ b) $12 \div (2 + 2)$ c) $7 - 3^2 \times 4 + 1$ d) $2 \times (3 + 4)^2$

3. Work out each of these calculations. Which answer is the odd one out?

   a) $3 \times (8 - 6) - 1$ b) $27 - 2 \div 1 + 4$ c) $20 \div 2 - 1 \times 5$ d) $(23 - 4 \times 2) \div 3$

4. Tamim says that $6 + 5 \times 8 = 88$ Explain why Tamim is wrong.

5. Put brackets where necessary to make these calculations true.

   a) $20 - 10 \div 2 = 5$ b) $2 \times 7 + 2 = 18$ c) $24 \div 8 - 2^2 = 6$ d) $10 \times 12 - 8 + 2 = 42$

6. Choose one of these symbols $<$, $>$, $=$ to make each statement correct.

   a) $4 + 6 \times 3$ ____ $(4 + 6) \times 3$ b) $5 \times 6 \div 2$ ____ $5 \times (6 \div 2)$

   c) $10 - 3 \times 2$ _____ $(10 - 3) \times 2$ d) $9 + 1^2$ _____ $(9 + 1)^2$

7. Jenna buys 3 notebooks costing £2 each and one calculator costing £5.
   Which of these statements show the change that she gets from £20?

   a) $20 - 3 \times 2 + 5$ b) $20 - 3 \times (2 + 5)$ c) $20 - (3 \times 2 + 5)$ d) $3 \times 2 + 5 - 20$

8. There are 18 students from 11A and 13 students from 11B at the marathon.

   The teacher takes two bottles of water for each student.

   To find the number of bottles needed the teacher does the calculation:

   $18 + 13 \times 2 = 44$

   She has made a mistake. Explain the mistake and show the correct calculation.

9. Find the missing numbers:

   a) $20 + (\_\_\_ - 3 \times 5) = 50$
   b) $(60 - 10 \times 2) \div \_\_\_\_ = 10$
   c) $13 - 3 \times \_\_\_ + 2 = 3$
   d) $14 + 3^2 - \_\_\_ \times 5 = -2$

10. Use only the symbols +, −, ×, ÷ to make each calculation correct.

   a) (3 __ 0) __ (10 __ 7) = 9
   b) 3 __ 1 __ (10 __ 7) = 1
   c) 3 __ ( 0 __ 10 __ 7) = 1
   d) 3 __ (0 __ 10 __ 7) = 10

SECTION 2

# PRIMES, FACTORS AND MULTIPLES

N3 N6 N7

## 2.1 SQUARE NUMBERS

### Objectives

**Work out and use squares of numbers**

When a number is multiplied by itself, it is being **squared**.

To square a number is the same as taking a number to the power of 2.

e.g. $3 \times 3 = 3$ squared $= 3^2$

A square number is the **answer** when a number has been squared.

e.g. $5^2 = 5 \times 5 = 25$ So **25** is a square number.

You should know the first 12 square numbers.

| | Square numbers | | Square numbers |
|---|---|---|---|
| $1^2$ | **1** | $7^2$ | **49** |
| $2^2$ | **4** | $8^2$ | **64** |
| $3^2$ | **9** | $9^2$ | **81** |
| $4^2$ | **16** | $10^2$ | **100** |
| $5^2$ | **25** | $11^2$ | **121** |
| $6^2$ | **36** | $12^2$ | **144** |

**Recognise square roots linked with known squares**

Squaring a number produces a square number.

The inverse operation to squaring is finding the square root.

The square root of a number finds the value which was originally squared.

e.g. Since $8 \times 8 = 64$, the square root of 64 is 8 .

$\sqrt{64} = 8$

**Understand the difference between positive and negative square roots**

Notice that $4 \times 4 = 16$ and $-4 \times -4 = 16$, so the square root of 16 could be 4 or −4

Square roots can be positive or negative.

Note – If you are asked for $\sqrt{16}$, a single answer of 4 is correct.

If you are asked for the positive and negative square roots of 16 or for the square roots of 16, you need to give the positive and negative values.

### Practice questions

1. Write down the value of:
   a) $5^2$ b) 7 squared c) $11 \times 11$ d) $13^2$

2. Write down the value of:
   a) $\sqrt{36}$ b) $\sqrt{81}$ c) $\sqrt{144}$ d) $\sqrt{100}$

3. Find:
   a) the square of 4
   b) the square root of 4
   c) the square root of 169
   d) a square number between 50 and 70

4. Which of these numbers is not a square number?

100 16 8 36 1 144

5. Place the correct symbol between each pair of numbers =, < , >

a) $14^2$ __ 144 b) $\sqrt{121}$ ___ $15^2$ c) $3^2$ ___ $\sqrt{81}$ d) −6 ___ the negative square root of 25

6. Match each square or square root to a single answer.

*Squares and square roots*

$(-2)^2$ $\sqrt{100}$ $(-3)^2$ $0^2$

*Answers*

−4 9 −10 0 −9 1 4

7. Which of these square roots are not integers?

a) $\sqrt{160}$ b) $\sqrt{196}$ c) $\sqrt{111}$ d) $\sqrt{1}$ e) $\sqrt{225}$

8. Write down an integer to go in each space and make correct number statements.

a) $3^2$ < ___ < $\sqrt{121}$ b) −3 > the negative square root of ___ > −5

c) ___ = square root of ___ = square of 3

9. Victor bends a piece of wire to make a square shape. The area of the square shape is $16cm^2$.

Victor does the following calculation to work out the dimensions of the square shape.

*length of the square = $\sqrt{16}$ = 8cm*

What mistake has Victor made? Calculate the length of the square and the perimeter of the square shape.

10. Say for each statement if it is always, sometimes or never true?

For each statement give an example which explains your answer.

a) The square of an integer is always greater than its square root.

b) All the square numbers are even numbers.

c) The square root of a number could be zero, positive or negative.

d) The square of a number could be positive or negative.

# 2.2 INDEX NOTATION

## Objectives

### Use index notation

**Index form**

The **index number** or **power** tells you how many times a number is multiplied by itself.

**e.g.** 3 to the power of 5 = $3^5 = 3 \times 3 \times 3 \times 3 \times 3$

6 to the power of 3 = $6^3 = 6 \times 6 \times 6$

When a number is written with a power or index number, it is written in index form.

**e.g.** $5^6$ is written in index form

### Recognise square and cube roots linked with known squares and cubes

**Cube numbers**

A **cube number** is an integer raised to the power of 3.

**e.g.** the cube of 8 = $8^3 = 8 \times 8 \times 8 = 512$

**Cubing** an integer produces a cube number.

**Cube root**

The inverse operation to cubing is finding the **cube root**.

The cube root of a number finds the value which was originally cubed.

**e.g.** Since $3 \times 3 \times 3 = 27$, the cube root of 27 is 3

The cube root of 27 = $\sqrt[3]{27} = 3$

### Recall the cubes of 1, 2, 3, 4, 5 and 10

### Recognise powers of 2, 3, 4, 5

**Powers of 2, 3, 4, 5 and 10**

The values of some powers are useful to remember.

**e.g.**

$2^2 = 2 \times 2 = 4$ | $2^3 = 2 \times 2 \times 2 = 8$ | $2^4 = 2 \times 2 \times 2 \times 2 = 16$

$3^2 = 3 \times 3 = 9$ | $3^3 = 3 \times 3 \times 3 = 27$ | $3^4 = 3 \times 3 \times 3 \times 3 = 81$

$4^2 = 4 \times 4 = 16$ | $4^3 = 4 \times 4 \times 4 = 64$ | $4^4 = 4 \times 4 \times 4 \times 4 = 256$

$5^2 = 5 \times 5 = 25$ | $5^3 = 5 \times 5 \times 5 = 125$ | $5^4 = 5 \times 5 \times 5 \times 5 = 625$

$10^2 = 10 \times 10 = 100$ | $10^3 = 10 \times 10 \times 10 = 1000$ | $10^4 = 10 \times 10 \times 10 \times 10 = 10\,000$

## Practice questions

1. Write each number in index form.
   a) $3 \times 3 \times 3$ b) $4 \times 4 \times 4 \times 4 \times 4$ c) $2 \times 2 \times 2 \times 2$ d) $5 \times 10 \times 10$

2. Write down the value of:
   a) $1^3$ b) $5^3$ c) $10^3$ d) $4^3$

3. Write down the value of:
   a) $\sqrt[3]{8}$ b) $\sqrt[3]{27}$ c) $\sqrt[3]{1}$ d) $\sqrt[3]{64}$

4. For each list of numbers find the odd one out. Give a reason for your answer.
   a) 196, 36, 200, 121, 1, 9, 81 b) 1, 64, 9, 27, 1000, 8, 125

5. Which of these numbers is larger? Show your working.
   a) $2^2$ or $5^2$ b) $10^3$ or $3^3$ c) 54 or $4^5$ d) $10^1$ or $3^1$

6. Match each power or root to its answer.

   *Power or root*
   $2^5$ $\sqrt[3]{1\,000}$ $\sqrt{196}$ $\sqrt{121}$ $\sqrt[3]{1}$

   *Answer*
   10 32 11 1 14 −1

7. Replace the * in each of these to make the statement correct.
   a) $\sqrt{\star} = 13$   b) $\sqrt[3]{\star} = 4$   c) $\star^3 = 27$   d) $5 \times 2^{\star} = 40$

8. Decide if each of the statements are true or false.
   Correct the statements that are false.
   a) $1^{10} = 10^1$   b) $2 \times 2 \times 2 = 2 \times 3$   c) $\sqrt[3]{1} = 1$ or $-1$   d) $4^2 = 2^4$

9. a) Jo says that if $n$ is a positive number then $\sqrt[3]{n} < \sqrt{n}$
   Give an example to show that she is not correct.
   b) Ben says that some numbers can be square numbers and cube numbers at the same time.
   Give an example to show that he is correct.

10. ■ is a cube number less than 200.
    ● is a square number less than 200.
    ● – ■ = $13^2$
    What is the value of the cube number?
    What is the value of the square number?

## 2.3 LAWS OF INDICES

N6 N7

### Objectives

Use index notation | Calculate with integer indices

**Know and use laws of indices with numbers**

**Laws of indices**

When **multiplying index numbers** with the same base, **add** the indices.

**e.g.** $2^3 \times 2^2 = 2^5$

When **dividing index numbers** with the same base, **subtract** the indices.

**e.g.** $2^6 \div 2^2 = 2^4$

When **raising a number** in index form to a power, **multiply** the indices

**e.g.** $(2^3)^2 = 2^6$

Any number to the **power of 0** equals 1.

**e.g.** $10^0 = 1$

**Negative powers** of a positive integer are the **reciprocals** of that number.

**e.g.** $10^{-2} = \frac{1}{10^2} = \frac{1}{100}$

### Practice questions

1. Simplify each expression. Leave your answer in index form.
   a) $4^3 \times 4^2$   b) $10^5 \times 10^1$   c) $5^2 \times 5^3 \times 5$   d) $11 \times 11^4 \times 11^3$

2. Simplify each expression. Leave your answer in index form.
   a) $3^5 \div 3^2$   b) $2^8 \div 2^3$   c) $\frac{11^{10}}{11^5}$   d) $\frac{7^2 \times 7^3}{7^5}$

3. Simplify each expression. Leave your answer in index form.
   a) $(2^3)^2$   b) $(2^3)^4$   c) $(3^3)^5$   d) $(3^5)^6$

4. Which of these numbers are not equal to $10^6$?
   a) $(10^2)^4$   b) $(10^3)^2$   c) $5^3 \times 2^3$   d) $(10^1)^6$

5. Simplify each expression. Leave your answer in index form.
   a) $2^3 \times 2^2 \times 5^2$  b) $3^3 \times 5^2 \times 5^2 \times 3^2$  c) $5^2 \times 2^3 \times 5^2 \times 2^2 \times 2^2$  d) $10 \times 3 \times 10^4 \times 3^{-1}$

6. Some of these statements are **not correct**.
   Which **one** is the odd one out? Give a reason for your answer.
   a) $\frac{(2 \times 2 \times 2 \times 2)}{(2 \times 2 \times 2 \times 2)} = 0$  b) $3^{-1} = -3$  c) $2^3 \times 2^5 = 2^{15}$  d) $(10^2)^3 = 1\,000\,000$

7. a) Ella says that $2^3 \times 2^2 = 4^5$ Is she correct? Give a reason for your answer.
   b) Tim says that $8^{10} \div 8^2 = 8^5$ Is he correct? Give a reason for your answer.

8. Replace the * in each statement to make the statement correct.
   a) $5^* \times 5^3 = 5^8$  b) $3^* \div 3^2 = 3^3$  c) $(2^5)^* = 2^{20}$

9. Decide which of these statements have mistakes in them.
   Correct the mistakes.
   a) $(2 + 1)^3 = 2^3 + 1^3 = 8 + 1 = 9$  b) $(3 \times 2^2)^2 = (3 \times 4)^2 = 12^2 = 144$
   c) $(2^2 + 3 \times 2^2)^2 = (4 + 12)^2 = 16^2 = (2^4)^2 = 2^8$  d) $(5^2)^3 \div (5^3)^2 = 5^6 \div 5^6 = 5^1 = 5$

10. Choose six different numbers from 0, 1, 2, 3, 4, 5 and 6 to replace the * in each of these statements.
   a) $10^* = 1^*$  b) $9^* = 3^*$  c) $2^* = 4^*$

## 2.4 PRIME NUMBERS

N4

### Objectives

**Use the concepts and vocabulary of prime numbers**

**Prime numbers**

A prime number is a number greater than 1 which is divisible only by itself and 1.

**e.g.** 2, 3, 5, 7 and 11 are prime numbers.

**Recognise prime numbers between 1 and 100**

**Use tests for divisibility to find factors**

**Divisibility tests**

A number will divide by 2 if it is even i.e. has a units digit which is even or 0.

**e.g.** 138, 2096 and 16 780 are all divisible by 2.

A number will divide by 3, if the sum of its digits is a multiple of 3.

**e.g.** 138 is divisible by 3 because 1 + 3 + 8 = 12 and 12 is divisible by 3.

A number will divide by 4, if the units and tens digits form a number divisible by 4.

**e.g.** 7532 is divisible by 4, because 32 is divisible by 4.

A number will divide by 5, if the units digit is 0 or 5.

**e.g.** 130, 265 are divisible by 5.

A number will divide by 6, if the number is even and it is a multiple of 3.

**e.g.** 5904 is divisible by 6 because it is an even number and it is a multiple of 3 because 5 + 9 + 4 = 18 which is a multiple of 3.

A number will divide by 8, if the last three digits are a multiple of 8.

**e.g.** 7104 is divisible by 8, because 104 is divisible by 8.

A number will divide by 9, if the sum of the digits is a multiple of 9.

**e.g.** 351 is divisible by 9, because 3 + 5 + 1 = 9 and 9 is divisible by 9.

## Practice questions

1. Which of these numbers are prime numbers?

   21, 37, 52, 63, 79, 81, 95

2. Place the numbers 20, 24, 36, 57, 65, 90 in the correct part of this Venn diagram.

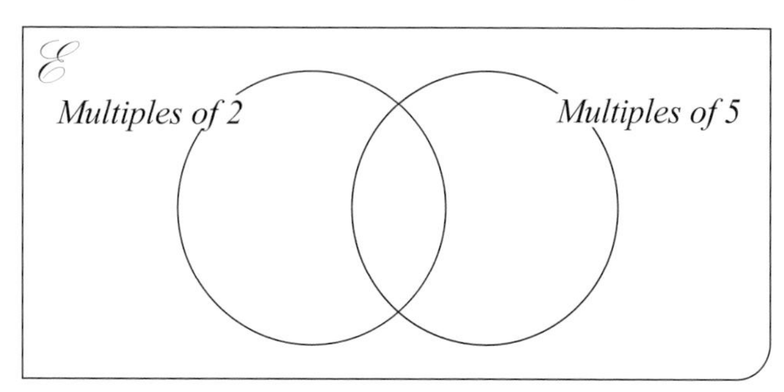


3. From each set of numbers, select the number which is the odd one out.

   Give a reason for your answer.

   a) 6, 28, 32, 49, 70 b) 10, 45, 51, 60, 85 c) 24, 28, 30, 42, 90 d) 3, 9, 27, 54, 72

4 Place these numbers in the correct place in the table 2, 3, 9, 10, 32, 45, 63

| | Divisible by 2 | Divisible by 3 |
|---|---|---|
| Prime number | | |
| Not prime number | | |

5. For each statement, decide if the statement is true or false.

   a) 414 is divisible by 4 b) 6248 is divisible by 8 c) 7590 is divisible by 5 d) 2304 is divisible by 9

6. For each statement decide if the statement is true or false.

   For false statements give a reason why they are false.

   a) All prime numbers are odd numbers
   b) 1 is the smallest prime number
   c) 99 is the largest prime under 100
   d) 2 is the only even prime number

7. By using divisibility tests, decide which of these answers is not an integer.

   a) $385 \div 5$ b) $4928 \div 9$ c) $3516 \div 6$ d) $9878 \div 3$

8. I am thinking of a number less than 100.

   It is divisible by 5 and 6. What is the biggest number that I could be thinking of?

9. Decide whether each statement is always true, sometimes true or never true.

   a) If $a$ and $b$ are prime numbers, then $a + b$ is a prime number.
   b) If $a$ and $b$ are prime numbers, then $a - b$ is a prime number.
   c) If $a$ and $b$ are prime numbers, then $ab$ is a prime number.
   d) If $a$ and $b$ are prime numbers, then $\frac{a}{b}$ is an integer.

10. Three prime numbers are all less than 20. Two of the numbers have a sum that is divisible by 8.

    The sum of all three numbers is divisible by 2, 3, 6 and 9. What are the three numbers?

# 2.5 FACTORS

## Objectives

### Use a factor tree to find factors

A factor tree is used to find the prime factors of a number.

e.g. This factor tree shows the prime factors of 60.

The **product** of prime factors of 60 is $2 \times 2 \times 3 \times 5$

The product of prime factors can be written in **index form** as $60 = 2^2 \times 3 \times 5$

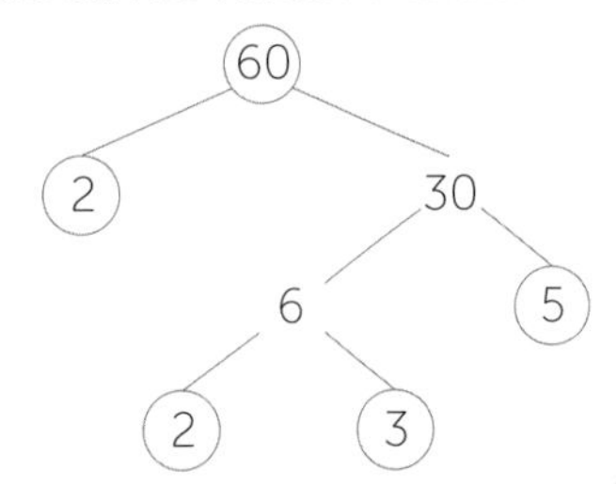


### Express a number as a product of prime factors

### Understand that an integer has a unique combination of prime factors (the unique factorisation theorem)

**Unique factorisation theorem**

Though the factor tree for a given number can be drawn in different ways, it always results in exactly the same prime factors.

e.g. Here are different factor trees for 60

But the prime factors of 60 remain the same;
$60 = 2 \times 2 \times 3 \times 5$

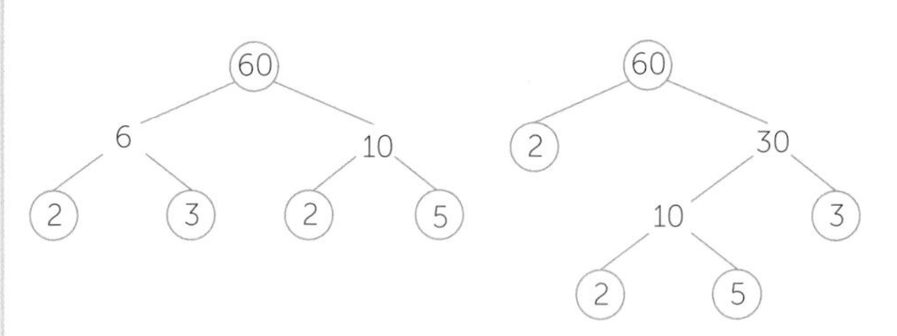


## Practice questions

1. Write each product of prime factors in index form.
   a) $64 = 2 \times 2 \times 2 \times 2 \times 2 \times 2$
   b) $72 = 2 \times 2 \times 2 \times 3 \times 3$
   c) $162 = 2 \times 3 \times 3 \times 3 \times 3$
   d) $300 = 2 \times 2 \times 3 \times 5 \times 5$

2. Write down the number that is represented by each of these products of primes.
   a) $2^2 \times 7$
   b) $2^2 \times 3^2$
   c) $5 \times 2^3$
   d) $2^2 \times 3^2 \times 5$

3. Draw factor trees for:
   a) 96
   b) 90

4. Amy has drawn a factor tree and written 24 as a product of prime factors.
   She has made a mistake.
   Explain her mistake and correct it.

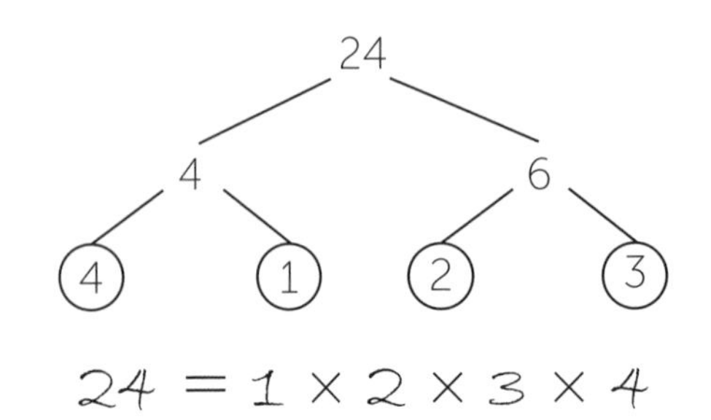


5. Write each number as the product of prime factors.
   a) 40
   b) 32
   c) 52
   d) 105

6. Three students are asked to express 50 as the product of prime factors in index form.
   Jake writes $2 \times 5 \times 5$
   Isabella writes $2 \times 5^2$
   Ben writes 2, 5, 5
   Who is correct?
   Give a reason for your answer.

7. Write 48 as a product of prime factors. Give your answer in index form.

8. Which of these is not the product of prime factors?
   a) $132 = 2 \times 2 \times 33$  b) $130 = 1 \times 2 \times 5 \times 13$  c) $65 = 5 \times 13$  d) $104 = 2^3 \times 13$

9. For each of these statements, decide if the statement is true or false.
   Give a reason for your answer.
   a) 6 is a factor of $2 \times 3 \times 5$  b) 4 is a factor of $2^2 \times 3^3$
   c) 120 is a factor of $2 \times 3 \times 5$  d) 40 is a factor of $2^3 \times 5$

10 Give an example of an integer that has:
   a) 5 prime factors and is less than 80
   b) 5 prime factors and is greater than 300
   c) Find a different answer for both parts a) and b).

## 2.6 MULTIPLES AND LCM

N4

### Objectives

#### Understand and find multiples

A multiple of a number is the product of the number with any integer.

**e.g.** Multiples of 5 are 5, 10, 15, 20, 25, ...

Multiples of 21 are 21, 42, 63, 84, ...

#### Find lowest common multiples

**Lowest common multiple**

The lowest common multiple (LCM) of two or more numbers, is the smallest multiple which the numbers share.

**e.g.** Find the lowest common multiple of 12 and 20.

List multiples of 12; 12, 24, 36, 48, **60**, 72, 84

List multiples of 20: 20, 40, **60**, 80, 100

60 is the first multiple which occurs in both lists. The LCM of 12 and 20 is 60.

#### Solve problems involving lowest common multiples

**Worked example**

Tom beats the drum every 3 seconds. Nala hits the cymbal every 5 seconds.

Tom and Nala both strike their instruments together at the start of a piece of music.

How many times will they strike their instruments together in a 4-minute piece of music?

Tom: 3, 6, 9, 12, **15**, 18, 21, ...

Nala: 5, 10, **15**, 20, 25, 30, ...

The LCM is 15. Tom and Nala will hit their instruments together every 15 seconds.

There are $4 \times 60 = 240$ seconds of music.

They will strike their instruments together once at the start, and then $240 / 15 = 16$ times during the piece, assuming they give a final drum beat or cymbal hit at the very end of the piece. Answer: 17 times.

### Practice questions

1. Write down the first 5 multiples of each of these numbers.
   a) 4  b) 10  c) 15  d) 30

2. Place the numbers 5, 6, 10, 18, 20, 30 in the correct part of this Venn diagram.

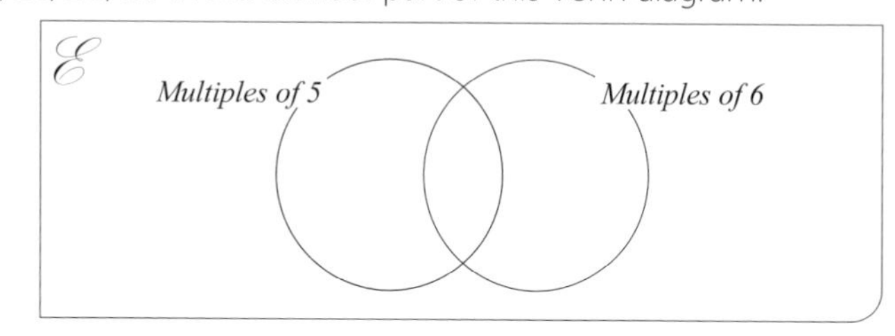

3. I am thinking of a number greater than 20.
   It is a multiple of 8 and a factor of 48. Its digits add up to 6.
   What is my number?

4. For the following.
   a) Express 18 and 24 each as a product of prime factors.
   b) Represent the prime factors on a Venn diagram

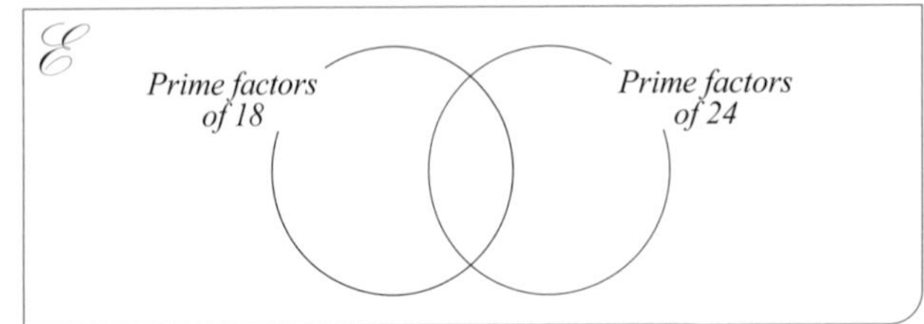


   c) Use this diagram to find the lowest common multiple of 18 and 24.

5. Find the lowest common multiple of each of these pairs of numbers.
   a) 6 and 8  b) 9 and 12  c) 10 and 25  d) 13 and 7

6. Find the lowest common multiple of 2, 3 and 5.

7. 10 and another number have a lowest common multiple of 90.
   Give three possible values for the other number.

8. The lowest common multiple of two numbers is 84.
   One of the numbers is 12.
   Give three possible values for the other number.

9. One church bell chimes every 15 minutes.
   Another church bell chimes every 12 minutes.
   They chime together at midnight.
   How many times in 12 hours will they chime together?

10. Three prime numbers are all less than 20.
    Two of them have a lowest common multiple of 91.
    A different pair from the three numbers have a lowest common multiple of 35.
    What are the three numbers?

11. Sandi can do 20 press ups in 2 minutes.
    Joel can do 24 press ups in 3 minutes.
    They start doing press ups at the same time.
    How many press ups will they start at the same time in a 5-minute session?

12. Mr Green bakes sets of 14 brownies. Miss Silver bakes sets of 30 cookies. They are bringing their baking to the village fete.
    There must be the same number of cookies and brownies at the fete and neither baker wants to bake more than they have to.
    How many batches of brownies does Mr Green have to bake?
    How many batches of cookies does Miss Silver have to bake?

## 2.7 FACTORS AND HCF

### Objectives

**Find the highest common factor of two or more numbers**

The **highest common factor (HCF)** of two numbers is the highest number that is a factor of both of them.

**e.g.** Common factors of 12 and 18 are 1, 2, 3 and **6**. The HCF of 12 and 18 is **6**.

**Solve problems involving LCM and HCF**

**Worked example**

The HCF or LCM of two numbers can be found by listing the factors or multiples of the numbers.

They can also be found by using the prime factors of the numbers.

**e.g.** A carpenter cuts two different lengths of wood into shelves of **equal length** with **none left over**.

One of the pieces of wood is 12 m long and the other piece is 8 m long.

What is the greatest length that the shelves can be?

$12 = \mathbf{2 \times 2} \times 3 \qquad 8 = \mathbf{2 \times 2} \times 2$

The HCF of 12 and 8 is $2 \times 2 = 4$.

The shelves are 4 m long.

### Practice questions

1. Find the HCF of each of these pairs of numbers.

   a) 4 and 6    b) 20 and 30    c) 12 and 18    d) 36 and 48

2. These numbers are expressed as a product of prime factors.
   Use the prime factors to work out the HCF and LCM of each pair.

   a) $30 = 2 \times 3 \times 5$    $50 = 2 \times 5 \times 5$    b) $44 = 2 \times 2 \times 11$    $66 = 2 \times 3 \times 11$

   c) $60 = 2 \times 2 \times 3 \times 5$    $84 = 2 \times 2 \times 3 \times 7$    d) $100 = 2 \times 2 \times 5 \times 5$    $150 = 2 \times 3 \times 5 \times 5$

3. Use the Venn diagrams to work out the HCF and LCM of each pair of numbers.

   a)

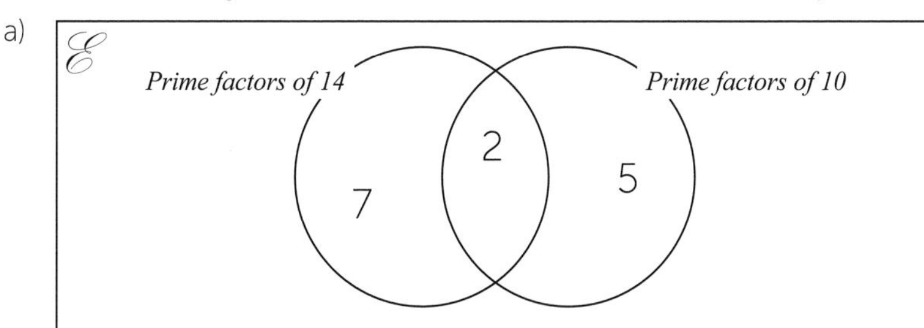


   b)

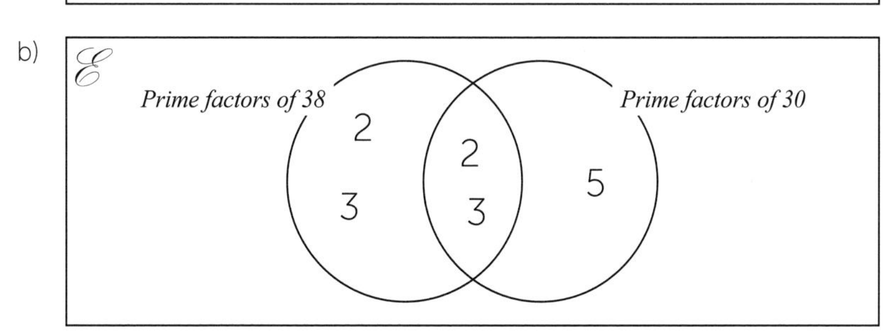

c)

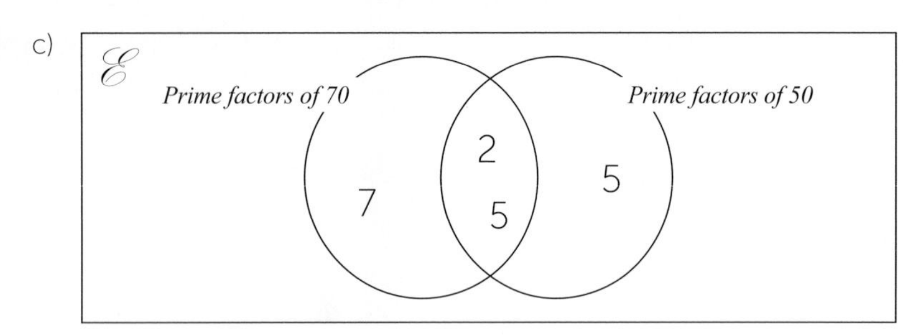


d)

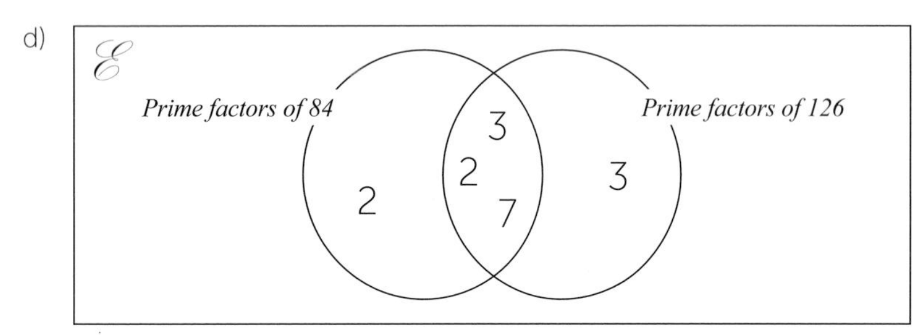


4. For the following.
   a) Express 54 and 45 each as a product of prime factors
   b) Work out the HCF of 54 and 45
   c) Work out the LCM of 54 and 45

5. Find the HCF and LCM of each of these pairs of numbers.
   a) 9 and 15 b) 32 and 12 c) 50 and 60 d) 45 and 75

6. Find the highest common factor and lowest common multiple of 10, 15, and 25.

7. 20 and another number have a highest common factor of 10.
   Give three possible values for the other number.

8. Archie has two pieces of ribbon.
   One piece is 1 m long and the other one is 75 cm long.
   He cuts them into smaller pieces of equal length, with no ribbon leftover.
   What is the greatest length in centimetres that each piece can be?

9. Mrs Chalk needs to buy workbooks and textbooks for her school.
   There are 18 textbooks in a pack and 30 workbooks in a pack.
   She needs exactly the same number of workbooks as textbooks.
   What is the minimum number of each pack she must buy?

10. Two numbers have an HCF of 2 and an LCM of 60.
    a) Jodie thinks that one of the numbers is 10. If so, what is the other number?
    b) George thinks that there is another pair of numbers with HCF = 2 and LCM = 60. Is he correct?
       Give a reason for your answer.

11. The product of two numbers is equal to the product of their HCF and LCM.
    Decide if this statement is true or false. Give a reason for your answer.

SECTION 3

# ALGEBRAIC EXPRESSIONS

## 3.1 ALGEBRAIC NOTATION

A1 A3 A21

### Objectives

#### Write multiplication and division of algebra using the correct notation

The letters in algebra represent unknown numbers; therefore, they work like numbers.

Algebra has a standard way of being written.

e.g. $a + a + a + a = 4 \times a$ which is written as $4a$

$5 \times b$ is written is $5b$

$c \div 8$ is written as a fraction as $\frac{c}{8}$

#### Know the difference between a term and an expression

An algebraic letter such as $x$ is a **variable** – the value of a variable can change.

A number on its own is called a **constant** – the value of this does not change.

A number in front of a letter is called a **coefficient** – so in $7x$, 7 is the coefficient.

An algebraic expression is a collection of algebraic terms with no equals sign.

$7x + 3$ is an **expression**

$7x$ and 3 are terms

#### Use indices correctly involving algebra

Multiplying something by itself is shown use powers or indices.

e.g. $3 \times 3$ is written as $3^2$ and in algebra $m \times m \times m = m^3$

#### Identify like terms

Like terms contain exactly the same variable.

e.g. In the expression $3v + 2w - w + 5v$...

$3v$ and $5v$ are like terms, $2w$ and $-w$ are also like terms

### Practice questions

1. In this expression $5g + 3h$, write down the:
   a) variables b) coefficients c) terms

2. Simplify these expressions using the correct algebraic notation.
   a) $5 \times p$ b) $y \div 2$ c) $a + a + a$ d) $n \times n$

3. Francis is asked to simplify $h \times h \times h \times h \times h$
   He writes down $5h$. Is he correct? Give a reason for your answer.

4. List the 'like terms' in each of these expressions.
   a) $a + a + b$ b) $p + q - p + p - q$ c) $5m + 3n - 2m + 3n$
   d) $v + 2w + v - x + 3v + 4x - w$ e) $y + 3 - 5y + 7$ f) $3g + 2g^2 - 5g$

5. Here is a rod made from six pieces of wood each $x$ centimetres long.
Write an expression for the total length of the rod in centimetres.

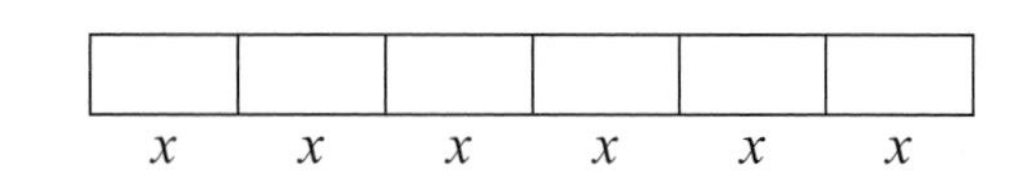


6. Write an expression for the total length of this rod.

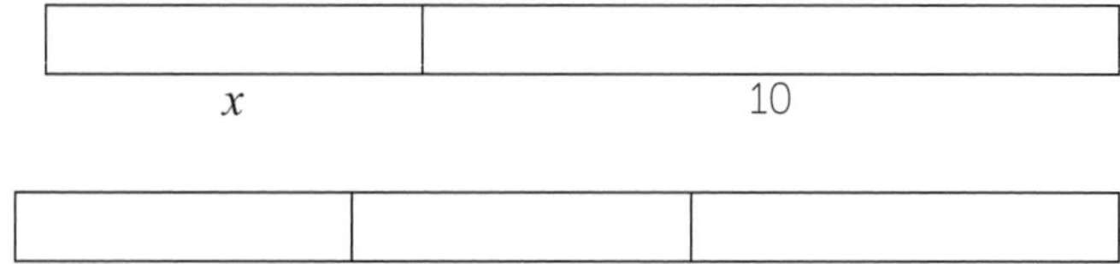


7. Write an expression for the total length of this rod.

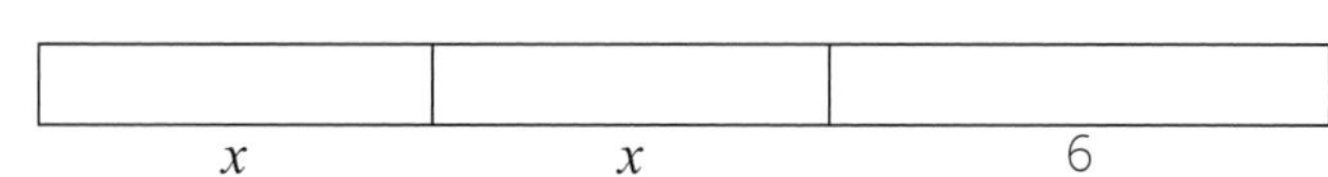


8. Write an expression for the perimeter of each shape.

a)

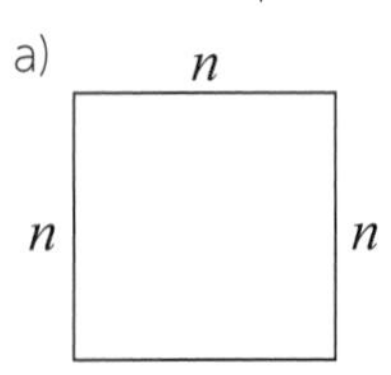


b)

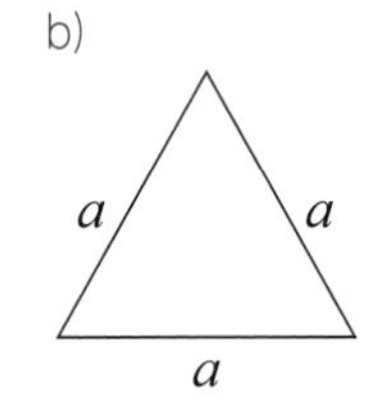


c)

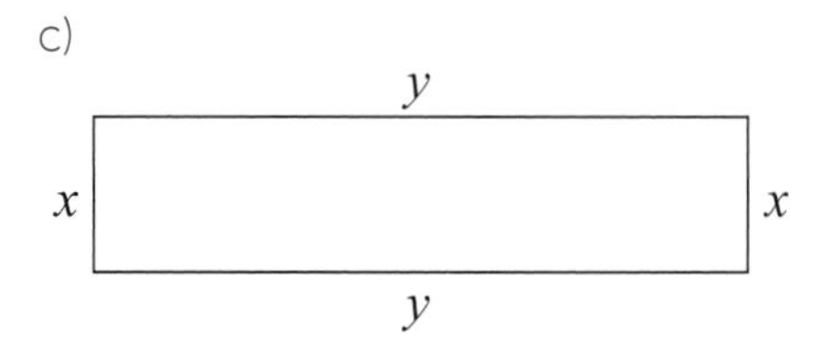


9. Richard has the expression $4x + 7y$
Richard says that $7y$ is the largest term in the expression.
Explain how Richard could be wrong.

10. Find an expression for the area of each shape below:

a)

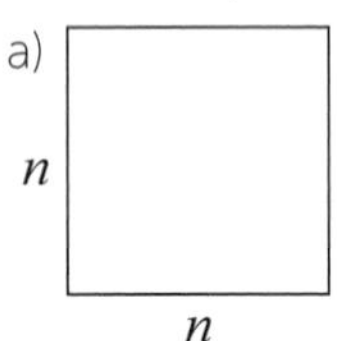


b)

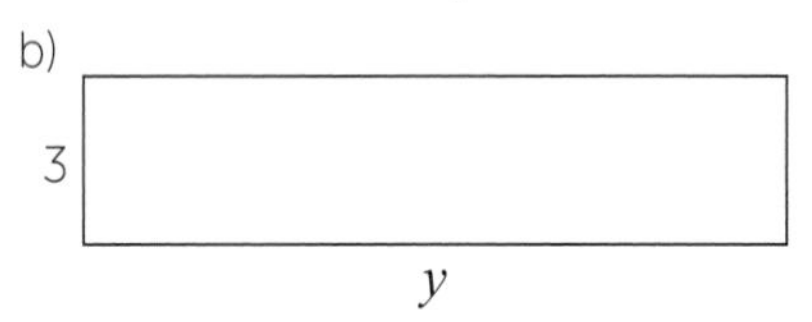


c)

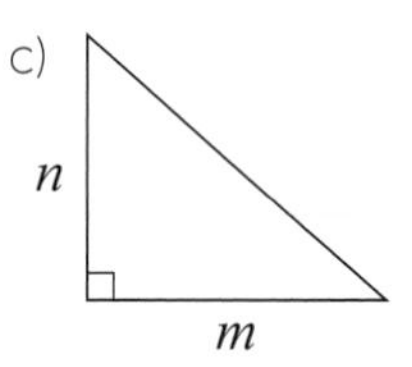


## 3.2 EXPRESSIONS AS FUNCTIONS

A2 A7

### Objectives

**Write an algebraic expression from a worded situation**

Decide which value could change; this is the **variable** that can be described as a letter

e.g. 'A number subtracted from 10' is written as **10 – $n$**

**Use function machines or input/output diagrams**

A **function machine** shows the operations and the order they come in.

e.g. This function machine multiplies by 3 and then adds on 5

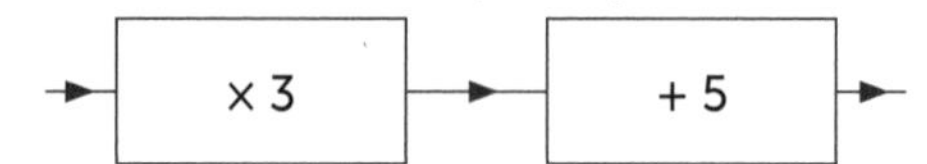


When the input is 2, the output is **11**

**Substitute values into an expression**

**Substitution** is when the letters are swapped for given numbers using the rules of algebra.

e.g. Substitute $x = 5$ into the expression $4x - 1$

$4 \times 5 - 1$ which gives the value **19**

## Practice questions

1. Write the following expressions using algebra.
   a) Four more than a number
   b) Three less than a number
   c) Five lots of a number
   d) A number subtracted from seven
   e) A number divided by five
   f) Twenty divided by a number
   g) A number added to ten
   h) A number shared between six

2. Fill in the letters and spaces in the function machines below.

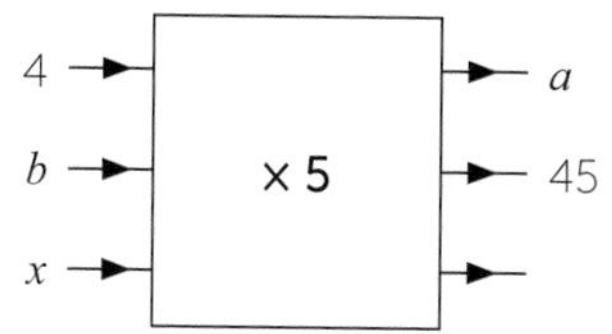


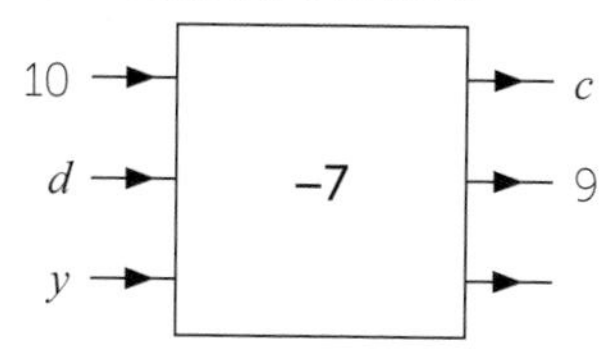


3. Find the value of each expression when n = 3
   a) $n + 7$ b) $3n$ c) $14 - n$ d) $\frac{12}{n}$ e) $n^2$

4. Choi is asked to "Write an expression for a number subtracted from 15"
   Her answer is $n - 15$
   Is Choi correct? Explain your answer.

5. Write the following using algebra.
   a) Three more than twice a number
   b) Divide a number by two then add three
   c) Multiply a number by three then subtract one
   d) Five less than a number multiplied by two

6. From the function machine:
   a) Work out the values of $a$, $b$ and $c$.
   b) The input is $n$, what is the output?

4, $b$, −3 → × 3 → + 2 → $a$, 20, $c$

7. Find the values of the expressions below when $n = 4$
   a) $2n + 5$ b) $3n - 2$ c) $\frac{n+5}{3}$ d) $20 - 3n$ e) $3n^2$

8. Yannick substitutes $p = -2$ into the expression $12 - 3p$
   Yannick gets the answer 6. Is he correct?

9. Look at the function machine.
   The input is $x$. What is the output?

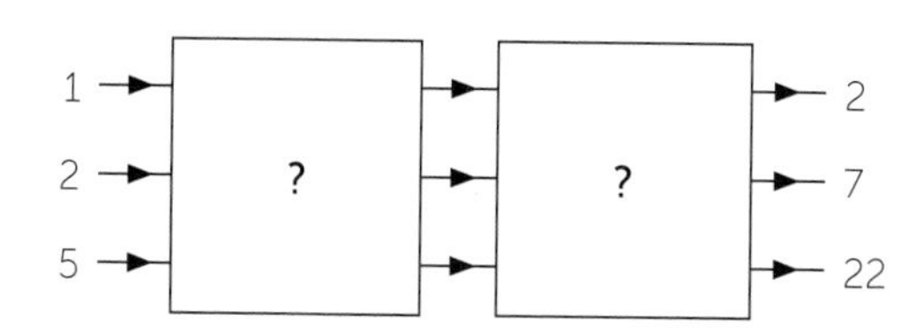

10. Sandip has been asked to substitute $y = 4$ into the expression $5y^2 + 1$
    Here is his answer: $5 \times 4^2 + 1 = 20^2 + 1$
    $= 400 + 1$
    $= 401$

    Explain why he is incorrect.

# 3.3 SIMPLIFYING ALGEBRAIC EXPRESSIONS

## Objectives

### Collect and simplify 'like terms'

'Like terms' are terms that have **exactly** the same variable.

**e.g.** In the expression $4a + 2b - a + 5b$

Like terms are:
$4a, -a$ and $2b, 5b$

Like terms can be collected together and simplified.

**e.g.** The expression $4a + 2b - a + 5b$ simplifies to $3a + 7b$

The expression $2n + 3n^2 - 4n + n^2$ simplifies to
$4n^2 - 2n$

### Recognise that powers can produce unlike terms

$x$ and $x^2$ are not like terms.

**e.g.** In the expression $2n + 3n^2 - 4n + n^2$

Like terms are: $2n, -4n$ and $3n^2, n^2$

### To multiply and divide together algebraic terms using index laws where appropriate

To simplify $2a^2 \times a^3$ rewrite it as $2 \times a \times a \times a \times a \times a$ which simplifies to $2a^5$

**e.g.** Simplify $3pq^3 \times 4p^3q^4$
$= 3 \times p \times q \times q \times q \times 4 \times p \times p \times p \times q \times q \times q \times q$
$= 12p^4q^7$

When dividing algebraic terms, common factors in the numerator and denominator can be cancelled

**e.g.** Simplify $\frac{g^6}{g^2} = \frac{g \times g \times g \times g \times g \times g}{g \times g}$

which simplifies to $g^4$

## Practice questions

1. Simplify the expressions below.
   a) $a + 3a$ b) $4y - y$ c) $6n + 3n - n$ d) $7p - 2p - p$

2. Write an expression for the perimeter of this rectangle.

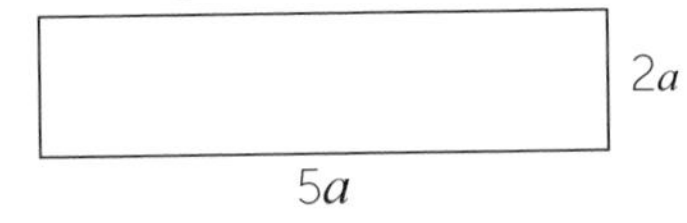


3. Dileep is asked to simplify $3q^2 + 4q - q^2$
   He got the answer $6q^2$. Explain why he is wrong.

4. Simplify the following giving your answers in index form.
   a) $n \times n^2$ b) $x^3 \times x^5$ c) $\frac{y^6}{y^2}$ d) $p^2 \times p^5 \times p$ e) $\frac{a^3 \times a^5}{a^4}$

5. Find an expression for the area of this square.

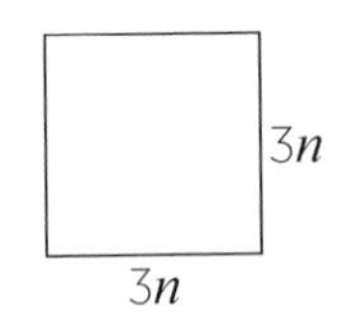

6. Simplify the following expressions.
   a) $2a + 3b - a + 4b$ b) $3p + 2q + p - 5q$ c) $6n^2 - 3n - 5n^2 + 3n$ d) $4x + 3x^2 - 2x + x^3$

7. Write expressions for the perimeters of the shapes. Simply your answers.

a) 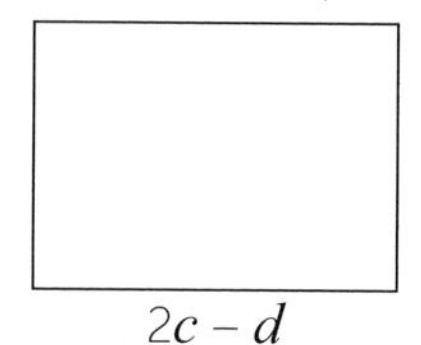


b) 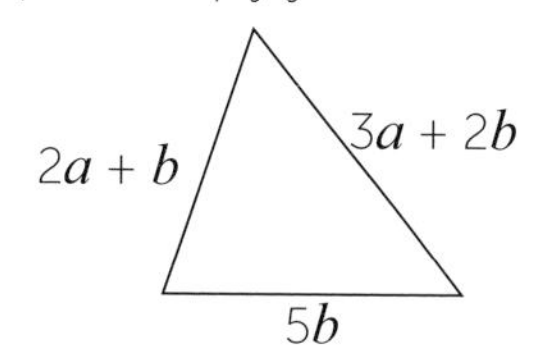


c) 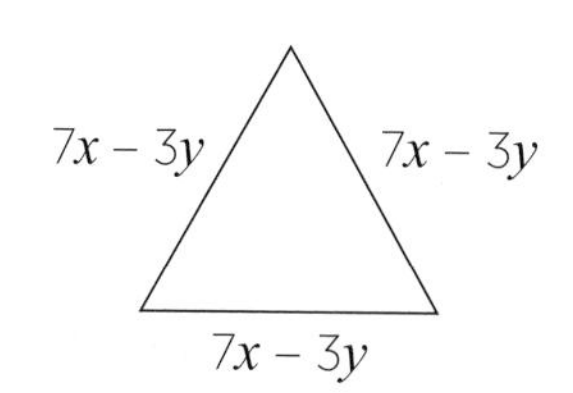

8. Which of the following expressions simplifies to $5x - 7y$?

a) $2x - 3y + 3x + 4y$ b) $8x - 9y - 3x + 2y$ c) $3y - x + 6x - 10y$

d) $3x - 3y + 2x^2 - 4y$ e) $7x + x^2 - y - x^2 - 6y - 2x$

9. Yolanda has to simplify $4a^2b^5 \times 7a^3b^4$

She arrives at the answer $28a^6b^{20}$

Is she correct? Give a reason for your answer.

10. Write an expression for the area of each of these shapes.

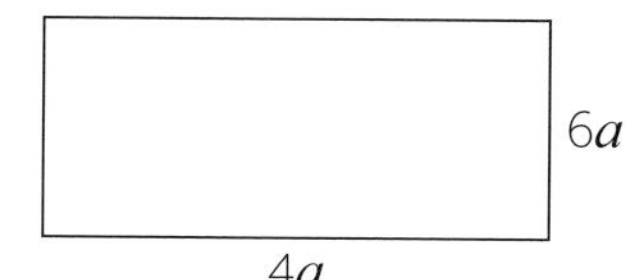

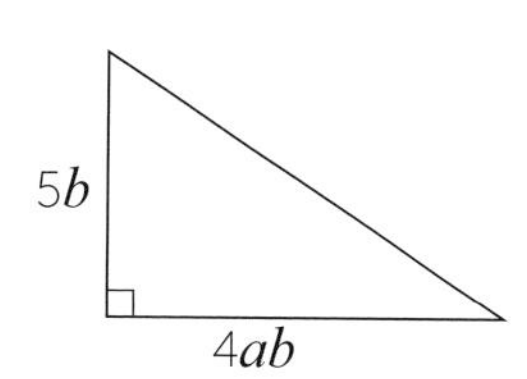


## 3.4 SIMPLE EQUATIONS

A17 A21

### Objectives

#### Recognise the difference between an expression, a formula and an equation

**Expressions**, **formulae** and **equations** all use algebra.

e.g. **Expression**: $5x - 3$; this is a mathematical statement using algebra

**Formula**: $T = 2n + 7$; one of the variables can be found, given values for the others

**Equation**: $3y - 4 = 14$; you can solve an equation to find the value of one variable.

#### Solve simple equations

**Solving equations**

Solving an equation means to find the value of the variable.

An equation must be balanced; this means that you must do the same to both sides of it.

e.g. Solve $2x + 3 = 11$

Subtract 3 from both sides $2x = 8$

Divide both sides by 2 $x = 4$

#### Form and solve simple equations from a worded situation

Make a statement about what you know. Change the words into algebra.

e.g. The perimeter of this triangle is 31 cm.

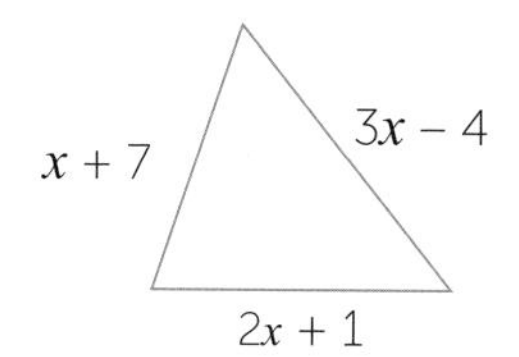


Form and solve an equation to find $x$

The perimeter of 31 cm equals the sum of the side lengths: $(x + 7) + (2x + 1) + (3x - 4) = 31$

Simplify the equation, by collecting the like terms

$6x + 4 = 31$

Solve the equation; subtract 4 from both sides $6x = 27$

Divide both sides by 6 $x =$ 4.5 cm

## Practice questions

1. State whether the each of these is an expression, a formula or an equation.
   a) $P = 2l + 2w$  b) $5x + 7 = 42$  c) $T = 30 - 2n$  d) $6a + 3b - 5c$

2. Solve the equations.
   a) $x + 6 = 11$  b) $y - 7 = 9$  c) $3n = 21$  d) $\frac{a}{3} = 15$  e) $15 - t = 11$

3. The total length of this rod is 19 centimetres.

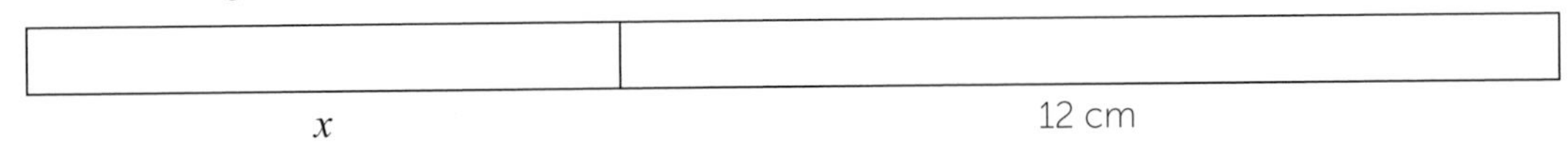


   Find the value of $x$ in centimetres.

4. Petra is 9 years old. Clem is $n$ years older than Petra. The sum of their ages is 31
   How much older is Clem than Petra?

5. Form an equation and solve it to find the length marked $y$ in the bar model.

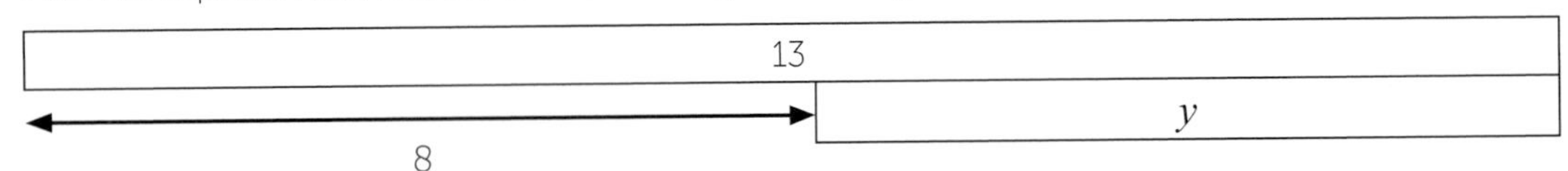


6. Solve the equations.
   a) $2a + 5 = 11$  b) $3y - 2 = 25$  c) $4t + 3 = 17$  d) $17 - 3q = 5$

7. The perimeter of this triangle is 34 centimetres.
   Set up an equation and solve it to find the value of $x$.

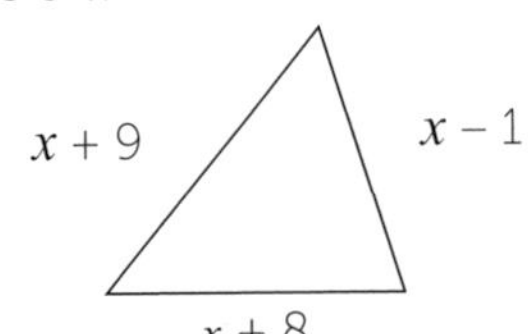


8. Florence has attempted to solve an equation. Here is her solution:

$$17 - 3x = 32$$

   Subtract 17 from both sides  $3x = 15$
   Divide both sides by 3  $x = 5$
   She has made a mistake. Find the mistake and correct it.

9. Form and solve an equation to find the value of $x$ in this diagram.

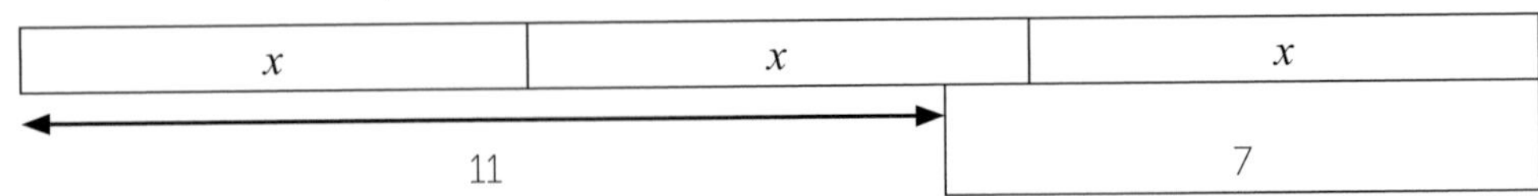


10. George is $n$ years old. Helen is 5 years older than George.
    Ian is twice as old as George. The sum of their ages is 49.
    How old is George?

# 3.5 SIMPLE FORMULAE

## Objectives

### Derive simple formulae from a worded situation

**Formulae** calculate the value of something given the other values of the variables.

e.g. A taxi firm charges £2 plus 30p per minute for a journey.

Write a formula to work out the total cost of a taxi journey.

The variable is time, which we will call $t$, measured in minutes

The total cost, in pounds, is what the formula will find, which we will call C

Think about how you would calculate the cost of journeys of certain lengths:

A journey of 1 minute: $C = 2 + 0.3 \times 1$

A journey of 2 minutes: $C = 2 + 0.3 \times 2$

A journey of $t$ minutes: $C = 2 + 0.3 \times t$

The formula is $C = 2 + 0.3t$

### Substitute values into a formula

When substituting into a formula, swap the variables for given numbers.

Then calculate using the rules of algebra and the correct order of operations

e.g. If $T = 3n^2 + 2$, find T when $n = 4$

Substitute $n = 4$: $T = 3 \times (4)^2 + 2$ $T = 3 \times 16 + 2$

$T = 50$

### Rearrange a simple formula to change its subject

Rearranging a formula involves using the rules of algebra to make a different variable the subject of the formula.

e.g. $T = 6x - 5$ is a formula which finds the value of T, so T is the subject.

Rearrange the formula to make $x$ the subject

Add 5 to both sides $T = 6x - 5$

Divide both sides by 6 $T + 5 = 6x$

Now $x$ is the subject $\frac{T+5}{6} = x$

## Practice questions

1. When organising a party for $p$ people, Louise uses the following rules:

   a) Two less apples than people $A = p - 2$

   b) Three biscuits for each person.

   c) One can of soda for each person plus five spares.

   d) A pizza between four people.

   Write a formula for each using $p$; the first one is done for you as an example.

2. Consider the formula $C = 4n + 7$; find C when:

   a) $n = 3$ b) $n = 7$ c) $n = -1$ d) $n = -3$ e) $n = 2.5$

3. Kelly has the formula $C = 10 - 3x$ and substitutes $x = -2$ into it.

   She gets the answer $C = 4$. Explain why Kelly is incorrect.

4. A bicycle hire company uses the following formula to calculate the total cost of hiring a bike: $C = 5 + h$

   C is the total cost of the hire and h is the number of hours the bike is hired for.

   a) Work out C when $h = 6$

   b) Work out C when $h = 8.5$

   c) Work out how many hours a bike can be hired for a cost of £43.

5. Rearrange the formulae to make $x$ the subject.

   a) $T = x + 5$ b) $T = 4x$ c) $T = 8 - x$ d) $T = \frac{x}{3}$ e) $T = \frac{24}{x}$

6. The formula to calculate the perimeter of this rectangle is $P = 2l + 2w$

$l$

$w$ $w$

$l$

Find:

a) P when $l$ = 5 cm and $w$ = 3 cm

b) P when $l$ = 12 cm and $w$ = 8 cm

c) $l$ when P = 30 m and $w$ = 4 m

d) $w$ when P = 29 m and $l$ = 9 m

7. Ice creams cost £1.50 each, ice lollies cost £1.80 each.

Write a formula to find the cost C, in pounds, of buying $a$ ice creams and $b$ ice lollies.

8. The equation of a line is $y = 5x - 2$. Rearrange the equation to make $x$ the subject.

9. Temperature conversions between Fahrenheit and Celsius use the formula:

$F = \frac{9}{5}C + 32$ F = temperature in Fahrenheit C = temperature in Celsius

a) Calculate the temperature in Fahrenheit when the temperature in Celsius is 15°C

b) Calculate the temperature in Celsius when the temperature in Fahrenheit is 77°F

10. A parcel delivery company has a fixed charge of £5 plus a charge of 40p per km.

a) What is the charge for a delivery with a distance of 50 km?

b) A delivery cost £8.20; how many kilometres away was the delivery?

c) Write a formula to work out the total cost in £s, C, of a delivery over $n$, kilometres.

d) Fred says that the cost of a delivery with distance 20 km will be double the cost of a delivery of 10km. Explain why he is wrong.

## 3.6 BRACKETS AND COMMON FACTORS

A4

### Objectives

#### Expand a single bracket

To expand a bracket means to **multiply out** the bracket.

A table can be used to help.

**e.g.** Expand $4(x + 3)$

$= 4x + 12$

| × | $x$ | + 3 |
|---|---|---|
| 4 | $4x$ | + 12 |

Expand $x(x - 5)$

$= x^2 - 5x$

| × | $x$ | − 5 |
|---|---|---|
| $x$ | $x^2$ | $-5x$ |

#### Simplify expressions involving brackets

Once brackets have been expanded, like terms can be collected and simplified.

**e.g.** Expand and simplify $2(x + 4) - 5(3x - 2)$

$= (2x + 8) - (15x - 10)$

$= 2x + 8 - 15x + 10$

$= 18 - 13x$

#### Factorise an expression into a single bracket

To factorise means to take out common factors from each term.

The factors are written outside a bracket.

An expression is fully factorised when every factor is outside the bracket.

**e.g.** Factorise fully $6x + 15$

$= 3(2x + 5)$

| × | $2x$ | + 5 |
|---|---|---|
| 3 | $6x$ | + 15 |

## Practice questions

1. Expand the brackets.

   a) $2(x + 1)$ b) $3(y - 2)$ c) $5(b + 3)$ d) $7(5 + n)$ e) $10(4 - m)$

2. Jeremy has been asked to expand $x(x + 6)$

   His answer is $x + 6x = 7x$

   Is he correct? Explain your answer.

3. Here is a rectangle.

   Which two of these are expressions for the area of the rectangle?

   | | |
   |---|---|
   | $2y + 10$ | $8(y - 3)$ |
   | $8y - 3$ | $y + 5$ |
   | $8y - 24$ | $8 + y - 3$ |

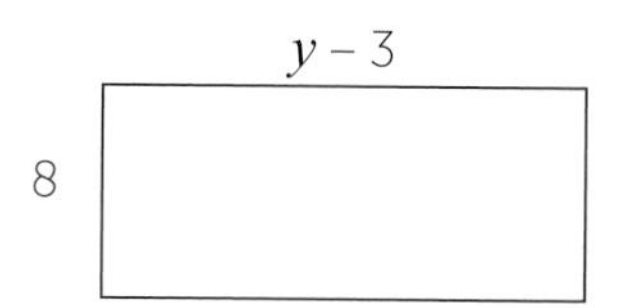


4. Factorise the following expressions.

   a) $2x + 4$ b) $3a - 12$ c) $5g - 30$ d) $6p + 9$ e) $8t - 20$

5. The expression for the perimeter of this equilateral triangle is $3x + 6$

   Write down the expression for the length of one side.

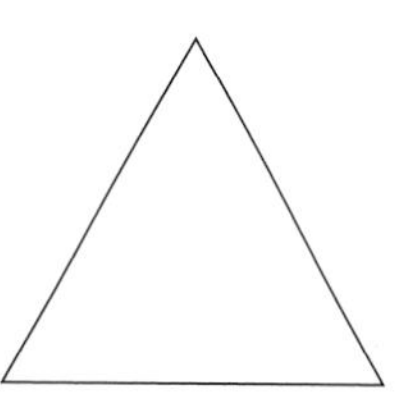

6. Expand and simplify the following.

   a) $2(x + 1) + 3(x + 2)$ b) $5(y + 3) + 4(y - 1)$

   c) $6(n + 2) - (n + 5)$ d) $4(p - 3) - 2(p + 6)$

7. Greg expands the bracket $3(4x + 5y)$ and writes down $7x + 8y$

   Explain what Greg has done wrong and correct his answer.

8. Expand the following brackets.

   a) $4(3x - 5)$ b) $y(y + 7)$ c) $p(q - 2)$ d) $n(n - 3m)$ e) $a(3a + 5b - 2c)$

9. Factorise fully these expressions.

   a) $ab - 5a$ b) $2xy - 7y$ c) $5pq + 15q$ d) $6n - 21mn$ e) $8xy + 16 - 20xz$

10. Write down an expression for the area of the compound shape.

    Simplify your answer.

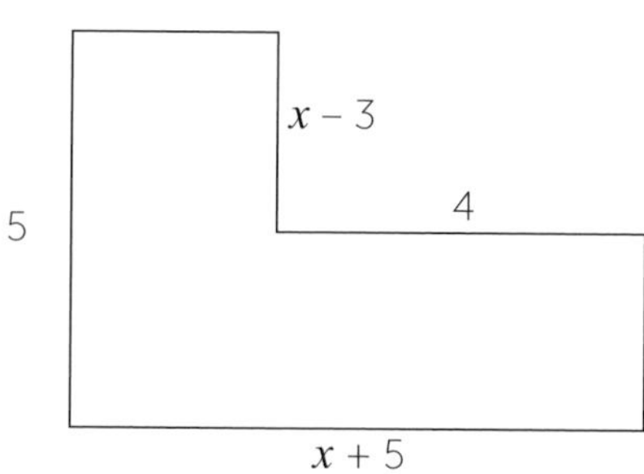

# 3.7 POWERS AND ROOTS

## Objectives

### Use index notation and the index laws when multiplying or dividing algebraic terms

**Laws of indices**

Algebraic terms follow the rules of indices.

e.g. $y^2 \times y^3 = y^5$ When multiplying index numbers with the same base, add the indices

$x^7 \div x^3 = x^4$ When dividing index numbers with the same base, subtract the indices

$(p^5)^2 = p^{10}$ When raising a number index form to a power, multiply the indices

**Surds**

A value which is expressed as a root, such as $\sqrt{7}$ is called a **surd**.

A surd cannot be simplified to remove the root.

e.g. $\sqrt{13}$ is a surd $\sqrt{25}$ is not a surd because it simplifies to 5

### Simplify and manipulate simple algebraic expressions involving powers and roots

In the expression $3x^2 + 2x - x^3$ there are no like terms.

The indices must match up to have like terms.

Like terms can be simplified.

e.g. Simplify $3b^2 + 7b^3 - 5b + 2b^2$
Like terms are $3b^2$ and $2b^2$
$= 5b^2 + 7b^3 - 5b$

e.g. Simplify $4x^2 \times 5x^4 = 4 \times 5 \times x^2 \times x^4$
Use rules of indices to combine the powers
$= 20x^6$

e.g. Simplify $10y^7 \div 5y^3 = \frac{10y^7}{5y^3}$
Use rules of indices to combine the powers
$= 2y^4$

e.g. Simplify $(3p^6)^2 = 3p^6 \times 3p^6$
Use rules of indices to combine the powers
$= 9p^{12}$

## Practice questions

1. Write down like terms and then simplify each of the expressions.
   a) $x + 3x^2 + 5x^2 + 4x$ b) $7y^2 - 3y^3 + y^2 - 2y^3 + y$

2. Kim and Laura are looking at the algebraic expression $4d^2 + 5d^4$
   Kim says that the expression cannot be simplified. Laura says it can. Who is correct? Explain your answer.

3. Find the total area of the compound shape below.
   The area of each section is given.

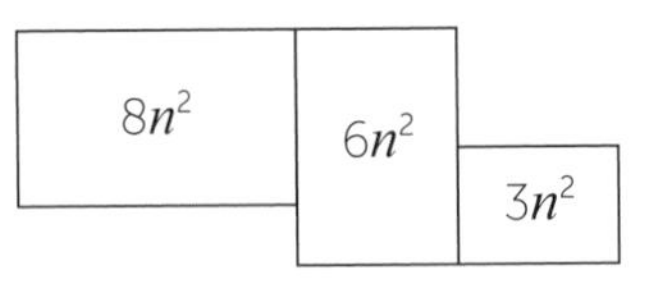


4. Simplify these algebraic expressions.
   a) $5c^3 \times c^4$ b) $2w^4 \times 4w^2$ c) $4p^5 \div 2p$ d) $2d \times 9d^4 \div 3d^2$

5. Identify which of these are surds.
   a) $\sqrt{4}$ b) $\sqrt{17}$ c) $\sqrt{1}$ d) $2\sqrt{3}$

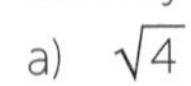

6. Simplify the expression: $p(p - 5) + p^2(4 - p)$

7. Write down an expression for the difference between the areas of the two rectangles.
   Simplify your answer.

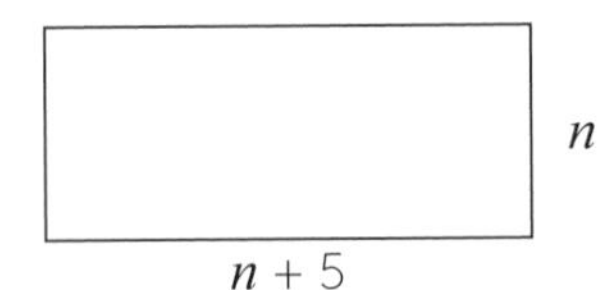


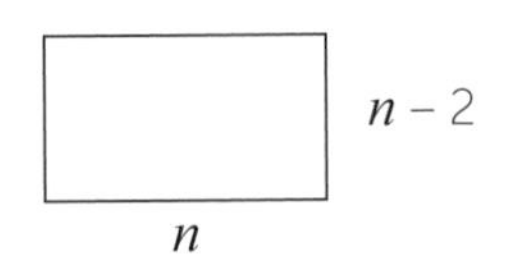


8. Expand and simplify the following.
   a) $2x(4 + 3x) - x(2 + x)$ b) $6pq^2(p - 3q)$ c) $2m^2(3m - 1) - 5m(4 - 7m)$

SECTION 4

# DECIMALS

## 4.1 DECIMAL PLACE VALUE

N1 N2

### Objectives

**Understand and use place value in decimals**

Decimal numbers have digits whose value depends on their place value or position

**e.g.** The number 4.603 can be shown as:

| units<br>1 | | tenths<br>0.1 | hundredths<br>0.01 | thousandths<br>0.001 |
|---|---|---|---|---|
| 4 | ● | 6 | 0 | 3 |

**e.g.** The number 0.346 has three tenths, four hundredths and six thousandths.

**Order positive and negative decimal numbers**

Decimal numbers can be written in ascending or descending order.

Write zero place holders when ordering decimals with different numbers of decimal places.

**e.g.** When ordering 8.307, 8.3, 8.07, 8.37, consider 8.370, 8.300, 8.070, 8.307.

In ascending order: 8.070, 8.300, 8.307, 8.370

**Use a number line**

### Practice questions

1. Draw out each number line and mark on each number.
   Use >, < or = to make a correct number statement.
   a) 0.13 ___ 0.2

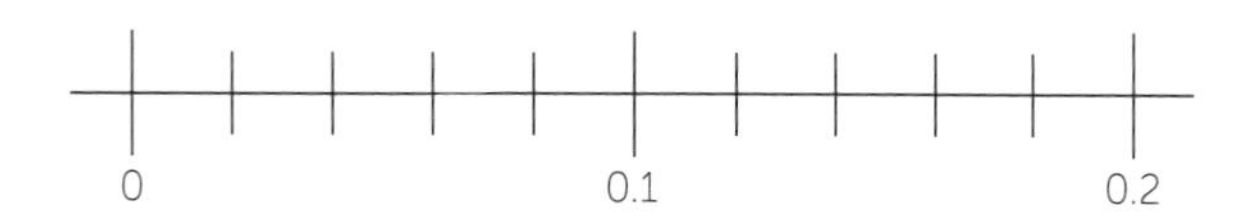


   b) 0.07 ___ 0.065

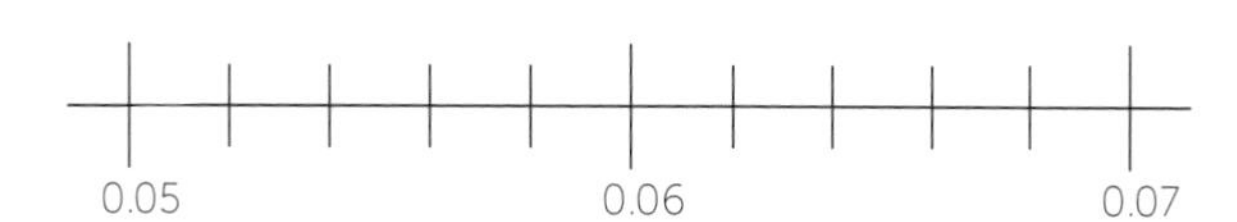


   c) 0.1000 ____ 0.1

   d) 1.329 ___ 1.4

2. Write down the smallest number in each set.
   a) 2.78, 2.709, 2.8
   b) −1.10, −1.4, −1.179
   c) −2.35, −2.204, −2.1

3. Write each set of numbers in ascending order.
   a) 4.8, 4.31, 4.2, 4.35, 4.12, 4.09
   b) 12.34, 14.85, 12.5, 13.61, 13.7
   c) −0.06, −0.11, −0.27, 0.35, −0.43
   d) −1.203, −1.281, 1.2, −1.29, −1.243

4. Write each set of numbers in descending order.
   a) 0.240 5, 0.24, 0.210 4, 0.214, 0.41
   b) 8.367, 8.850 1, 8.52, 8.7, 8.145 7
   c) −0.16, −0.8, −0.27, −0.25, −0.63
   d) −0.76, −0.713, −0.74, −0.701, −0.72

5. Write down a decimal number in each space to make a correct statement.
   a) 0.6 < ____ < 0.7
   b) 0.87 < ____ < 0.88
   c) 0.31 < ____ < 0.329
   d) 0.4 < ____ < 0.479

6. Here are the top three results in a race.

| Name | Time (seconds) |
|---|---|
| Josephine Rosberg | 38.2 |
| Rosie Dixon | 38.15 |
| Keshena Mustafa | 38.176 |

   Who came first, second and third in the race?

7. Owen has written a decimal number in his notebook.
   Use these clues to find his decimal number.
   - There is a two in the units (ones) place.
   - The digit in the tenths place is four times the digit in the hundredths place.
   - The digit in the thousandths place is two more than the digit in the hundredths place.
   - The number is less than 2.5

8. The table shows the average minimum temperature in 7 cities in December.

| City | Moscow | Sofia | Milan | Athens | Oslo | Tirana | Budapest |
|---|---|---|---|---|---|---|---|
| Temperature | −4.8 | −4.79 | −4.3 | −3.5 | −7.8 | −1.3 | −4.22 |

   a) True or false?
      - The coldest city is Tirana.
      - The warmest city is Tirana.
      - Oslo was colder than Moscow.
      - The temperature in Budapest is higher than the temperature in Milan.
   b) Write the city names in descending order of temperature.

9. Rebecca says that there is only one decimal number that could make this number statement true. 1.23 < ____ < 1.24
   Rebecca is not correct. Explain why.

10. Decimal numbers are written on four cards. Two of the cards are left blank.

| −6.9 | −6.07 | −6.84 | −6.528 | | |
|---|---|---|---|---|---|

   a) Choose two different numbers to write on the cards, between −6.84 and −6.528
   b) Arrange all the cards in descending order.

# 4.2 CALCULATING WITH DECIMAL NUMBERS

## Objectives

### Add and subtract decimal numbers

When adding or subtracting decimals:

- Set out the calculations in columns so that the decimal points line up.
- Always start from the column on the right.
- Use zeros as placeholders.

e.g.

```
   1
   3.67        3 1
 + 1.80        4.35
 ------      - 2.70
   5.47      ------
               1.65
```

### Multiply and divide decimal numbers by powers of 10

When multiplying by 10, move all the digits one place to the left.

When dividing by 10, move all the digits one place to the right.

e.g. $56.09 \times 10 = \mathbf{560.9}$

$14.92 \div 10 = \mathbf{1.492}$

When multiplying by 100, move all the digits two places to the left.

When dividing by 100, move all the digits two places to the right.

e.g. $4.589 \times 100 = \mathbf{458.9}$

$13.052 \div 100 = \mathbf{0.130\ 52}$

### Use place value when calculating

Knowing the answer to one calculation can be used to find the answer to similar questions, by using place value.

e.g. Given that $123 \times 37 = 4551$ work out $0.123 \times 37$

| | |
|---|---|
| Given: | $123 \times 37 = 4551$ |
| Divide both sides by 10 | $12.3 \times 37 = 455.1$ |
| Divide both sides by 10 | $1.23 \times 37 = 45.51$ |
| Divide both sides by 10 | $\mathbf{0.123 \times 37 = 4.551}$ |

## Practice questions

1. Work out

   a) $4.37 + 6.79$ b) $12.63 + 7.509$ c) $9.31 - 2.4$ d) $10.23 - 3.4$

2. Work out:

   a) $3.47 \times 10$ b) $0.891 \times 100$ c) $7.25 \div 10$ d) $0.431 \div 100$

   e) $4.895 \times 1000$ f) $0.006\ 78 \times 1000$ g) $348 \div 1000$ h) $78.3 \div 1000$

3. Complete the following calculations.

   a) 0.28 + ___ = 0.5 b) 0.743 − ___ = 0.25

   c) ___ + 0.6 = 3.459 d) ___ − 0.37 = 2.703

4. Work out:

   a) $23.45 + 17.6 - 6.185$ b) $13.79 - 7.661 + 3.4$

   c) $10.67 - (1.306 + 3.8)$ d) $(8.53 + 7.9) - (2.561 + 4.9)$

5. Given that $36 \times 11 = 396$, work out:

   a) $37 \times 11$ b) $36 \times 9$ c) $36 \times 22$ d) $198 \div 11$

6. Match each calculation to the missing number.

   **10 100 1000 10 000**

   a) 0.58 × __ = 5.8 b) 9.24 ÷ __ = 0.092 4 c) 0.336 7 × __ = 3.367

   d) 7.25 ÷ __ = 0.007 25 e) 0.039 × __ = 3.9 f) 43.1 ÷ __ = 0.043 1

7. Write your own decimal addition and subtraction questions that give the following answers, using only the numbers 0.38, 6.265, 11, 3.735.

   a) 10 b) 10.62 c) 2.53 d) 4.115

8. Given that 18 × 17 = 306 work out:

   a) 0.18 × 1.7 b) 0.018 × 1.7 c) 0.018 × 170 d) 30.6 ÷ 18

9. You are told that 15 × 160 = 2400.

   The answers to some related calculations are given below.

   Write down one possible calculation which would give each answer.

   a) 2560 b) 4800 c) 2415 d) 120

10. a) 100 pens have a mass of 0.8 kg. What is the mass of one pen in grams?

    b) George runs 1 km in 6.43 minutes.

    How long does he take to run 10 m if he maintains a constant speed?

11. Given that 40 × 25 = 1000, Nora calculates 39 × 25 as follows:

    39 × 25 = 1000 − 40 = 960

    Is she correct? Give a reason for your answer.

12. Selim buys 1 umbrella for £ 7.99, 2 bottles of water for £0.85 each and 1 magazine for £2.30.

    He pays with a £20 note. How much change does he get?

## 4.3 ROUNDING AND DECIMAL PLACES

N15

### Objectives

**Understand the purpose of rounding**

Rounding can be used in real-life to calculate quickly or to estimate.

Rounding can be used when numbers are very large, and accuracy is not critical.

**e.g.** When shopping, rounding the prices helps you to estimate overall spending.

**e.g.** The population of a city is 3 000 000 to the nearest million.

### Objectives

**Round a number to a given number of decimal places**

When rounding a number to a given number of decimal places (d.p.), look at the digit in that place and the digit to its right.

**e.g.** Round 56.0871 to 2 decimal places

Look at the digit in the 2nd decimal place and the digit to its right: 56.0**87**1

'87' rounds to '90'; all other digits after the 2nd decimal place go: **56.09**

### Understand the purpose of rounding

When rounding a number to the nearest 1, 10, 100, 1000 etc. look at the digit in that particular place in the number and at the digit to its right.

**e.g.** Round 145 679 to the nearest 100

Look at the digit in the hundred's column and the digit to its right 145 **67**9

'67' is closest to '70'; any other digits to the right become zero **145 700**

**e.g.** Round 1559.4 to the nearest 1000

Look at the digit in the thousand's column and the digit to its right **15**59.4

'15' rounds up to '20'; any other digits to the right become zero **2000.0**

## Practice questions

1. Round each number to the degree of accuracy shown.
   a) 184 (nearest 100) b) 2361 (nearest 10) c) 1631 (nearest 100) d) 16 328 (nearest 1000)
   e) 34 006 (nearest 10) f) 20 091 (nearest 1000)

2. Round each number to the degree of accuracy given.
   a) 107.53 (nearest unit) b) 8.453 (1 d.p.) c) 79.761 (nearest unit) d) 4.371 (1 d.p.)

3. Round each length given in metres, to 1 d.p.
   a) 34.709 2 m b) 100.289 1 m c) 78.067 1 m d) 9.890 3 m

4. Use a calculator to complete these calculations.
   Write the answer correct to 2 decimal places.
   a) $2 \div 3$ b) $5 \div 11$ c) $8 \div 9$ d) 10/17

5. Use a calculator to convert these fractions into decimals.
   Write the decimal answer correct to 2 decimal places.
   a) $\frac{5}{7}$ b) $\frac{5}{8}$ c) $\frac{13}{6}$ d) $\frac{11}{9}$

6. Jake works out $\sqrt{7}$ on his calculator, which gives the number 2.645 751 311
   Round this number to 3 decimal places.

7. Jessie rounds 0.234 7 to 2 decimal places.
   She rounds 0.2347 to 0.235 and then rounds 0.235 to 0.24
   Is Jessie correct? Give a reason for your answer.

8. Two lion cubs Aina and Simba are weighed. Aina weighs 1.751 8 kg. Simba weighs 1.745 7 kg.
   a) Which lion cub weighs the most?
   b) Give both of the weights to 2 decimal places. Does the result change?
   c) Give both of the weights to 3 decimal places.

9. I am thinking of a decimal number.
   If I round my number to the nearest unit, the answer is 10.
   If I round my number to 1 decimal place the answer is 10.5.
   Write 5 possible numbers that I could be thinking of.

10. Use the digits 3, 4, 5, 6, 7 once to complete these statements.
    a) 0. .... .... .... rounds to 0.75 to 2 decimal places.
    b) 0. .... .... .... rounds to 0.6 to 1 decimal place.

## 4.4 SIGNIFICANT FIGURES

### Objectives

#### Round a number to a given number of significant figures (s.f.)

Numbers can be rounded to a given number of **significant figures**

The first significant figure is the first non-zero digit with the highest place value.

**e.g.** Consider 380; the first significant number is 3 which has the value of 300.

**e.g.** Consider 0.005 009: the first significant figure is 5, the second significant figure is 0.

When rounding a number to a given number of significant figures, look at the digit in that place and the digit to its right.

**e.g.** Round 52.047 to 4 significant figures.

Look at the 4th significant figure and the digit to its right
52.0**47**

'47' rounds to '50'; all the other digits after this are removed.
**52.047 = 52.05 (4 s.f.)**

**e.g.** Round 0.0058621 to 3 significant figures.

Look at the 3rd significant figure and the digit to its right
0.005 8**62** 1

'62' rounds to '60'; all other digits after this are removed.
**0.005 862 1 = 0.005 86 (3 s.f.)**

**e.g.** Round 25 784 to 2 significant figures.

Look at the 2nd significant figure and the digit to its right.

2**5 7**84 rounds to **26 000** to 2 significant figures.

#### Understand and express error intervals

The error interval shows the range of values which an approximated number can take.

**e.g.** The length of a garden is 3.5 m given accurate to 2 significant figures.

The smallest value that the length of the garden can be is 3.45

The largest value that the length of the garden can be is 3.549 999.... m

The error interval for the length of the garden is **3.45 ≤ 3.5 (2 s.f.) < 3.55**

### Practice questions

1. Round each number to one significant figure.
   a) 3247 b) 18 211 c) 0.032 1 d) 0.003 718

2. Round each number to the accuracy given in the brackets.
   a) 13 478 (2 s.f.) b) 102 489 (3 s.f.) c) 4 671 083 (2 s.f.) d) 2.678 1 (3 s.f.)
   e) 0.035 781 8 (3 s.f.) f) 0.070 413 (2 s.f.)

3. Write down the error interval for each of these approximated values.
   a) 12.5 km (1 d.p.) b) 10.4 kg (3 s.f.) c) 3.00 m (2 d.p.) d) 0.043 kg (2 s.f.)

4. Write each fraction as a decimal correct to 3 significant figures.
   a) $\frac{2}{11}$ b) $\frac{13}{30}$ c) $\frac{29}{6}$

5. The population of Russia in 2019 was 143 902 736.
   Round this population figure to:
   a) 1 significant figure b) 2 significant figures c) 3 significant figures

6. Ali says that 7.956 7 rounded to 2 significant figures is 8
   Explain why Ali is wrong.
   What should the answer be?

7. By rounding each number to 1 significant figure, estimate the answers to these calculations.
   a) $6459 \div 3.17$ b) $0.032\,7 \times 99.85$

8. a) The width of an aeroplane $a$ is 947.8 cm correct to 1 decimal place.
   Write down an inequality to show the error interval for the width of the aeroplane.
   b) The speed $v$ of an aeroplane is 878 km/h correct to the nearest whole number.
   Write down an inequality to show the error interval for the speed of the aeroplane.

9. 19 878 tickets are sold to watch the Summer Concert.
   The average price of a ticket is £58.50. 48.75% of the ticket money goes to charity.
   By using approximations, estimate how much money the charities will receive.

10. The table shows the radius of seven planets.

| **Planet** | Earth | Mercury | Venus | Mars | Neptune | Jupiter | Uranus |
|---|---|---|---|---|---|---|---|
| **Radius (km)** | 6371 | 2439.7 | 6051.8 | 3389.5 | 24 622 | 69 911 | 25 362 |

   a) Round each planet's radius to 1 significant figure.
   b) Copy and complete the following:
   'The radius of _______is approximately 10 times bigger than the radius of _______.'

11. There is a gap of 752 mm (measured to 3 significant figures) in Mrs Smith's office.
    Mrs Smith buys a bookshelf to fit in the gap.
    The width of the bookshelf is given as 75 cm correct to 2 significant figures.
    Will the bookshelf definitely fit in the gap?

## 4.5 MULTIPLYING DECIMAL NUMBERS

N2 N14

### Objectives

**Multiply decimal numbers**

When multiplying decimals, first complete the multiplication ignoring the decimal point.

Then put the decimal point back into the answer, considering the powers of 10 involved.

e.g. Apples cost £1.35 per kilogram. Find the cost of 1.4 kg of apples.

First work out $135 \times 14 = 1890$

```
  135
×  14
  540
+1350
 1890
```

Then replace the decimal point

$1.35 \times 1.4 = (135 \times 14) \div 1000 = 1890 \div 1000 = 1.89$

**Cost of apples = £1.89**

**Estimate answers to calculations using approximations**

To check your answer is correct, approximate the numbers in the calculation and estimate the answer.

e.g. Use approximation to determine whether $2.3 \times 12.67 = 291.41$ is correct.

$2.3 \approx 2$ and $12.67 \approx 13$ so $2.3 \times 12.67 \approx 2 \times 13$

**Estimate = 26**

So, the answer of 291.41 is **not correct**.

## Practice questions

**Do not use a calculator for these questions**

1. Given that 176 × 54 = 9504; work out:

   a) 17.6 × 5.4    b) 1.76 × 0.54    c) 0.176 × 0.54    d) 0.017 6 × 5.4

2. Work out:

   a) 8 × 12    b) 0.8 × 1.2    c) 0.08 × 0.12    d) 1.8 × 1.2

3. Work out:

   a) 2.4 × 1.1    b) 1.63 × 0.53    c) 15.8 × 1.15    d) 3.26 × 0.278

4. Copy and complete the multiplication grid.

| × | 20 | 2 | 0.2 |
|---|---|---|---|
| 9 | | | |
| 0.9 | | | |
| 0.09 | | | |

5. Match each multiplication with its answer.

   a) 0.7 × 0.6    b) 1.4 × 0.3    c) 0.02 × 2.1    d) 0.12 × 3.5    e) 0.06 × 6.7

   **0.42    0.042    0.042    4.02    4.20**

6. Determine if each statement is true or false.

   a) 16 × 1.3 > 16 × 0.976 8    b) 13.8 × 0.78 > 13 × 1.1    c) 1.56 × 1.02 > 1.56 × 0.976 8

7. Estimate the answers to these calculations

   a) 14.25 × 0.976 8    b) 1.438 × 15.071    c) 2.68 × 4.768    d) 0.027 × 498
   e) 5.123 × 3.009    f) 120 × 0.21    g) 3.69 × 4.53    h) 9.08 × 1.297 8

8. A giraffe sleeps 15.75 hours each day. How many hours does it sleep in a week?

9. Mrs Smith changes £120 into Euros.
   The exchange rate is £1 = 1.06 Euros.
   How many Euros does Mrs Smith receive?

10. a) Use rounding to estimate the area of each of these picture frames.

    A: 3.05 cm by 4.32 cm    B: 1.98 cm by 5.95 cm

    b) Now calculate the exact area of each of the frames.
    Was your estimation correct?

# 4.6 DIVIDING DECIMAL NUMBERS

## Objectives

### Divide decimal numbers

When dividing decimal numbers, find equivalent divisions which have an integer on the denominator

**e.g.** $55.2 \div 1.2 = \frac{55.2}{1.2} = \frac{55.2 \times 10}{1.2 \times 10} = \frac{552}{12}$

Now divide 552 by 12

$$12\overline{)552} = 46$$

$55.2 \div 1.2 = 46$

### Estimate answers to calculations using approximations

To check your answer is correct, approximate the numbers in the calculation and estimate the answer.

**e.g.** Use approximation to estimate the answer to $178.84 \div 3.4$

$178.84 \approx 180$ and $3.4 \approx 3$

So $178.84 \div 3.4 \approx 180 \div 3$

Estimate = 60

## Practice questions

1. Work out:
   a) $0.48 \div 0.06$ b) $200 \div 0.05$ c) $0.035 \div 0.7$ d) $0.021 \div 0.7$

2. Which of these divisions are equivalent to $23.9 \div 1.4$?
   a) $\frac{239}{14}$ b) $\frac{0.239}{0.14}$ c) $\frac{2.39}{0.14}$ d) $\frac{23.9}{14}$

3. Work out these calculations.
   a) $0.96 \div 0.8$ b) $2.43 \div 0.09$ c) $5.52 \div 0.12$ d) $30.78 \div 0.19$

4. Use >, < or = to make these calculations true.
   a) $5.7 \div 1.5$ ___ $5.7$ b) $1.54 \div 2.8$ ___ $\frac{7.7}{14}$ c) $6.24 \div 0.6$ ___ $6.24$ d) $0.204 \div 0.4$ ___ $5$

5. Copy and complete the division pyramids. The first one has been done for you.

   a)

   b) 2.28 / 0.24 | [ ]

   c) 0.078 / [ ] | 0.65

   d) 13.975 / [ ] | 3.25

6. Two shops sell writing pens.
   Work out the cost of 1 pen at Shop A and 1 pen at Shop B.
   Give your answers in pounds.

   | **Shop A** | **Shop B** |
   |---|---|
   | 5 for £1.80 | 3 for £1.05 |

7. Lena works out $4.05 \times 0.48$ on her calculator. Her answer is 4. 52222....
   Use an approximate calculation to show that Lena has made a mistake.

8. Use > or < to make these statements true.
   a) $35 \div 1.21$ ___ $35$ b) $13.8 \div 0.89$ ___ $13.8$ c) $38.75 \div 1.52$ ___ $20$ d) $27.89 \div 2.93$ ___ $9$

9. A bag of 2.58 kg of sand costs £1.25.
   Velma thinks that 1 kg of sand costs just under 50p. Tom thinks that 1 kg of sand costs just over £2.
   Who is correct? Use approximations to decide. Show your working.

10. 'Dividing one number by the another always makes it smaller.'
    Is this statement always, sometimes or never true?
    Use examples to support your answer.

SECTION 5
# MEASURES

## 5.1 ESTIMATING ANSWERS

N14 N15

### Objectives

**Estimate answers to calculations by rounding numbers to 1 significant figure**

The answer to a calculation can be estimated by approximating each of the numbers in the calculation.

Often the numbers are approximated to 1 significant figure.

**e.g.** Estimate the answer to $\frac{0.95 \times 3.207}{1.12 + 2.8}$ by rounding each number to 1 significant figure.

$0.95 \approx 1$ $3.207 \approx 3$ $1.12 \approx 1$ $2.8 \approx 3$

So $\frac{0.95 \times 3.207}{1.12 + 2.8} \approx \frac{1 \times 3}{1 + 3}$ Estimate = $\frac{3}{4}$

**Round to a sensible degree of accuracy**

When giving a final answer to a calculation, give a degree of accuracy which is sensible.

**e.g.** A room of total area 44 $m^2$ is to be painted. A tin of paint covers 15 $m^2$

The number of tins required would be $44 \div 15 = 2.933...$ tins.

It is sensible to round this to **3 tins**, since only whole tins of paint can be bought.

**Estimate lengths by comparing with known lengths**

We can use known measurements to approximate unknown ones.

**e.g.** A handspan is approximately 20 cm

This can be used to estimate the length of the diagonal of a TV screen.

**Use inequality notation to indicate error intervals due to truncation or rounding**

A **truncated number** is approximated by removing all decimal digits to the right of the required accuracy level.

**e.g.** Truncate the number 3.509 73 to 3 d.p.

Look at the digit in the third decimal place.
3.509 73

Remove all digits to the right of the third decimal place
3.509

Notice how the 7 does not make the 9 round up.

Error intervals can be written for truncated numbers too.

**e.g.** The width of an envelope is given as 14.2 cm truncated to 1 d.p.

The error interval of the width of the envelope is
$14.2 \text{ cm} \leq \text{actual width} < 14.3 \text{ cm}$

## Practice questions

1. Write each of the following numbers rounded to 1 significant figure.

   a) 18.5　b) 7.3　c) 135　d) 0.029　e) 0.006 62

2. Write each of the following numbers truncated to 1 decimal place.

   a) 14.59　b) 3.084　c) 0.271　d) 13.829　e) 0.006 62

3. By approximating each number, estimate the answer to each of these calculations.

   a) $\frac{6.6 \times 9.4}{7.4}$　b) $\frac{11.6 \times 3.87}{5.7}$　c) $\frac{14.2 \times 9.6}{19.7}$　d) $\frac{18.5 \times 2.8}{6.42}$

4. Use approximation to estimate the answer to each of these calculations.

   a) $\frac{(43.2 - 8.9)}{2.8}$　b) $(6.7 + 4.8) \times (10.45 - 2.7)$　c) $\frac{(58 - 19)}{(1.2 \times 3.9)}$

5. Mark's handspan is approximately 11 cm wide.
   The length of his guitar is approximately 8 handspans.
   Estimate the length of his guitar in centimetres.

6. Rachel has an arm span of 1.8 m.
   She estimates the length of a garden hedge as 7 of her arm spans.
   Estimate the length of the garden in metres.

7. The reaction time for a science experiment is 35.7 seconds, rounded to 1 d.p.
   Give the error interval of the time of the experiment.

8. Ann measures the length of her dog walk as 4.8 km truncated to 1 decimal place.
   Give the error interval of the length of her dog walk.

9. Work out the each of these calculations on a calculator.
   (i) Write down all the digits in the answer on your calculator display.
   (ii) Write your answer to a suitable degree of accuracy.

   a) $\frac{5.3^3}{2.87 \times 1.76}$　b) $\frac{1.94 \times 5.89}{2.6^2}$　c) $\frac{6.22 + \sqrt{141}}{29}$　d) $\frac{19.7 \times 1.6^3}{\sqrt{39}}$

10. Here is a formula $v^2 = u^2 + 2as$, where:

    $v$ is the final velocity,
    $u$ is the initial velocity,
    $a$ is acceleration and
    $s$ is the distance.

    A car starts from rest and accelerates at 4 m/s$^2$ over a distance of 100 m.
    Work out the final velocity of the car.
    Give your answer to a sensible degree of accuracy.

# 5.2 SCALE DIAGRAMS

## Objectives

### Make an accurate scale drawing from a diagram

A scale diagram represents a place or object.

The distances and lengths of the real places and objects are usually 'scaled down'.

Angles stay the same in a scale diagram.

### Know that scale diagrams, including bearings and maps, are 'similar' to the real-life examples

A map is a scale diagram of an area showing various physical features

### Use and interpret scale drawings

The scale is used to work out actual distances or distances on the scale diagrams.

**e.g.** A scale of 1 cm to 50 cm is used to build a model of a helicopter.
The rotor blades on the model are 8 cm long.
Using the scale, the length of the real blades is $8 \times 50 =$ **400 cm or 4 m long**

The real length of the helicopter is 30 m.

Using the scale, the length of the model helicopter is $3000 \div 50 =$ **60 cm**

*Explain: Why is the real length converted into centimetres first?*

**Find the scale**

The scale can be found by comparing a distance on the scale drawing with the same distance in real life.

**e.g.** The distance between two towns is 50 km in real life. On a map it is 2.5 cm.

Work out the scale of the map.

2.5 cm is equivalent to 50 km

1 cm is equivalent to $50 \div 2.5 = 20$ km. **Scale is 1 cm to 20 km**

## Practice questions

1. Triangle ABC is right-angled at B. AB is 6 m and BC is 10 m.
   a) Draw a scale diagram of the triangle using a scale of 1 cm to 1 m.
   b) Measure the length of AC on the scale drawing
   c) Write down the actual length of AC on the triangle.

2. The scale of a drawing is 12 mm to 3.5 m. Copy and complete the table.

| Scale length (mm) | Real-life length (m) |
|---|---|
| 24 | |
| | 17.5 |
| | 28 |
| 84 | |

3. A model of a yacht is built to a scale of 1 cm to 32 cm.
   a) The model is 60 cm long. What is the actual length of the yacht in metres?
   b) The yacht has a mast which is 6.8 m high. How high is the mast on the model?

4. A scale drawing of an office building shows the building as 24 cm tall. The scale of the drawing is 1 cm to 5 m. Work out how tall the real office building is.

5. A map is drawn with a scale of 1 cm to 2.5 km. Jasmine says this is the same as using a scale of 5 cm to 15 km. Is she correct? Give a reason for your answer.

6. On a scale drawing, a roof beam is shown as 13 mm long. The real beam measures 2.6 m long. What is 1 cm on the scale drawing equivalent to, in real-life?

7. The real distance between two towns is 16 km. On a map, the distance between the towns is 40 cm. What is the scale of the map?

8. A garden is 45 m long. A scale drawing of the garden shows the length as 18 cm. What is the scale of the drawing?

9. An architect builds a model of a proposed new athletics stadium. Her model is built on a scale of 1 cm to 5 m.
   a) On the model, how long will the 100 m running straight be?
   b) On the model, what is the distance around the inside of the 400 m track?
   c) The model long jump run-up measures 5 cm. How long is the real run-up?

10. A rectangular wall measures 18 m wide by 8 m high. The wall has a door in the centre measuring 2 m by 1 m.
    a) Make a scale drawing of the wall and the door. Use a scale of 1 cm to 2 m.
    b) Use your scale drawing to work out the real length of the diagonal of the door. Give your answer to the nearest centimetre.

## 5.3 BEARINGS

G15 | R2

### Objectives

**Give bearings between points on a scale diagram or map**

A **bearing** describes a direction.

All bearings are measured clockwise from North and are given with 3 digits.

e.g. The bearing of B from A is 075°.

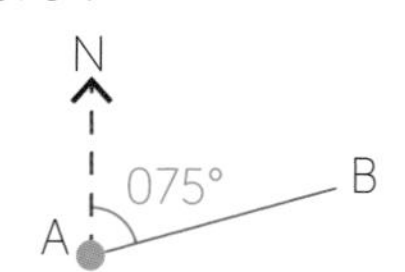


**Using bearings**

Bearings can be used in scale diagrams.

e.g. The bearing of a ship from Falmouth is 124°.

The bearing of the ship from Plymouth is 206°.

Here is a diagram to show the position of the ship.

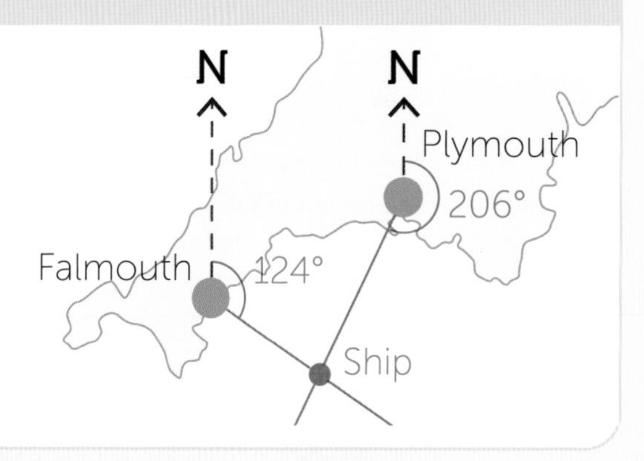


**Understand clockwise and anticlockwise**

An angle can be measured with a turn in a clockwise or anticlockwise direction.

e.g. 🡅 rotated through 90° clockwise looks like this 🡆

**Make an accurate scale drawing from a diagram, including for solving bearings problems**

## Practice questions

1. Measure the bearing of B from A in the following diagrams:

a)

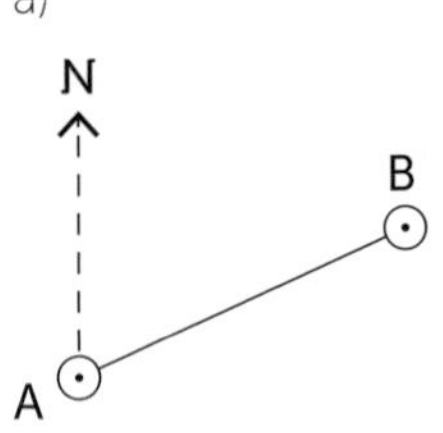


b)

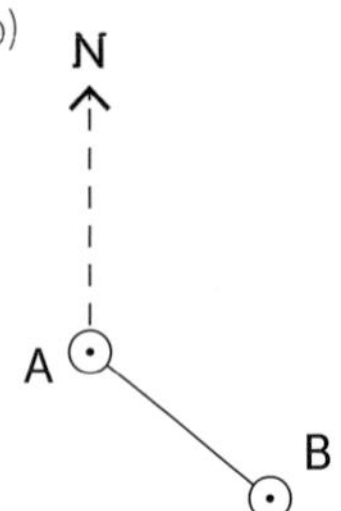


c)

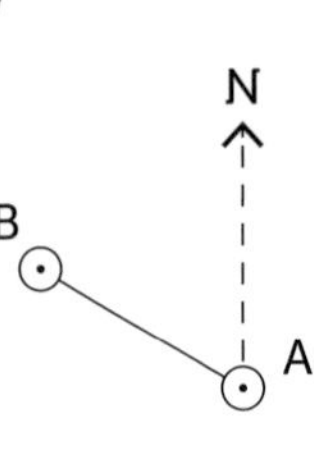


2. Draw diagrams like those in Q1 to show the bearing of B from A as:

a) 057° b) 098° c) 204° d) 314°

3. The church is 300 m east of the shop. The shop is 300 m south of the school.
What is the bearing of the church from the school?

4. The bearing of the lake from the forest is 180°.
What is the bearing of the forest from the lake?

5. Town A is 5 km directly north of Town C.
Town B is directly east of town A and is on a bearing of 045° from Town C.
   a) Using a scale of 1 km to 1 cm, draw an accurate diagram to show the positions of the three towns.
   b) By measuring on your scale diagram and using the scale, work out the real distance of Town B to Town C.

6. A, B and C are three towns as shown.

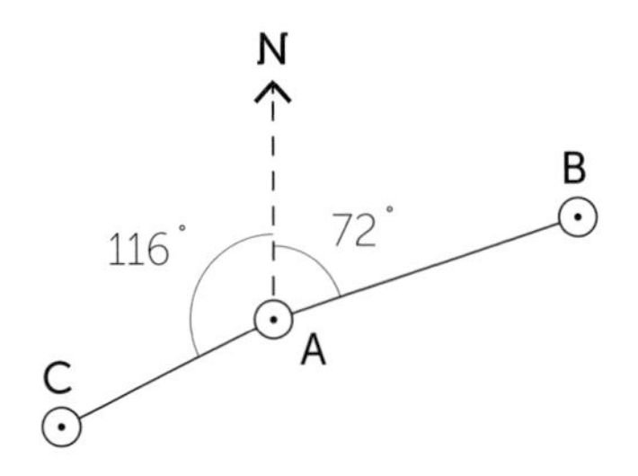


   a) What is the bearing of A from B?
   b) What is the bearing of C from A?

7. The bearing of Y from X is 069°. The bearing of Z from Y is 180°.
The distance from X to Y is the same as from X to Z.
What is the bearing of Z from X?

8. The bearing of the station from the library is 084°
The superstore is directly east of the station.
The station is the same distance from both the library and the superstore.
Draw a diagram to show the relative positions of the station, the library and the superstore and work out the bearing of the superstore from the library.

# 5.4 PLANS AND ELEVATIONS

## Objectives

**Construct and interpret plans and elevations of 3D shapes**

### Plans and elevations

The diagram shows a step block, used for aerobics and fitness.

The view from above the block is called the plan view.

The view from the front is called the front elevation.

The view from the side is called the side elevation.

Here are each of those views:

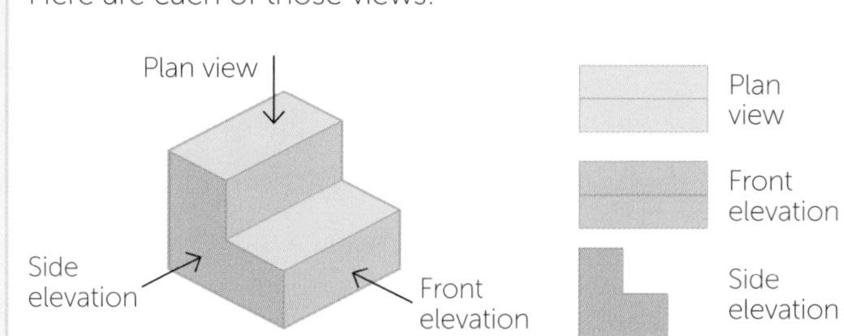


### Isometric drawings

Isometric drawings

Solid (3D) shapes can also be drawn on isometric paper, which has a pattern of dots arranged in a triangular lattice.

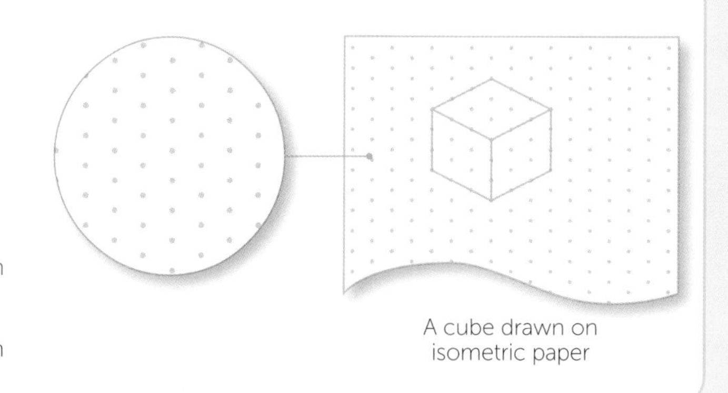
A cube drawn on isometric paper

## Practice questions

1. Draw the plan, front and side elevations of a cube measuring 5 cm.

2. A cuboid measures 8 cm long, 5 cm wide and 4 cm high.
   Draw to scale, the plan view, the side elevation and the front elevation of the cuboid.

3. On isometric paper, draw a 3 cm cube.

4. A wooden block is made from joining eight 1 cm cubes together to make the shape as shown here.
   a) Draw this shape on isometric paper.
   b) Draw the plan, front and side elevations of the shape.

5. A shape is built as shown from linking cubes.
   a) Draw its plan, front and side elevations.
   b) Draw the shape on isometric paper.

6. The 3D shape and side elevation of a metal block are shown.
   a) Draw the front elevation of the block.
   b) Draw the plan view of the block.

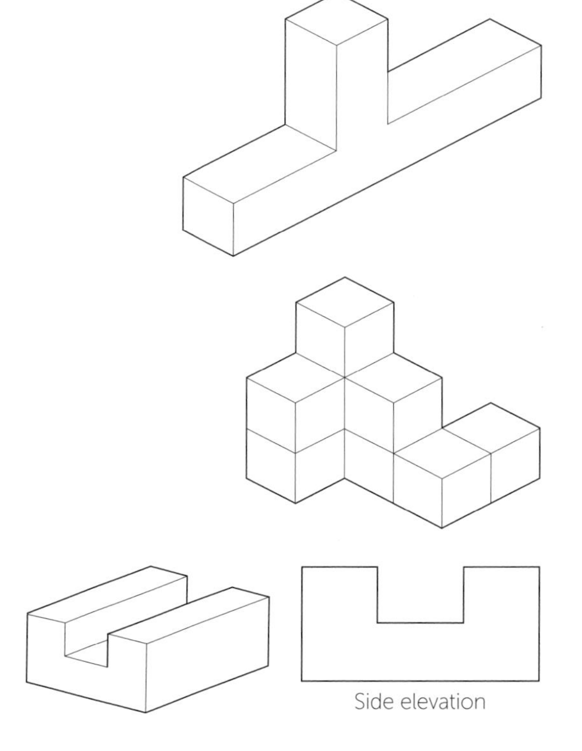

7. Here are the plan, front and side elevations of a solid shape. Draw the solid shape on isometric paper.

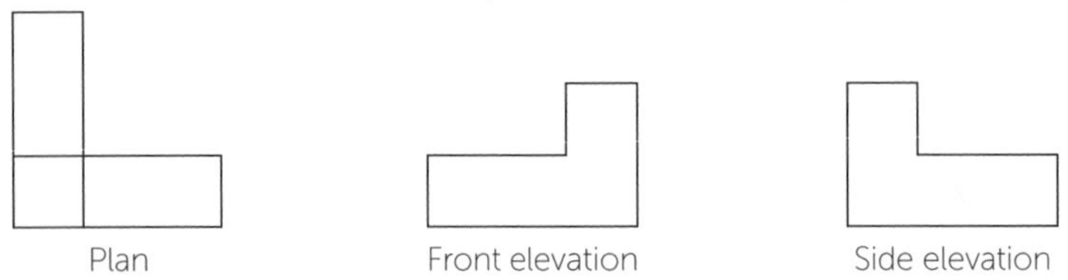

8. Draw a sketch of the following 3D shapes.

a) Cuboid  b) Triangular prism  c) Square-base pyramid

9. For each solid shape in Q8, draw and label the plan, front and side elevation.

10. For each solid in Q8 and 9, draw the solid on isometric paper.

## 5.5 PERIMETER

G4 G14 A5

### Objectives

**Find the perimeters of rectangles, triangles, parallelograms, trapezia and compound shapes**

The **perimeter** of a 2D shape is the total length of all its sides.

**e.g.** The perimeter of the rectangle
$= 6 + 4 + 6 + 4$
$= 20$ cm

**e.g.** The perimeter of the parallelogram $= 23 + 7 + 23 + 7$
$= 2 \times (23 + 7)$
$= 60$ m

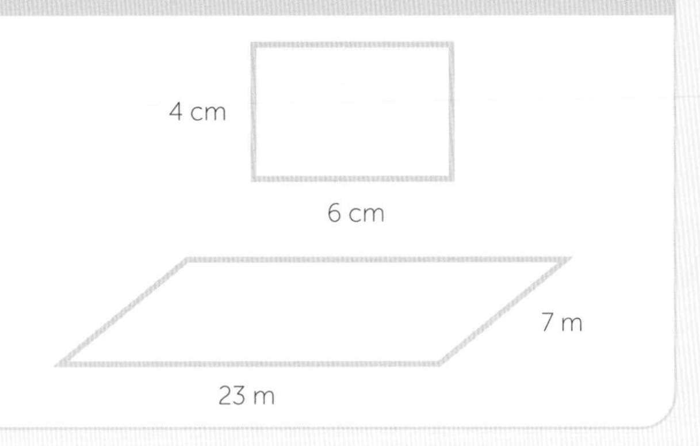


**Solve perimeter problems using algebra**

**Perimeters of compound shapes**

The perimeter of a compound shape is the sum of all the outside lengths.

If some measures are missing, these can be calculated before finding the perimeter.

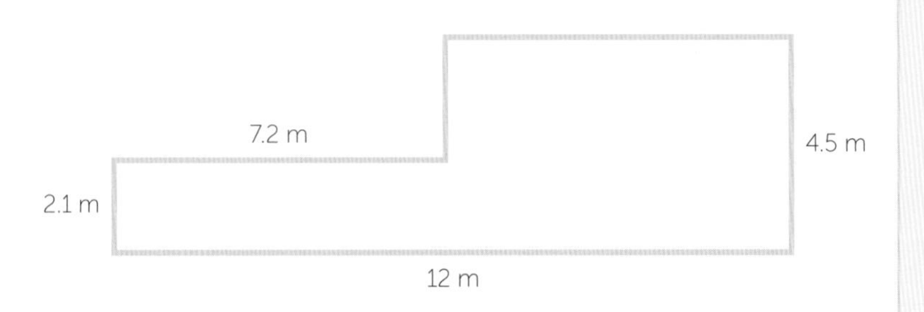


**Example:**

Find the perimeter of the shape shown.

Missing horizontal length is $12 - 7.2 = 4.8$ m

Missing vertical length is $4.5 - 2.1 = 2.4$ cm

Perimeter is $12 + 4.5 + 4.8 + 2.4 + 7.2 + 2.1 = 33$ m

*Explain: How can the perimeter be found without finding the missing lengths?*

## Practice questions

1. A rectangular garden is 24 m long and 15 m wide. Work out the perimeter.

2. What is the perimeter of a square with a side length of 45 cm?

3. The perimeter of a rectangle is 50 cm. If the width is 8 cm, how long is the rectangle?

4. A rectangle is 22 m long and 6 m wide. A square has the same perimeter as the rectangle.
   How long is each side of the square?

5. This diagram shows the plan of a floor. What is its perimeter?

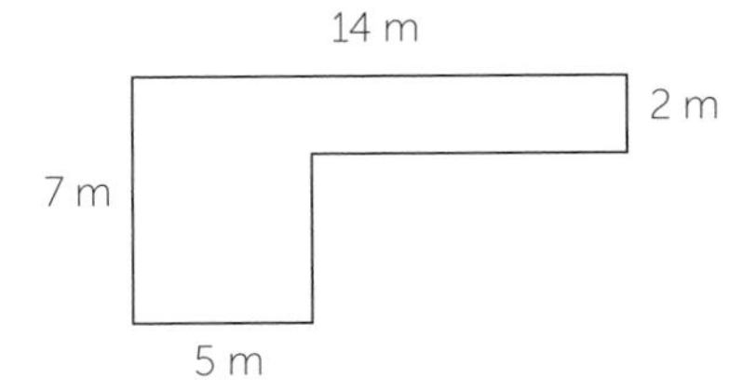


6. A square has side length 5 cm. A rectangle has length 7 cm and width 4 cm.
   A triangle has sides of lengths 5 cm, 7 cm and 8 cm.
   Luke says the shapes all have the same perimeter. Becky says they do not all have the same perimeter.
   Who is correct? Show your working.

7. Draw 3 different rectangles that each have a perimeter of 18 cm. Label the rectangles with their dimensions.

8. A right-angled triangle has sides of 10 cm, 24 cm and 26 cm. What is its perimeter?

9. A square has an area of 144 $m^2$. What is its perimeter?

10. A football manager warms up the players before a match by making them run around the edge of the pitch 3 times.
    The pitch is 118 m long and 74 m wide and the warm up run takes 6 minutes.
    What is the average speed of the players?

11. A rectangle has a length of $(4x + 2)$ and a width of $(5x - 1)$.
    Its perimeter is 56 cm.
    Work out the dimensions of the rectangle.

## 5.6 AREA OF SIMPLE SHAPES

### Objectives

**Recall and use the formulae for the area of a triangle, rectangle, trapezium and parallelogram**

**Area of simple shapes**

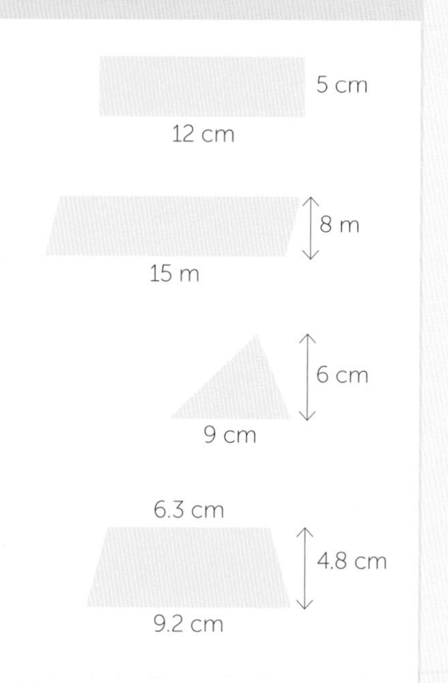


A **rectangle** with length $l$ and width $w$ has an area $A = l \times w$

e.g. Area of rectangle $= 5 \times 12$
$= 60\text{ cm}^2$

A **parallelogram** with a base of $b$ and a vertical height of $h$ has an area $A = b \times h$

e.g. Area of parallelogram $= 15 \times 8$
$= 120\text{ m}^2$

A **triangle** with a base of $b$ and a vertical height of $h$ has area $A = \frac{1}{2} \times b \times h$

e.g. Area of triangle $= \frac{1}{2} \times 9 \times 6$
$= 27\text{ cm}^2$

A **trapezium** with parallel lengths $a$ and $b$ and vertical height $h$ has area
$A = \frac{1}{2}(a + b) \times h$

e.g. Area of trapezium $= \frac{1}{2} \times (6.3 + 9.2) \times 4.8$
$= 37.2\text{ cm}^2$

**Units of area**

Area is measured in square units. e.g. $\text{cm}^2$, $\text{m}^2$ or $\text{mm}^2$

**Calculate areas of compound shapes made from triangles and rectangles**

### Practice questions

1. What is the area of a rectangle measuring 13 m by 17 m?

2. A rectangle has a length of 24 cm and an area of 456 $\text{cm}^2$. What is the width of the rectangle?

3. This parallelogram has a base length of 15 cm and an area of 127.5 $\text{cm}^2$.
Work out h, the perpendicular height.

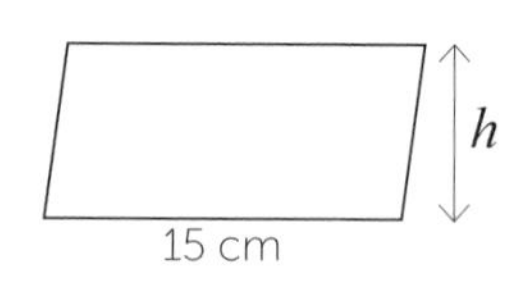


4. A triangle has a base of 10 cm and an area of 100 $\text{cm}^2$ What is the perpendicular height of the triangle?

5. A trapezium has parallel sides of 36 cm and 12 cm and perpendicular height of 9 cm.
Work out the area of the trapezium.

6. Work out the area of this triangle.
Give your answer in $\text{cm}^2$

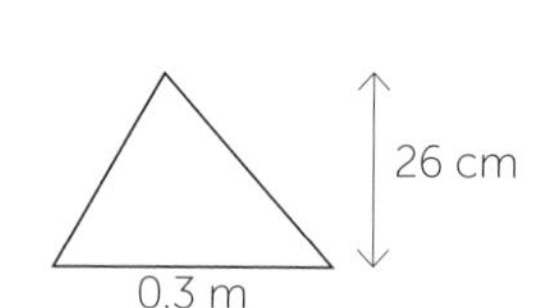


7. This trapezium has an area of 890 $\text{m}^2$.
Work out its perpendicular height.

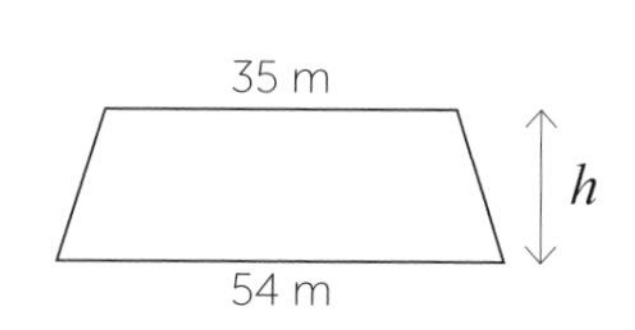

8. A parallelogram has a base of 9 cm and perpendicular height of 11 cm. A triangle has a base of 25 cm and a perpendicular height of 8 cm. Mia says the parallelogram has a greater area than the triangle. Is she correct?
   Show your working.

9. Draw three different rectangles with integer side lengths and areas of 24 $cm^2$. Which of your rectangles has the smallest perimeter? Show your working.

10. This large square has side length of 3 units. Inside the square is a smaller 1 unit square, a rectangle, a triangle with a base of 1 unit, a parallelogram and a trapezium.
    Work out the area of all the shapes.

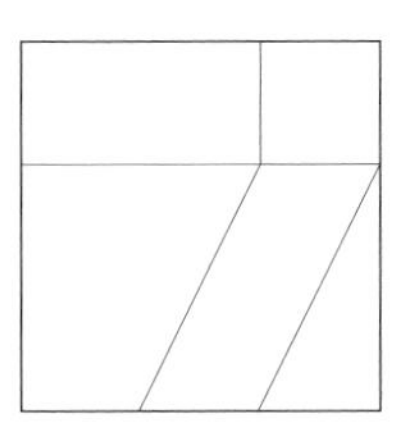

## 5.7 VOLUME OF SIMPLE SHAPES

G14 G16 A5 A17 A21

### Objectives

**Recall and use the formula for the volume of a cuboid**

A cuboid has side lengths of length ($l$), width ($w$) and height ($h$).

The volume of this cuboid is given by $V = l \times w \times h$

**e.g.** Volume of cuboid $= 6 \times 3 \times 2$
$= 36$ $cm^3$

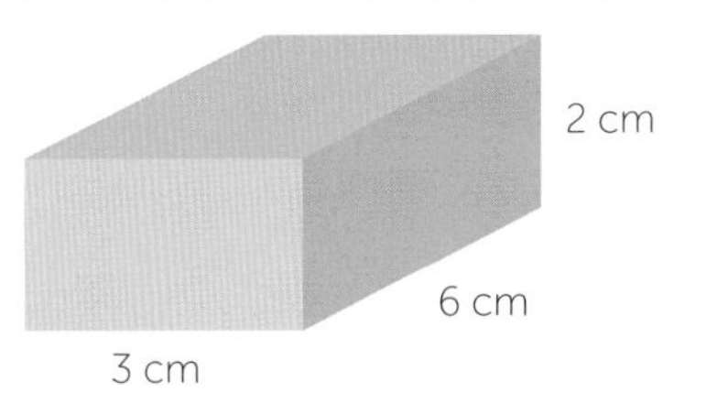


**Volume is measured in cubed units.** e.g. **$cm^3$, $m^3$ or $mm^3$**

**Find the volume of a prism, including a triangular prism, cube and cuboid**

A **prism** is a 3D shape with the same cross-sectional area.

**e.g.** A cuboid is an example of a prism with a rectangular cross-section.

**Cuboid**

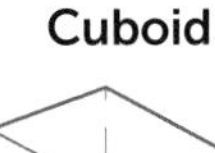
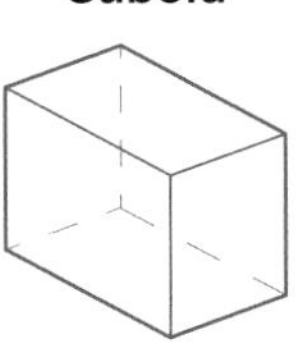

**Triangular prism**

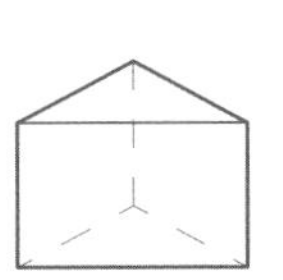

**Hexagonal prism**

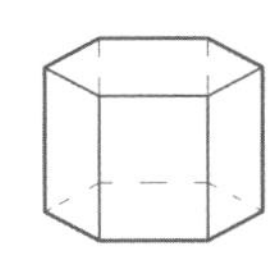

The volume of a prism is found by multiplying the area of the cross-section ($A$) and the length ($l$).

For any prism, Volume $= A \times l$

**e.g.** Volume of triangular prism $= 12 \times 28$
$= 336$ $cm^3$

Area of triangular end = 12 $cm^2$

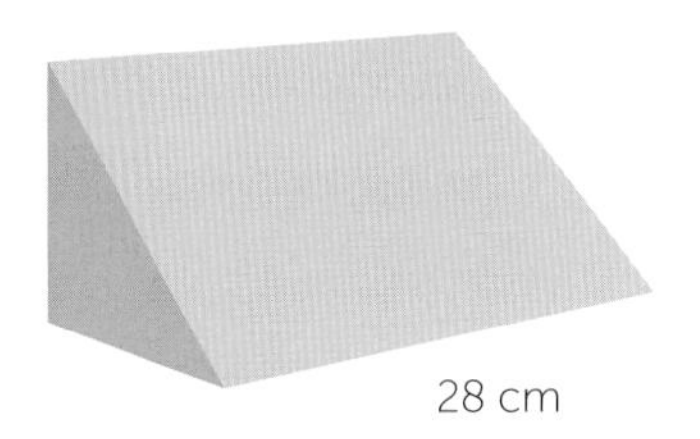

## Practice questions

1. Work out the volume of a cuboid measuring 50 cm by 30 cm by 15 cm.

2. A cuboid measures 5 m by 3 m by 20 cm. Judie says the volume is 300 $m^3$.
   Is she correct? Give a reason for your answer.

3. Work out the volume in of a cube of side length 60 cm. Give your answer in $m^3$.

4. A cuboid has a volume of 320 $cm^3$. Its length is 10 cm, and its width is 8 cm. What is the height of the cuboid?

5. An L-shaped box has dimensions as shown.
   Work out the volume of the box.

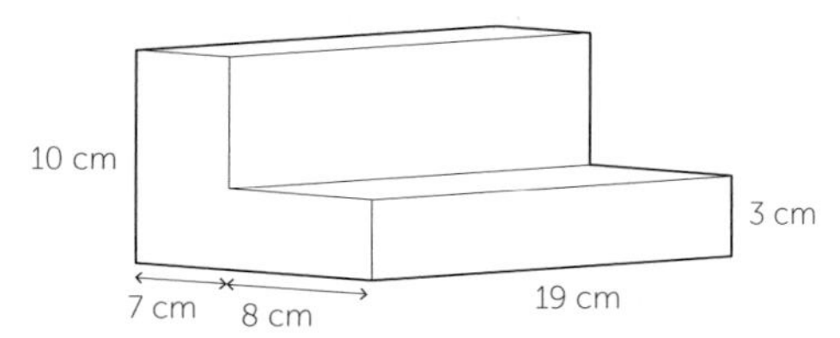


6. A shipping company packs crates into a large metal container.
   The container measures 6 m in length, 2.5 m in width and is 2.6 m high.
   A crate measures 50 cm by 50 cm by 130 cm.
   How many crates will fit into a container?

7. A chocolate bar is packed in a triangular prism-shaped box of length 40 cm and with a triangular cross-sectional area of 6 $cm^2$.
   Work out the volume of the box.

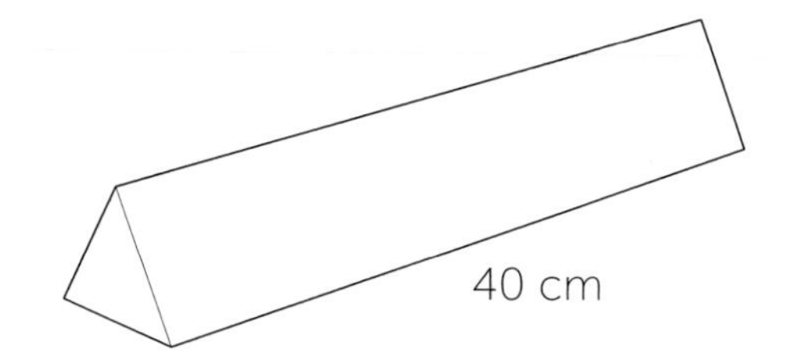


8. The diagram shows the cross-section of a prism.
   The prism is 16 cm long.
   What is the volume of the prism?

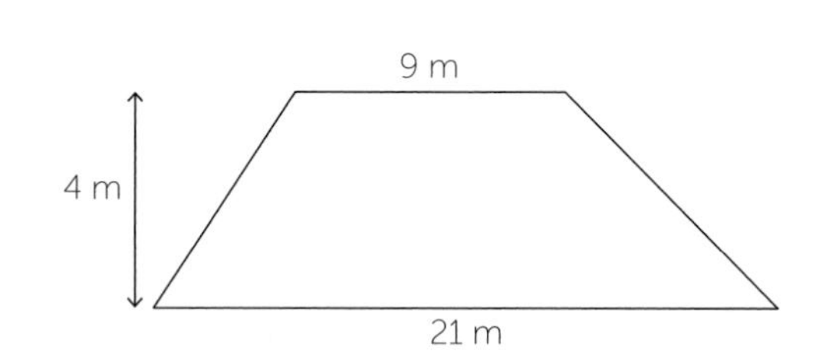


9. Two containers have exactly the same volume.
   One box is a cuboid measuring 10 cm by 9 cm by 8 cm.
   The other box is a prism of length 12 cm.
   What is the cross-sectional area of the prism?

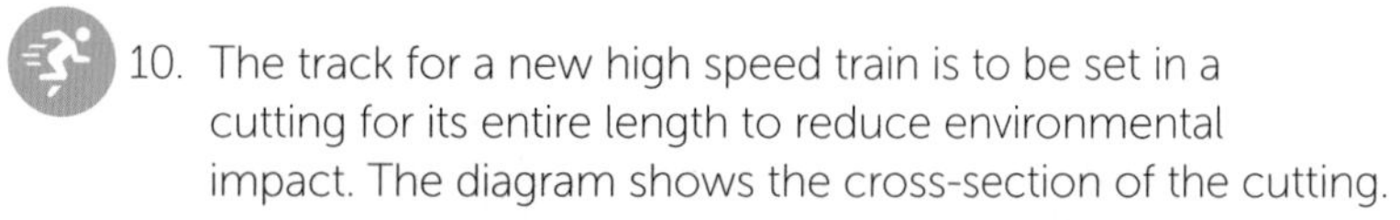

10. The track for a new high speed train is to be set in a cutting for its entire length to reduce environmental impact. The diagram shows the cross-section of the cutting.
    a) Calculate the area of the cross section.
    b) Use your answer to calculate the volume of soil which will be dug out for every kilometre of the railway.
    c) Soil has a mass of 1200 $kg/m^3$.
    Work out the mass of the soil removed for 1 km of track.

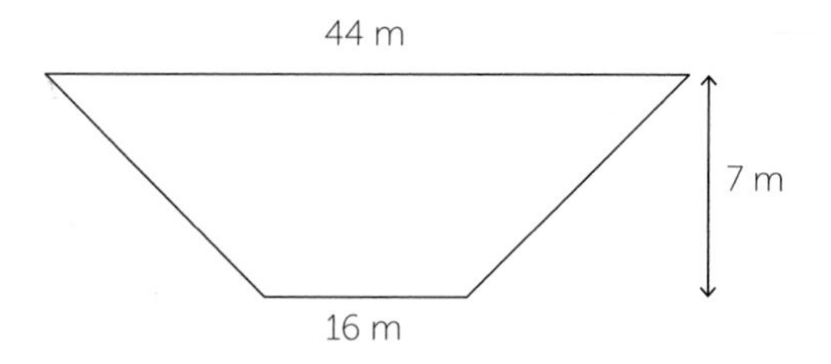

SECTION 6

# FRACTIONS

## 6.1 EQUIVALENT FRACTIONS

N1 R3

### Objectives

#### Understand the terms numerator and denominator

$\frac{3}{8}$ of the circle is shaded. 3 is the **numerator** (the top number). It shows the number of the parts that are shaded.
8 is **denominator** (the bottom number). It shows the number of equal parts the whole has been divide into.

**e.g.** $\frac{5}{8}$ of the circle is not shaded.

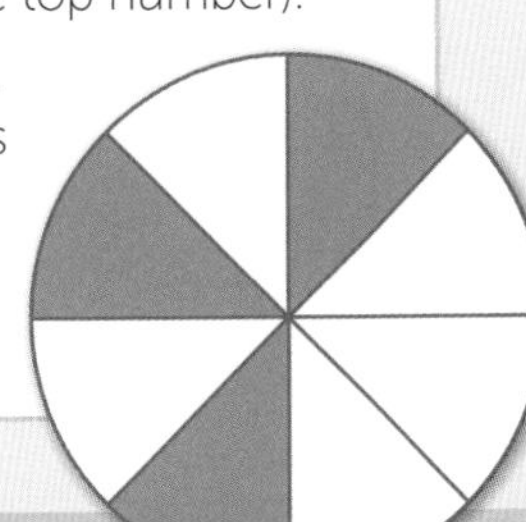

#### Find equivalent fraction

You can find equivalent fractions by multiplying or dividing the numerator and the denominator by the same number.

**e.g.** $\frac{1}{2} = \frac{3}{6}$

$$\frac{1}{2} \underset{\times 3}{\overset{\times 3}{=}} \frac{3}{6}$$

$\frac{4}{8} = \frac{2}{4} = \frac{1}{2}$

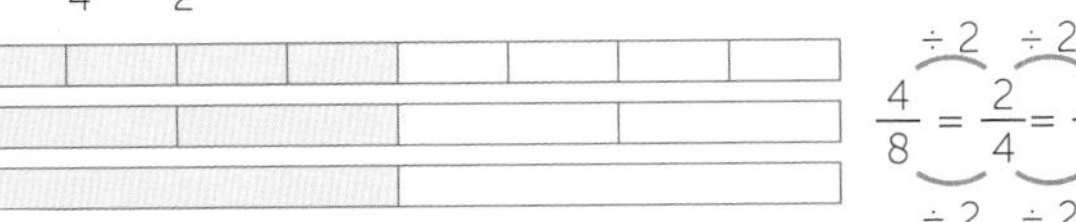

$$\frac{4}{8} \underset{\div 2}{\overset{\div 2}{=}} \frac{2}{4} \underset{\div 2}{\overset{\div 2}{=}} \frac{1}{2}$$

#### Express fractions in their simplest form

To simplify a fraction, divide the numerator and the denominator by a common factor.

The fraction is in its simplest form when it cannot be simplified any further.

**e.g.** Write the fraction of each part of the shape that is shaded.
Give your answer in its simplest form.

$\frac{4}{8} = \frac{1}{2}$ $\frac{4}{12} = \frac{1}{3}$

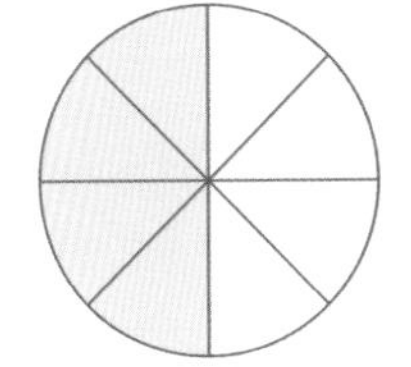

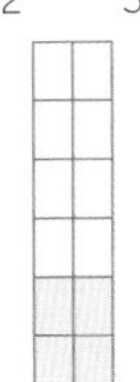

### Practice questions

1. Write each of these fractions into its simplest form.
   a) $\frac{21}{28}$ b) $\frac{45}{72}$ c) $\frac{42}{126}$ d) $\frac{36}{90}$

2. What fraction of each shape is shaded?
   Give your answer in its simplest form.
   a)

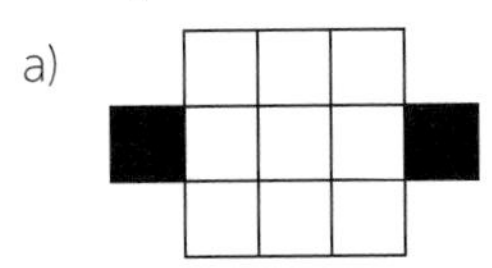

   b)

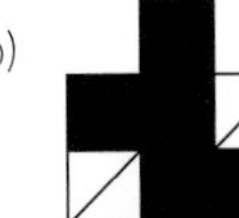

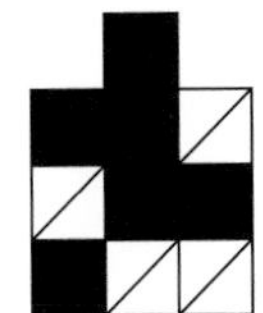

   c)

   d)

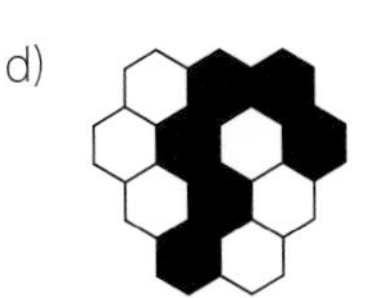

3. Circle the odd one out. Explain your reasoning.
   a) $\frac{9}{3}$ $\frac{8}{24}$ $\frac{11}{33}$ $\frac{7}{21}$ b) $\frac{8}{12}$ $\frac{12}{18}$ $\frac{9}{12}$ $\frac{20}{30}$ c) $\frac{12}{15}$ $\frac{8}{10}$ $\frac{100}{125}$ $\frac{40}{100}$ d) $\frac{60}{140}$ $\frac{12}{14}$ $\frac{72}{84}$ $\frac{30}{35}$

4. Valma thinks that $\frac{32}{36} = \frac{32-30}{36-30} = \frac{2}{6} = \frac{1}{3}$.
   Explain why she is wrong.

5. Find the missing numbers in each of these pairs of equivalent fractions.
   a) $\frac{?}{18} = \frac{20}{24}$ b) $\frac{16}{24} = \frac{20}{?}$ c) $\frac{27}{36} = \frac{15}{?}$ d) $\frac{45}{72} = \frac{30}{?}$

6. Use three numbers from 1, 4, 8, 12, 16, 20 to make this statement correct.
   $\frac{1}{?} = \frac{3}{?} = \frac{75}{300} = \frac{?}{32}$

7. Tom thinks that $\frac{2}{5}$ of the shape is shaded.
   Tom is wrong. Explain why.

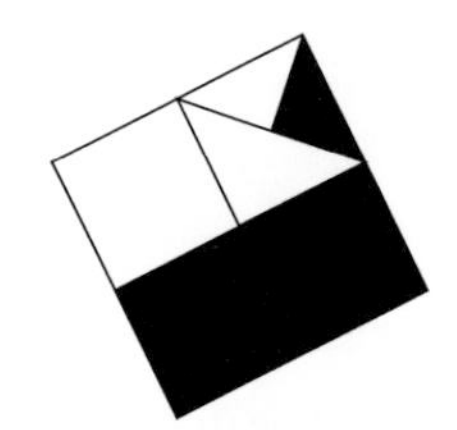

8. Choose integers from 1 to 15 to make each number statement true.
   $\frac{1}{?} = \frac{3}{?}$ $\frac{2}{?} = \frac{8}{?} = \frac{10}{?} = \frac{6}{9}$ $\frac{1}{?} = \frac{?}{8} = \frac{?}{22}$ $\frac{5}{?} = \frac{?}{14}$

9. These are the results that Jessie gets in her latest tests.
   She got the same fraction of answers correct in two subjects.
   Which subjects are they?

| Subject | Marks gained | Total marks |
|---|---|---|
| English | 24 | 30 |
| Science | 16 | 25 |
| French | 15 | 18 |
| Drama | 12 | 18 |
| Spanish | 20 | 24 |

10. There are 30 students in a class.
    18 of the students in the class are girls and 3 of the girls study Spanish.
    a) What fraction of the girls study Spanish?
    b) What fraction of the students are girls that study Spanish?
    Give your answers in the simplest form.

11. Always, sometimes or never true?
    a) To simplify a fraction, multiply or divide the numerator and denominator by the same number.
    b) Equivalent fractions are fractions that have the same denominator.
    c) The numerator and denominator of a fraction in its simplest form are prime numbers.

12. Fill in the missing numbers to make the simplified fraction statements correct.
    a) $\frac{45}{18+?} = \frac{5}{6}$ b) $\frac{30-?}{30+?} = \frac{3}{7}$ c) $\frac{12}{21-?} = \frac{?}{4}$ d) $\frac{?+12}{50-?} = \frac{11}{13}$

## 6.2 PROPER AND IMPROPER FRACTIONS

### Objectives

**Understand the terms proper and improper fractions**

In a **proper fraction**, the numerator is smaller than the denominator.

In an **improper fraction**, the numerator is greater the denominator.

**e.g.** $\frac{1}{3}$ is a proper fraction

$\frac{4}{3}$ is an improper fraction

**Convert between mixed numbers and improper fractions**

**Mixed numbers and improper fractions**

Mixed numbers have a whole number part as well as a fractional part. They convert to improper fractions. Improper fractions can be changed back to mixed numbers.

One whole

One third of one whole

**e.g.** $1\frac{1}{3} = \frac{4}{3}$ or $\frac{4}{3} = 1\frac{1}{3}$

**Compare fractions and write fractions in order of size**

**e.g.** There are three trees in a garden which are $\frac{1}{5}$ m, $\frac{1}{6}$ m and $\frac{3}{10}$ m tall.

Which of the trees is the tallest?

Compare $\frac{1}{5}$, $\frac{1}{6}$ and $\frac{3}{10}$. The LCM of 5, 6 and 10 is 30. Change all the denominators to 30.

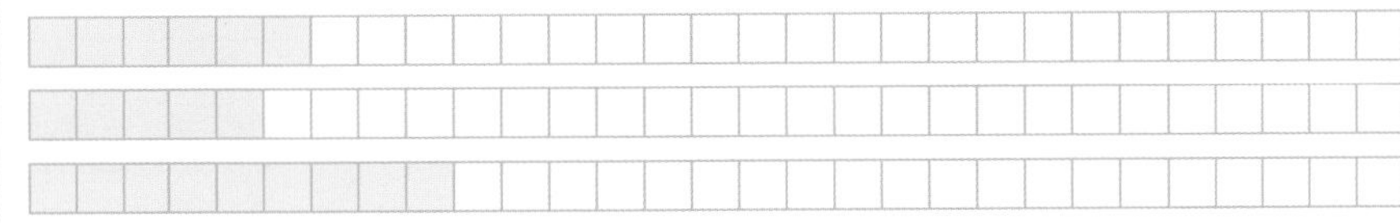

$\frac{1}{5} = \frac{6}{30}$, $\frac{1}{6} = \frac{5}{30}$, $\frac{3}{10} = \frac{9}{30}$ $\Rightarrow$ $\frac{5}{30} < \frac{6}{30} < \frac{9}{30}$ or $\frac{1}{6} < \frac{1}{5} < \frac{3}{10}$

The tallest tree is $\frac{3}{10}$ m

### Practice questions

1. State which fraction is the odd one out in each list and give a reason for your answer.
   a) $\frac{3}{5}$ $\frac{1}{9}$ $\frac{7}{8}$ $\frac{11}{4}$ $\frac{1}{6}$ b) $\frac{10}{3}$ $\frac{13}{9}$ $\frac{6}{7}$ $\frac{11}{4}$ $\frac{3}{2}$ c) $1\frac{3}{11}$ $2\frac{1}{9}$ $3\frac{7}{8}$ $1\frac{1}{5}$ $\frac{4}{7}$ d) $\frac{7}{6}$ $1\frac{1}{2}$ $\frac{11}{8}$ $\frac{2}{5}$ $2\frac{1}{8}$

2. Match each mixed number to the equivalent improper fraction.

   $3\frac{3}{8}$ $4\frac{1}{4}$ $1\frac{1}{5}$ $2\frac{1}{7}$ $1\frac{1}{3}$

   $\frac{6}{5}$ $\frac{5}{3}$ $\frac{34}{8}$ $\frac{10}{7}$ $\frac{4}{3}$ $\frac{7}{5}$ $\frac{15}{7}$ $\frac{27}{8}$

3. Convert these improper fractions to mixed numbers.
   a) $\frac{11}{3}$ b) $\frac{14}{5}$ c) $\frac{23}{4}$ d) $\frac{6}{5}$
   e) $\frac{23}{8}$ f) $\frac{30}{7}$

4. Put the correct sign <, > or = between each pair of fractions.
   a) $\frac{13}{5}$ $1\frac{3}{5}$ b) $\frac{7}{3}$ $3\frac{1}{3}$ c) $2\frac{1}{10}$ $\frac{13}{10}$ d) $\frac{25}{2}$ $12\frac{1}{2}$
   e) $\frac{16}{9}$ $2\frac{2}{9}$ f) $2\frac{1}{13}$ $\frac{16}{13}$

5. Choose one of the symbols, <, >, or = to make these statements true.

   a) $\frac{5}{8}$ $\frac{7}{16}$ b) $\frac{4}{9}$ $\frac{5}{12}$ c) $\frac{9}{24}$ $\frac{3}{24}$ d) $\frac{5}{6}$ $\frac{3}{4}$

   e) $\frac{5}{8}$ $\frac{7}{10}$ f) $\frac{11}{14}$ $\frac{18}{21}$

6. Put each set of fractions in order from smallest to largest.

   a) $\frac{5}{12}$ $\frac{5}{6}$ $\frac{3}{4}$ b) $\frac{7}{10}$ $\frac{3}{5}$ $\frac{5}{8}$ c) $\frac{7}{9}$ $\frac{2}{3}$ $\frac{5}{6}$ d) $\frac{2}{9}$ $\frac{1}{6}$ $\frac{2}{11}$

7. State which is the largest in each of these sets of three fractions.

   a) $1\frac{5}{6}$ $1\frac{3}{4}$ $1\frac{11}{12}$ b) $2\frac{5}{12}$ $2\frac{3}{8}$ $2\frac{1}{2}$ c) $\frac{3}{2}$ $\frac{23}{12}$ $\frac{27}{13}$

8. Three children work out what fraction of the day they spend playing in the holidays.

   Jake: $\frac{7}{9}$ Sam: $\frac{3}{5}$ Polly: $\frac{7}{10}$

   Which child spends the most time playing?

9. Gwen says that $\frac{8}{15} > \frac{6}{11}$ because 8 > 6 and 15 > 11. Is she correct?
   Explain your answer.

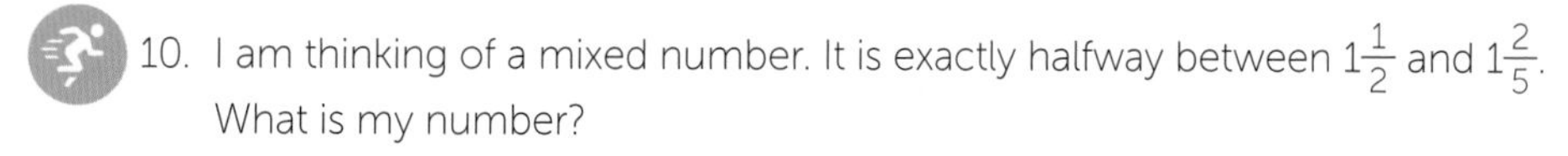

10. I am thinking of a mixed number. It is exactly halfway between $1\frac{1}{2}$ and $1\frac{2}{5}$.
    What is my number?

11. Find the missing numbers to make these statements true.

    a) $2\frac{3}{4} < 2\frac{?}{8}$ b) $4\frac{?}{6} > 4\frac{5}{8}$ c) $3\frac{?}{4} = \frac{45}{12}$ d) $\frac{1}{2} < \frac{?}{8} < \frac{?}{20} < \frac{2}{3}$

## 6.3 ADDING AND SUBTRACTING FRACTIONS

**N2** **N8**

### Objectives

**Add and subtract proper fractions**

#### Adding fractions

Make sure that the denominators are the same, by making equivalent fractions.

Add the numerators.

**e.g.** Show that $\frac{1}{3} + \frac{1}{6} = \frac{1}{2}$

The lowest common multiple of 3 and 6 is 6:

$\frac{1}{3} = \frac{?}{6}$ ⇒ $\frac{1 \times 2}{3 \times 2} = \frac{2}{6}$

$\frac{2}{6} + \frac{1}{6} = \frac{3}{6}$ which simplifies to $\frac{1}{2}$

#### Subtracting fractions

Make sure that the denominators are the same, by making equivalent fractions.

Subtract the numerators.

**e.g.** Show that $\frac{3}{4} - \frac{2}{5} = \frac{7}{20}$

The lowest common multiple of 4 and 5 is 20:

$\frac{3}{4} = \frac{?}{20}$ and $\frac{2}{5} = \frac{?}{20}$

$\frac{3 \times 5}{4 \times 5} = \frac{15}{20}$ and $\frac{2 \times 4}{5 \times 4} = \frac{8}{20}$

$\frac{15}{20} - \frac{8}{20} = \frac{7}{20}$

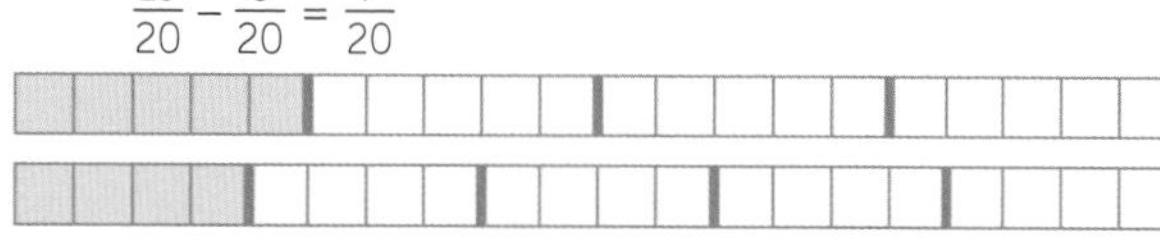

**e.g.** Faryah completed $\frac{1}{2}$ of her work on Monday, $\frac{1}{3}$ of her work on Tuesday and the rest on Wednesday. What fraction of the work did she complete on Wednesday?

The fraction of the work completed on Monday and Tuesday $= \frac{1}{2} + \frac{1}{3}$

$= \frac{3}{6} + \frac{2}{6}$

$= \frac{5}{6}$

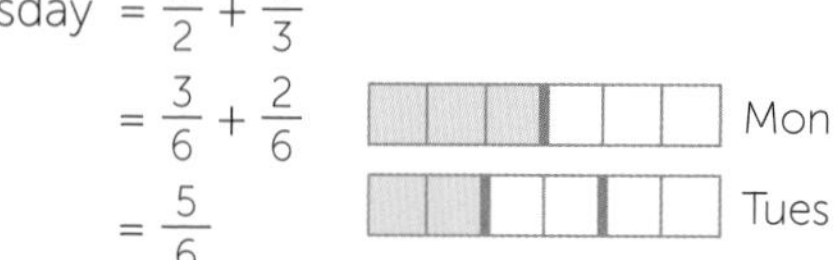


The fraction of the work completed on Wednesday

$= 1 - \frac{5}{6}$

$= \frac{1}{6}$

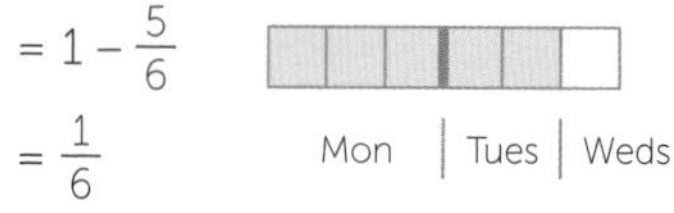

## Practice questions

1. Work out the following, simplifying your answer where possible.

   a) $\frac{3}{5}+\frac{1}{5}$ b) $1-\frac{1}{3}$ c) $\frac{1}{6}+\frac{5}{6}$ d) $\frac{7}{10}-\frac{3}{10}$ e) $\frac{7}{8}-\frac{1}{8}$ f) $\frac{5}{12}+\frac{1}{12}$

   g) $\frac{3}{10}+\frac{1}{10}$ h) $\frac{8}{9}-\frac{1}{9}$

2. Work out the following, simplifying your answer where possible.

   a) $\frac{3}{8}+\frac{1}{4}$ b) $\frac{2}{3}-\frac{5}{9}$ c) $\frac{1}{2}+\frac{1}{20}$ d) $\frac{11}{12}-\frac{3}{6}$ e) $\frac{3}{4}+\frac{1}{8}$ f) $\frac{1}{2}-\frac{5}{12}$

   g) $\frac{2}{5}+\frac{7}{15}$ h) $\frac{7}{3}-\frac{11}{6}$

3. Match each calculation to its answer. Two of the answers do not match.

   a) $\frac{3}{8}+\frac{1}{5}$ b) $\frac{5}{6}-\frac{1}{9}$ c) $\frac{1}{10}+\frac{1}{4}$ d) $\frac{1}{2}-\frac{2}{9}$

   i) $\frac{23}{40}$ ii) $\frac{23}{80}$ iii) $\frac{7}{20}$ iv) $\frac{13}{18}$ v) $\frac{5}{18}$ vi) $\frac{4}{13}$

4. Sam's homework has some mistakes. Find the mistakes and correct them.

   a) $\frac{2}{3}+\frac{1}{5}=\frac{10}{15}+\frac{3}{15}=\frac{13}{30}$ b) $1-\frac{9}{12}=\frac{1}{3}$ c) $\frac{1}{4}+\frac{1}{4}=\frac{1}{2}$ d) $\frac{2}{10}-\frac{1}{8}=\frac{1}{2}$

5. Find the missing numbers.

   a) $?+\frac{1}{10}=\frac{3}{5}$ b) $\frac{5}{8}-?=\frac{5}{16}$ c) $?-\frac{1}{3}=\frac{1}{18}$ d) $1-?=\frac{7}{9}$

6. Here is a list of fractions. $\frac{5}{15}$ $\frac{5}{6}$ $\frac{1}{10}$ $\frac{7}{9}$ $\frac{3}{8}$ $\frac{1}{4}$

   Choose from the list:

   a) Two fractions that have the largest sum. What is their sum?

   b) Two fractions that have the smallest difference. What is their difference?

7. Ben spent $\frac{3}{5}$ of an hour on his Maths homework. He spent $\frac{1}{3}$ of an hour on his English homework. What fraction of an hour did Ben spend on his homework altogether?

8. Work out

   a) $\frac{3}{5}+\frac{1}{10}+\frac{1}{15}$ b) $\frac{11}{12}-\frac{1}{4}-\frac{1}{6}$ c) $\frac{9}{10}-\frac{1}{3}+\frac{3}{5}$ d) $\frac{8}{9}-\frac{1}{2}+\frac{1}{4}$

9. I read $\frac{1}{5}$ of a book on Monday. By Tuesday, I had read $\frac{3}{4}$ of the book.

   What fraction of the book did I read on Tuesday?

10. Find the missing numbers.

    a) $\frac{?}{3}+\frac{5}{18}=\frac{17}{18}$ b) $\frac{3}{4}-\frac{?}{16}=\frac{5}{16}$ c) $\frac{1}{?}+\frac{1}{?}=\frac{1}{30}$ d) $\frac{1}{?}-\frac{1}{?}=\frac{1}{3}$

11. Kate, Jo and Will shared a pizza.

    Kate said that she ate $\frac{1}{4}$ of the pizza. Jo said that she ate $\frac{1}{3}$ of the pizza. Will said that he ate $\frac{1}{2}$ of the pizza.

    Explain why this cannot be true.

12. Noah added two fractions with different denominators to get $\frac{5}{6}$.

    He also subtracted two fractions with different denominators to get $\frac{5}{6}$.

    Write down the fractions he might have used for each calculation.

# 6.4 MIXED NUMBERS

## Objectives

### Add and subtract mixed numbers and improper fractions

Change the fractions to improper fractions first, and then add them.

Change back to a mixed number.

**e.g.** Work out, without a calculator $2\frac{2}{3} + 1\frac{1}{2}$

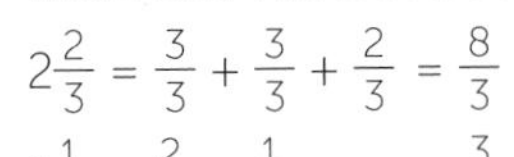

$2\frac{2}{3} = \frac{3}{3} + \frac{3}{3} + \frac{2}{3} = \frac{8}{3}$

$1\frac{1}{2} = \frac{2}{2} + \frac{1}{2} \quad = \frac{3}{2}$

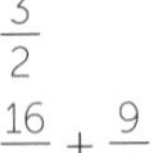

$\frac{8}{3} + \frac{3}{2} \quad = \frac{16}{6} + \frac{9}{6}$

$= \frac{25}{6}$

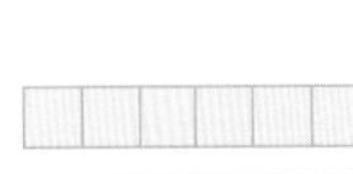

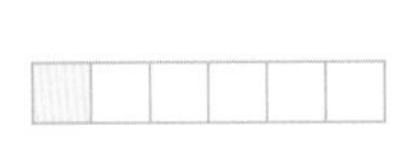

$\frac{25}{6} \quad = 4\frac{1}{6}$

### Convert between mixed numbers and improper fractions

Change the fractions to improper fractions first, and then add them.

Change back to a mixed number if necessary.

**e.g.** Work out, without a calculator $3\frac{1}{2} - 1\frac{2}{5}$

$3\frac{1}{2} = \frac{2}{2} + \frac{2}{2} + \frac{2}{2} + \frac{1}{2} \quad = \frac{7}{2}$

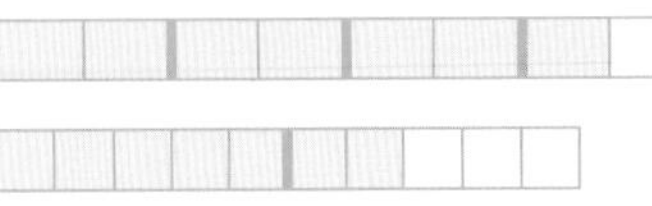

$1\frac{2}{5} = \frac{5}{5} + \frac{2}{5} \quad = \frac{7}{5}$

$\frac{7}{2} - \frac{7}{5} \quad = \frac{35}{10} - \frac{14}{10}$

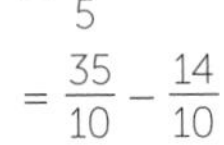

$= \frac{21}{10}$

$= 2\frac{1}{10}$

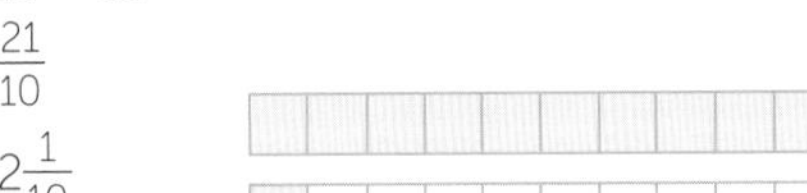

## Practice questions

1. Work out the following, simplifying your answer where possible. Give your answer as a mixed number.
   a) $\frac{8}{9} + \frac{1}{2}$ b) $\frac{7}{10} + \frac{3}{4}$ c) $1\frac{1}{2} + \frac{5}{6}$ d) $2\frac{7}{10} + \frac{3}{5}$ e) $1\frac{7}{12} - \frac{1}{4}$ f) $1\frac{5}{12} - \frac{1}{2}$
   g) $2\frac{1}{4} - \frac{1}{2}$ h) $2\frac{1}{3} - 1\frac{1}{9}$

2. Work out the following, simplifying your answer where possible. Use a calculator to check your answers.
   a) $3\frac{7}{8} + 2\frac{1}{2}$ b) $1\frac{1}{2} + \frac{5}{6}$ c) $3\frac{1}{4} + 1\frac{1}{3}$ d) $2\frac{7}{10} - \frac{3}{5}$ e) $2\frac{8}{15} - 1\frac{4}{5}$ f) $2\frac{3}{20} - 1\frac{1}{10}$
   g) $2\frac{5}{9} - 1\frac{5}{6}$ h) $3\frac{1}{7} - \frac{4}{5}$

3. Here is a list of fractions. $1\frac{3}{5}$, $2\frac{10}{15}$, $7\frac{1}{3}$, $1\frac{1}{7}$, $2\frac{1}{10}$, $1\frac{1}{2}$. From the list choose:
   a) two fractions that have the sum of 10.
   b) two fractions that have the difference of $\frac{1}{2}$.
   c) two fractions such that one of them is $\frac{5}{14}$ more than the other one.
   d) two fractions such that one of them is $6\frac{4}{21}$ less than the other one.

4. Work out these calculations. Which is the odd one out?

a) $1\frac{2}{3} + \frac{4}{5}$ b) $5 - 1\frac{8}{15}$ c) $2\frac{2}{5} + 1\frac{1}{15}$ d) $4\frac{4}{5} - 1\frac{1}{3}$

5. Use >, < or = to make these statements true:

a) $2\frac{1}{2} + 1\frac{4}{5}$ ... $5\frac{1}{2} - 1\frac{1}{5}$ b) $\frac{2}{3} + 3\frac{1}{11}$ ... $5 - 1\frac{8}{33}$ c) $2\frac{1}{9} - 1\frac{5}{6}$ ... $1\frac{1}{6} + \frac{2}{3}$ d) $1\frac{1}{8} + \frac{5}{12}$ ... $1\frac{5}{8} + \frac{1}{12}$

6. Add or subtract to complete these addition pyramids. Give the answers as mixed numbers.

a)

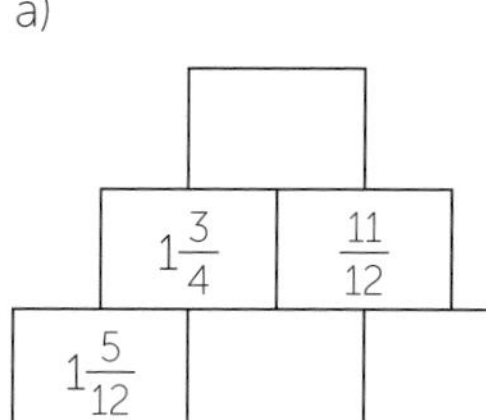


b)

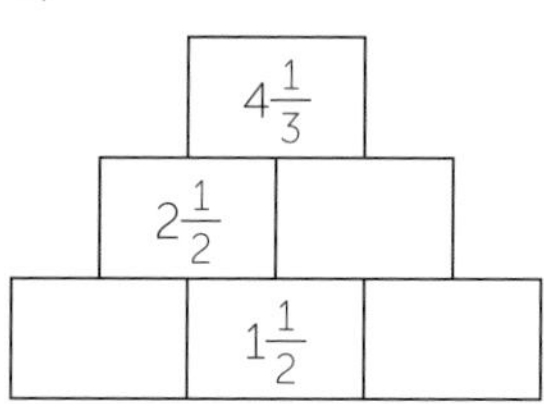


7. Ben mixed $10\frac{1}{2}$ kg of soil and $4\frac{3}{5}$ kg of soil improver.
He wants to fill in a pot which takes 16 kg. How much more soil mixture does he need?

8. Work out each of these calculations. Which one has the largest answer?

a) $2\frac{1}{4} + 3\frac{7}{20} - 1\frac{1}{5}$ b) $2\frac{7}{10} - 1\frac{1}{4} + 1\frac{1}{5}$ c) $3\frac{1}{8} - 1\frac{9}{16} + 1\frac{1}{4}$ d) $3\frac{8}{10} - 1\frac{1}{2} - 1\frac{1}{3}$

9. The fence around the garden is $5\frac{1}{2}$ m.
Calculate the missing length.

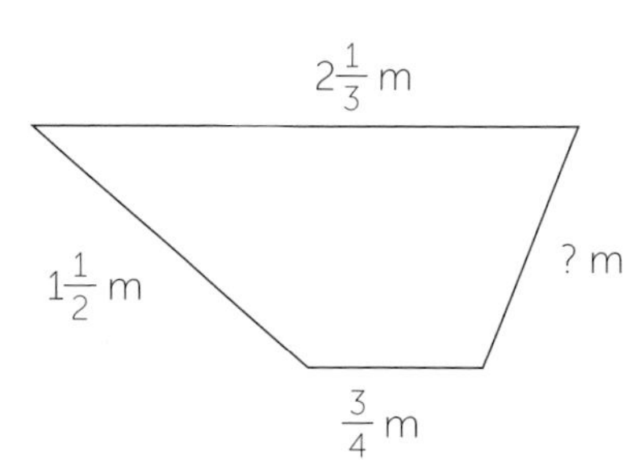


10. Fiona walks $1\frac{9}{10}$ km on Monday. On Tuesday she walks $1\frac{1}{3}$ km further than on Monday.
From Monday to Wednesday, Fiona walks 6 km in total.

a) How many km does she walk on Tuesday?

b) How many km does she walk on Wednesday?

# 6.5 MULTIPLYING FRACTIONS

## Objectives

### Multiply proper and improper fractions

**Multiplying proper fractions**

Find common factors in the numerators and denominators, and cancel.

Multiply the numerators together and multiply the denominators together.

Simplify where possible.

*(Cancelling and simplifying could be done at the end.)*

**e.g.** Work out: $\frac{3}{5} \times \frac{4}{3}$

$= \frac{\cancel{3}^1 \times 4}{5 \times \cancel{3}^1}$

$= \frac{4}{5}$

or by doing: $\frac{3}{5} \times \frac{4}{3}$

$= \frac{3 \times 4}{5 \times 3}$

$= \frac{\cancel{12}^4}{\cancel{15}^5} = \frac{4}{5}$

When multiplying by an integer, change the integer to a fraction. Use 1 as the denominator.

**e.g.** $\frac{2}{3} \times 9 \Rightarrow \frac{2 \times \cancel{9}^3}{\cancel{3}^1 \times 1}$

$= \frac{2 \times 3}{1 \times 1} = 6$

**Multiplying mixed numbers and improper fractions**

Change the mixed numbers to improper fractions before performing the multiplication.

**e.g.** Work out, without a calculator $2\frac{1}{7} \times 1\frac{2}{5}$

$2\frac{1}{7} = \frac{7}{7} + \frac{7}{7} + \frac{1}{7} = \frac{15}{7}$

$1\frac{2}{5} = \frac{5}{5} + \frac{2}{5} = \frac{7}{5}$

$\frac{\cancel{15}^3 \times \cancel{7}^1}{\cancel{7}^1 \times \cancel{5}^1} = \frac{3 \times 1}{1 \times 1}$

$= 3$

### Find the reciprocal of a fraction

**Reciprocals**

The reciprocal of a number $= \frac{1}{number}$. If the number is a fraction, turn it upside down.

Mixed numbers must be turned into improper fractions first.

**e.g.** the reciprocal of $\frac{2}{5} = \frac{5}{2} = 2\frac{1}{2}$

the reciprocal of 10 or $\frac{10}{1} = \frac{1}{10}$

the reciprocal of $1\frac{1}{8}$ or $\frac{9}{8} = \frac{8}{9}$

## Practice questions

1. Work out the following, simplifying your answer where possible. Give your answer as a mixed number.

a) $\frac{4}{5} \times \frac{1}{2}$ b) $\frac{7}{10} \times 20$ c) $\frac{1}{2} \times \frac{4}{9}$ d) $\frac{15}{21} \times \frac{3}{5}$ e) $8 \times \frac{3}{16}$ f) $\frac{3}{4} \times 28$

2. Work out the following, simplifying your answer where possible.

Use a calculator to check your answers.

a) $1\frac{1}{16} \times \frac{8}{9}$ b) $1\frac{1}{6} \times \frac{10}{11}$ c) $\frac{7}{10} \times 1\frac{7}{28}$ d) $15 \times 1\frac{1}{10}$ e) $15 \times 2\frac{2}{9}$ f) $2\frac{1}{3} \times \frac{5}{14}$

3. Work out the following, simplifying your answer where possible.
   Use a calculator to check your answers.
   a) $1\frac{3}{7} \times 1\frac{1}{2}$  b) $2\frac{1}{3} \times 3\frac{5}{14}$  c) $3\frac{1}{5} \times 4\frac{1}{6}$  d) $2\frac{4}{15} \times 3\frac{1}{3}$  e) $3\frac{1}{4} \times 2\frac{2}{13}$  f) $3\frac{1}{8} \times 1\frac{3}{5}$

4. Match each fraction with its reciprocal.
   Two of the numbers do not match.

| | | | |
|---|---|---|---|
| $\frac{1}{2}$ | $-2$ | $\frac{2}{9}$ | $\frac{3}{5}$ |
| $1\frac{2}{3}$ | $2$ | $4\frac{1}{2}$ | $1\frac{3}{5}$ |

5. Work out the following:
   a) $\frac{7}{8} \times \frac{1}{2} =$  b) $\frac{3}{10} \times 1\frac{1}{2} =$  c) $\frac{5}{11} \times 4\frac{2}{5} =$  d) $1\frac{1}{3} \times \frac{9}{16}$

6. This is a plan of a children's playing area.

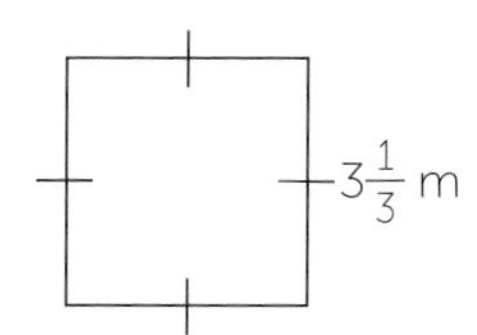


   It is square. George thinks that the number of metres for the perimeter is greater than the number of square metres for the area.
   Is he correct?

7. Mrs Kott used $1\frac{5}{9}$ kg of oranges every day.
   How many kg of oranges did Mrs Kott use in 10 days?

8. Are the following statements true, sometimes true or never true?
   a) The product of two fractions less than 1 is smaller than the fractions multiplied.
   b) The product of a number with its reciprocal is 1.
   c) The reciprocal of a fraction is smaller than the fraction.

9. Explain the mistake in this calculation. Rewrite the correct solution.

$$1\frac{3}{4} \times 4\frac{1}{3} =$$

$$1 \times 4 \times \frac{1^{3}}{4} \times \frac{1}{3_{1}}$$

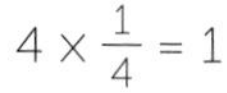

$$4 \times \frac{1}{4} = 1$$

10. In a race Jake covered a total distance of $\frac{14}{15}$ km.
    He ran $\frac{5}{6}$ of the distance and walked the rest.
    How many kilometres did he run?

11. $\frac{2}{5}$ of the school garden is used to grow vegetables.
    $\frac{1}{3}$ of the remaining garden is used to grow strawberries.
    Owen thinks that $\frac{2}{15}$ of the school garden is used to grow strawberries.
    Is Owen correct? Explain your answer.

12. There are pink and yellow marbles in a jar.
    Some are small, and the rest are large.
    $\frac{3}{4}$ of the marbles are yellow and $\frac{1}{5}$ of the pink marbles are large. What fraction of the marbles are pink and small?

# 6.6 DIVIDING FRACTIONS

## Objectives

### Divide proper and improper fractions

**Dividing proper fractions**

Dividing by a fraction is the same as multiplying by its reciprocal.

**e.g.** Show that $\frac{2}{5} \div \frac{3}{4} = \frac{8}{15} \Rightarrow \frac{2}{5} \times \frac{4}{3}$

$= \frac{2 \times 4}{5 \times 3} = \frac{8}{15}$

When dividing by an integer, change the integer to a fraction. Use 1 as the denominator.

*(Cancelling could be done at the end.)*

**e.g.** Work out $\frac{2}{3} \div 6$

$= \frac{2}{3} \div \frac{6}{1}$

$= \frac{2}{3} \times \frac{1}{6}$

$= \frac{\cancel{2}^1 \times 1}{3 \times \cancel{6}^3}$

$= \frac{\cancel{2}^1}{\cancel{18}^9} = \frac{1}{9}$

**Dividing mixed numbers and improper fractions**

Change the mixed numbers to improper fractions before performing the division.

**e.g.** Work out, without a calculator $1\frac{2}{3} \div \frac{2}{5}$

$1\frac{2}{3} = \frac{3}{3} + \frac{2}{3} = \frac{5}{3}$

$\frac{5}{3} \div \frac{2}{5} = \frac{5}{3} \times \frac{5}{2}$

$= \frac{5 \times 5}{3 \times 2}$

$= \frac{25}{6}$ or $4\frac{1}{6}$

## Practice questions

1. Work out the following, simplifying your answer where possible. Give answers as mixed numbers.

a) $\frac{4}{5} \div \frac{1}{2}$ b) $\frac{7}{10} \div 3$ c) $\frac{1}{2} \div \frac{5}{6}$ d) $\frac{7}{10} \div \frac{1}{5}$ e) $\frac{5}{12} \div \frac{1}{6}$ f) $12 \div \frac{3}{4}$

g) $\frac{1}{4} \div 12$ h) $\frac{1}{3} \div \frac{5}{9}$

2. Work out the following, simplifying your answer where possible.

a) $1\frac{1}{8} \div 9$ b) $1\frac{1}{2} \div \frac{5}{6}$ c) $2\frac{1}{3} \div 7$ d) $2\frac{7}{10} \div \frac{3}{5}$ e) $18 \div 1\frac{4}{5}$ f) $\frac{3}{20} \div 1\frac{1}{10}$

g) $\frac{1}{2} \div 1\frac{10}{11}$ h) $\frac{1}{10} \div 1\frac{4}{25}$

3. Work out the following, simplifying your answer where possible.
Use a calculator to check your answers.

a) $1\frac{1}{8} \div 1\frac{1}{2}$ b) $2\frac{1}{2} \div 1\frac{5}{6}$ c) $2\frac{1}{3} \div 1\frac{1}{9}$ d) $1\frac{7}{30} \div 4\frac{1}{3}$ e) $2\frac{4}{15} \div 1\frac{1}{3}$ f) $3\frac{5}{12} \div 4\frac{1}{10}$

g) $10\frac{1}{2} \div 1\frac{3}{4}$ h) $15\frac{5}{8} \div 1\frac{1}{4}$

4. Match the calculations that have the same answer.

$5 \div 1\frac{8}{15}$ $1\frac{8}{15} \div 5$ $1\frac{3}{5} \div 5$ $4\frac{3}{5} \div 15$

5. Which is bigger?

a) $\frac{4}{9} \div 2$ or $\frac{4}{9} \div \frac{1}{2}$ b) $2\frac{7}{10} \div 1\frac{1}{2}$ or $2\frac{7}{10} \div 1\frac{1}{3}$ c) $3\frac{1}{8} \div 1\frac{1}{4}$ or $2\frac{5}{6} \div 1\frac{1}{4}$

6. A ribbon is $4\frac{5}{8}$ m long. How many pieces of $\frac{1}{2}$ m can be cut from the ribbon?

7. $\frac{3}{4}$ m of a water pipe has a mass of $\frac{2}{3}$ kg. To find the mass of 1 m of the water pipe
Anushka does the following calculation:
$$\frac{3}{4} \div \frac{2}{3} = \frac{2}{4}$$
$$= \frac{1}{2} \text{ kg}$$
Write down **two** mistakes that Anushka made.

8. Kleo bought a plot of land with an area of $1\frac{3}{5}$ km$^2$. She divided the land into smaller plots of equal area of $\frac{16}{45}$ km$^2$.
How many smaller plots did Kleo have?

9. Bethan buys $1\frac{9}{10}$ kg of mushrooms. She divides them into 6 equal portions.
   a) Find the mass in kg of one portion of mushrooms.
   b) Find the total mass in kilograms of 5 portions.

10. Mary uses $1\frac{3}{4}$ kg of flour to make 20 bread rolls.
How many kg of flour are used for:
   a) 12 bread rolls?
   b) 8 bread rolls?

SECTION 7

# STRAIGHT LINE GRAPHS

## 7.1 WORKING WITH COORDINATES

A8

### Objectives

**Plot and read coordinates in all four quadrants**

All coordinates are written in the form $(x, y)$ where:

$x$ is the horizontal movement from (0, 0) and
$y$ is the vertical movement.

**e.g.** Point A is at (−2, −2), B is at (3, 4), C is at (−3, 3) and D is at (2, 0)

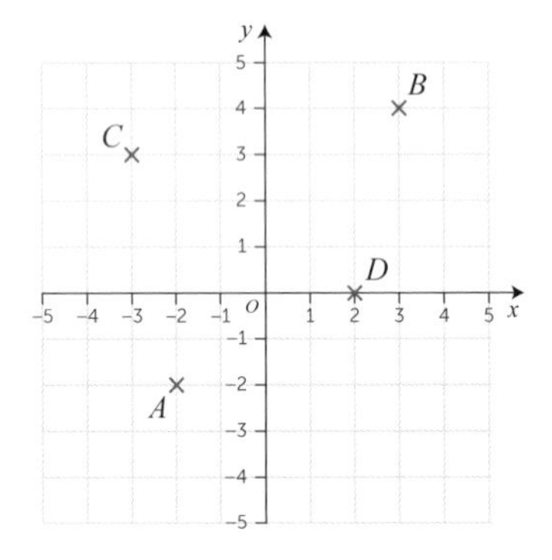


**Plot and read coordinates in all four quadrants**

**Midpoint of a line segment**

The midpoint has coordinates exactly in the middle of the coordinates of the two end points.

We calculate the midpoint between $(x_1, y_1)$ and $(x_2, y_2)$ by using $\left(\frac{x_1 + x_2}{2}, \frac{y_1 + y_2}{2}\right)$

**e.g.** Midpoint of AB is $\left(\frac{-4 + 2}{2}, \frac{-2 + 5}{2}\right) = (-1, 1.5)$

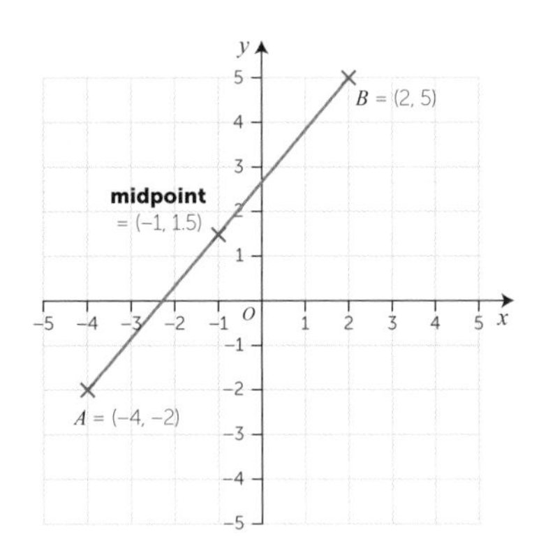


**Solve problems involving coordinates**

### Practice questions

1. Look at the coordinate grid.
   a) Write down the coordinates of points A, B and C.
   b) Write down the coordinates of a fourth point D, which forms a square ABCD.

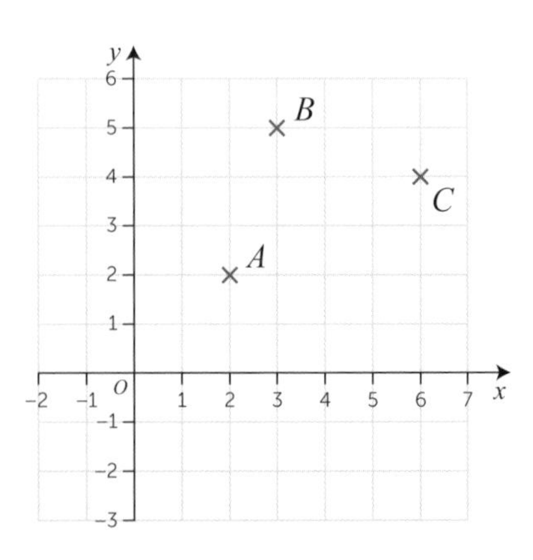

2. Using this letter grid, write down the letters at these points.
   a) (2, 3) b) (−6, −1)
   c) (−9, 0) d) (1, −6)
   e) (−5, 8) f) (−4, −9)

3. Write down the coordinates for these letters.
   a) W b) X
   c) D d) E
   e) B f) C

4. Write down the name of each line segment on the letter grid, which has these midpoints.
   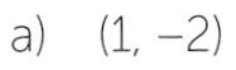
   a) (1, −2) b) (−6, 1.5)
   c) (4, 5) d) (−5, 5.5)

5. Find the coordinates of the midpoint of each of these line segments on the letter grid.

   a) OX b) SE c) AH d) BR

6. (9, 4) and (1, 2) are connected with a line segment.
   a) Find the coordinates of the midpoint of the line segment.
   b) Write down the coordinates of two more points, with positive integer coordinates which have the same midpoint.

7. Find the midpoint of the following pairs of coordinates.
   a) (0, 0) and (6, 8) b) (−4, 10) and (0, 0) c) (2, 6) and (−2, 4) d) (−1, 3) and (−7, 4)

8. Look at this grid.
   a) Write down the coordinates of the point D that forms a rhombus ABCD.
   b) Work out the midpoint of AC and of BD
   c) Comment on what you notice.

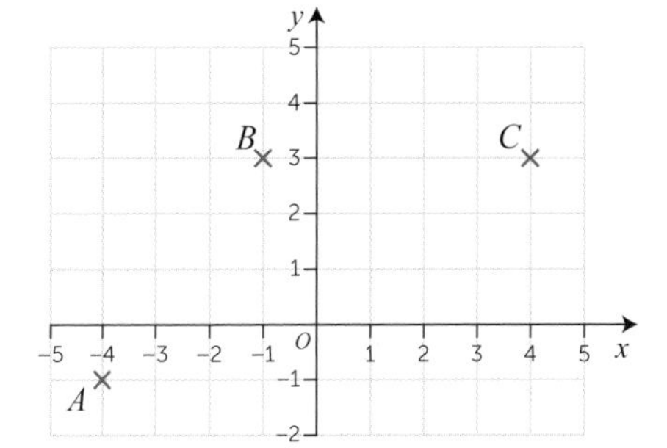


9. Gary says that the midpoint of (3, 4) and $(p, q)$ is (−1, 1.5)
   Find the values of $p$ and $q$.

# 7.2 EQUATIONS OF LINES

## Objectives

### To recognise and state equations of horizontal and vertical lines

Here is a set of coordinates: (2, 0) (2, 5) (2, 11) and (2, −3)

All these coordinates have an **$x$-coordinate** of 2

A line which passes through these points has the equation $x = 2$

**Horizontal lines** have the equation $y = a$

**Vertical lines** have the equation $x = a$

($a$ can equal any number).

**e.g.** The green horizontal line has the equation $y = 3$

The pink vertical line has the equation $x = 4$

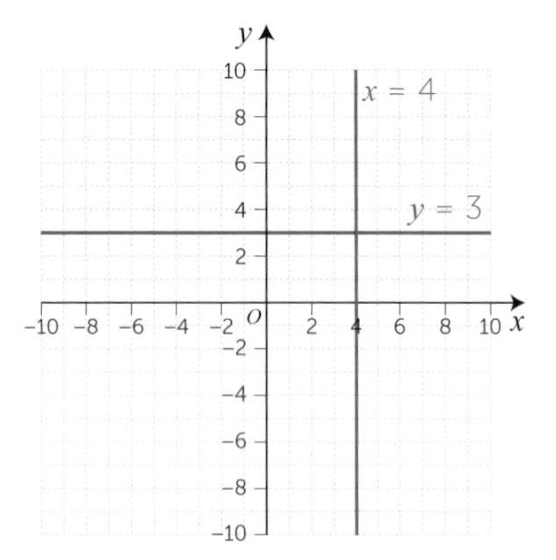


### To recognise the lines $y = x$ and $y = -x$

**Diagonal lines** have equations that involve both $y$ and $x$.

The equation describes the relationship between the $x$ and $y$ coordinates.

**e.g.** Equation of the pink line is $y = x$

Equation of the green line is $y = -x$

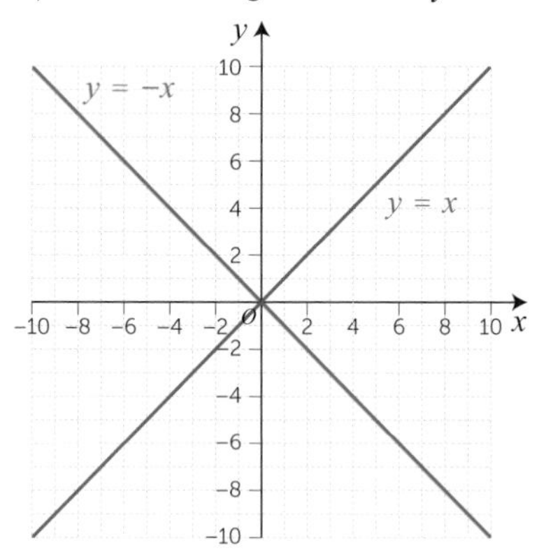


## Practice questions

1. Copy and complete the sentences below using one of these words.

   horizontal vertical parallel perpendicular

   a) The line with equation $x = 4$ is ____________.
   b) The line with equation $y = -3$ is ____________.
   c) The line with equation $x = 10$ is _________ to the $x$-axis.
   d) The line with equation $y = 1$ is _________ to the $x$-axis

2. State three sets of coordinates on each of these lines.

   a) $x = 3$ b) $y = 7$ c) $y = -2$ d) $x = -1$

3. Hattie is looking at this line.

   She says that the equation of the line is $x = 4$ because the $x$-axis is also horizontal.

   Explain why she is not correct.

   Write down the equation of the line.

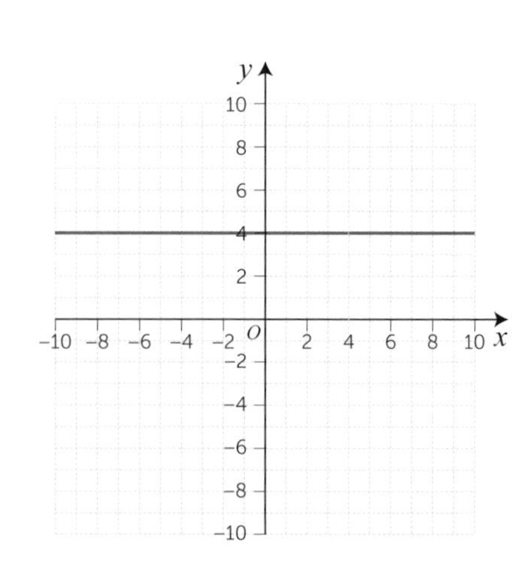

4. Copy this grid and draw on it, the following lines.

   a) $x = 2$ b) $y = 5$

   c) $y = -7$ d) $x = -3$

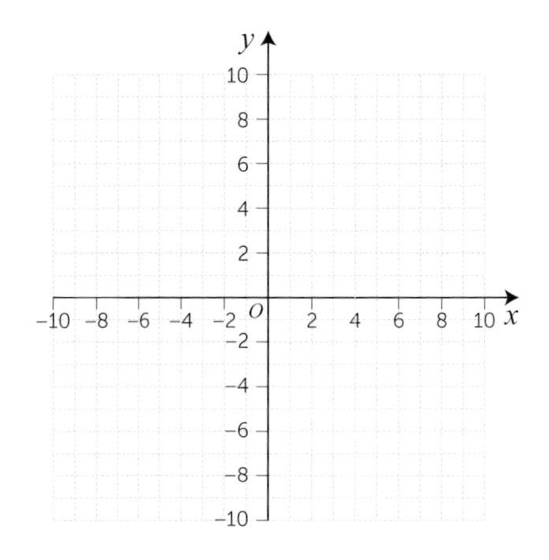

5. Look at the coordinate grid.

   Write down the equations of the lines A, B, C and D.

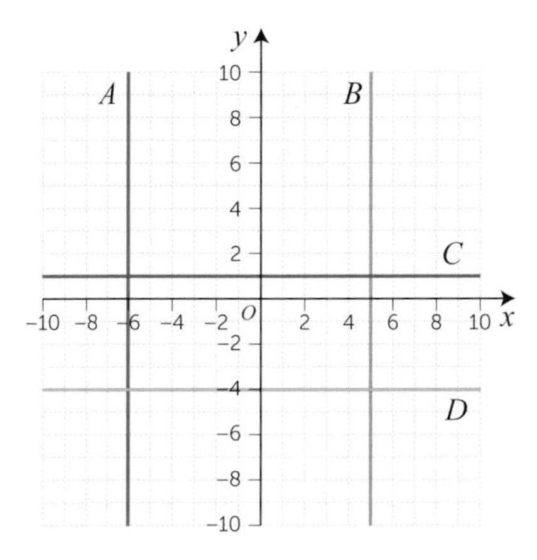


6. P is plotted at (3, −4). Write down the equations of two lines which pass through P.

7. George says that the points (0, 0), (−4, 4) and (4, −4) all lie on the same line.

   Is he correct? Give a reason for your answer.

8. Choose the odd one out of these sets of coordinates. Explain why it does not fit the pattern.

   a) (3, 6), (3, −5) (−3, 8) (3, 0) b) (0, 0), (2, 2), (−4, 4) and (−6, −6)

9. State the equations of the lines that contain each of these sets of coordinates.

   a) (−6, 3), (−2, 3), ( 1, 3) and (3, 3) b) (2, −2), (0, 0), ( −5, 5) and (14, −14)

10. A company is designing a logo using a coordinate grid.

    The design they require looks like this.

    Write down the equations of all the lines used to make the logo.

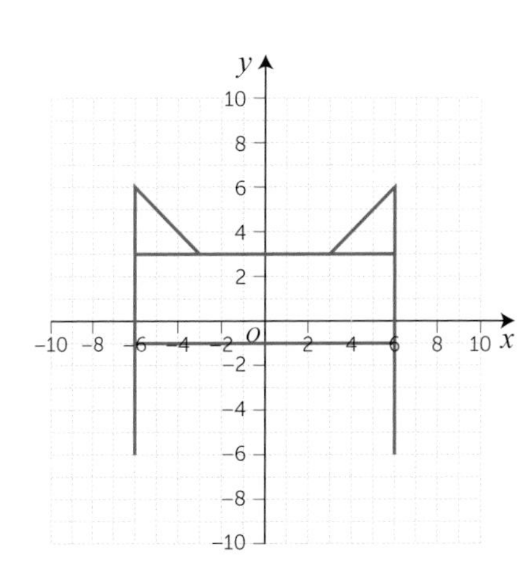

# 7.3 PLOTTING GRAPHS 1

## Objectives

### Understand that the general equation of a straight line is $y = mx + c$

All straight lines that are not horizontal or vertical can be written in the form $y = mx + c$

**e.g.** $y = 3x + 1$, $y = 5 - 2x$ and $y = \frac{1}{2}x + 6$ are all equations of straight lines.

The equation shows the relationship between the $x$-coordinates and the $y$-coordinates of the points on the line.

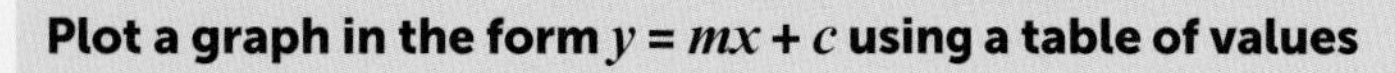

### Plot a graph in the form $y = mx + c$ using a table of values

To draw a graph from the equation, work out the $y$-coordinates from the $x$-coordinates using a table of values.

**e.g.** Plot the line $y = 3x - 1$ for $x$ values between −2 and 2.

Draw a table of values; put in the $x$-coordinates and work out the $y$-coordinates.

| $x$ | −2 | −1 | 0 | 1 | 2 |
|---|---|---|---|---|---|
| Calculation | 3 × (−2) − 1 | 3 × (−1) − 1 | 3 × (0) − 1 | 3 × (1) − 1 | 3 × (2) − 1 |
| $y$ | −7 | −4 | −1 | 2 | 5 |

Plot the $x$- and $y$-coordinates on a grid. Join the points up with a line.

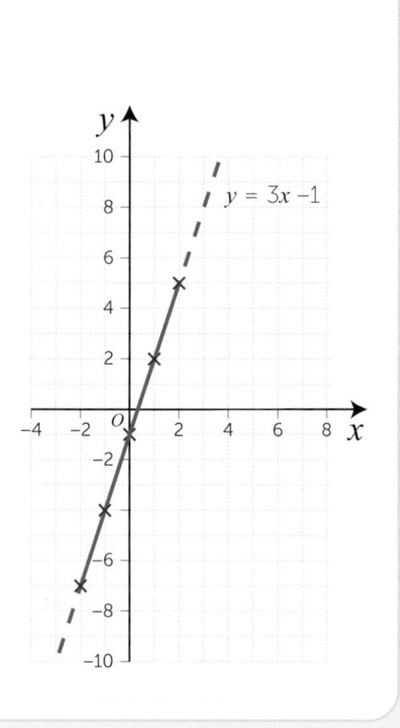


## Practice questions

1. State which of these is the equation of a straight line.

   a) $y = x + 5$  b) $y = x^2 - 1$  c) $y = \frac{2}{x} + 3$  d) $y = 5x - 6$

2. Point A lies on the line $y = 2x + 1$. The $x$-coordinate of A is 3. Work out the $y$-coordinate of A.

3. Copy and complete the table of values for the equation $y = x + 2$. Draw axes for $x$ from 0 to 4 and $y$ from 0 to 10 and plot the graph.

| $x$ | 0 | 1 | 2 | 3 | 4 |
|---|---|---|---|---|---|
| $y$ | | | | | |

4. Copy and complete the table of values for the equation $y = 2x - 1$. Draw axes for $x$ from 0 to 4 and $y$ from − 4 to 10 and plot the graph.

| $x$ | 0 | 1 | 2 | 3 | 4 |
|---|---|---|---|---|---|
| $y$ | | | | | |

5. Gary says that the coordinate (2, 5) lies on the line $y = 2x + 1$. Is he correct? Give a reason for your answer.

6. Copy and complete the table of values for the equation $y = 3x - 2$. Draw axes for $x$ from −2 to 2 and $y$ from −10 to 10 and plot the graph.

| $x$ | −2 | −1 | 0 | 1 | 2 |
|---|---|---|---|---|---|
| $y$ | | | | | |

7. Copy and complete the table of values for the equation $y = 8 - x$. Draw axes for $x$ from – 2 to 2 and $y$ from 0 to 10 and plot your graph.

| $x$ | −2 | −1 | 0 | 1 | 2 |
|---|---|---|---|---|---|
| $y$ | | | | | |

8. Draw and complete a table of values to work out the coordinates for the line $y = 4x + 1$.
Use integer values of $x$ from −2 to 2. Draw a suitable set of axes and plot your graph.

9. Kiera says that the point with coordinates (−2, −7) lies on the line $y = 2x - 3$
Bobby disagrees and says that it lies on the line $y = 6x + 5$
Who is correct? Give a reason for your answer.

10. Find the equations of the lines that contain the following sets of coordinates.
a) (0, 1), (3, 4) (8, 9) and (12, 13)
b) (3, 1), (4, 2), (8, 6) and (−1, −3)
c) (0, 3), (1, 5), (2, 7) and (3, 9)
d) (0, 6), (1, 5), (−1, 7) and (−2, 8)

## 7.4 PLOTTING GRAPHS 2

A9 A17

### Objectives

**Recognise that equations of straight lines can be rearranged to the form $ax + by = c$**

Equations of straight lines can be written in the form $ax + by = c$ with $a$, $b$ and $c$ as numbers

To draw graphs of equations in this form; find the $x$-intercept and $y$-intercept by substituting values of zero in for $x$ and for $y$

**e.g.** Plot a graph of the equation $4x + 3y = 12$

When $x = 0$ $3y = 12$ so $y = 4$ The $y$-intercept is (0, 4)

When $y = 0$ $4x = 12$ so $x = 3$ The $x$-intercept is (3, 0)

Plot the two intercepts. Join them up to give the graph.

Check a third point lies on the line.

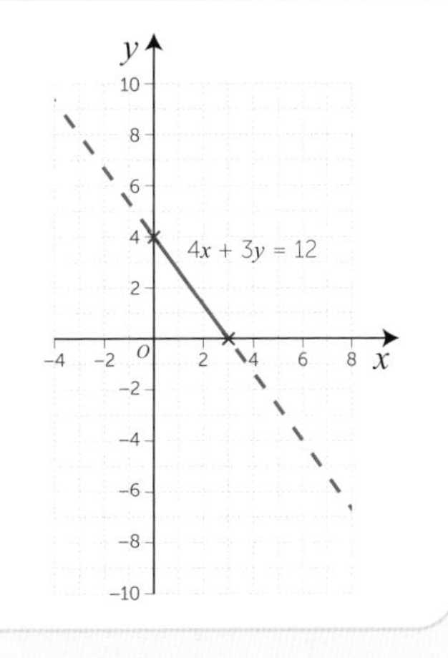


**Plot graphs in the form $ax + by = c$ using a table of values**

The equation shows how the $x$ and the $y$ values are related.

Use a table of values to work out values of $y$ from given values of $x$.

**e.g.** Complete the table of values for the equation $2x + y = 5$ and draw the graph.

| $x$ | −1 | 0 | 1 | 2 | 3 |
|---|---|---|---|---|---|
| Calculation | $2 \times (-1) + y = 5$<br>$-2 + y = 5$ | $2 \times (0) + y = 5$<br>$0 + y = 5$ | $2 \times (1) + y = 5$<br>$2 + y = 5$ | $2 \times (2) + y = 5$<br>$4 + y = 5$ | $2 \times (3) + y = 5$<br>$6 + y = 5$ |
| $y$ | 7 | 5 | 3 | 1 | −1 |

Plot the coordinates from the table of values.

Join up the points in a straight line.

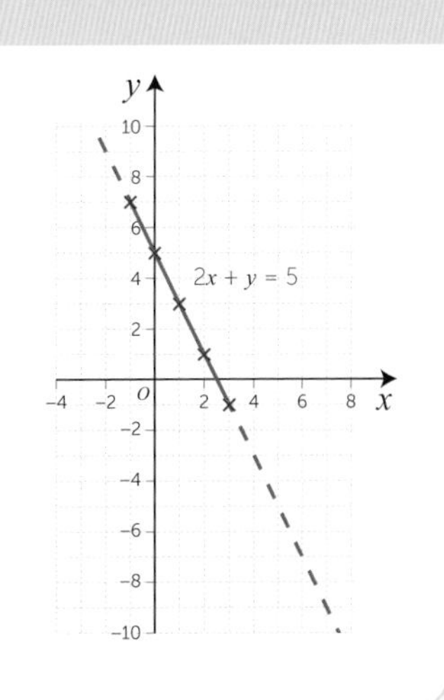

## Practice questions

1. State which of these is the equation of a straight line.
   a) $x + y = 5$ b) $y = 2x^2 + 1$ c) $y - \frac{3}{x} = 8$ d) $2x + y = -6$

2. State the coordinates of the $x$-intercept and the $y$-intercept for each of these graphs.
   a) $x + y = 3$ b) $x + y = -2$ c) $2x + y = 10$ d) $3x + 4y = 12$

3. Copy and complete the table of values for the equation $x + y = 4$

| $x$ | 0 | 1 | 2 | 3 | 4 |
|---|---|---|---|---|---|
| $y$ | | | | | |

   Draw axes for $x$ from 0 to 10 and $y$ from 0 to 10 and plot your graph.

4. Copy and complete the table of values for the equation of $x + y = 7$

| $x$ | 0 | 1 | 3 | 5 | 7 |
|---|---|---|---|---|---|
| $y$ | | | | | |

   Draw axes for $x$ from 0 to 10 and $y$ from 0 to 10 and plot your graph.

5. Write down **three** coordinates that lie on the line $x + y = 10$
   Draw axes for $x$ from 0 to 10 and $y$ from 0 to 10 and plot your graph.

6. From the equation $2x + 5y = 10$
   a) State where the line intersects the axes.
   b) Use your answer to a) to sketch a graph of the equation $2x + 5y = 10$

7. Copy and complete the table of values for the equation $2x + y = 4$

| $x$ | 0 | 1 | 2 | 3 | 4 |
|---|---|---|---|---|---|
| $y$ | | | | | |

   Draw axes for $x$ from 0 to 5 and $y$ from 0 to 5 and plot your graph.

8. Copy and complete the table of values for the equation $2x + 3y = 12$

| $x$ | 0 | 3 | 6 | 9 |
|---|---|---|---|---|
| $y$ | | | | |

   Draw axes for $x$ from −2 to 10 and $y$ from −2 to 8 and plot your graph.

9. Copy and complete the table of values for the equation $x - 2y = 8$

| $x$ | 0 | 2 | 4 | 6 | 8 |
|---|---|---|---|---|---|
| $y$ | | | | | |

   Draw axes for $x$ from 0 to 10 and $y$ from −5 to 5 and plot your graph.

10. Give the equation of each line upon which each set of coordinates lie.
    a) (0, 3), (2, 1), (−1, 4) and (6, −3)
    b) (−3, 8), (2, 3), (0, 5) and (7, −2)
    c) (0, 8), (1, 6), (3, 2) and (−1, 10)

# 7.5 GRADIENTS OF STRAIGHT LINES

## Objectives

### Understand how to calculate gradient

Gradient is a measure of steepness of a line.

It is found by dividing the vertical change by the horizontal change, between two points on a line.

Gradient = $\frac{\text{change in } y}{\text{change in } x}$ or

Gradient = $\frac{y_2 - y_1}{x_2 - x_1}$ between two points $(x_1, y_1)$ and $(x_2, y_2)$

**e.g.** Gradient = $\frac{\text{change in } y}{\text{change in } x} = \frac{6}{3} = 2$

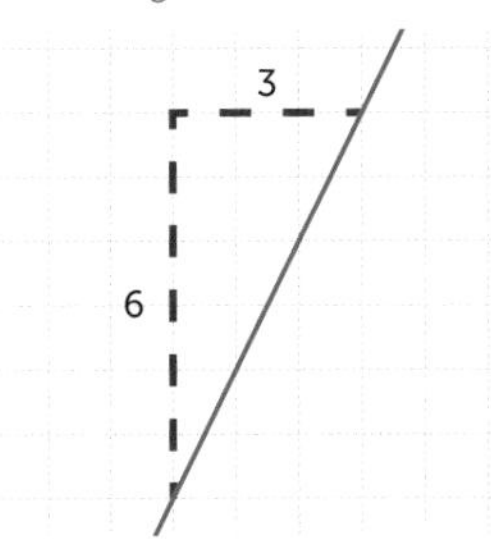


### Recognise the difference between positive and negative gradients

Gradient can be positive or negative.

**e.g.** Line A has a positive gradient

gradient of A = $\frac{4}{4} = 1$

Line B has a negative gradient

gradient of B = $\frac{-5}{5} = -1$

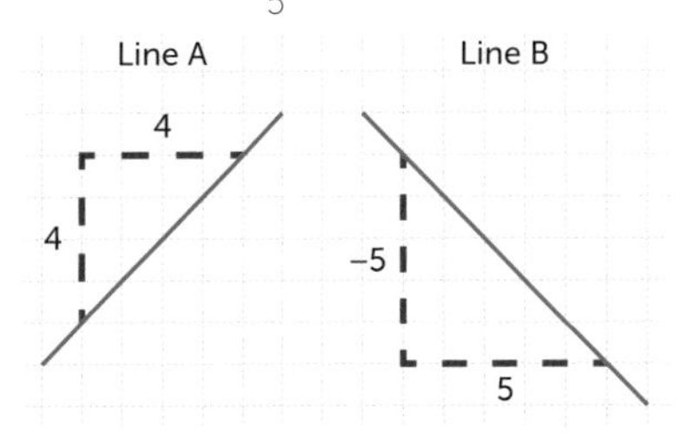


### Identify lines with different gradients and parallel lines

## Practice questions

1. Find the gradients of these lines.
   Give your answers as fractions where appropriate.

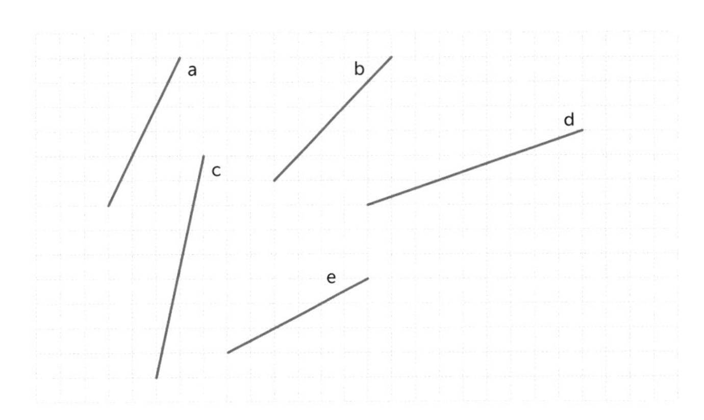


2. Greg has a line with a gradient of 3. Harriet has a line with a gradient of $\frac{7}{2}$.
   Whose line is the steepest? Give a reason for your answer.

3. Find the gradients of these lines, giving answers as fractions where appropriate.

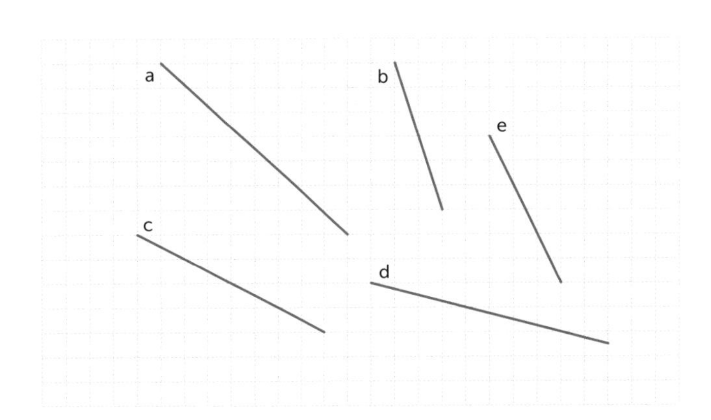

4. Sally has a line with a gradient of −4. Tim has a line with a gradient of −3
   Whose line is steeper? Give a reason for your answer.

5. Find the gradients of these lines.
   Give your answers as fractions.

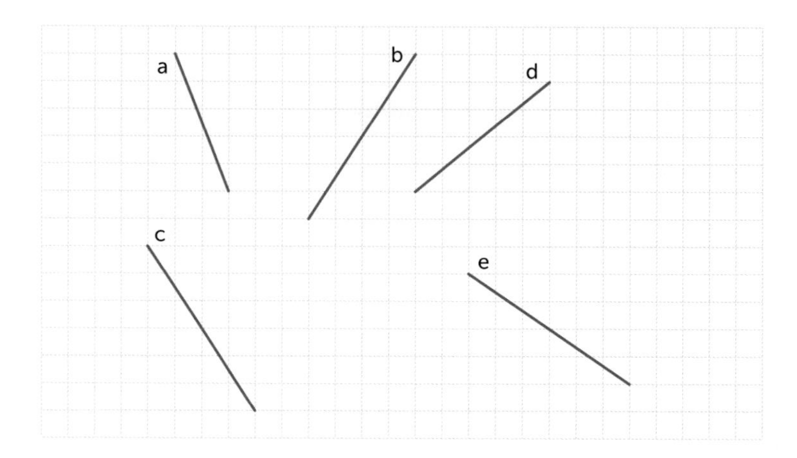


6. Plot each pair of points and join them up with a line.
   Calculate the gradient of each line.
   a) (0, 0) and (3, 6) b) (0, 0) and (−2, 8) c) (0, 1) and (5, 6)
   d) (3, −1) and (7, 3) e) (−1, 2) and (−7, −1)

7. From the graph:
   a) Work out the gradient of the pink line.
   b) Work out the gradient of the green line.

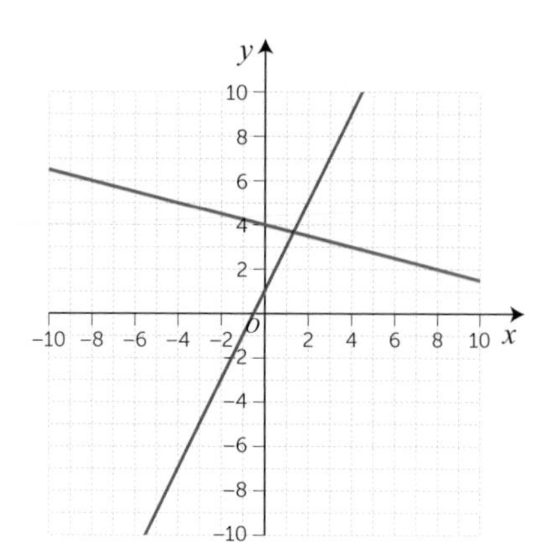


8. David is trying to work out the gradient between the coordinates (1, −2) and (−3, −14)
   This is David's solution:
   Gradient = $\frac{-14 - 2}{-3 - 1} = \frac{-16}{-4} = 4$
   David has made a mistake.
   Find the mistake and correct it.

9. Edie and Stan both have lines passing through the point (2, −3).
   Edie's line also passes through (10, 9). Stan's line also passes through (−4, −12).
   Which is the steeper line? Show your working.

10. Alex's line passes through (2, 3) and (8, 11).
    Chloe's line passes through (1, −2) and (−7, −12).
    Whose line has a gradient that is closer to 1?

# 7.6 EQUATION OF A STRAIGHT LINE 1

## Objectives

### Plot a line given its equation without a table

**Plotting without a table**

When an equation is written in the form $y = mx + c$;
$m$ is the gradient and $c$ is the $y$-intercept.

This information can be used to draw a graph of an equation.

**e.g.** Draw the graph of the equation $y = 3x - 2$

We know the $y$-intercept is –2 and so can plot the point (0, –2)

We know the gradient is 3, which means the line rises 3 for every 1 to the right..

Use this to move in steps from (0, –2) up 3, right 1.

Join the points up to give the line.

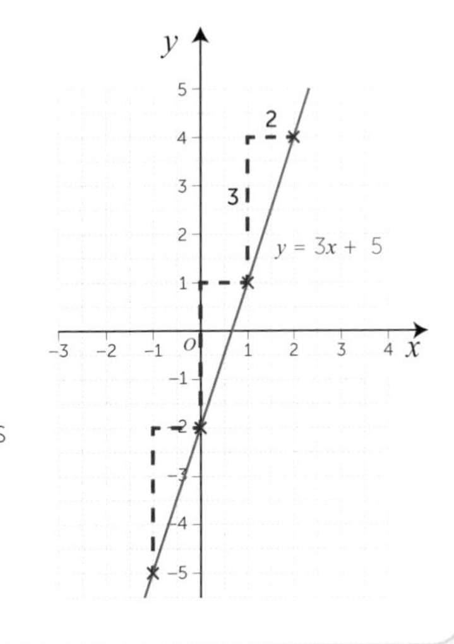


**Sketch a graph using the concept of $y = mx + c$**

### Find the equation of a line given its graph

**Finding the equation of a line**

When a graph is shown, we can read off the $y$-intercept and the gradient from the graph.

This information can be put into the equation $y = mx + c$ to give the equation of the line.

**e.g.** Find the equation of this line.

The $y$-intercept is at (0, 3) so the value of $c = 3$

The gradient of the line is –2, as it goes down 2 for every 1 it goes right.

Hence $m = -2$

Put the values of $m$ and $c$ into the equation $y = mx + c$, giving:

Equation of line: $y = -2x + 3$

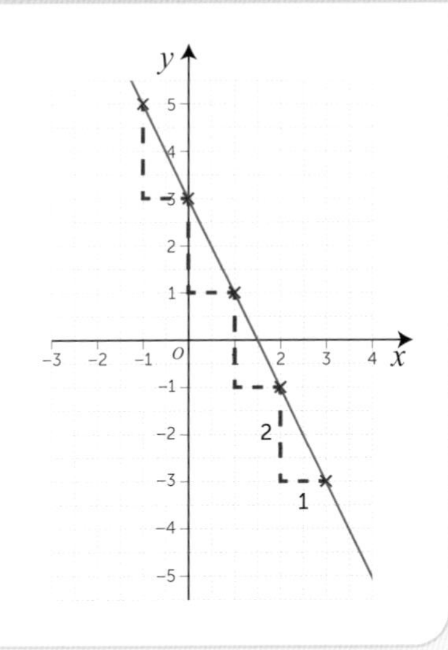


## Practice questions

1. State the coordinates of the y-intercept and gradient of each of these lines.
   a) $y = 3x - 2$ b) $y = 2x - 9$ c) $y = 4x + 5$ d) $y = 8 + 3x$

2. Matilda is plotting the line $y = 4x$. She has plotted (0, 0) and (4, 1). Has Matilda plotted her coordinates correctly? Give a reason for your answer.

3. Draw a coordinate grid with $x$ from –5 to 5 and $y$ from –10 to 10. Draw the graph of $y = 2x$ using $y = mx + c$.

4. Draw a coordinate grid with $x$ from –3 to 3 and $y$ from –10 to 10. Draw the graph of $y = 3x - 2$ using $y = mx + c$.

5. Work out the equation of this line.

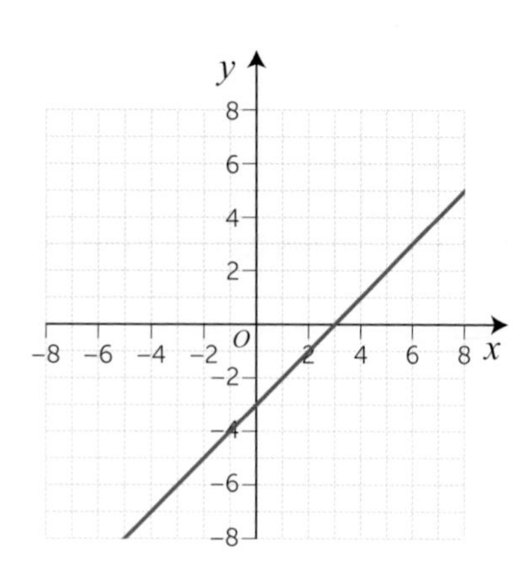


6. Harry has been asked to plot $y = 2x + 3$ on a coordinate grid.
   He has plotted (0, 2) and (1, 5)
   Has Harry plotted his coordinates correctly? Give a reason for your answer.

7. Work out the equation of this line.

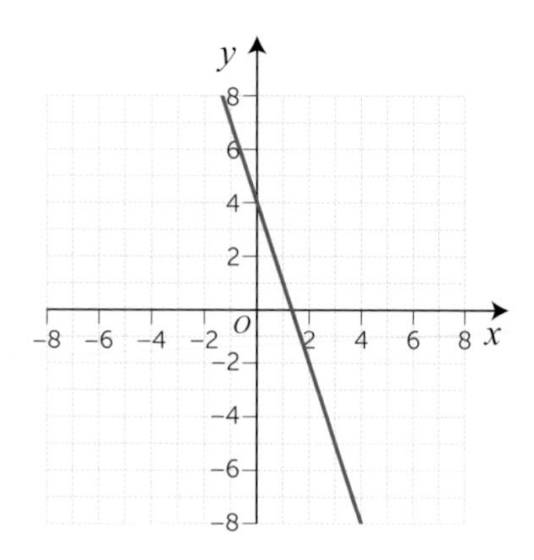


8. Draw a coordinate grid with $x$ from −3 to 3 and $y$ from −5 to 15.
   Draw the graph of $y = 5 - 2x$ using $y = mx + c$.

9. Draw a coordinate grid with $x$ from −3 to 3 and $y$ from −5 to 5.
   Draw the graph of $y = \frac{3}{2}x - 1$ using $y = mx + c$.

10. Match equations for lines A, B and C from the table.

| Line 1 | $y = 3x + 1$ |
|---|---|
| Line 2 | $y = -0.5x - 1$ |
| Line 3 | $y = 2 - x$ |
| Line 4 | $y = 0.5x + 1$ |
| Line 5 | $y = 0.5x - 2$ |

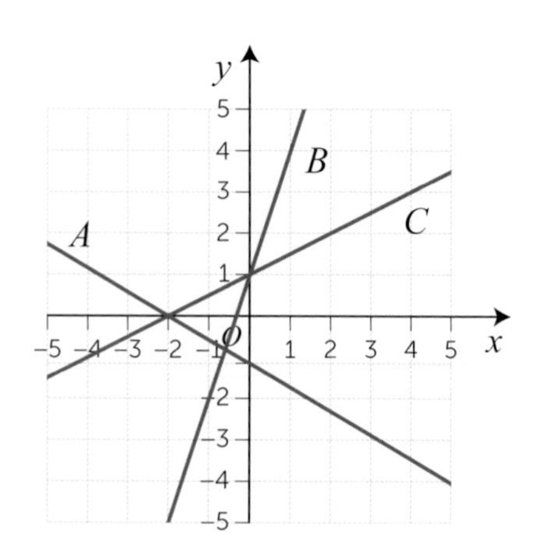

# 7.7 EQUATION OF A STRAIGHT LINE 2

## Objectives

### Find the equation of a line given any coordinate on the line and its gradient

Given the gradient of a line and the coordinates of a point, we can work out the equation.

e.g. Find the equation of a line with gradient 5 which passes through the point (2, −3)

The general equation of a straight line is $y = mx + c$

The gradient is 5 so we know that $m = 5$

Substitute $m = 5$ and the coordinates $x = 2$ and $y = -3$ into the general equation.

$-3 = 5 \times 2 + c$ $\quad\quad$ $-3 = 10 + c$ $\quad\quad$ $c = -13$

Equation of line is $y = 5x - 13$

### Find the equation of a line given two points on the line

Given two points, we can find the gradient of a line between them.

Use the gradient and one of the coordinates to find the equation of the line.

e.g. Find the equation of the line that passes through (−1, 3) and (1, −5)

First find the gradient.

Gradient $= \frac{-8}{2} = -4$ $\quad\quad$ so $m = -4$

Substitute $m = -4$ and the coordinates $x = 1$ and $y = -5$ into the general equation.

$-5 = -4 \times 1 + c$ $\quad\quad$ $-5 = -4 + c$ $\quad\quad$ $c = -1$

The equation of the line is $y = -4x - 1$

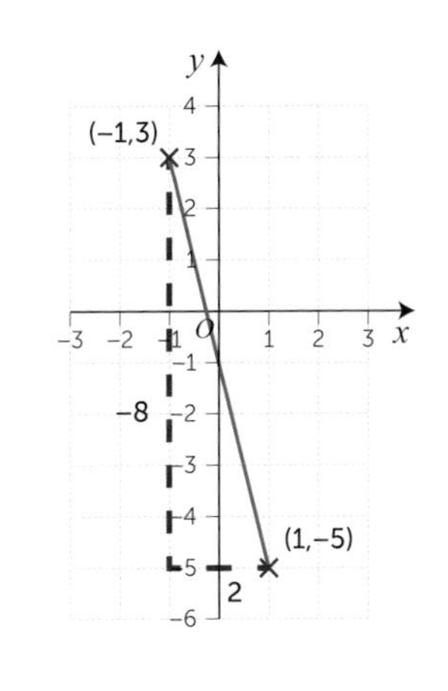


## Practice questions

1. State whether the gradients between the pairs of points below is positive or negative.

   a) (0, 0) and (−2, 6) $\quad$ b) (0, 0) and (−3, −9) $\quad$ c) (2, 2) and (6, 0)

   d) (3, 5) and (0, 8) $\quad$ e) (1, 8) and (−2, −4)

2. The two points (1, 3) and (5, 15) are drawn on a coordinate grid. Jamie says that the gradient between the coordinates is 3. Kiera says that the gradient between the coordinates is $\frac{1}{3}$. Who is correct? Show your working.

3. Work out the equation of this line.

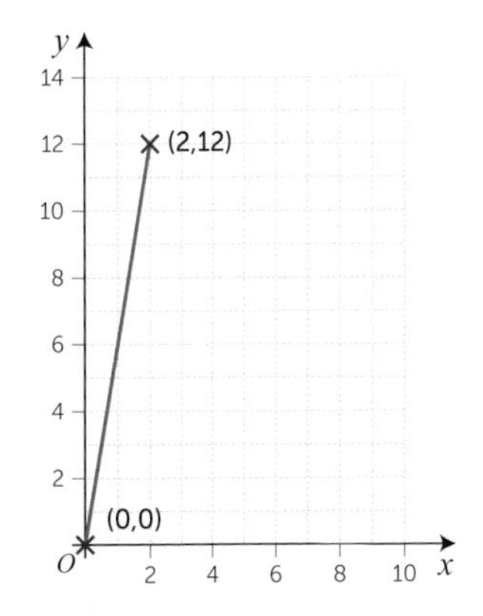

4. Use $y = mx + c$ to write the equation of each line.
   a) Gradient = 2, point = (3, 5)
   b) Gradient = 5, point = (2, −4)
   c) Gradient = −3, point = (5, 1)
   d) Gradient = $\frac{1}{2}$, point = (−4, −6)

5. Work out the equation of the line which passes through each pair of points.
   a) (0, 0) and (2, 6)
   b) (0, 0) and (3, −9)
   c) (0, 2) and (4, 10)
   d) (0, −1) and (5, – 16)

6. A line passes through (0, 3) on the $y$-axis and through (3, 0) on the $x$-axis.
   Toby says that the equation of the line is $y = x + 3$
   Explain why Toby is wrong.

7. Work out the equation of this line.

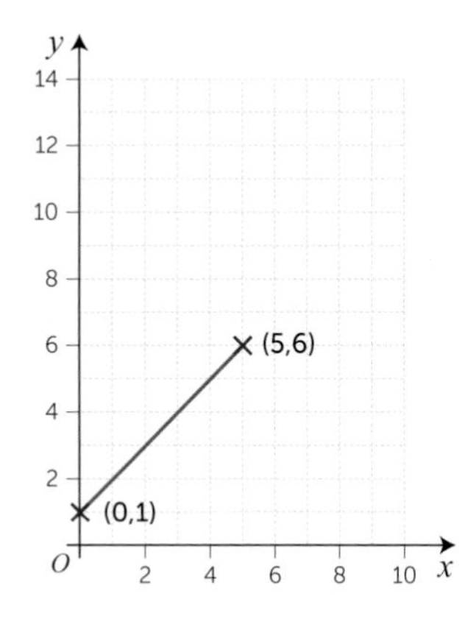


8. David is finding the gradient of the line between (1, 8) and (5, 6).
   He does: Gradient = $\frac{4}{2} = 2$
   He has made a mistake. Find the mistake and correct it.

9. Work out the equation of this line.

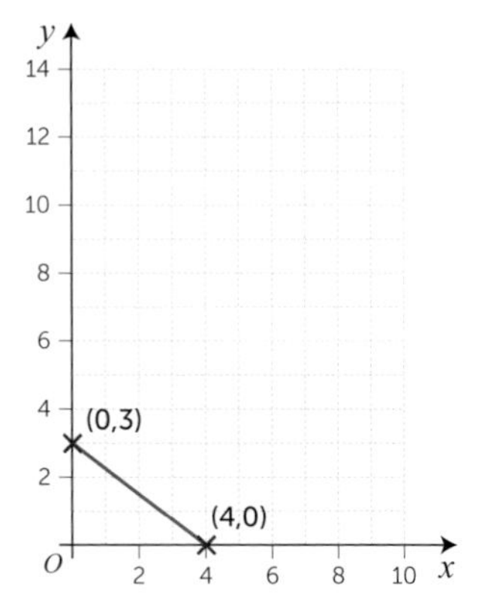


10. Work out the equation of the line that passes through each set of points.
    a) (2, 5) and (4, 7)
    b) (1, 10) and (3, 4)
    c) (−2, −8) and (3, 2)
    d) (4, 2) and (12, 4)
    e) (−2, 7) and (−5, 13)

SECTION 8

# FRACTIONS, DECIMALS, AND PERCENTAGES

## 8.1 DECIMALS TO FRACTIONS

N10

### Objectives

**Convert decimals to fractions**

Work out the place value of the digit that is furthest to the right of the decimal number.

This is the denominator of the fraction.

The numerator is the number formed by the sequence of digits after the decimal point.

Where possible, simplify the fraction.

| Decimal number | Units (ones) | | $\frac{1}{10}$ | $\frac{1}{100}$ | $\frac{1}{1000}$ | Fraction | |
|---|---|---|---|---|---|---|---|
| 0.7 | 0 | . | 7 | | | $\frac{7}{\mathbf{10}}$ | |
| 0.37 | 0 | . | 3 | 7 | | $\frac{37}{\mathbf{100}}$ | |
| 0.037 | 0 | . | 0 | 3 | 7 | $\frac{37}{\mathbf{1000}}$ | |
| 0.205 | 0 | . | 2 | 0 | 5 | $\frac{205}{\mathbf{1000}}$ | $= \frac{41}{\mathbf{200}}$ |

**e.g.** Match these decimals and fractions. One of the fractions cannot be matched.

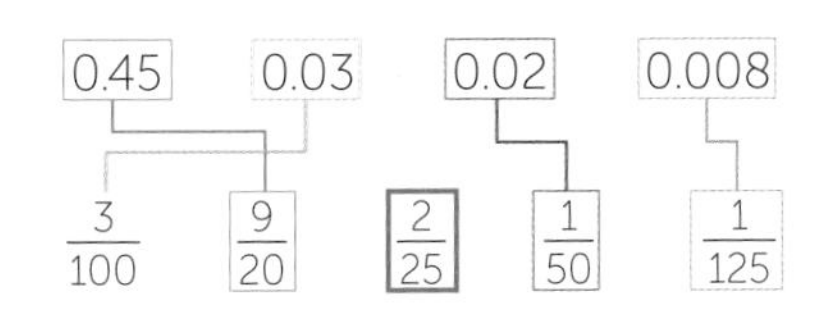


$0.45 = \frac{\cancel{45}^{9}}{\cancel{100}_{20}} \Rightarrow \frac{9}{20}$ $\qquad 0.03 = \frac{3}{100}$

$0.02 = \frac{\cancel{2}^{1}}{\cancel{100}_{50}} \Rightarrow \frac{1}{50}$ $\qquad 0.008 = \frac{\cancel{8}^{1}}{\cancel{1000}_{125}} \Rightarrow \frac{1}{125}$

### Practice questions

1. Write these decimals as fractions.

a) 0.3 b) 0.03 c) 0.003

d) 0.09 e) 0.009 f) 0.001 9

2. Write these decimals as fractions in their simplest form.

a) 0.6 b) 0.32 c) 0.150

d) 0.8 e) 0.45 f) 0.052

3. Match each picture with its corresponding fraction.

a) $\frac{7}{100}$ b) $\frac{3}{5}$ c) $\frac{9}{50}$ d) $\frac{1}{100}$

i) 

ii) 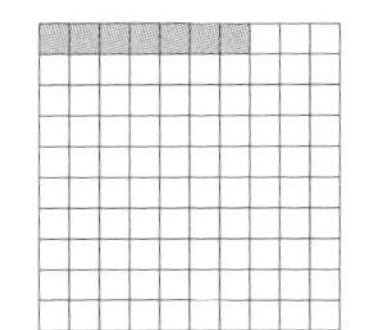

iii) 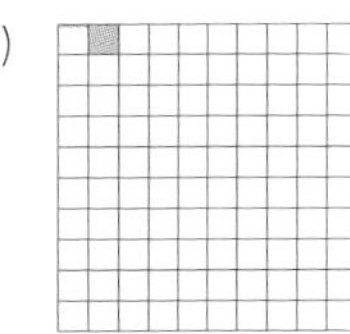

iv) 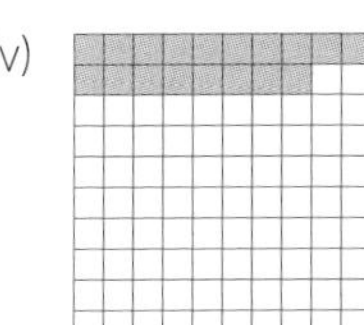

4. Write down which fraction, decimal or division does not belong to each group.

   a) 0.3 $\frac{3}{10}$ $\frac{1}{3}$ $3 \div 10$  b) 0.02 $2\frac{1}{10}$ $2 \div 100$ $\frac{1}{50}$

   c) 0.040 $\frac{1}{250}$ $4 \div 1000$ $\frac{4}{1000}$  d) 0.006 $\frac{6}{1000}$ $6 \div 10\,000$ $\frac{3}{500}$

5. Which of these statements are untrue? Correct them, to make them true.

   a) $0.70 = \frac{7}{10}$  b) $0.200 = \frac{1}{5}$  c) $0.050 = \frac{1}{2}$

   d) $0.008 = \frac{1}{125}$  e) $0.04 = 0.4 \div 10$  f) $0.006 = 6 \div 100$

6. Write each as fractions in the simplest form. Match the ones that are equivalent.

   a) $3 \div 50$  b) 0.02  c) $\frac{6}{50}$

   d) 0.006  e) $\frac{12}{50}$  f) 0.12  g) $0.6 \div 10$

7. Complete the table. The first one has been done for you.
   Simplify your answers where appropriate.

| 0.4 | 0.40 | 0.04 | 0.004 | 0.008 | |
|---|---|---|---|---|---|
| $\frac{2}{5}$ | | | | | $\frac{1}{50}$ |

8. Matilda ran for 0.2 km and cycled for $\frac{1}{6}$ km. She thinks she cycled further than she ran. Is she correct?

9. On a cycling track, there is a notice saying '0.675 km to the finish line'.
   What fraction of the last kilometre is left?

10. Lily spends 0.15 hours doing her homework. Which statement is correct?
    Lily spends $\frac{3}{20}$ of an hour doing her homework.
    Lily spends $\frac{1}{4}$ of an hour doing her homework.
    Lily spends $\frac{15}{60}$ of an hour doing her homework.

11. Tom thinks that $0.005 = 0.5 \div 1000$ or $\frac{1}{2} \div 1000 = \frac{1}{200}$.
    Explain why Tom is wrong.

12. Jake thinks of a decimal number. It is less than $\frac{1}{2}$ but greater than $\frac{2}{5}$. The tenths digit is half of the hundredths digit.
    What decimal number did Jake think of? Write it as a fraction.

# 8.2 FRACTIONS TO DECIMALS

## Objectives

### Convert fractions to decimals

If possible, change the fraction to its equivalent with a power of 10 as its denominator.

Write the numerator after the decimal point, with the right hand digit in the place value represented by the denominator.

| Fraction | Multiplier | Equivalent fraction | Units | | $\frac{1}{10}$ | $\frac{1}{100}$ | $\frac{1}{1000}$ | Decimal number |
|---|---|---|---|---|---|---|---|---|
| $\frac{3}{5}$ | × 2 | $\frac{6}{\mathbf{10}}$ | 0 | . | 6 | | | 0.6 |
| $\frac{7}{50}$ | × 2 | $\frac{14}{\mathbf{100}}$ | 0 | . | 1 | 4 | | 0.14 |
| $\frac{3}{250}$ | × 4 | $\frac{12}{\mathbf{1000}}$ | 0 | . | 0 | 1 | 2 | 0.012 |

Write zero in any empty place value spaces to the right of the decimal point.

If changing the denominator to a power of 10 is too complex, or a calculator is not allowed, divide the numerator by the denominator.

**e.g.** Work out the decimal equivalent of $\frac{7}{40}$

$7 \div 40$

$$40\overline{)7\,.\,{}^{7}0\,{}^{30}0\,{}^{20}0} \quad \text{quotient } 0\,.\,1\;7\;5$$

= **0.175**

### Work interchangeably with terminating decimals and their corresponding fractions

Change all the numbers to decimals.

Compare their values.

**e.g.** Put these numbers in descending order: $\frac{14}{25}$ 0.57 $\frac{14}{20}$ 0.6 $\frac{7}{8}$ 0.599

First change all the fractions to decimals.

$\frac{14}{25} = \frac{56}{100} = 0.56$ $\quad \frac{14}{20} = \frac{7}{10} = 0.7$ $\quad \frac{7}{8} = 0.875$

Now compare all the decimal numbers and put them into descending order.

0.875 0.7 0.6 0.599 0.57 0.56

Write the final answer in the form given in the original question.

$\frac{7}{8}$ $\frac{14}{20}$ 0.6 0.599 0.57 $\frac{14}{25}$

## Practice questions

1. Write the fraction and the decimal number that each shape represents.

a) 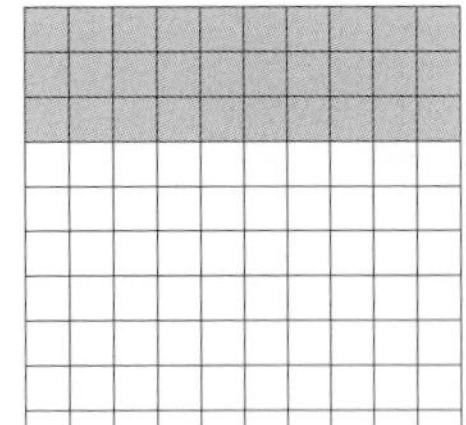

b) 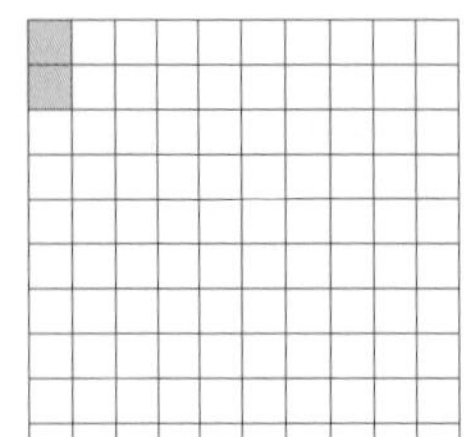

c) 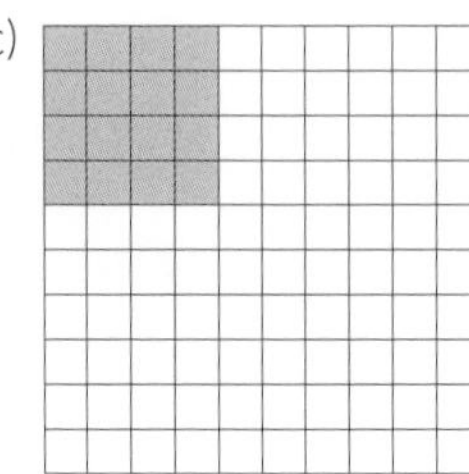

d) 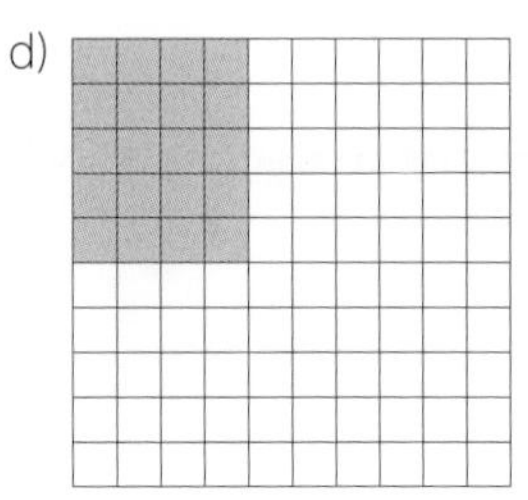

2. Use equivalent fractions to change these fractions to decimals.

a) $\frac{3}{5} = \frac{?}{10}$  b) $\frac{11}{50} = \frac{?}{100}$  c) $\frac{7}{25} = \frac{?}{50} = \frac{?}{100}$  d) $\frac{27}{30} = \frac{?}{?}$

3. Write each fraction as a decimal.

a) $\frac{23}{25}$  b) $\frac{7}{8}$  c) $\frac{9}{40}$  d) $\frac{1}{25}$  e) $\frac{15}{16}$

4. Which decimal numbers cannot be matched with one of the fractions below?

0.164  0.0164  0.04  0.022  0.052  0.02

$\frac{1}{25}$  $\frac{13}{250}$  $\frac{28}{400}$  $\frac{11}{500}$  $\frac{41}{2500}$

5. Convert the fractions to decimals and list them in ascending order.

a) 0.27  0.615  $\frac{3}{8}$  $\frac{13}{20}$  0.07  $\frac{11}{25}$

b) $\frac{3}{5}$  0.46  $\frac{7}{10}$  0.66  $\frac{17}{20}$  $\frac{32}{40}$

6. Dan says that $\frac{7}{20} > \frac{3}{8}$ because 7 > 3 and 20 > 8. Is Dan correct?
Explain your answer.

7. 35% of the students in Year 10 like football. 37 out of 100 students like basketball and 8 out of 24 students like hockey.
   a) Which of the sports is the most popular?
   b) Which of the sports is the least popular?

8. Which number is exactly halfway between $\frac{3}{5}$ and $\frac{4}{5}$. Write it as a decimal.

9. Eamon has a bag of 40 balloons. 0.45 of the balloons are red, $\frac{2}{5}$ of the balloons are yellow, the rest are blue. What fraction of the balloons are blue?

10. Chiara uses a written method to change $\frac{5}{8}$ to a decimal number. Look at her method. Is Chiara correct?

$$8\,\overline{)\,5.{}^{6}00{}^{4}0}\quad \text{quotient } 0.775$$

11. Which of the following fractions are between 0 and 0.3? Show how you know.

a) $\frac{3}{8}$  b) $\frac{11}{200}$  c) $\frac{13}{25}$  d) $\frac{1}{20}$  e) $\frac{7}{40}$

12. Use numbers 3, 6, 7 and 9 exactly once each to complete this number statement.

$\frac{??}{125} = 0.2??$

# 8.3 PERCENTAGES 1

## Objectives

### Interpret percentages as 'number of parts per hundred'

**e.g.** This is 1%.

1 out of 100 squares is shaded.

$1\% = \frac{1}{100}$

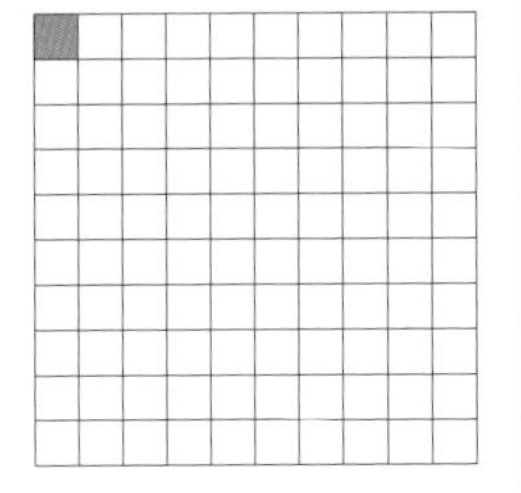

**e.g.** This is 5%.

5 out 100 squares are shaded.

$5\% = \frac{5}{100}$

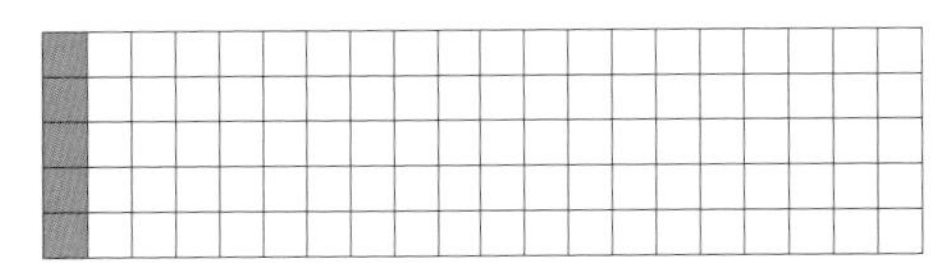

### Express one quantity as a percentage of another

Make a fraction from the two numbers.

Multiply the fraction by 100 to find the percentage.

**e.g.** Write 14 as a percentage of 40 .

$\frac{14}{40} = \frac{7}{20}$ (÷ numerator and denominator by 2)

$= \frac{35}{100}$ (× numerator and denominator by 5)

$= 0.35$ (× 100)

$= 35\%$

### Compare two quantities using percentages

**e.g.** Debbie achieved 28 out of 30 in her French test and 91% in her Spanish test. She thinks that her test result in French is better than in Spanish.

Is she correct?

French result: $\Rightarrow \frac{28}{30} = \frac{28}{30} \times 100\%$

$= 93\%$

Spanish result: $\Rightarrow$ 91%

The French result is better than Spanish result.
Debbie is correct.

## Practice questions

1. On squared paper, draw four 10 by 10 squares. Shade in:
   a) 28% b) 85% c) $\frac{11}{25}$ d) $\frac{3}{20}$

2. Use equivalent fractions to change these fractions to percentages:
   a) $\frac{13}{25} = \frac{?}{100} = ?\%$ b) $\frac{56}{80} = \frac{?}{10} = \frac{?}{100} = ?\%$ c) $\frac{72}{120} = \frac{?}{20} = \frac{?}{100} = ?\%$ d) $\frac{100}{125} = \frac{4}{?} = \frac{?}{100} = ?\%$

3. Work out:
   a) 13 as a percentage of 20 b) 39 as a percentage of 50
   c) 54 as a percentage of 90 d) 33 as a percentage of 60

4. Rewrite these statements as percentages. Use a calculator.
   Give your answer to 2 decimal places.
   a) 21 out of 47 students use iPads b) 13 out of 31 students are vegans
   c) 12 out of 14 girls achieved grade 9 d) 127 out of 150 men like cats

5. True or false? Use percentages to explain your answer.
   a) 26 out of 65 > 21 out of 35 b) 20 out of 50 < 10 out of 40
   c) 36 out of 48 = 75 out of 100 d) 55 out of 60 > 23 out of 24

6. Use >, < or = to complete the following statements

a) 25% ...... $\frac{1}{4}$ b) 8% ...... $\frac{8}{10}$ c) $\frac{3}{20}$ ...... 15% d) $\frac{1}{50}$ ...... 2% e) $\frac{2}{3}$ ...... 60%

7. Kim knows that 1 as a percentage of 25 = 4%. To find 1 as a percentage of 50, Kim multiplies 4% by 2.
Is Kim correct? Explain your answer.

8. Below is the number of people attending two different shows in Leeds.

| | **Total number of people** | **Number of occupied seats** |
|---|---|---|
| Strictly Comedy | 1200 | 960 |
| Leeds' got talent | 1500 | 1130 |

Work out which show has the greater percentage of occupied seats.

9. This table shows the ages of laptops owned by two group of people. Use percentages to compare the ages of laptops owned by doctors and accountants.

| **Age of laptop** | **Less than 3 years** | **3 years or more** |
|---|---|---|
| Doctors | 43 | 39 |
| Accountants | 50 | 78 |

10. 85 people took a driving test for the first time on Monday and 61 passed.
250 people took the test for the second time. 214 people passed.
Based on these results, are people more likely to pass the test the first or the second time?

11. Use numbers 4, 5, 6, 7 or 8 once only to make this statement true.
$\frac{?}{8} < ??\% < \frac{?}{?}$

## 8.4 PERCENTAGES 2

R9 N12

### Objectives

#### Interpret fractions and decimals as operators

To find a percentage of a number, convert the percentage to a decimal (divide it by 100) then multiply the decimal by that number.

**e.g.** Find 3% of £26.

3% = 3 ÷ 100
= 0.03
0.03 × 26 = 0.78 ⇒ 78p

To find a fraction of a number, divide it by the denominator then multiply by the numerator.

**e.g.** Amy had £72. She spent $\frac{5}{8}$ of her money on food and saved the rest. How much money did Amy save?

Amy saves $1 - \frac{5}{8}$ = $\frac{3}{8}$ of £72
$\frac{3}{8}$ of £72 = 72 ÷ 8 × 3
= 27
Money saved = **£27**

#### Convert between fractions, decimals and percentages

To change a percentage to a decimal, divide the number by 100.

**e.g.** 35% = 35 ÷ 100 ⇒ **0.35**

To change a decimal to a percentage, multiply the number by 100.

**e.g.** 0.04 = 0.04 × 100 ⇒ **4%**

To change a percentage to a fraction, write the percentage as the numerator of a fraction with denominator 100. Leave the answer as a fraction in its simplest form.

**e.g.** 2% = $\frac{2}{100}$ ⇒ $\mathbf{\frac{1}{50}}$

To change a fraction to a percentage, multiply the number by 100.

**e.g.** $\frac{1}{25}$ = $\frac{1}{25} \times 100$
= 4 ⇒ **4%**

## Practice questions

1. Write down the percentage, decimal and fraction that the shaded region represents.

   a) 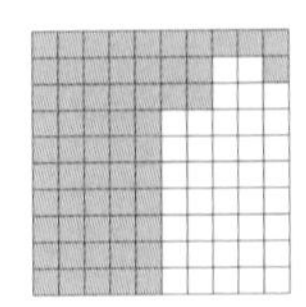
   b) 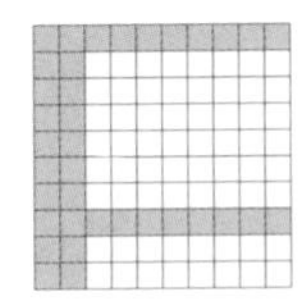
   c) 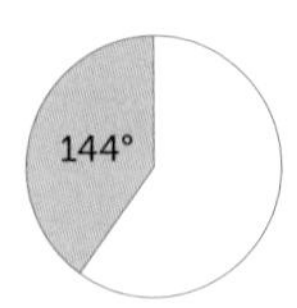

   d) 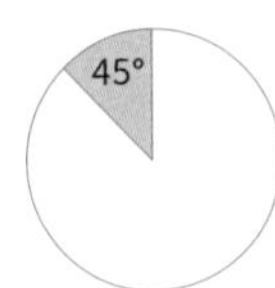


2. Change the following percentages to decimals:

   a) 35% b) 21% c) 18.6% d) 4% e) 0.2% f) $2\frac{1}{2}\%$

3. Change the following decimals to percentages:

   a) 0.06 b) 0.8 c) 0.009 d) 0.85 e) 1.2 f) 12.5

4. Calculate:

   a) 20% of 160 m b) 25% of £16 c) 16% of 25 kg d) 12.5% of 200 ml
   e) 41% of 420 cm f) 0.5% of 300 g

5. Copy this table. Complete the missing boxes.

| **Fraction** | $\frac{1}{5}$ | | | $\frac{3}{4}$ | | |
|---|---|---|---|---|---|---|
| **Decimal** | | 0.5 | | | 0.8 | |
| **Percentage** | | | 25% | | | 100% |

6. Place in order starting with the smallest.

   a) $\frac{18}{20}$, 0.18, 0.18% b) 34%, $\frac{3}{4}$, 0.8 c) 0.7, 69%, $\frac{3}{7}$ d) $\frac{3}{5}$, 38%, 3.8

7. Write down the calculation in each set that gives a different answer to the other two.

   a) 25% of 128 $\frac{1}{2}$ of 256 0.125 × 256 b) 24% of 600 95% of 160 150% of 96
   c) 2.5% of 400 100% of 20 $\frac{2}{11}$ of 110

8. Here is some of the nutrition information from a tin of soup. Write as a fraction, decimal and percentage the proportion of:

   a) protein b) carbohydrate
   c) fibre d) salt

| **Per 50g** |
|---|
| Protein - 4.0g |
| Carbohydrate - 12.6g |
| Saturated fat - 6.4g |
| Fibre - 3.9g |
| Salt - 1.2g |

9. A right-angled triangle has been cut from the rectangular card shown.

   a) What fraction of the card is left?
   b) What percentage of the card is left?

35 cm
2.4 cm
20 cm
10 cm

10. Freddie took part in a triathlon that was 24 km long. He cycled 35% of the distance, ran $\frac{5}{6}$ of the rest and swam the remainder. How many kilometres did Freddie swim?

11. Anthony, Beth, Charlie, Dina and Ella ran the following distances in 2 minutes. Write their results in ascending order.

| **Anthony** | **Beth** | **Charlie** | **Dina** | **Ella** |
|---|---|---|---|---|
| $\frac{2}{5}$ of a km | 0.25km | 30% of 1km | $\frac{3}{8}$ of 1km | 35% of 1km |

12. a) Jordan uses the fact that $\frac{1}{100} = 0.01$ to work out $\frac{1}{200}$, $\frac{1}{50}$ and $\frac{1}{25}$ as decimals.
   i) What could Jordan's method be?
   ii) Change $\frac{1}{200}$, $\frac{1}{50}$ and $\frac{1}{25}$ to decimals.

   b) Helen uses the percentage equivalent of $\frac{1}{4}$ to work out $\frac{1}{8}$, $\frac{3}{8}$ and $\frac{7}{8}$ as percentages.
   i) What could Helen's method be?
   ii) Change $\frac{1}{8}$, $\frac{3}{8}$ and $\frac{7}{8}$ to percentages.

13. Fill in some possible percentages, fractions and decimals on the spider diagram below so that the answer is 6 km. Use different numbers each time. One has been done for you.

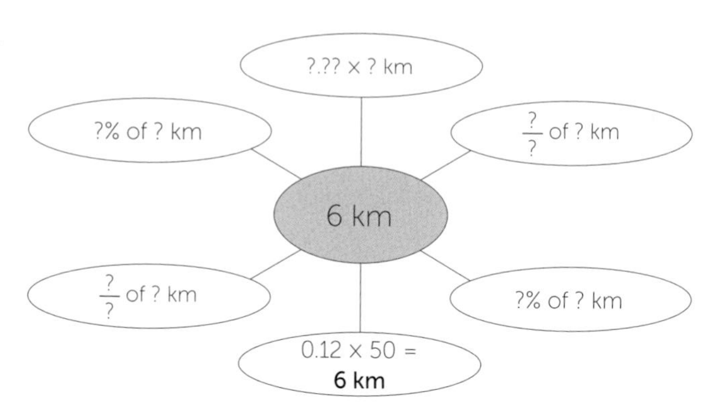


## 8.5 PROBLEMS INVOLVING PERCENTAGES

R9 N12

### Objectives

**Solve problems involving percentages**

**e.g.** There are 1200 patients in a dental surgery. 624 of the patients are male.
What percentage of the patients are female?

Female patients $= 1200 - 624$
$= 576$

% of female patients $= \frac{576}{1200} \times 100\% \Rightarrow$ **48%**

**e.g.** There are 120 pupils in Year 11.

- 15% of the students are studying French and 36 students are studying German.
- The rest are studying Spanish.
- 10 students who study French, 75% of German students and 50% of Spanish students are expected to achieve a grade 5.
- What percentage of all the students are not predicted a grade 5?

**Solution:** Use a table to fill in the information given in the question.

| | French | German | Spanish | **Total** |
|---|---|---|---|---|
| Grade 5 | **10** | 27 | 33 | 70 |
| Not Grade 5 | 8 | (9) | (33) | 50 |
| **Total** | 18 | **36** | 66 | **120** |

| | | |
|---|---|---|
| 15% of 120 study French: | $0.15 \times 120$ | = 18 |
| Number not gaining grade 5: | $18 - 10$ | = 8 |
| Number studying Spanish: | $120 - (18 + 36)$ | = 66 |
| 75% of 36 German students predicted grade 5: | $0.75 \times 36$ | = 27 |
| 50% of 66 Spanish students predicted grade 5: | $0.5 \times 66$ | = 33 |
| Total predicted Grade 5 : | $10 + 27 + 33$ | = 70 |
| Total not predicted grade 5: | $120 - 70$ | = 50 |
| Percentage not predicted a grade 5: | $\frac{50}{120} \times 100\%$ | = **41.7%** (1 d.p.) |

## Practice questions

1. $\frac{1}{2}$ kg of a cake mix contains 180 g of dark chocolate. What percentage of the cake mix is dark chocolate?

2. Linda's mark in a test was 36 out of 50. What percentage is that?

3. When full, James' phone battery lasts 12 hours. The battery is 75% full. How many more hours will it last?

4. The table shows the number of calculators a shop sold last spring.

| March | April | May |
|---|---|---|
| 150 | 165 | 175 |

What percentage of the calculators were sold in May?

5. There are 65 planes that leave each morning from the airport. 15 of them were late. What percentage of the planes were on time?

6. Alistair sold 1650 books in the last sales period. Sales information is shown in the table.

| Books | Number of Sales | Percentage of Sales (nearest whole) |
|---|---|---|
| Crime | 579 | 35% |
| Romance | 300 | 18% |
| Biographies | 391 | 24% |
| Short stories | 380 | 33% |
| **Total** | **1650** | **110%** |

a) Without doing any further calculations, how can you tell that Alistair has made a mistake in recording his data?

b) One of the values in the percentage column is wrong. Use a calculator to check the values and find the mistake. Correct it.

7. Evelyn scored 18 out of 20 in her Maths test and 21 out of 24 in her Physics test. Did she do better in Maths or Physics? Explain your answer.

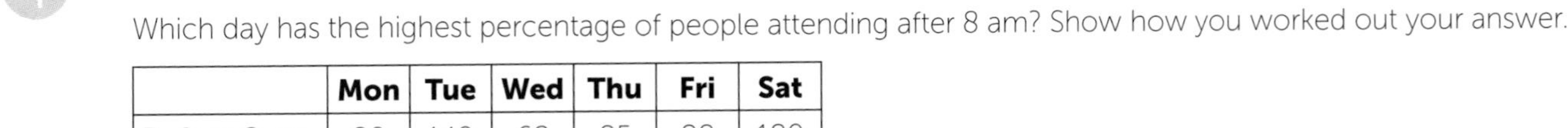

8. The table shows the number of people who went swimming in a city swimming pool last week.
Which day has the highest percentage of people attending after 8 am? Show how you worked out your answer.

| | Mon | Tue | Wed | Thu | Fri | Sat |
|---|---|---|---|---|---|---|
| **Before 8 am** | 80 | 140 | 68 | 85 | 88 | 180 |
| **After 8 am** | 90 | 160 | 32 | 65 | 92 | 220 |

9. There are 40 students in the school orchestra. 40% of the students are girls. 75% of the girls in the orchestra are in Year 11 and the rest of the girls are in Year 10. What percentage of the students in the orchestra are girls from Year 10?

10. 40% of the people who went to a concert were adults, the rest were teenagers. $\frac{2}{5}$ of the people that went to the concert were male, 60% of the people were teenagers. What % of concert attendees were teenage girls?

11. Ben and Sophie are each thinking of a number.

a) 30% of Ben's number is 93. What is his number?

b) One fifth of Sophie's number is 350. What is 80% of her number?

12. Izzy spent 20% of her salary on travelling. She spent $\frac{1}{2}$ of what was left on food and bills and saved the rest. If she saved £200, how much is Izzy's salary?

# 8.6 PERCENTAGE INCREASE

## Objectives

### Work with percentages greater than 100

To find a number as a percentage of another one, calculate $\frac{\text{1st number}}{\text{2nd number}} \times 100\%$

**e.g.** Find 27 as a percentage of 180:

$\frac{27}{180} \times 100\% \quad = \mathbf{15\%}$

**e.g.** What is 36 as a percentage of 24?

$\frac{36}{24} \times 100\% \quad = \mathbf{150\%}$

### Interpret percentages multiplicatively as decimals

### Solve original value problems

To find the value before a percentage increase (the original value), divide by the original multiplier.

**e.g.** A gold coin has increased in value by 5%.
It is now worth £661.50
What was the original value?

| Multiplier: | (100 + 5)% | = 105% |
|---|---|---|
| | 105 ÷ 100 | = 1.05 |
| | 661.50 ÷ 1.05 | = **£630** |

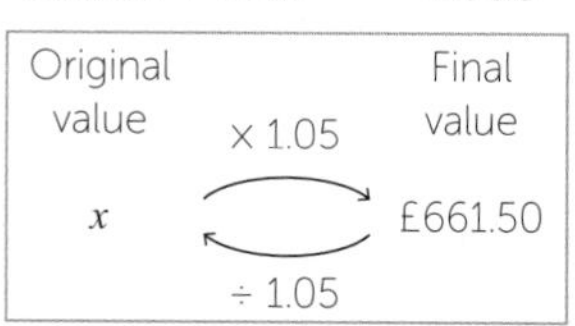


### Solve problems involving percentage increase

The multiplier needed to increase a number by $a\%$ is the decimal equivalent of $(100 + a)\%$.

**e.g.** Increase £35 by 60%.

| Multiplier: | (100 + 60)% | = 160% |
|---|---|---|
| | 160 ÷ 100 | = 1.6 |
| | 1.6 × 35 | = **£56** |

To find a percentage increase, calculate $\frac{\text{actual increase}}{\text{original}} \times 100\%$.

**e.g.** Jake planted a sunflower seed and measured the height of the plant after a few weeks. It was 50 cm. A week later it was 60 cm. Find the percentage increase in height.

Actual increase: 60 − 50 = 10

Percentage increase: $\frac{10}{50} \times 100\% \quad = \mathbf{20\%}$

## Practice questions

1. Increase these numbers by the percentages given:
   a) £120 by 20% b) 300 kg by 18% c) 5 km by 20% d) 150 kg by 25%
2. Holiday prices increased by 18% this year. Last year a holiday cost £800. How much does it cost this year?
3. The price of the latest mobile phone has increased from £150 to £180. Work out the percentage increase.
4. Work out the multiplier required for the following percentage increases.
   a) 5% b) 80% c) 150% d) 0.5%
5. I am thinking of three different numbers $x$, $y$ and $z$. Find my numbers from the clues.
   a) 10% of $x$ is 5. b) 20% of $y$ is 18. c) 30% of $z$ is 18.
6. The number of students in Year 11 last year was 90. This year the number of students has increased by 10%. Will says that there are 100 students this year in Year 11. Is Will correct? Explain why.

7. Tom and Jessie want to buy a printer that costs £80 + VAT. Considering VAT = 20%, Tom and Jessie calculate the price of the printer using two different methods.

| **Tom's method** | | |
|---|---|---|
| Multiplier: | (100 + 20)% | = 120% |
| | | = 1.2 |
| | 80 × 1.2 | = **£96** |

| **Jessie's method** | | |
|---|---|---|
| 20% of 80 | 0.2 × 80 | = 16 |
| | 80 + 16 | = **£96** |

a) Which method do you prefer? Why?

b) Use your preferred method to calculate the price after VAT is added for a laptop costing £360.

8. The population of the world has increased from 6.42 billion to 7.52 billion from 2005 to 2017.

Calculate the percentage increase.

9. Amy invests some money in a new business. After one year, her investment has increased by 110%. She now has £1575.

How much did she invest at the start of the year?

10. A plant grows by 5% each year. This year it is 1.2 m tall. How tall was it last year?

11. Matilda's savings increased by 2% last year. This year they decreased by 2%. Evelina says that Matilda has the same amount now as she had at the start.

Evelina is wrong. Explain why.

## 8.7 PERCENTAGE DECREASE

R9 N12

### Objectives

#### Solve problems involving percentage decrease

The multiplier needed to decrease a number by $a\%$ is the decimal equivalent of $(100 - a)\%$.

**e.g.** A new motorbike **depreciates** by 15% in a year. When new, the motorbike costs £12 000. Calculate the value of the motorbike after one year.

Multiplier: (100 – 15)% = 85%
= 0.85
12 000 × 0.85 = **£10 200**

To find a percentage decrease, calculate

$$\frac{\text{actual decrease}}{\text{original}} \times 100\%$$

**e.g.** When a car is new, it costs £10 000. After two years its value is £6500. Calculate the percentage decrease in value after 2 years.

Actual decrease:
10 000 – 6500 = 3500

Percentage decrease:
$\frac{3500}{10\,000} \times 100\%$ = **35%**

#### Solve original value problems

To find the value before a percentage decrease (the original value), divide by the original multiplier.

**e.g.** A jumper in the sale costs £24 after a 60% reduction. Calculate its original value.

Multiplier: (100 – 60)% = 40%
= 0.4
24 ÷ 0.4 = **£60**

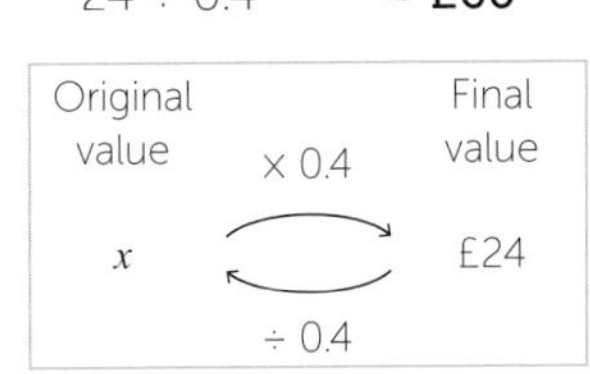

## Practice questions

1. Decrease:
   a) £150 by 20%  b) 250 cm by 25%  c) 80 kg by 40%  d) 120 km by 35%

2. Work out the multipliers given each of the following percentage changes:
   a) 12% increase  b) 8% decrease  c) 15% increase  d) 27% decrease
   e) 135% increase  f) 0.1% increase  g) 22.5% decrease  h) $3\frac{3}{4}$% decrease

3. A shirt originally costs £25. It is reduced in a sale by 30%. Work out the sale price of the shirt.

4. Cars depreciate by 22% each year. A new car costs £15 000. Calculate its value after 1 year.

5. For each of these statements, state whether they are true or false. Rewrite any false statements to make them true.
   a) 60 decreased by 40% = 60 – (0.4 × 22.8)  b) 0.35 decreased by 90% = 0.35 × 0.9
   c) 110 decreased by 22.5% = 11.5 × 0.775

6. The value of a house could increase or decrease each year.
   Work out the percentage change after the following valuations:
   a) £120 000 reduced to £108 000  b) £130 000 reduced to £104 000
   c) £132 000 reduced to £120 000  d) £110 000 increased to £137 000

7. In 2000, there were 230 000 wild orangutans. In 2010, there were 104 700 orangutans in the wild.
   Calculate the percentage decrease between 2000 and 2010.

8. During the summer sale, all holiday packages have been reduced by 15%.
   Callum's family bought a holiday package and saved £231.
   a) What was the original price?  b) How much did they pay for their holiday?

9. All items in the School Shop are reduced by 25%. A graphic calculator costs £80.
   What was the price before the sale?

10. Tameem wants to buy a new bike, costing £800. He can pay the full price now, and receive 10% off. He could pay a deposit of 30%, then ten monthly instalments of £58.
    Which method is cheaper and by how much?

11. Part of the print on a sales sticker is smudged.
    Give three examples of what the original price and percentage off could have been.

12. The original price of an item is £100. The price is decreased by 10%, then decreased by another 10%.
    Oliver says 'The new price will be £80'
    Explain why Oliver is wrong.

SECTION 9

# PROBABILITY

## 9.1 MEASURING PROBABILITY

P3

### Objectives

**Distinguish between outcomes which are impossible, unlikely, even chance, likely, and certain to occur**

The probability of something happening is a measure of how likely it is.

It can be described using words such as; likely, unlikely, certain, impossible and even chance.

**e.g.** the chance of the sun rising tomorrow is **certain**

the chance of rolling a 2 on a fair normal dice is **unlikely**

the chance of getting a head on a fair coin is an **even chance**

**Mark probabilities on a probability scale of 0 to 1**

**Measuring probability**

Probability can be measured using numbers on a scale from 0 to 1

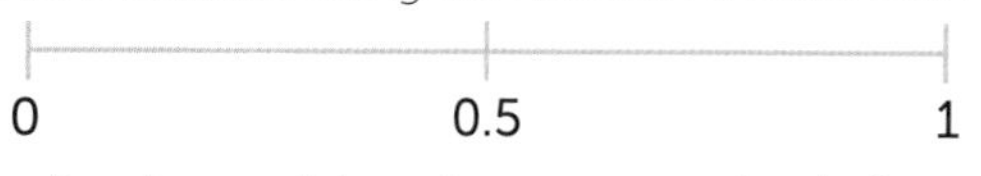


The probability of an impossible outcome occurring is 0
The probability of a certain outcome occurring is 1

All other measures of probability lie in between 0 and 1. Probability values can be fractions, decimals, or percentages.

**e.g.** The probability of each outcome is shown on the probability scale.

**A** Getting a 'head' when a fair coin is thrown
**B** Rolling a 4 with a fair normal dice
**C** Being given homework this week
**D** Rolling two normal dice and getting a total score of 13
**E** The sun setting tomorrow evening

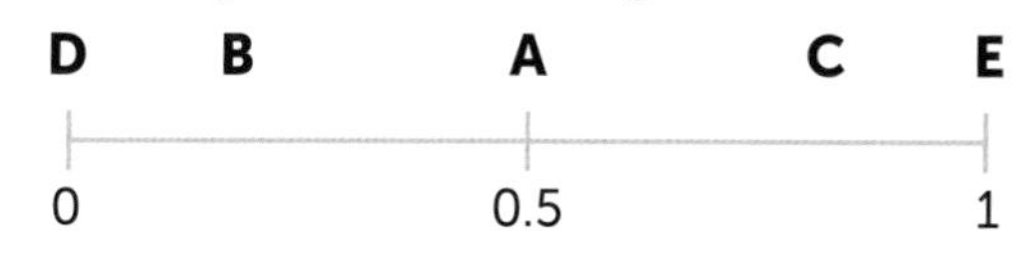


### Practice questions

1. Describe the likelihood of each of these events occurring.
   Choose words from: 'impossible', unlikely', 'equally likely', 'likely' and 'certain'.
   a) Saturday will follow Friday.
   b) Someone in your class is left-handed.
   c) You will live to be 110 years old.
   d) You roll a 7 using a normal 6-sided dice.

2. Draw a probability scale from 0 to 1.
   Mark it with each letter to indicate the approximate probability of each event.
   a) The next car you see will have been made in Japan.
   b) It will rain where you live tomorrow.
   c) At the next Olympic Games, someone will run 1500 m in less than 3 minutes.
   d) A manned flight will reach Mars in the next 10 years.

3. The top prize in a raffle is a new television.
   Jasmine wishes to enter the raffle for a chance to win the television. Draw a probability scale from 0 to 1. Mark on it the letters estimating the probability of her winning if she buys:
   a) all the tickets  b) half the tickets  c) no tickets  d) most of the tickets

4. Which is the more likely outcome in each of the following pairs?
   a) From a pack of 52 standard playing cards, you will pick a red card, or you will throw a six with a fair dice.
   b) 'Two fair dice will both land on 5' or 'two fair coins will both land on heads'.

5. Write down 3 events which have a probability of 0. Give reasons for your answers.

6 Write down 3 events which have a probability of 1. Give reasons for your answers.

7. In a family with 3 children, the eldest child is a girl. What is the probability of this?

8. In a family with 2 children, both children are boys. What is the probability of this?

9. Arrange the following events in order from least likely to most likely.
   a) It will rain during the next 30 days.
   b) You will talk to your Headteacher next week.
   c) The next two cars you see will both be red.
   d) Someone you know will win 'Britain's Got Talent'.

10. Amy and Emma are watching the National Lottery draw. The first ball drawn is 51, the second ball is 48 and the third ball is 54. Emma says that since these are all high numbers, the next ball drawn will be a low number. Amy says the next ball could be any of the other numbers.
    Who is correct? Give a reason for your answer.

## 9.2 LISTING SYSTEMATICALLY

P2 P6 P7 N5

### Objectives

#### List all outcomes for single events systematically

An **event** is an activity which leads to different **outcomes**.

e.g. Throwing a coin is an event. The outcomes of an event are the possible results following the event.

e.g. When a coin is thrown, there are two outcomes – either heads, or tails.

When listing all the possible outcomes be **systematic** to ensure no outcomes are missed.

e.g. A fair 8-sided spinner has 8 equal sections numbered 1 to 8. The possible outcomes of this event are: 1, 2, 3, 4, 5, 6, 7, 8.

A bag contains 3 coloured counters – 2 blue and 1 green. Two counters are taken together from the bag at random. The possible outcomes are: (blue, blue), (blue, green) and (green, blue).

#### List all outcomes for combined events systematically

Two different events can take place and the outcomes are a combination of the outcomes of the separate events.

e.g. A fair six-sided dice and a fair coin are thrown. The outcomes from these combined events are: 1-heads, 2-heads, 3-heads, 4-heads, 5-heads, 6-heads, 1-tails, 2-tails, 3-tails, 4-tails, 5-tails and 6-tails.

#### Use and draw sample space diagrams

The sample space shows all the outcomes from one or more events.

A sample space could be a list of outcomes or could be shown in a table.

e.g. Two fair coins are thrown, and each can land heads (H) or tails (T). The sample space of outcomes is: HH, HT, TH, TT

A committee choose 2 representatives, one male, one female. The men are Nico, Asif and Rob and the women are Tia, Sonia, Usha and Katie.

The sample space of outcomes is:

| | | | |
|---|---|---|---|
| Nico, Tia | Nico, Sonia | Nico, Usha | Nico, Katie |
| Asif, Tia | Asif, Sonia | Asif, Usha | Asif, Katie |
| Rob, Tia | Rob, Sonia | Rob, Usha | Rob, Katie |

## Practice questions

1. List the possible outcomes when rolling a fair 20-sided dice numbered from 1 to 20.

2. Muffins come in lemon, blueberry or chocolate flavour. They can have sprinkles or not.
   List the possible types of muffin.

3. Akua, Bianca, Carly, Dora and Ebba are in the same class. Their teacher needs to choose two of them to do a job.
   Write down the possible pairs that the teacher could choose.

4. Three fair coins, each showing a head and a tail, are thrown.
   List the possible outcomes.

5. A restaurant sells pizzas. There are three types; pepperoni, margherita and Napoli.
   The pizza comes with one extra topping; mushrooms, anchovies or extra cheese.
   List the possible types of pizza.

6. A bag contains 4 coloured balls; 1 blue, 1 yellow, 1 green and 1 red.
   A ball is taken from the bag at random, its colour noted, then it is replaced in the bag. A ball is again taken from the bag and its colour noted.
   Draw a table to show the sample space of possible outcomes.

7. a) Two normal fair six-sided dice are rolled together.
   List the possible outcomes as pairs of numbers.
   b) The scores on the two dice are added together.
   Write the sample space for possible outcomes.

8. Gurdeep has a normal fair six-sided dice and four cards showing clubs, hearts, diamonds and spades.
   He rolls the dice and picks a card at random.
   Draw a table to show the sample space of possible outcomes.

9. Liz has a selection of scarves; 4 are flowery, 3 are plain and 2 are spotty. She picks out 3 scarves at random to lend to a friend.
   List the possible combinations of scarves the friend could receive.

10. There are two sets of numbered cards.
   Set A has numbers 1, 2, 3 and 4, and set B has numbers 1, 3, 5, 7.
   One card is drawn at random from each set and the difference between the numbers on the chosen cards is found.
   Complete the table to show the sample space of possible outcomes.
   The first one has been done for you.

| | | **Set A** | | | |
|---|---|---|---|---|---|
| | | 1 | 2 | 3 | 4 |
| **Set B** | 1 | 1 − 1 = 0 | | | |
| | 3 | | | | |
| | 5 | | | | |
| | 7 | | | | |

# 9.3 THEORETICAL PROBABILITY

## Objectives

**Find the probability of an outcome occurring using theoretical probability**

The **theoretical probability** is the probability which is expected to happen.

Theoretical probability is the ratio of successful outcomes to all possible outcomes.

Theoretical probability $= \dfrac{\text{number of successful outcomes}}{\text{total number of possible outcomes}}$

**e.g.** The probability of getting a five on a fair six-sided dice is $P(5) = \frac{1}{6}$

The probability of tossing a head using a fair coin is $P(H) = \frac{1}{2}$

In a bag of 20 sweets, 10 are red, 3 are green and the rest are yellow. The probability of a taking a yellow sweet at random is found by dividing the number of yellow sweets by the total number of sweets.

Number of yellow sweets is $20 - (10 + 3) = 7$ $\quad P(\text{yellow}) = \frac{7}{20}$

**Estimate the number of times an outcome will occur, given the probability and the number of trials**

We can estimate the number of times an outcome will occur based on theoretical probability.

An estimate of the number of times is found by multiplying the theoretical probability of the outcome by the number of trials.

**e.g.** The theoretical probability of winning a game, based entirely on chance is $\frac{3}{5}$.

Anna plays the game 10 times. The number of times she should expect to win is

$\frac{3}{5} \times 10 =$ **6 times**.

**Use theoretical models to include outcomes using dice, spinners and coins**

## Practice questions

1. A fair coin is thrown. What is the theoretical probability that it will land as tails?

2. If a fair coin is thrown 800 times, estimate the number of heads.

3. Two fair coins are thrown together.
   a) What are the possible outcomes?
   b) What is the theoretical probability of getting 2 heads?
   c) What is the theoretical probability of getting a head and a tail?
   d) If two fair coins are thrown 360 times, estimate the number of times would you expect to get 2 tails.

4. Yan has a pencil case with 8 pens in it. Three pens are red, and the rest are blue. He picks a pen out of the case at random.
   a) What is the probability of getting a blue pen?
   b) He has 6 lessons in a day and each time he draws out a pen at random. Estimate the number of times he will draw a red pen.

5. Carrie has a biased coin which lands on heads with a probability of 0.65 If she throws the coin 100 times, estimate the number of tails she can expect.

6. A spinner is made from a square of card and coloured as shown.
   a) What is the probability it will land on yellow?
   b) What is the probability it will land on blue?
   c) It is spun 80 times. Estimate the number of times it lands on red.

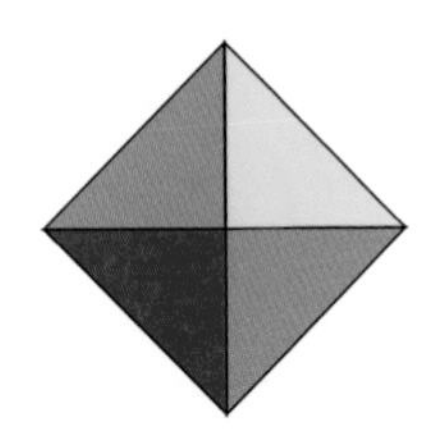

7. A spinner is made from a regular hexagon and coloured as shown.
   a) What is the probability it will land on red?
   b) What is the probability it will land on yellow?
   c) In 300 spins, estimate the number of times it will land on blue.

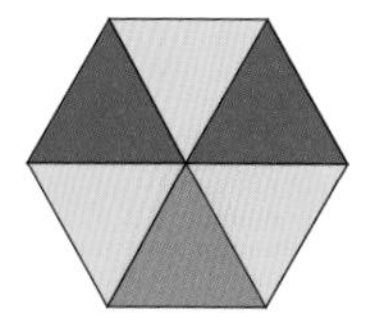

8. Two fair six-sided dice are rolled, and the scores are added to give a total.
   a) What are the possible totals?
   b) What is the probability of scoring 12?
   c) What is the probability of scoring 4?
   d) What is the probability of scoring 7?
   e) Dixie says the probability of scoring 3 is the same as scoring 11.
   Seb thinks the probability of scoring 3 is less than the probability of scoring 11.
   Who is correct? Give a reason for your answer.

9. A dice is made up of some red faces and some blue faces. The probability of it landing on a red face is 0.7
   It is rolled many times and lands on red 70 of the times. Estimate the number of times it has been rolled.

10. A dice is rolled 600 times and lands on three, 200 times. Is the dice fair? Give a reason for your answer.

11. Heather has a biased twelve-sided dice numbered from 1 to 12. The probability of the dice landing on 12 is 0.15
    She rolls the dice many times and gets 12, 60 times and 11, 20 times.
    a) Estimate the number of times she has rolled the dice.
    b) Estimate the probability of getting 11.

## 9.4 EXPERIMENTAL DATA

P1 P6

### Objectives

#### Record outcomes of probability experiments in tables

When an experiment is carried out, it is helpful to record the outcomes in a frequency table.

**e.g.** A fair dice is rolled 60 times and the outcomes recorded.

| Score | 1 | 2 | 3 | 4 | 5 | 6 |
|---|---|---|---|---|---|---|
| Frequency | 10 | 7 | 12 | 9 | 11 | 11 |

The frequency shows the number of times a particular score occurred.

If the dice is fair, the frequency of each score should be similar. Explain why.

#### Estimate probability using relative frequency

The frequency can be used to find the relative frequency of an outcome.

$$\text{Relative frequency} = \frac{\text{frequency of outcome}}{\text{total frequency}}$$

Relative frequency gives an estimate of the probability of an outcome occurring.

**e.g.** From the frequency table above, the relative frequency of getting a 2 on the dice is $\frac{7}{60}$.

An estimate of the probability of a getting a 2, from this data, is $\frac{7}{60}$.

#### Record outcomes of probability experiments in a frequency tree

A frequency tree is another way of displaying data from an event.

**e.g.** This frequency tree shows the number of girls and boys in a group of 40 students.

It also shows how many girls and boys are vegetarian or not.

A student is chosen at random. The probability of selecting a student who is a boy and a vegetarian is $\frac{4}{40} = \frac{1}{10}$

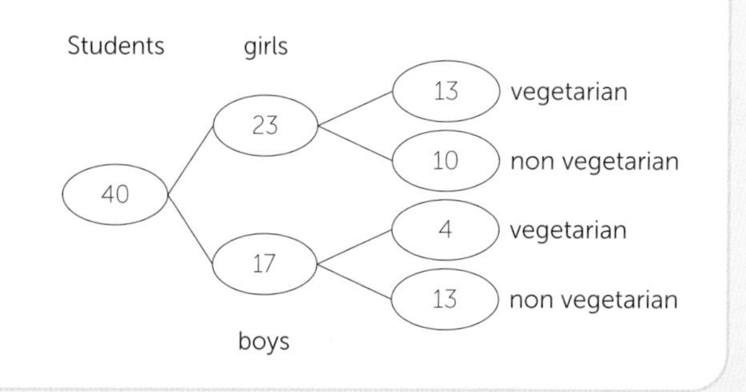

## Practice questions

1. Throw a fair coin 100 times.
   a) Record the outcomes in a frequency table.
   b) Using your data, calculate the relative frequency of throwing a head.

2. Roll a fair six-sided dice 30 times.
   a) Record the results in a frequency table.
   b) Write down the relative frequency of getting a 4 from your data.
   c) If you repeated your experiment would you expect to get the same relative frequency for 4?

3. A bag has red, blue and green counters in it.
   A counter is drawn out, its colour noted and then replaced.
   The experiment is repeated 60 times and the results put into this frequency table.

| **Colour** | Red | Blue | Green |
|---|---|---|---|
| **Frequency** | 20 | 10 | 30 |

   a) Write down the relative frequency of getting a blue counter.
   b) Estimate the probability of getting a green counter, if the experiment is repeated.
   c) The experiment is repeated a further 30 times. Estimate the number red counters drawn out this time.

4. 80 men sat a test and 34 passed. 50 woman sat the test and 19 did not pass.
   a) Put this information in a frequency tree.
   b) One person is chosen at random from those who took the test. Write down the relative frequency of selecting a man who did not pass the test.

5. This frequency tree shows the chosen sport for some students.

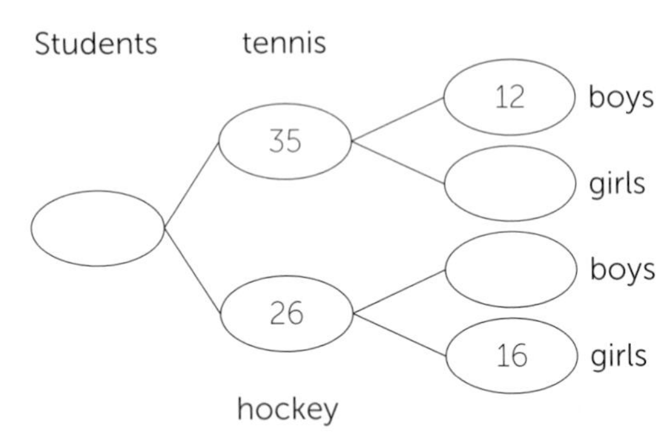


   a) Copy and complete the frequency table.
   b) A student is chosen at random. What is the probability that the student plays tennis?
   c) What is the probability that the student is a boy who plays hockey?

6. Customers at a café sometimes buy food (with or without a drink) and sometimes just a drink.
   The frequency table shows how people shopped at the cafe on a typical day.

| | Food | Just drink |
|---|---|---|
| Sit in | 45 | 15 |
| Take out | 24 | 62 |

   a) Draw a frequency tree displaying this data. Start with the total number of customers surveyed.
   b) Use this data to estimate the probability that a customer will take out their purchase.
   c) Use this data to estimate the probability that a customer who buys just a drink will sit in.

## 9.5 PROBABILITY EXPERIMENTS

### Objectives

**Understand that a larger sample size will produce a more reliable estimate of probability**

Relative frequency can be found from the frequencies given by a probability experiment.

The relative frequency of an outcome is also an estimate of the probability of that outcome occurring in further experiments.

**e.g.** A square spinner is coloured as shown. The spinner is spun 100 times and the results recorded in this frequency table.

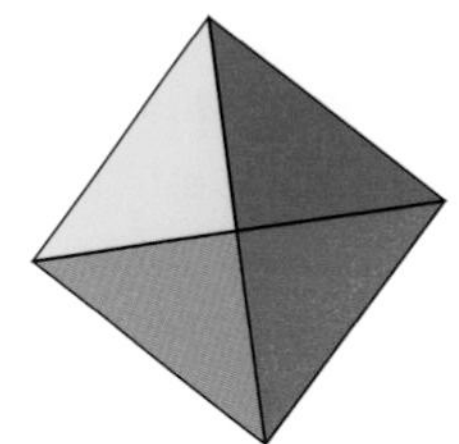

| **Colour** | Yellow | Green | Red | Blue |
|---|---|---|---|---|
| **Frequency** | 24 | 26 | 35 | 15 |
| **Relative frequency** | 0.24 | 0.26 | 0.35 | 0.15 |

The relative frequency of getting a red is 0.35; so, an estimate of the probability of getting a red next on the spinner is 0.35.

**Compare experimental data and theoretical probabilities**

**Comparing probabilities**

Comparison can be made between an estimated probability from experimental data and a theoretical probability.

**e.g.** In the spinner example shown above, the theoretical probability of getting a red (if the spinner is fair) = $\frac{1}{4}$ or 0.25.

This is much lower than the estimated probability of 0.35. It looks likely that the spinner is not fair.

**Sample size**

The more trials carried out, the more reliable is the data collected.

**e.g.** In the spinner example shown earlier, a further 900 spins are carried out and the results shown here.

| **Colour** | Yellow | Green | Red | Blue |
|---|---|---|---|---|
| **Frequency** | 236 | 242 | 267 | 255 |
| **Relative frequency** | 0.236 | 0.242 | 0.267 | 0.255 |

Now all the relative frequencies are much closer to the theoretical probability of 0.25. It now looks as if the spinner is fair.

### Practice questions

1. A coin is thrown 1000 times and lands on heads 485 times.
   Based on this experiment, estimate the probability of getting a head on the next throw.

2. A **biased** coin is thrown 400 times and lands on heads 267 times.
   a) How many times does it land on tails in this trial?
   b) What is the relative frequency of landing on heads?
   c) If the coin is tossed again, estimate the probability it will land on tails.

3. A sack contains 20 coloured balls. Each ball is either white, black or red. Dania chooses a ball from the sack, notes its colour and replaces it. She does this a total of 400 times and records her results in this table:

| White | Black | Red |
|---|---|---|
| 240 | 30 | 130 |

a) What is the relative frequency of a red ball?

b) Which colour ball is least likely to be drawn at random next?

4. Out of the last 1000 candidates who took their driving test, 750 passed.

a) Estimate the probability that the next student who takes the test will pass.

b) Out of the **next** 1000 candidates only 640 passed.

Using **all the data**, estimate the probability that the next candidate will pass.

5. Sammy rolls a six-sided dice 100 times and it lands on 6, 15 times.

a) Estimate the probability of getting a 6 next on this dice.

b) Write down the theoretical probability for getting a 6 on a fair dice.

c) Compare your answers to parts a) and b).

6. Courtney has a box of 100 letter tiles. She takes a tile at random, notes the letter and returns the tile to the box. Here are her results.

| **Letter** | I | S | L | U | O | Z | V | E | Other |
|---|---|---|---|---|---|---|---|---|---|
| **Frequency** | 14 | 21 | 10 | 8 | 17 | 1 | 3 | 24 | 102 |

a) How many times does she take a tile from the box?

b) What is the relative frequency of choosing the letter S?

c) Which of these letters is three times as likely to be picked as Z?

d) There are four tiles with the letter L in the box. Does she draw L as often as expected? Give a reason for your answer.

7. A triangular spinner can land on blue (B), yellow (Y) or green (G).

The spinner is spun 20 times giving: B Y Y G Y B G G Y G B Y Y G B G Y G G B

a) Write down the relative frequency of getting each colour.

b) After a total of 100 spins, the relative frequency of a blue is now 0.19. Which of the relative frequencies for blue is the best estimate for the probability of the spinner landing on blue next? Give a reason for your answer.

8. Harry throws a coin 10 times and finds the relative frequency of tails.

He repeats this 6 times. The results are shown in the table.

| **Total number of throws** | 10 | 20 | 30 | 40 | 50 | 60 |
|---|---|---|---|---|---|---|
| **Relative frequency of tails** | 0.4 | 0.45 | 0.4 | 0.45 | 0.44 | 0.4 |

a) How many tails in total has Harry got after 40 throws?

b) Estimate the probability of getting a tail on the sixty-first throw.

# 9.6 MUTUALLY EXCLUSIVE OUTCOMES

## Objectives

### Identify different mutually exclusive outcomes and know that the sum of the probabilities of all outcomes is 1

Outcomes that **cannot** happen together are called **mutually exclusive**.

**e.g.** An ordinary coin is thrown.

The outcome of a head and the outcome of a tail are **mutually exclusive**. It is not possible to get both together.

A card is chosen from an ordinary pack of cards.

The outcome of a King and the outcome of a Heart are **not mutually exclusive**. It **is possible** to get both together – the King of Hearts!

### Use 1 – p as the probability of an outcome not occurring where p is the probability of the outcome occurring

The probability of an outcome occurring + the probability of the outcome not occurring = 1

This can be used to find the missing probabilities.

**e.g.** The probability of it raining tomorrow is 0.3

This means the probability of it **not raining** tomorrow is 1 – 0.3 = 0.7

### Find a missing probability from a list or table including algebraic terms

The sum of the probabilities of all possible outcomes of an event equals 1.

This can be used to find missing probabilities.

**e.g.** A team plays a hockey match. The probability of a win is 0.25 and of a draw is 0.1

This means that the probability of losing is 1 – (0.25 + 0.1) = **0.65**

## Practice questions

1. Identify which of these are mutually exclusive outcomes:
   a) From an ordinary pack of cards, picking a card that is red and a Queen.
   b) Rolling a three and a four at the same time on an ordinary dice.
   c) Picking a number that is a prime number and an even number.
   d) Picking a number that is a prime factor of 13 and a prime factor of 24.
   e) Choosing student from a class who is a girl and who wears glasses.

2. Work out each of the following.
   a) P(**not** a 1) when rolling a fair six-sided dice.
   b) P(**not** a tail) when throwing a fair coin.
   c) P(**not** an Ace) when choosing a card from an ordinary pack of cards.

3. The probability that a team wins a rugby match is 0.72.
   Is the probability of losing 0.28? Give a reason for your answer.

4. Jay and Kim play a game of noughts and crosses. The probability that Jay wins is 0.34 and the probability that Kim wins is 0.47. What is the probability that the game is a draw?

5. Gil buys a comic book once a month. The probability that he buys 'Dark Knight' is 0.55. The probability that he buys 'The Panther' is 0.25.
   a) What is the probability that he does **not** buy 'The Panther'?
   b) What is the probability that he does **not** buy 'The Dark Knight'?
   c) What is the probability that he buys a comic which is neither of these two?

6. Tahir travels to a monthly meeting by either; bus, car, train or tram, with the following probabilities.

   P(bus) = 0.1 P(car) = 0.2 P(train) = 0.5

   a) What is the probability that he travels to the meeting by tram?
   b) What is P(**not** by car)?
   c) How many times in a year would Tahir travel by train?

7. Four athletes take part in a cross-country race.
   The probabilities of each of them winning are shown in the table.

| Dee | Joy | Kit | Sally |
|---|---|---|---|
| 0.4 | | 0.15 | 0.1 |

   a) What is P(Joy wins)?
   b) What is P(Kit does not win)?

8. Ruby has 3 hamsters. She chooses one to play with each day.

| Pip | Hammy | Nibbles |
|---|---|---|
| | $x$ | $x$ |

   She is equally likely to choose Hammy as Nibbles. She chooses Pip 3 times out of every 10 days.

   a) What is the probability that she chooses Pip?
   b) What is the probability that she chooses Hammy?
   c) What is the probability that she does not choose Nibbles?
   d) In the month of June, how many times can Pip be expected to be chosen?

## 9.7 COMBINED EXPERIMENTS AND OUTCOMES

P2 P7

### Objectives

#### Add simple probabilities

**Combined outcomes**

The probability of combined outcomes of an event can be found by adding together the probability of each single outcome.

This bar model shows the three outcomes possible on a triangular spinner.

| P(red) = 0.4 | P(green) = 0.3 | P(blue) = 0.3 |
|---|---|---|

The probability of the spinner landing on red or green is 0.7.

It is found by adding the probability of landing on red and the probability of landing on green.

For a single event:

**P(A or B) = P(A) + P(B)**

e.g. The captain of a team is to be chosen from four students A, B, C and D.

The probability of each being chosen is given by:

P(A) = 0.2, P(B) = 0.15, P(C) = 0.4, P(D) = 0.25

P(A or B) = P(A) + P(B)
= 0.2 + 0.15
= 0.35

P(B or C or D) = P(B) + P(C) + P(D)
= 0.15 + 0.4 + 0.25
= 0.8

#### Find the probability of a given outcome for multiple independent events or experiments, such as several throws of a single dice

e.g. A dice is rolled three times.
To find the probability of an outcome for successive events, multiply the probability of the separate outcomes. (Note: This assumes that the probability of the second event is unaffected by the first event.)

**P(A and B) = P(A) × P(B)**

e.g. A fair six-sided dice is thrown twice. The probability of getting 3 on one throw is $\frac{1}{6}$

The probability of getting 3 on two throws = P(3) × P(3)

$= \frac{1}{6} \times \frac{1}{6}$

$= \frac{1}{36}$

## Practice questions

1. What is the probability of throwing a prime number on a fair six-sided dice?

2. A bag has red, blue and yellow counters in. The probability of choosing each is given by:
   P(red) = 0.4  P(blue) = 0.15  P(yellow) = 0.45
   a) Work out the probability of choosing red or blue.
   b) Work out the probability of choosing yellow or blue.
   c) Work out the probability of choosing red, blue or yellow. Explain your answer.

3. A fruit bowl contains 2 apples, 3 bananas and 5 oranges.
   Parveen chooses a piece of fruit at random.
   a) What is the probability she chooses an apple or a banana?
   b) What is the probability she chooses an orange or an apple?
   c) What is the probability she does **not** choose an orange or a banana?

4. Zane has three favourite lunches; pizza, pasta or panini.
   The probability that he chooses each one is given by:
   P(pizza) = 0.4  P(pasta) = 0.25  P(panini) = 0.35
   a) Work out P(pizza or pasta).
   b) Work out P(panini or pizza or pasta). Explain your answer.

5. Mr Tick the teacher owns lots of ties. He chooses a tie at random to wear each day.
   The probability of him choosing a tie is given by:
   P(striped) = 0.3  P(plain) = 0.25  P(spotty) = 0.1  P(flowery) = 0.35
   a) Work out P(striped or plain).
   b) Work out P(flowery or spotty or plain).

6. Work out the probability, that in a family of three children, all the children are girls.

7. A fair ten-sided dice numbered 1 to 10, is rolled twice.
   Work out the probability that the dice will land on 9 both times.

8. A sack contains 8 blue counters and 2 white counters.
   A counter is chosen at random and then replaced. A counter is chosen again.
   What is the probability that both of the counters chosen are blue?

9. The probability of a particular seed germinating is 0.85.
   Three seeds are planted. Work out the probability that all three seeds germinate.

10. Luke has a drawer full of socks; 12 are black, 6 are blue and 2 are brown.
    He takes out a sock, notes the colour and then replaces it.
    Then he takes out a sock again.
    a) What is the probability that the two socks he picks out are both black?
    b) Luke says he has a better than even chance of picking out two socks of the same colour.
       Is he correct? Show your working.

SECTION 10

# RATIO

## 10.1 RATIO NOTATION

R4 R6 R8 N11

### Objectives

**Write ratios in their simplest form, and in the form 1 : m or m : 1**

To **simplify** a ratio is to divide both quantities by a common factor.

A ratio is in its **simplest form** when the quantities do not share any prime factors.

**e.g.** 30 : 50 can be simplified by dividing both sides by 10 to give 3 : 5

**Equivalent ratios** express the same proportion of quantities but with different numbers.

**e.g.** 30 : 50, 6 : 10 and 3 : 5 are all equivalent ratios.

Use division of both sides to express a ratio in the form 1 : m or m : 1

**e.g.** Dividing both sides by 3 gives

$3 : 5 = 1 : \frac{5}{3}$

Dividing both sides by 5 gives

$3 : 5 = \frac{3}{5} : 1$

**Understand and express the division of a quantity into a number of parts as a ratio**

A **ratio** compares the sizes of two quantities.

A ratio is written using a colon ( : ) between each quantity in the order they are given.

**e.g.** There are 3 oranges and 2 apples.

The ratio of oranges to apples is **3 : 2**

The ratio of apples to oranges is **2 : 3**

**Write a ratio as a fraction**

A ratio can also be expressed as a fraction.

Find the total number of parts and express each quantity out of this total.

**e.g.** In the diagram, the ratio of white to grey squares is 20 : 4 or 5 : 1

The total number of parts is 5 + 1 = **6** and 1 part is grey

The proportion of the rectangle that is grey is $\frac{1}{6}$.

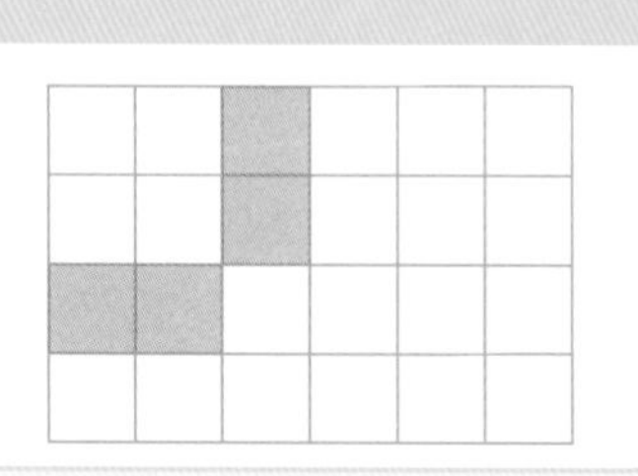

### Practice questions

1. Write the ratio of grey to white parts for each diagram.
   Give your answers in the simplest form.

   a)  b)

2. Jane bought 4 kg of flour, 12 kg of sugar, 28 kg of rice and 36 kg of beans. Write each of these ratios in their simplest form.

   a) The mass of rice to beans
   b) The mass of flour to rice
   c) The mass of sugar to flour
   d) The mass of beans to sugar

3. Write each ratio in its simplest form.

   a) 15 : 20 b) 24 : 36 c) 28 : 35 d) 20 : 70

4. Write each ratio in the form 1 : m

a) 25 : 50  b) 30 : 10  c) 64 : 40  d) 60 : 72

5. Tomas bought 15 kg of strawberries and 18 kg of plums to make jam.

a) Write the ratio of strawberries to plums in its simplest form.

b) What fraction of the fruit is strawberries?

6. The ratio of red pens to black pens in a packet of pens is 2 : 5 There are 10 red pens in the packet.

a) How many black pens are in the packet?

b) What fraction of the packet are red pens?

7. There are 12 boys and 17 girls in a class. 3 new boys joined the class and 2 girls left.

a) Write the new ratio of the number of boys to the number of girls.

Give your ratio in its simplest form.

b) What fraction of the class are boys now?

8. The table shows the number of students that study different languages in a school.

| **Language** | Spanish | Latin | German | Italian | French |
|---|---|---|---|---|---|
| **Number of students** | 48 | 28 | 70 | 35 | 56 |

Copy and complete the missing words:

a) Latin : ...................... = 1 : 2

b) ...................... : ...................... = 4 : 5

c) ...................... : ...................... = 5 : 2

d) ...................... : ...................... = 6 : 7

e) ...................... : ...................... = $1 : \frac{7}{12}$

f) ...................... : ...................... = $\frac{4}{5} : 1$

9. Copy and complete the missing numbers to form equivalent ratios

a) 16 : ....... = 2 : 3  b) 24 : 60 = 2 : .......  c) 42 : ....... = 6 : 7

d) ....... : 56 = 3 : 2  e) ....... : 7 = $1 : \frac{7}{3}$  f) ....... : 4 = $\frac{5}{4} : 1$

10. The ratio of the number of stamps Grace has to the number of stamps Josh has is $1 : \frac{5}{6}$

What is the smallest number of stamps that Gracie and Josh could have in total?

## 10.2 DIVIDING IN A GIVEN RATIO

R5

### Objectives

**Write a ratio to describe a situation**

**Share a quantity in a given ratio including three-part ratios**

#### Find one quantity when the other is known

Knowing a ratio and one quantity, means the ratio can be scaled up or down to find the other quantity.

**e.g.** Dan and Bex share the cost of a holiday in the ratio of 4 : 5.

Dan pays £320.

Dan has 4 parts and has paid £320, so 1 part costs 320 ÷ 4 = **£80**

Bex has 5 parts so she pays 5 × £80 = **£400**

#### Share a quantity in a given ratio including three-part ratios

Quantities can be shared in a ratio.

Divide the total number of parts of the ratio into the amount to be shared to find the value of one part. Then multiply the value of one part by the ratio.

**e.g.** Anton and Bella share £540 in the ratio 5 : 1

The total number of parts is 5 + 1 = 6

| 90 | 90 | 90 | 90 | 90 | 90 |
|---|---|---|---|---|---|

1 part is £540 ÷ 6 = £90

Anton has 5 parts = 5 × £90 = **£450**

Bella has 1 part = 1 × £90 = **£90**

## Practice questions

1. Share the quantities in the given ratios.
   a) 20 kg in the ratio 3 : 7
   b) 180 g in the ratio 4 : 5
   c) 120 cm in the ratio 3 : 5 : 4
   d) £450 in the ratio 2 : 9 : 4

2. Fiona and Will have 126 stickers in total. They share them in the ratio 3 : 4
   How many stickers does Will get?

3. The ratio of windy days : sunny days : rainy days one September was 3 : 1 : 2
   a) How many days were sunny?
   b) How many more days were windy than rainy?

4. There need to be 3 teachers for every 14 students on any school trip. One school trip has 210 students.
   How many teachers need to go?

5. A cereal box contains a mixture of nuts, oats and raisins in the ratio 3 : 10 : 1
   There is 350 g of cereal in the box. Find the mass of oats in the cereal box.

6. There are twice as many red roses as there are yellow roses in a garden. There are three times as many yellow roses as there are pink roses in the garden.
   a) Write the ratio of red : yellow : pink roses.
   b) How many red roses will there be if there are 60 roses altogether in the garden?

7. A rope is split into two lengths in the ratio 3 : 8. The longer piece is 120 cm longer than the shorter piece.
   How long was the original rope?

8. Rae, Mo and Lara collect 60 shells. The ratio of small shells to big shells is 1 : 2. They share the small shells in the ratio 2 : 1 : 2 and the big shells in the ratio 2 : 3 : 5 Work out the number of shells each girl gets.

9. The ratio of the width of a rectangle to the length of a rectangle is 5 : 7
   The perimeter of the rectangle is 48 cm. Work out the area of the rectangle.

   Perimeter = 48 cm

10. There are pink, yellow and blue marbles in a box. The ratio of pink marbles to yellow marbles to blue marbles is 3 : 7 : 4
    There are 434 marbles in the box.
    a) Lily says that there are 90 pink and 170 yellow marbles. Without doing any detailed calculations, how can you tell that Lily is wrong?
    b) Work out how many more blue marbles than pink marbles there are in the box.

## 10.3 RATIO AND PROPORTION

### Objectives

**Solve word problems involving direct proportion, including recipes**

If two quantities are in **direct proportion**, the numbers stay in the same proportion when scaled up or down.

**e.g.** 250 g of flour are needed to make 10 cupcakes. To make 5 cupcakes we halve the number of cupcakes, so we halve the amount of flour too.

We need 250 ÷ (10 ÷ 5) = **125 g of flour**

If we have 750 g of flour, we can make 750 ÷ 250 = 3 batches of cupcakes.

So, we can make 10 × 3 = **30 cupcakes**

**Solve proportion problems using the unitary method**

When working with direct proportion problems, one approach is to work out the **value of 1** in the relationship. Then this can be scaled up as needed.

**e.g.** The cost of hiring a car is £322 for 2 weeks.

The cost is directly proportional to the number of days of hire.

To work out the cost of hiring the car for 30 days, first work out the cost for 1 day.

14 days cost £322

**1 day** costs £322 ÷ 14 = **£23**

30 days cost £23 × 30 = **£690**

**Compare ratios**

To compare ratios, it can help to express the ratio as a fraction.

**e.g.** The ratio of copper to silver in Metal A is 7 : 5

The ratio of copper to aluminium in Metal B is 3 : 2

Which of the metals has the higher quantity of copper?

Metal A is $\frac{7}{12}$ copper. Metal B is $\frac{3}{5}$ copper.

$\frac{7}{12} < \frac{3}{5}$ so **Metal B has the higher quantity of copper in it.**

**Express a multiplicative relationship between two quantities as a ratio or fraction**

### Practice questions

1. The cost of 14 textbooks is £196. Work out the cost of 1 textbook.

2. Twelve pens cost £4.80. Work out:
   a) the cost of 1 pen
   b) The cost of 8 pens
   c) the cost of 11 pens
   d) how many pens can be bought with £6

3. A packet of 8 chocolate bars costs 96p. Work out:
   a) the cost of 2 bars
   b) the cost of 6 bars
   c) the cost of 12 bars
   d) how many bars can be bought with £4.80

4. The ages of 3 cousins are in the ratio 2 : 3 : 4. The eldest cousin is 24 years old. How old are the other two cousins?

5. The ratio of the height of the Eiffel Tower and the London Eye is 20 : 9
   The London Eye is 135 m tall. Work out the height of the Eiffel Tower.

6. The ratio of the width of a picture frame to its length is in the ratio 2 : 5
   The length of the picture frame is 70 cm. Which of these lengths represent its width?
   a) 20 cm b) 50 cm c) 28 cm d) 35 cm e) 175 cm

7. Amin, Bora and Chee raised money for charity in the ratio 10 : 11 : 7
   Chee raised £98. How much money did they raise altogether?

8. One week Joe is paid £192 for working 16 hours in the week and £128 for working 8 hours at the weekend.
The next week he works 12 hours in the week and 10 hours at the weekend.
How much will he get paid for this week?

9. A garden shop sells soil in bags. 8 bags of soil weigh 360 kg.
How many bags are needed for a customer who wants 1710 kg of soil?

10. The ratio of the number of teachers : the number of students in five schools is shown below.

| School 1 | School 2 | School 3 | School 4 | School 5 |
|---|---|---|---|---|
| 3 : 27 | 1 : 17 | 2 : 23 | 1 : 14 | 7 : 43 |

Place the schools in descending order according to the proportion of teachers.

11. The ratio of bike sales : scooter sales is 7 : 3
After 6 more bikes are sold, the ratio of bike sales to scooter sales changes to 3 : 1
Work out the total number of scooters which have been sold.

12. A, B and C are three numbers such that A : B = 1 : 6 and B : C = 3 : 3
Work out the ratio of A : B : C

## 10.4 SOLVING PROBLEMS WITH RATIO

R1 R2 R5 R10

### Objectives

**Solve problems involving direct proportion**

**Write and interpret a ratio to describe a situation**

**Solve a ratio problem in context**

**e.g.** The table shows the number of visitors to a museum on Friday and Saturday.

| | Adults | Children |
|---|---|---|
| **Saturday** | 120 | 180 |
| **Sunday** | 160 | 360 |

Work out the day which has the largest proportion of children visiting the museum.

To solve this problem, work out the ratio of adults to children on each day.

Saturday Adults to children ratio is 120 : 180 = 2 : 3

Sunday Adults to children ratio is 160 : 360 = 4 : 9

Now scale the ratios up until they have a matching value.

Saturday Adults to children ratio is 2 : 3 = **4 : 6**

Sunday Adults to children ratio is **4 : 9**

On Saturday for every 4 adults visiting there were 6 children.

On Sunday for every 4 adults visiting there were 9 children.

There was a greater proportion of children visiting on Sunday.

## Practice questions

1. In a bowl of fruit, there are 10 bananas, 6 apples and 8 pears. Some fruit is eaten but the ratio of banana: apple: pear stays the same.
   How many pieces of each fruit have been eaten?

2. There are 3 toffees for every 2 mints in a pack of sweets. Jordan counts 12 toffees and 11 mints in her pack of sweets.
   Explain why Jordan cannot be correct.

3. Ned wins a packet of 75 lollies. He gives $\frac{4}{5}$ of the lollies to Dan, Kesh and Polly who shared them in the ratio 1 : 2 : 3
   Work out the number of lollies each person gets.

4. A recipe states that 320 g of flour is mixed with 200 ml of water to make 8 bread rolls.
   a) Alex has 400 g of flour. How much water does she need?
   b) How many rolls will she make?

5. Julian mixes cordial with water in the ratio 2 : 7 to make a fruit drink.
   a) How much water should he mix with 100 ml of cordial?
   b) How much cordial should he add if he uses 2.1 litre of water?

6. Two different types of biscuit have these ratios of ginger to nut in them.
   Nutty ginger biscuits 2 : 3 Ginger and nut cookies 3 : 4
   Lena thinks that both biscuits contain the same proportion of ginger.
   She is not correct. Give a reason why. Show your working.

7. Keira and Fred share £200 in the ratio 3 : 5. Kiera gives 20% of her money to Joe. Fred gives $\frac{3}{5}$ of his share to Joe.
   Work out the ratio of the money that Keira, Fred and Joe have.

8. Max mixes 2 units of squash with 5 units of water. Lars mixes 5 units of squash with 2 units of water.
   Do their drinks have the same proportion of squash? Give a reason for your answer.

9. Sandi wants to make 360 kg of concrete mix. She needs to mix cement, sand and gravel in the ratio 1 : 3 : 5 by mass.
   She has 30 kg of cement, seven 25 kg bags of sand and three 70 kg bags of gravel.
   Does she have enough to make the concrete mix?

10. A tin contains 40 biscuits. The biscuits are either square or circular.
    The ratio of square biscuits to circular biscuits is 3 : 2. All biscuits are covered with either pink or yellow icing. 25% of the square biscuits are covered in pink icing. Half of the circular biscuits have yellow icing.
    What fraction of all the biscuits have yellow icing?

11. Tom says that the three angles in his triangle are in the ratio 2 : 5 : 5
    The smallest angle in his triangle is 60°. Explain why Tom cannot be correct.

12. The ratio of boys to girls in the drama club and science club are shown below.

    | **Drama club** | **Science club** |
    |---|---|
    | Boys to girls = 4 : 3 | Boys to girls = 3 : 4 |

    The same number of girls go to both drama and science club.
    Are there more boys in the drama club or in the science club?

# 10.5 RATIO AS A LINEAR FUNCTION

## Objectives

**Recognise when values are in direct proportion by reference to a graph**

Two values are in **direct proportion** when they increase or decrease in the same ratio.

A graph of a direct proportion relationship is a **straight line** passing through the **origin**.

The gradient of the graph represents the quantity of $y$ for 1 unit of $x$.

**e.g.** A decorator uses 2 tins of paint to paint 3 chairs.

He uses 6 tins of paint to paint 9 chairs.

To the right there is a graph representing this information. It is a straight line passing through the origin because the variables are in direct proportion. The graph can be used to read off values:

4 tins of paint are needed to paint 6 chairs.

The gradient of the line = $\frac{\text{change in } y}{\text{change in } x} = \frac{3}{2}$ or 1.5

The gradient shows that 1.5 chairs can be painted with 1 tin of paint.

The ratio of tins of paint to the number of chairs painted is 1 : 1.5 or 2 : 3

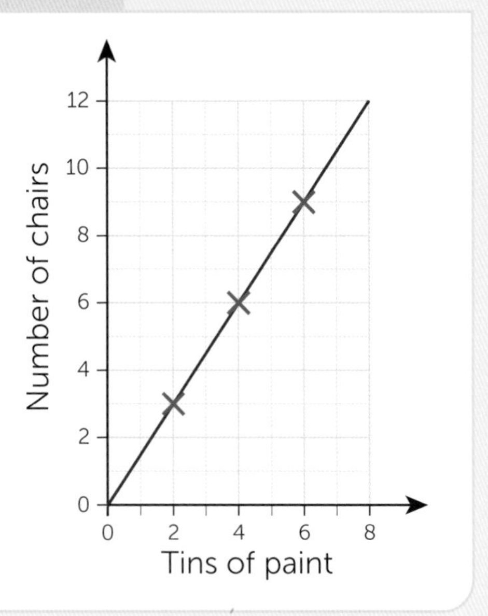


**Write a ratio as a linear function, linking to direct proportion**

**Recognise when values are in direct proportion by reference to a graph**

## Practice questions

1. Which of these proportions is the odd one out?
   a) The number of people in the house and the cost of the weekly food shop.
   b) The number of builders working and the time it takes to build a house.
   c) The length of the side of an equilateral triangle and its perimeter.
   d) The price of an apple and the cost of buying a basket of apples.

2. Which of these tables shows two variables in direct proportion?

a)

| $x$ | 3 | 15 | 75 |
|---|---|---|---|
| $y$ | 4 | 20 | 100 |

b)

| $a$ | 4 | 8 | 16 |
|---|---|---|---|
| $b$ | 3 | 6 | 9 |

c)

| $w$ | 108 | 36 | 12 |
|---|---|---|---|
| $z$ | 180 | 60 | 20 |

3. The ratio of the number of cards bought ($x$) to their total cost ($y$) are in direct proportion. One card costs 25p. Look at the statements below. Decide if each statement is true or false.
   a) $x = 25y$ b) $y = 25x$ c) $x = 25$ d) When $x = 10$, $y = 35$

4. Which of these graphs show two quantities in direct proportion to another?

a) 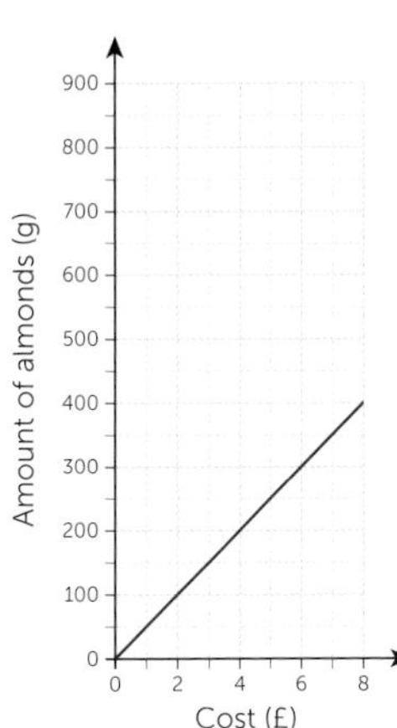


b) 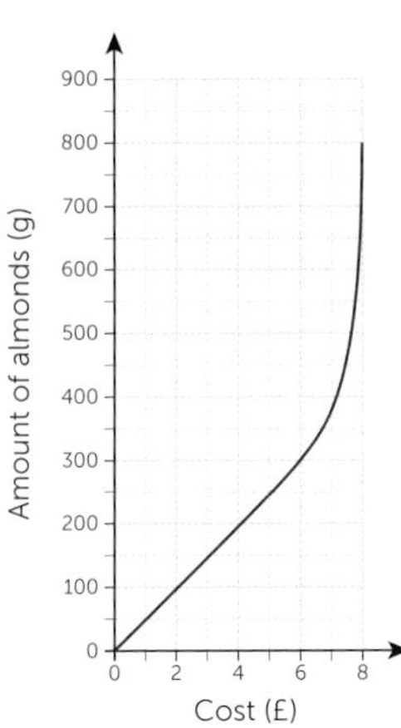


c) 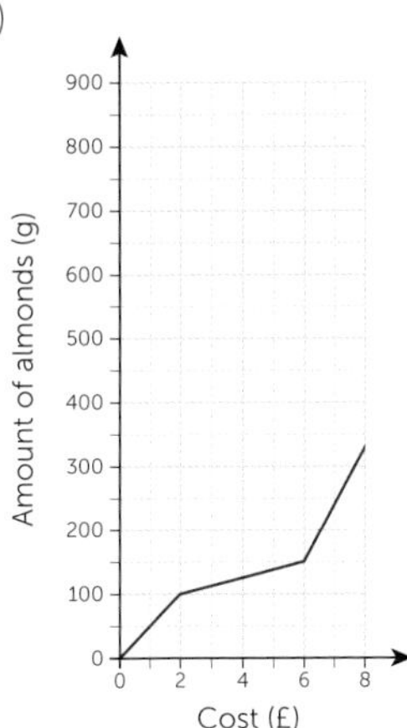


d) 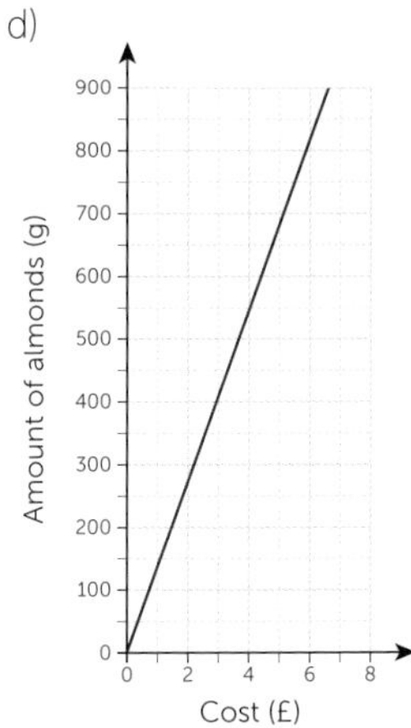


5. Daphne says that this graph shows that $y$ is directly proportional to $x$.
Daphne is not correct. Give a reason why.

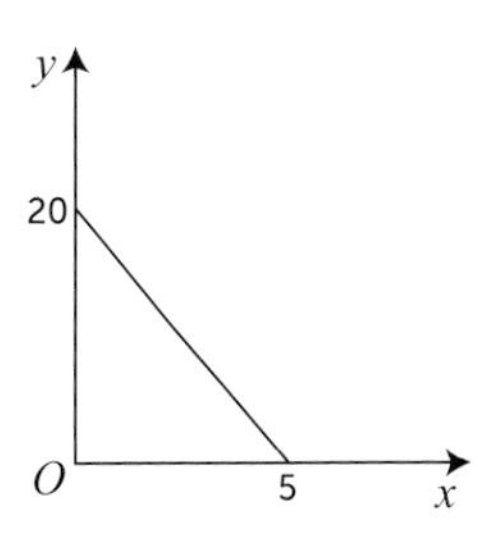


6. Match each ratio to the correct equation.

$x:y = 1:4$ $\quad$ $y:x = 1:4$ $\quad$ $y:x = 1:2$

a) $x = 4y$ $\quad$ b) $y = 4x$ $\quad$ c) $x + 4 = y$ $\quad$ d) $y = 0.5x$

7. Which of these ratios of number of packets to cost is represented by the graph?

a) 1 : 4
b) 4 : 9
c) 2 : 4
d) 2 : 4.5

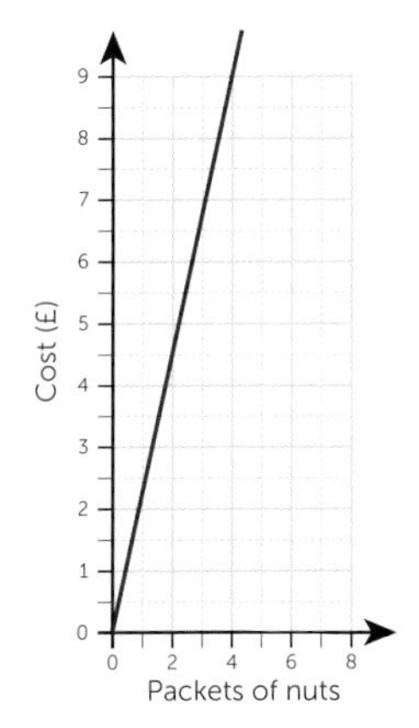


8. A cook makes jam with strawberries ($x$) and raspberries ($y$) in the ratio of 2 : 5
a) Draw a graph to represent the ratio.
b) Work out the gradient of the graph and explain what the gradient represents.

9. This graph shows the cost of a dinner party plotted against the number of people at the party.

| **Number of people** | 2 | 4 | 6 | 8 | 10 |
|---|---|---|---|---|---|
| **Cost per meal (£)** | | | | | |

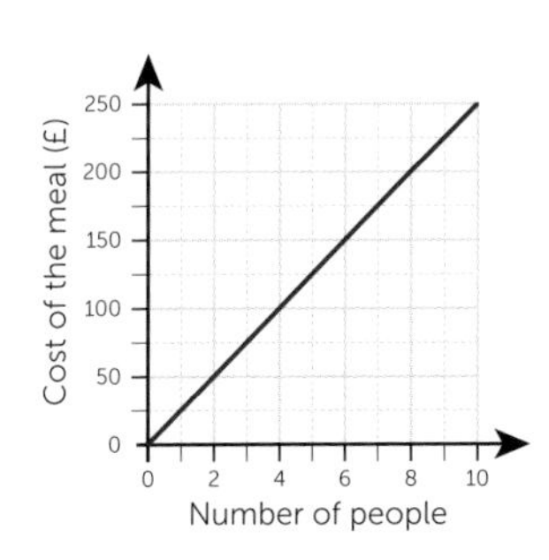


a) Copy and complete the table.
b) Write down the ratio of the number of people at the party to the cost of the meal.

# 10.6 MAP SCALES

## Objectives

### Use and interpret maps

The scale of a map or scale drawing can be given using ratio.

The ratio shows how many units in real life one map unit represents.

**e.g.** 1 : 200 means that 1 unit on the map represents 200 units in real life.

**e.g.** 1 : 100 000 means that 1 unit on the map represents 100 000 units in real life.

This can be converted to actual units. Take centimetres as the units.

**1 cm : 100 000 cm   1 cm : 1000 m   1 cm : 1 km**

### Convert scales as ratios to units

The ratio showing the scale on a map can be used to find real or map distances.

**e.g.** The scale of a map is 1 : 200 000. The real distance between two cities is 156 km. Find the map distance.

The scale is 1 : 200 000 or 1 cm : 2 km

The map distance is 156 ÷ 2 = **78 cm**

**e.g.** The scale of a map is 1 : 10 000

The map distance between two villages is 16 cm. Find the real distance.

The scale is 1 : 10 000 or 1 cm : 100 m

The real distance is 16 × 100 = **1600 m or 1.6 km**

### Estimate distances on maps

## Practice questions

1. On a map, 1 cm represents 8 km.
   What distance on the map will represent a real distance of:
   a) 16 km   b) 32 km   c) 4 km   d) 12 km

2. On a map, 1 cm represents 5 km.
   Work out the real distance between two places if their map distance is:
   a) 2 cm   b) 4.8 cm   c) 6.5 cm   d) 12.2 cm

3. The scale of a map is 1 : 500 000
   On the map the distance between two towns is 6.2 cm.
   Work out the real distance between the two towns. Give your answer in km.

4. The scale of a map is 1 : 100 000
   Work out the map distance between two towns which are 19 km apart.

5. Match the map scales with the correct map ratio.

   a) 1 : 35 000
   b) 1 : 350 000
   c) 1 : 100 000
   d) 1 : 10 000

   i) 1 cm to 3.5 km
   ii) 1 cm to 350 m
   iii) 1 cm to 10 km
   iv) 1 cm to 100 m
   v) 1 cm to 1 km

6. Write each scale as a map ratio.
   a) 1 cm to 2 km   b) 3 cm to 1.5 km   c) 1 cm to 300 m   d) 5 cm to 1 km

7. The distance between two towns is 6 km. On a map the distance between the towns is 30 cm.
   Work out the ratio scale of the map.

8. This is a map of the UK.
   It has a scale of 1 cm to 50 km.
   Estimate the real distance between:
   a) Exeter and Aberdeen
   b) Belfast and London
   c) Cardiff and Norwich

9. Georgia wants to draw a map of her school which covers an area of 1 km by 1.3 km.
   Her map must fit onto a piece of paper measuring 28 cm square.
   Suggest a suitable scale for her map. Give your answer as a ratio.

10. The dimensions of a park on the city map are 12 cm by 15 cm. The area of the park is 1.8 $km^2$.
    Which **one** of these map ratios has been used: 1 : 10   1 : 100   1 :1000   1 : 10 000
    Give reason for your answer.

SECTION 11

# SHAPES AND TRANSFORMATIONS

## 11.1 ANGLE PROPERTIES OF LINES

G1 G3

### Objectives

**Use two-letter notation for a line and three-letter notation for an angle**

Figure 1 shows two line segments AB and DM. M lies on segment AB.

Angle $x$ can be written as ∠AMD or ∠ DMA.

Angle $y$ can be written as ∠DMB or ∠BMD.

**Angles that meet at a point on a straight line add up to 180°**. ($x + y = 180°$)

Figure 1

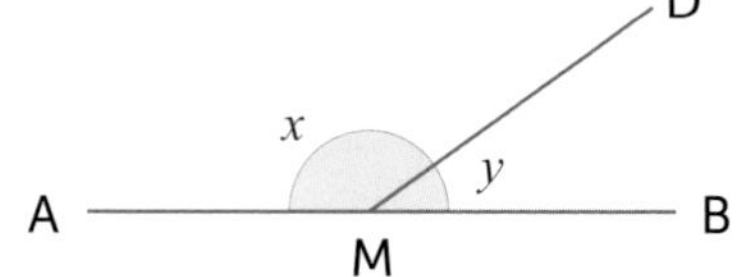


Figure 2 shows two line segments PQ and RS, which intersect at point X.

**Angles that meet around a point add up to 360°** ($a + b + c + d = 360°$)

Angles that are opposite each other when two straight lines intersect are called **vertically opposite angles**.

**Vertically opposite angles are equal:** $a = b$ and $c = d$

Figure 2

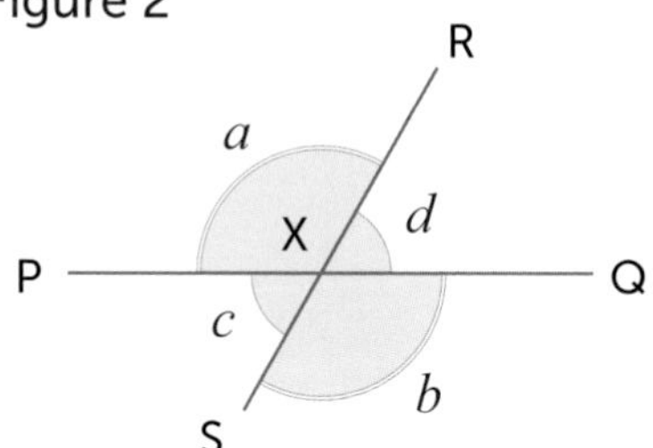


When a straight line crosses a pair of parallel lines, it makes a set of **corresponding angles**, and a set of **alternate angles**.

Corresponding angles lie in corresponding positions on each parallel line.

**Corresponding angles are equal to each other**:

In Figure 3: $w = u$, $x = v$, $z = t$ and $y = s$

Alternate angles lie inside the parallel lines, on opposite sides of the line that crosses them.

**Alternate angles are equal to each other**.

In Figure 3: $y = u$ and $x = t$

Figure 3

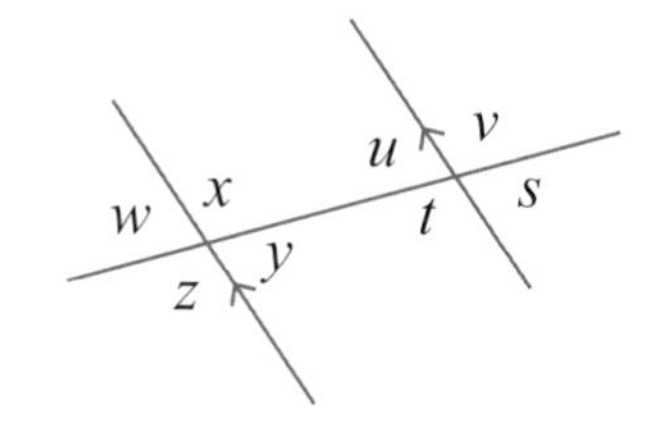


e.g. In Figure 4, find the angles marked with a letter in this diagram:

$a$ = 180 – 66
 = 114° (angles on a straight line add up to 180°)
$b$ = 66° (vertically opposite to 66°)
$c$ = 114° (vertically opposite to $a$)
$d$ = 114° (alternate to $c$)
$e$ = 66° (corresponding to $b$)
$f$ = 114° (corresponding to $c$)
$g$ = 66° (alternate to $b$)

Figure 4

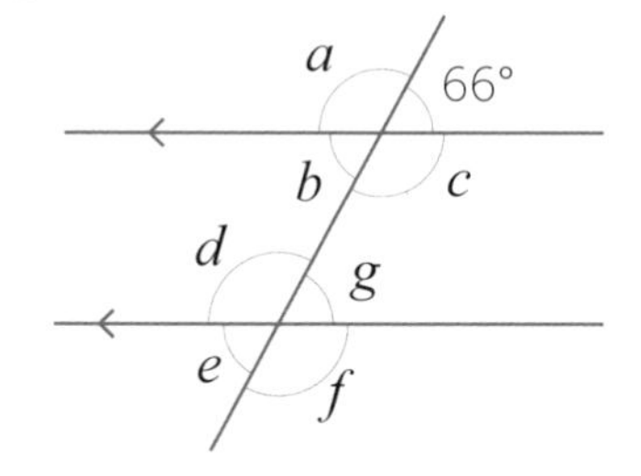


**Recall and use properties of angles at a point, angles at a point on a straight line, right angles and vertically opposite angles**

**Find missing angles in parallel lines using properties of corresponding and alternate angles**

## Practice questions

1. Use the three-letter notation to describe angles $t$, $u$, $v$, $w$, $x$ and $y$ in the following diagrams:

a)

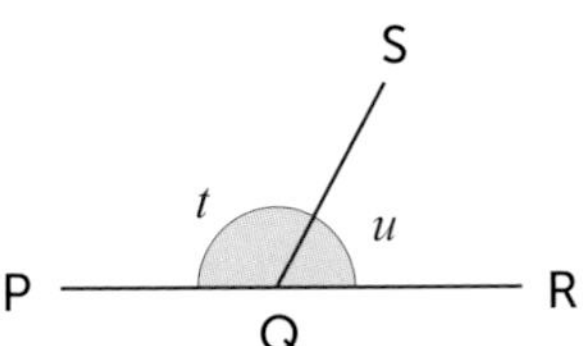


b)

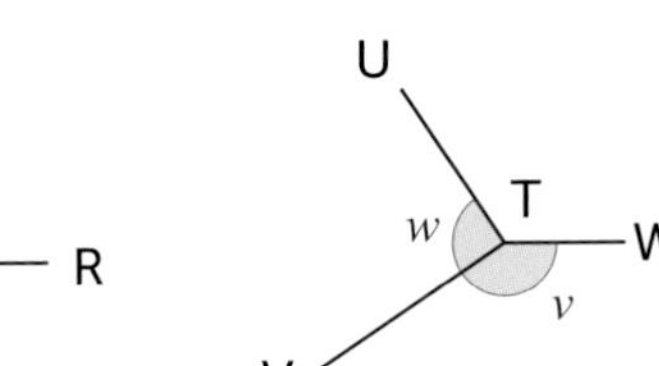


c)

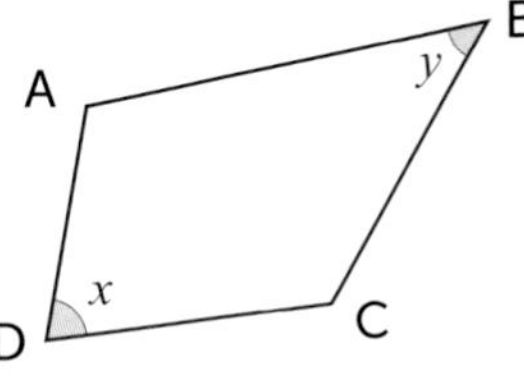


2. Work out the value of each angle marked with a letter.
Give reasons for your answers.

a)

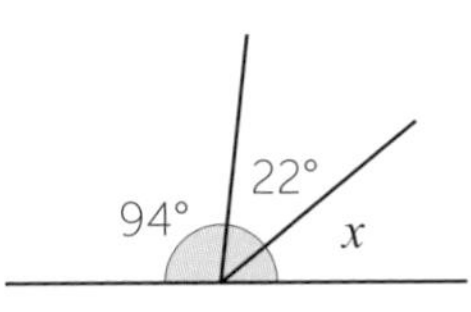


b)

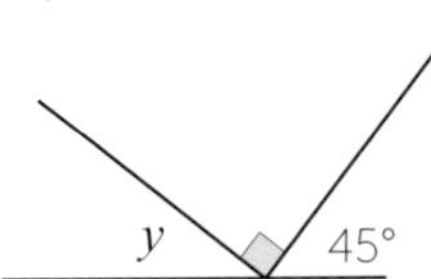


c)

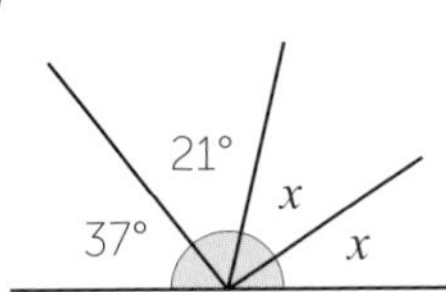


d)

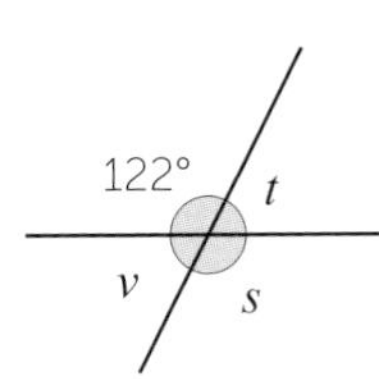


e)

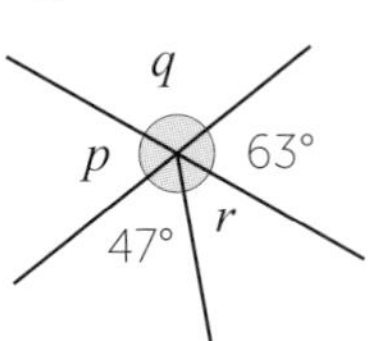


f)

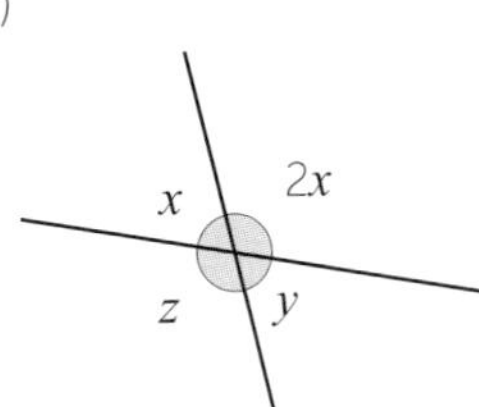


3. For each diagram, find the lettered angles .Give reasons for your answers.

a)

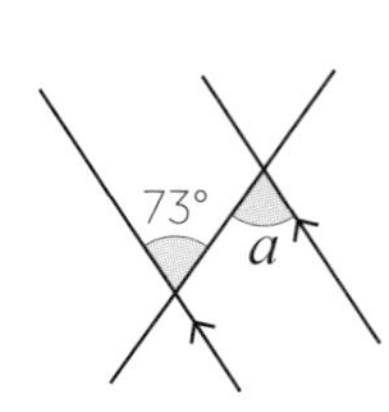


b)

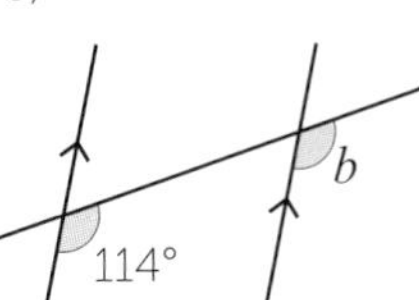


c)

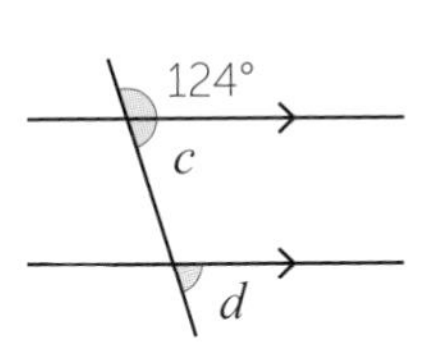


d)

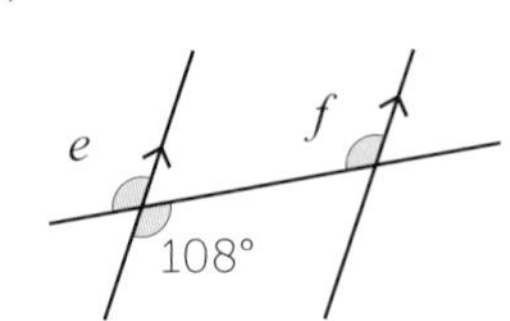


e)

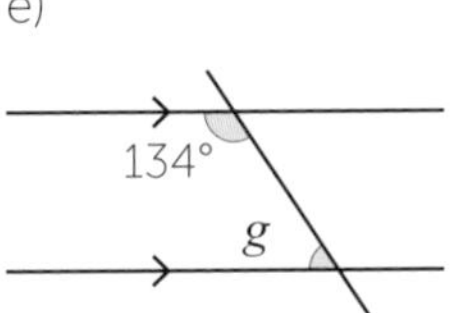


f)

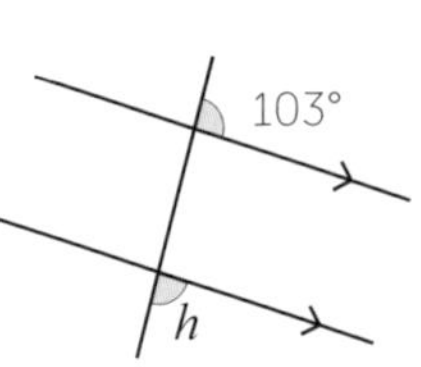


4. For each of the following diagrams, work out the size of the lettered angles.
Give reasons for your answers.

a)

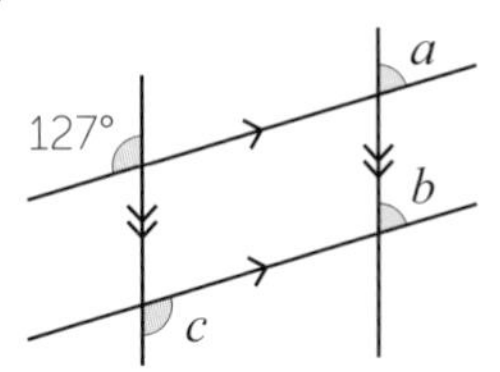


b)

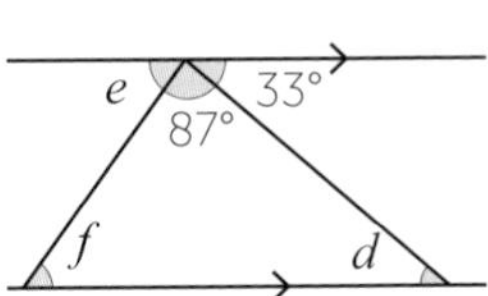


c)

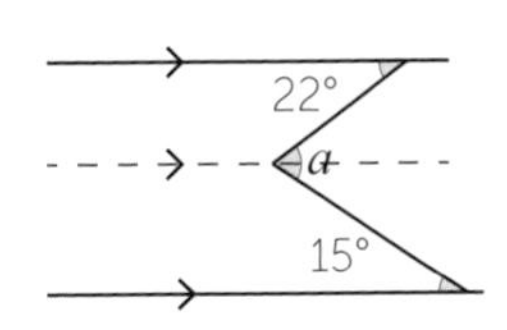

## 11.2 ANGLE PROPERTIES OF SHAPES

### Objectives

#### Use the sum of angles in a triangle, knowing the properties of special triangles

The angle sum of any triangle is 180°.

An isosceles triangle has two equal angles, and two equal sides.

An equilateral triangle has three angles of 60°, and three equal sides.

The exterior angle of any triangle is equal to the sum of the two interior opposite angles.

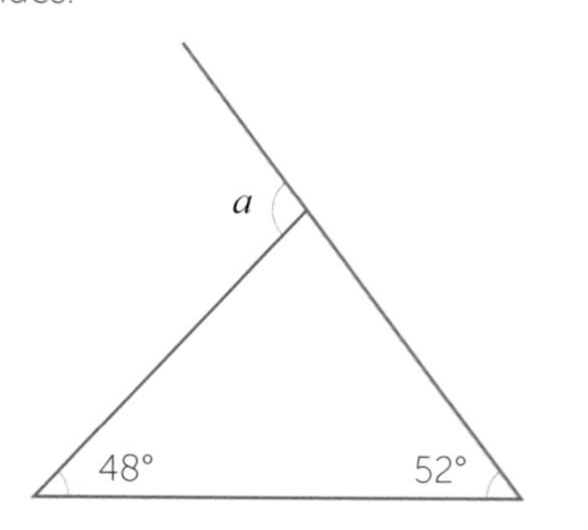


**e.g.** Work out the value of angle $a$.

Give reasons for your answer.

$a = 48 + 52$

$= $ **100°** (Exterior angle of a triangle is equal to the sum of the two interior opposite angles)

#### Name polygons and understand 'regular' and 'irregular' polygons

Polygons are named as follows:

| **Name** | Quadrilateral | Pentagon | Hexagon | Heptagon | Octagon | Nonagon | Decagon |
|---|---|---|---|---|---|---|---|
| **Sides** | 4 | 5 | 6 | 7 | 8 | 9 | 10 |
| | 4 | 5 | 6 | 7 | 8 | 9 | 10 |

Polygons with 4 or more sides can be split into triangles using straight lines from a single vertex to each of the other vertices.

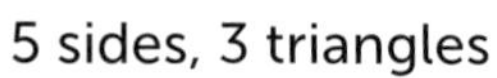


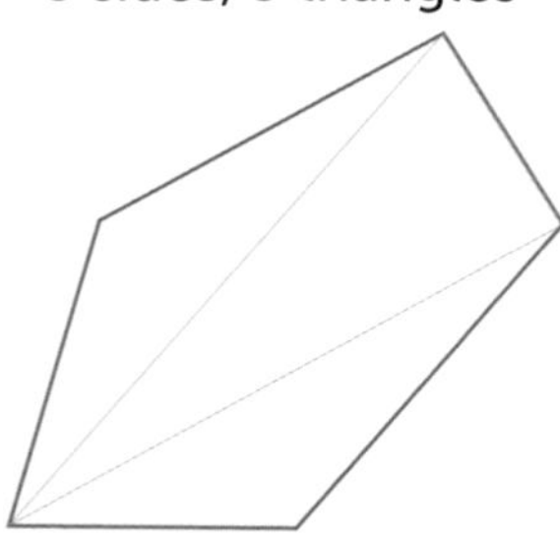

For an $n$-sided polygon, there are always $(n - 2)$ triangles.

Sum of **interior** angles of any n-sided polygon $= (n - 2) \times 180$

Sum of **exterior** angles of any polygon $= 360°$

Interior angle + exterior angle $= 180°$

**e.g.** What is the sum of the interior angles of a 23-sided polygon?

Sum of interior angles $= (23 - 2) \times 180$ $=$ **3780°**

#### Calculate and use the sums of the interior and exterior angles of polygons

Regular polygons have all sides equal in length, and all angles equal.

For any regular polygon: Size of one interior angle $= \frac{(n - 2) \times 180}{n}$

Size of one exterior angle $= \frac{360}{n}$

**e.g.** A regular polygon has one interior angle 162°.

How many sides does it have?

Exterior angle $= 180 - 162$

$= 18°$

Number of sides $= \frac{360}{18}$

$=$ **20**

## Practice questions

1. Find the missing angles in these triangles:

a)

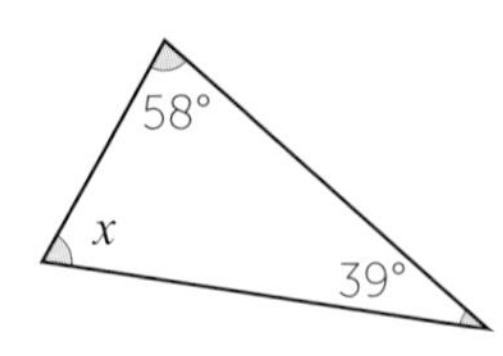


b)

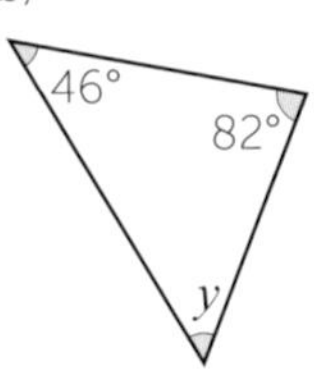


c)

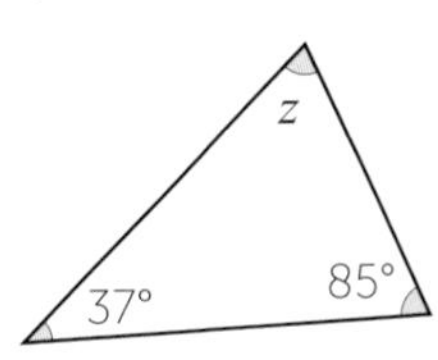


2. Calculate the angles marked with letters in these diagrams:

a)

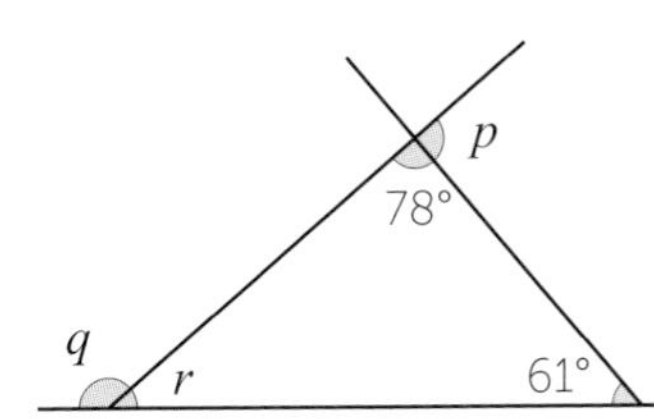


b)

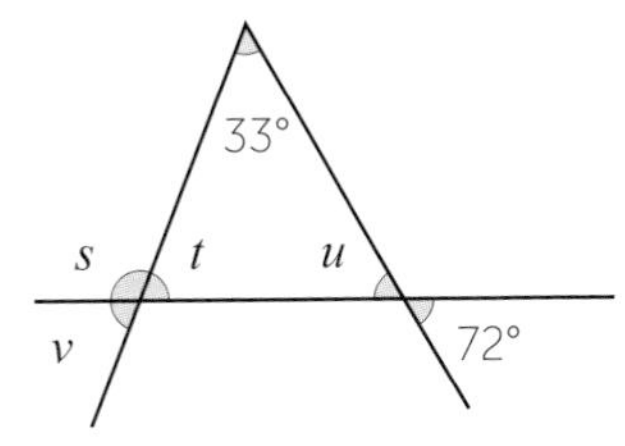


3. Find the angles marked with letters in these isosceles triangles:

a)

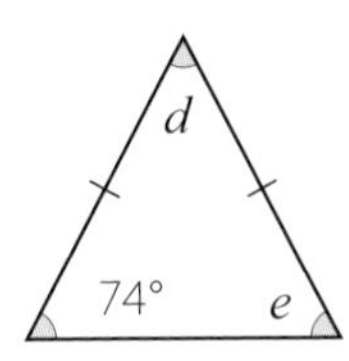


b)

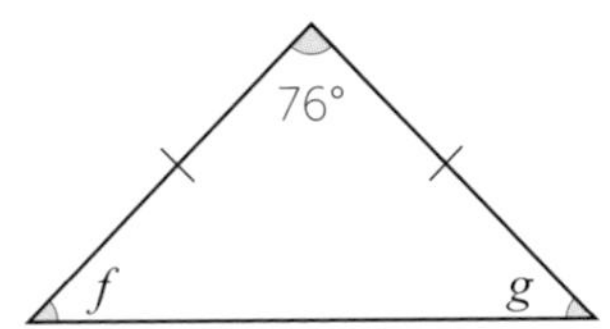


c)

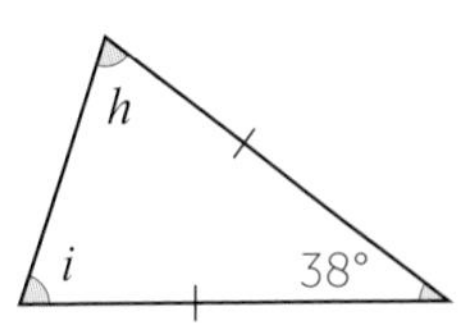


4. Three of the angles in a quadrilateral are 72°, 81° and 103°.
What size is the other angle?

5. Two angles of a quadrilateral are identical, and the others are 117° and 135°. What size are the identical angles?

6. In this diagram, the lengths of the sides are all equal. Is this a regular pentagon?
Explain your answer. Work out the size of the missing angles.

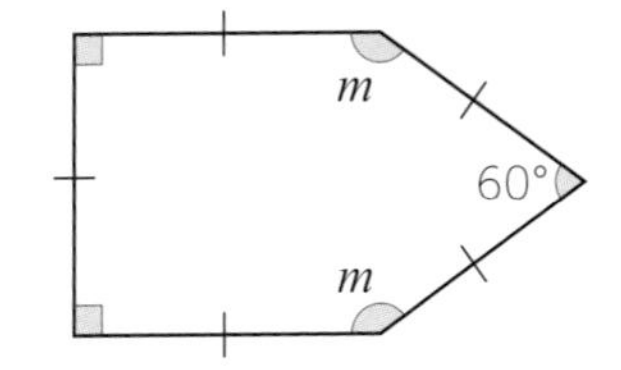


7. What is the size of each angle in a regular octagon?

8. In a kite, two of the angles are 46° and 114° What could the other angles be? Give your reasons.

9. What is the sum of the angles in a decagon?
If the decagon is regular, what size is each angle?

10. The angles in a hexagon are $x°$, $(x + 10)°$, $(x + 20)°$, $(x + 30)°$, $(x + 40)°$ and $(x + 50)°$.
What size is the largest angle in the hexagon?

11. A cake is made in the shape of a regular dodecagon (12-sided polygon).
A slice is cut from each vertex of one side to the centre.
What is the angle of the slice at the centre? How do you know?

12. The angles in a regular polygon are each 160°.
How many sides does the polygon have?

# 11.3 TRANSLATIONS AND ROTATIONS

## Objectives

### Translate a given shape by a vector

A **translation** can:

- move a point in the $x$ direction or in the $y$ direction, or both.
- be represented by an arrow, from starting point to finishing point.
- be represented by a vector $\begin{pmatrix}\text{Units moved in } x \text{ direction}\\ \text{Units moved in } y \text{ direction}\end{pmatrix}$

**e.g.** Triangle A has coordinates (-4, 1), (−2, 1) and (−2, 3).

Translate A through vector $\begin{pmatrix}4\\-2\end{pmatrix}$ and label the image B.

Describe fully the transformation that maps B back to A.

Translation through vector $\begin{pmatrix}-4\\2\end{pmatrix}$

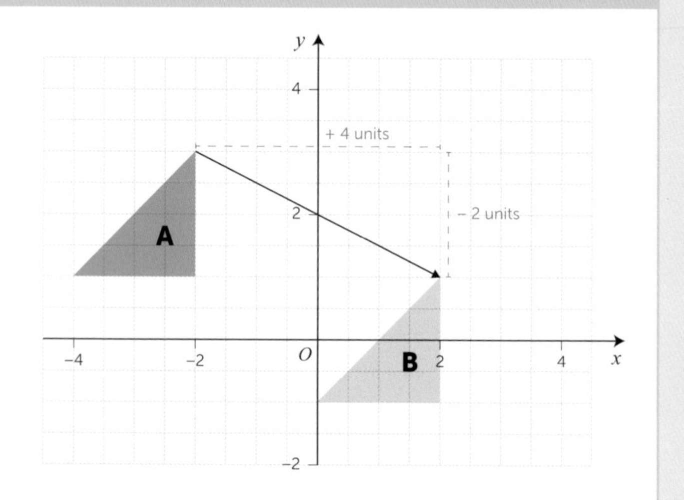


### Describe fully and transform 2D shapes after rotation about the origin or any other point

A rotation turns a shape about a given point.

Its size does not change but its position and orientation can.

A rotation is described by its centre, the angle of rotation and the direction of rotation.

**e.g.** Triangle C has coordinates (1, 2), (4, 2) and (2, 5).

Rotate C through 90° clockwise about point (0, 0) and label its image D.

Describe fully the transformation that maps D back to C.

Rotation 90° anticlockwise about (0, 0)

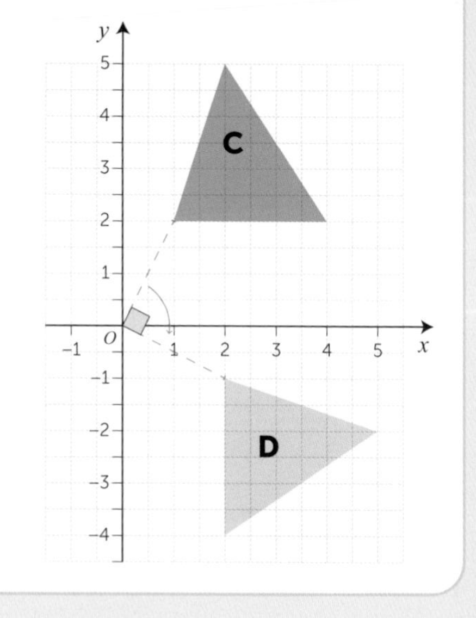


## Practice questions

1. Write the following translations as vectors:
   a) 4 units in the $x$ direction and 3 units in the $y$ direction.
   b) −3 units in the $x$ direction and −5 units in the $y$ direction.
   c) 7 units in the $x$ direction.

2. Draw arrows to represent the following translations.
   a) $\begin{pmatrix}6\\-3\end{pmatrix}$ b) $\begin{pmatrix}-8\\0\end{pmatrix}$ c) $\begin{pmatrix}-3\\4\end{pmatrix}$ d) $\begin{pmatrix}0\\5\end{pmatrix}$

3. A point A is translated through vector $\begin{pmatrix}1\\4\end{pmatrix}$ and then by vector $\begin{pmatrix}-2\\3\end{pmatrix}$, to point B. Write down the vector that maps A straight onto B.

4. A point is translated by vector $\begin{pmatrix} 5 \\ -2 \end{pmatrix}$. Which translation will return it to where it started?

5. Write down the vectors that will translate shape A onto each of the shapes B, C and D:

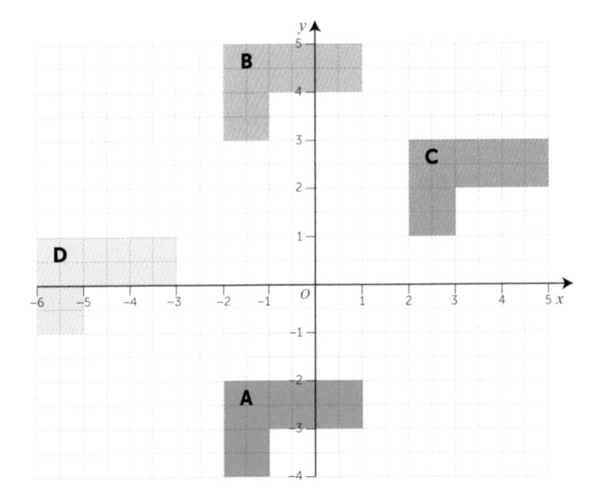


6. The coordinates of point M on trapezium A are (–2, 2)
   The trapezium is translated so that point M moves to M' (4, –3)
   a) Draw the trapezium after the translation and label it B.
   b) Describe fully the vector that moves all points on trapezium A to their images on trapezium B.

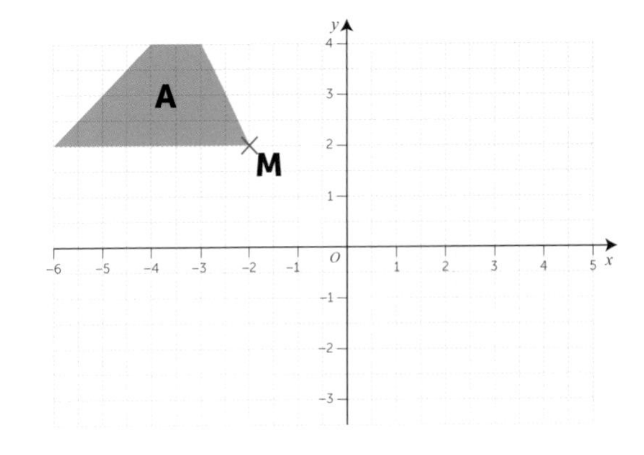

7. A clockwise rotation maps rectangle E onto rectangle F.
   a) State the angle of rotation.
   b) Write down the coordinates of the centre of rotation.
   An anticlockwise rotation maps rectangle E onto rectangle F.
   c) State the angle of rotation.
   d) Write down the centre of rotation.

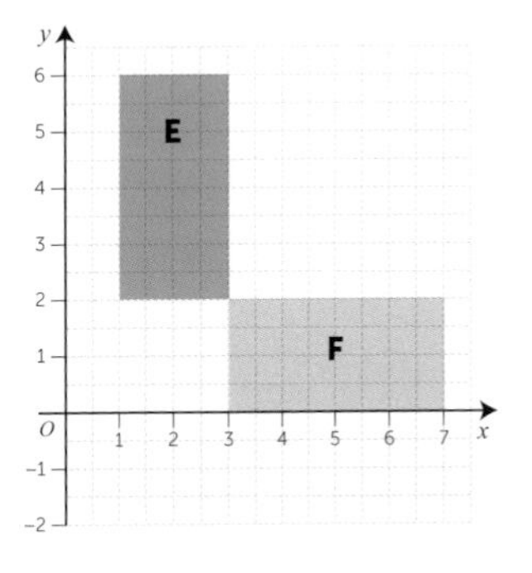

8. A rectangle G has vertices at (1, 1), (4, 1), (4, 3) and (1, 3).
   a) Plot the rectangle on a set of axes.
   b) Rotate it through 90° anticlockwise about (1, 3) and write down the coordinates of the vertices of the image of G.

9. Four shapes A, B, C and D are shown in the diagram opposite:
   a) Describe fully the rotation that maps A onto B.
   b) Describe fully the rotation that maps B onto C.
   c) Describe fully the rotation that maps C onto D.
   d) Which single rotation will map D back onto A?

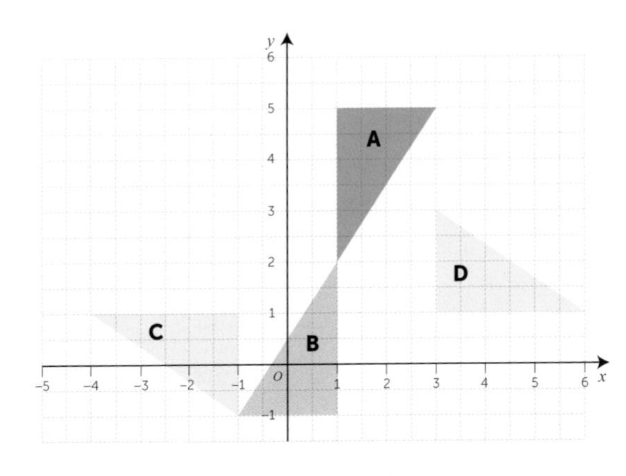

# 11.4 REFLECTIONS

## Objectives

**Describe and transform 2D shapes using single reflections with vertical, horizontal and diagonal mirror lines**

A **reflection** maps a shape from one side of a mirror line to the other.

The reflected shape is the same size and shape as the original and the same distance from the mirror line.

As **Figure 1** shows, horizontal mirror lines have equations $y = a$ where $a$ is the intercept with the $y$-axis.

Vertical mirror lines have equations $x = b$ where $b$ is the intercept with the $x$-axis.

Diagonal mirror lines in the exam will either have equation $y = x$ or $y = -x$

Reflections are described fully by stating

- The word reflection
- The equation of the mirror line

**e.g.** In **Figure 2**

a) Shape A is reflected in the line $y = 3$ creating the image B.

b) Describe fully the transformation that maps shape C onto shape D.

Reflection in the line $y = x$

Figure 1

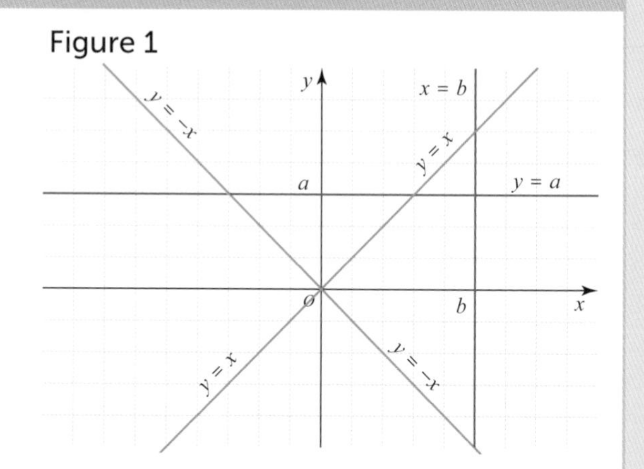


Figure 2

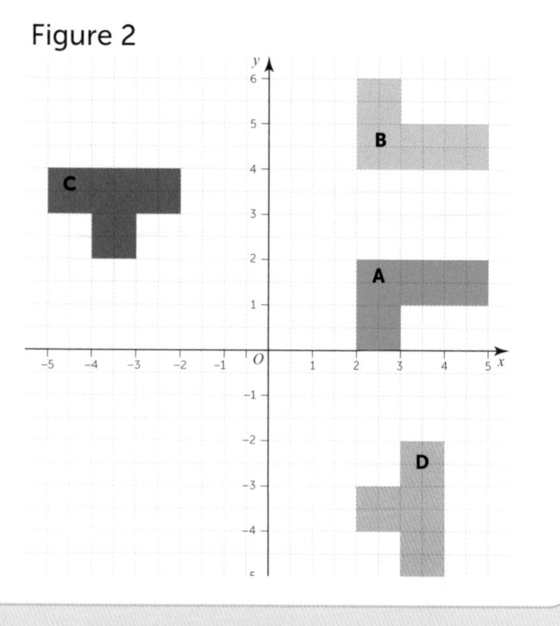


**Understand that distances and angles are preserved under rotations, translations and reflections**

**Understand that any figure is congruent under any of these transformations**

## Practice questions

1. Draw a triangle with vertices at coordinates (−1, 2), (−5, 2) and (−4, 4). Label it A.
   Reflect it in the mirror line $x = 0$. Label the image A'.
   What are the coordinates of the reflected triangle A'?
   What do you notice about the $x$ and $y$ coordinates of the two triangles?

2. Draw a trapezium which has vertices at coordinates (3, 1), (4, 5), (6, 5) and (8, 1). Label it A.
   Reflect it in the mirror line $y = 0$. Label the image A'.
   What are the coordinates of the reflected trapezium A?
   What do you notice about the $x$ and $y$ coordinates of the two trapeziums?

3. Reflect A in the mirror line $y = 2$. Label the image B.
   Reflect B in the mirror line $x = 1$. Label the image C.
   Describe fully the transformation that will map A onto C.

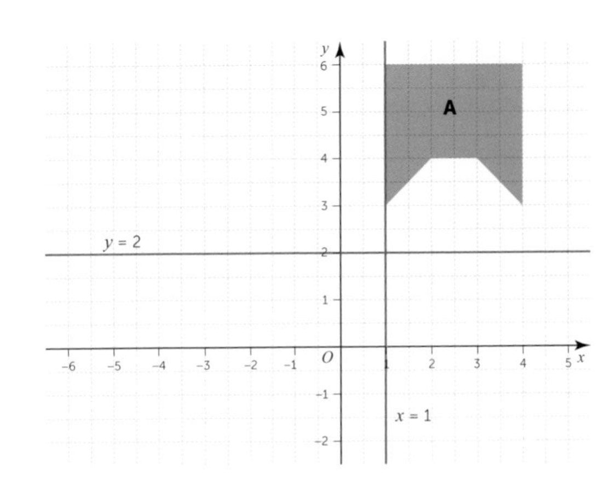


4. Rotate shape E 90° anticlockwise about point (–3, 2)
   Label the new triangle F.
   Translate triangle F through vector $\begin{pmatrix} 1 \\ -5 \end{pmatrix}$. Label the image G.
   Describe fully the transformation that will map triangle E onto triangle G.

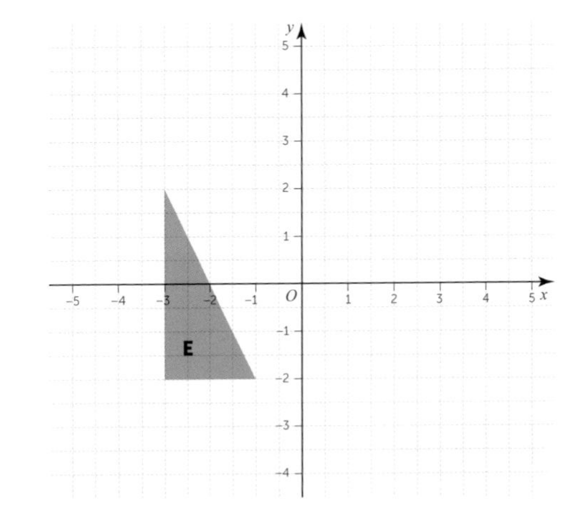


5. Reflect shape H in the mirror line $y = -x$. Label the image J.
   Rotate shape J 180° about centre (2, 0). Label the image K.
   Write down the coordinates of the vertices of shape K.

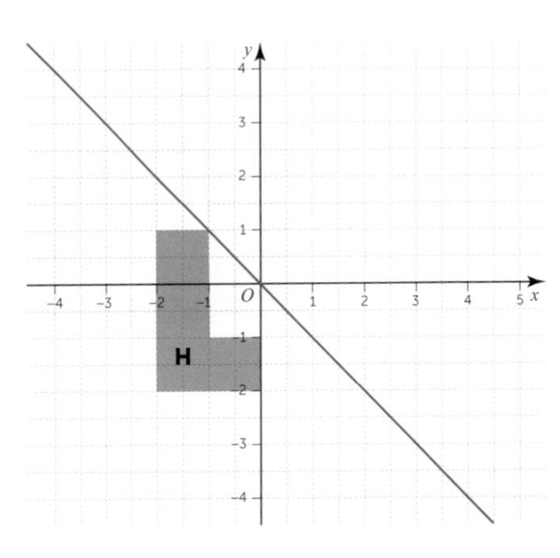


6. Translate quadrilateral L through vector $\begin{pmatrix} -4 \\ -2 \end{pmatrix}$. Label the image M.
   Rotate M 180° about (0, 0). Label the image N.
   Reflect N in the line $y = 3$. Label the image P.
   Write down the coordinates of the vertices of P.

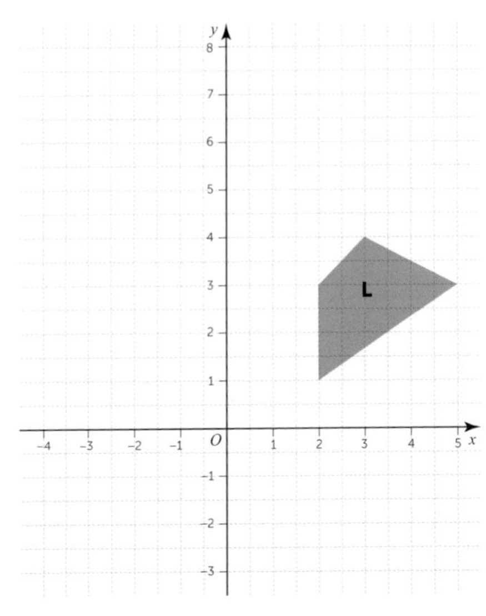

# 11.5 ENLARGEMENTS AND SIMILARITY

## Objectives

### Understand that similar shapes are enlargements of each other and angles are preserved

**Similarity**

When two shapes have the same shape, including the same angles, but are different sizes by length then they are **similar**. A shape and its enlargement are always similar.

### Enlarge objects using simple integer scale factors

Here is Radcliffe Camera in Oxford. Every length on the second picture is three times that of the first picture, so the second picture is an enlargement of the first by scale factor 3.

In an enlargement, lengths and areas change, but angles do not.

1 unit

3 units

### Identify the scale factor of an enlargement of a shape as the ratio of the lengths of two corresponding sides:

To find the scale factor of an image, find the ratio of one length to the corresponding length on the object.

$$\text{Scale Factor} = \frac{\text{Length of side on image}}{\text{Length of corresponding side on object}}$$

**e.g.** Shape B is an enlargement of shape A.

To find the scale factor enlargement from A to B.

The base of A is 2 units, the base of B is 4 units.

The scale factor from A to B $= \frac{4}{2}$

$= 2$

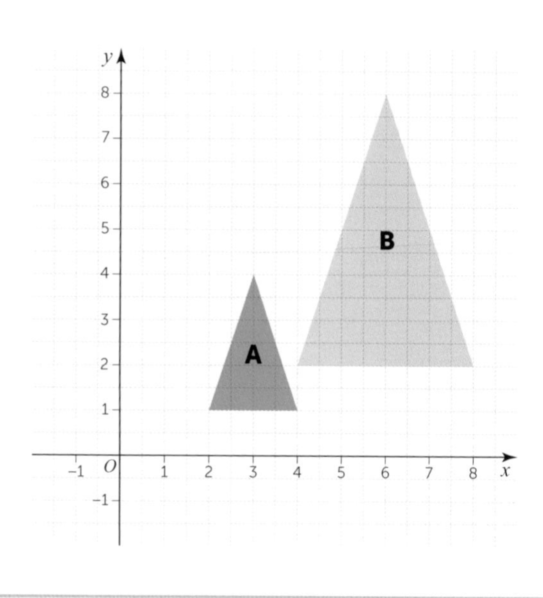


## Practice questions

1. A rectangle has length 24 cm and width 15 cm. If it is enlarged by scale factor 4, what will its new length and width be?

2. A photo which is 12 cm x 18 cm is enlarged by scale factor 3.
   What are the dimensions of the enlarged photo?

3. Work out the scale factor of the enlargement that transforms shape A onto shape B.

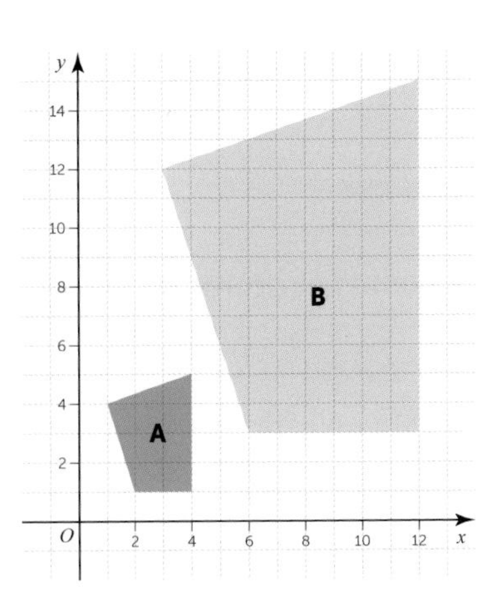


4. Here are four triangles, C, D, E and F.
   Work out the scale factor of:
   a) D from C.

   b) E from C.
   c) F from C.
   d) F from D.

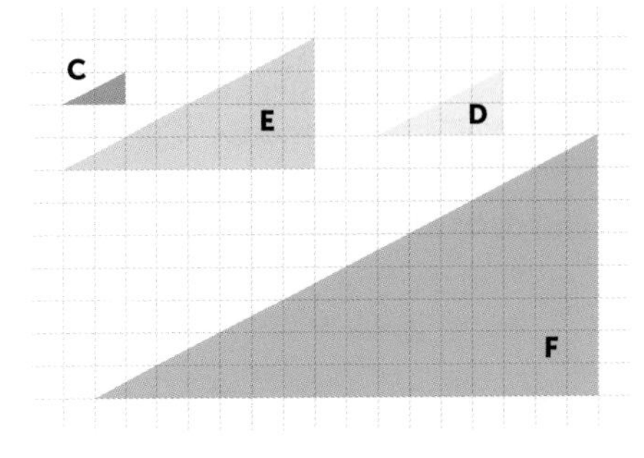


5. A parallelogram has a base of 42 cm, a perpendicular height of 16 cm and its smaller angles are each 47°.
   If the parallelogram is enlarged by a scale factor 5, what will its new measurements be?

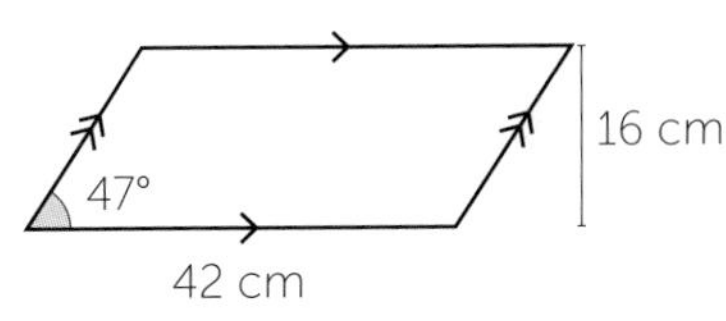


6. Shape G is as shown.
   Draw an enlargement, scale factor 4 and label it H.
   Enlarge H by a scale factor of 2 and label it J.
   What scale factor would enlarge shape G to get shape J?

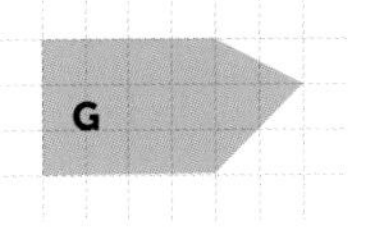


7. An enlarged shape has dimensions as shown.
   It was enlarged from a smaller shape by a scale factor of 6.
   What were the dimensions of the original shape?

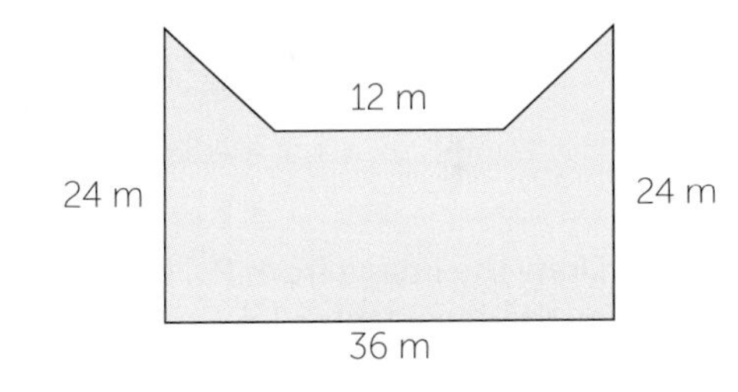

8. For each pair of shapes, say if they are similar and give your reasons.
   a)

8 cm 11 cm

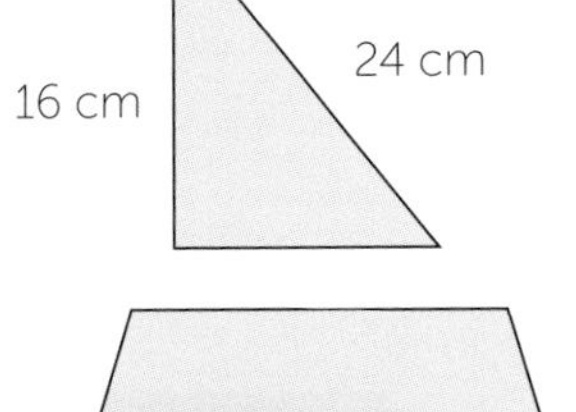


   b)

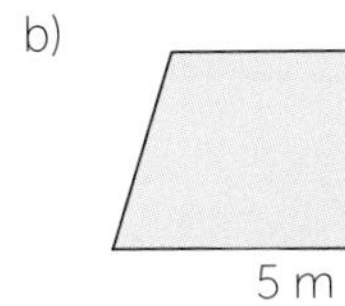


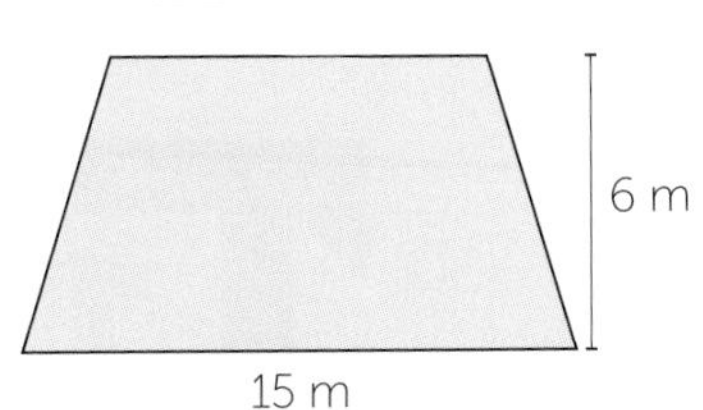

## 11.6 FURTHER ENLARGEMENTS

### Objectives

**Describe and transform 2D shapes using enlargements by a positive fractional scale factor**

A fractional scale factor (less than 1) makes the image smaller than the object.

It is still called an enlargement.

e.g. Enlarge shape A by a scale factor of $\frac{1}{2}$. Label the image B.

All sides of shape A are multiplied by the scale factor $\frac{1}{2}$.

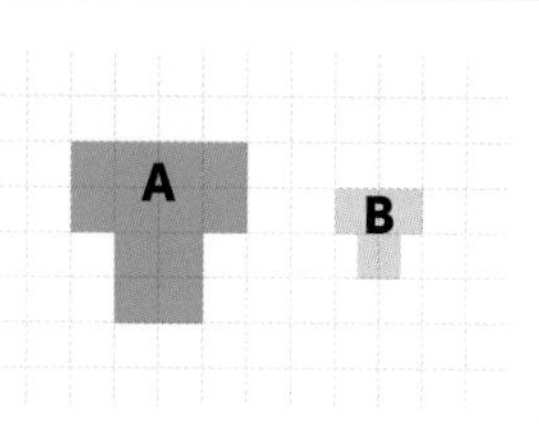


**Describe and interpret enlargements using positive integer scale factors, positive fractional scale factors and centres of enlargement**

**Centre of enlargement**

e.g. Find the scale factor of the enlargement that maps shape P onto shape Q:

Scale factor = $\frac{2}{4}$ ⇒ $\frac{1}{2}$

State the coordinates of the centre of enlargement.

Draw two straight lines that join corresponding points on object and image.

Centre of enlargement is at (0, 0)

**Enlarging through a given centre**

e.g. Enlarge shape R by scale factor 1.5 through centre C (1, 1).

Horizontal distance from C to P = 3 units

Vertical distance from C to P = 1 unit

P′ is at position: (**5.5, 2.5**)

Horizontal: 3 × 1.5 = 4.5 units from C

Vertical: 1 × 1.5 = 1.5 units from C

Draw the image from P′, multiplying each side by the scale factor 1.5.

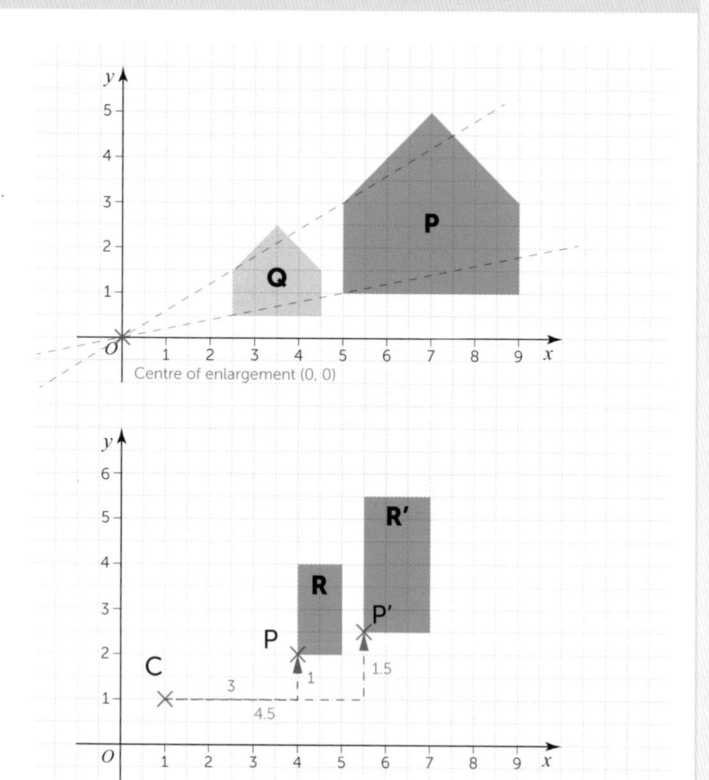


### Practice questions

1. State the scale factor of each of the following enlargements, of P to P′.

a)

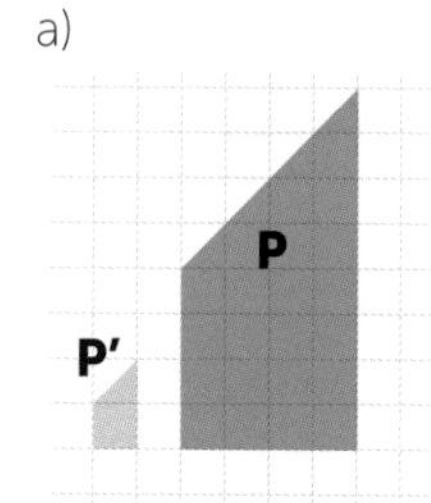


b)

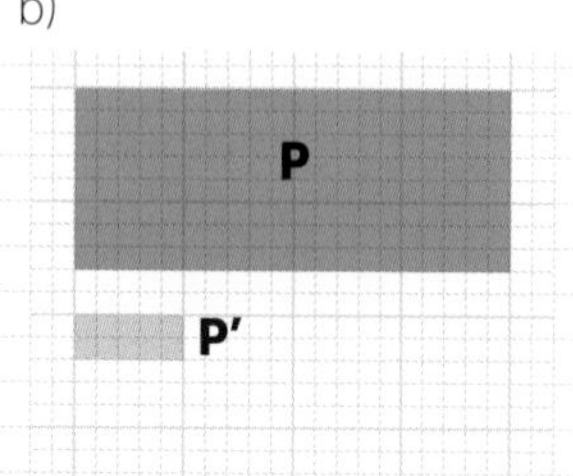


c)

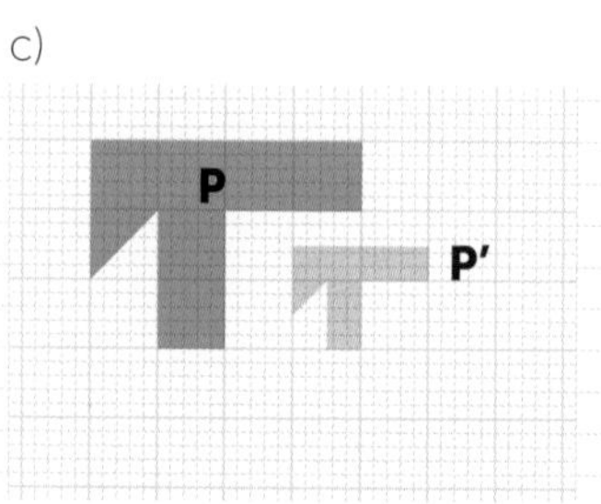

2. Copy each diagram below and enlarge the shape with the given centre of enlargement and scale factor.
   a) Scale factor 2
   b) Scale factor 3
   c) Scale factor 2

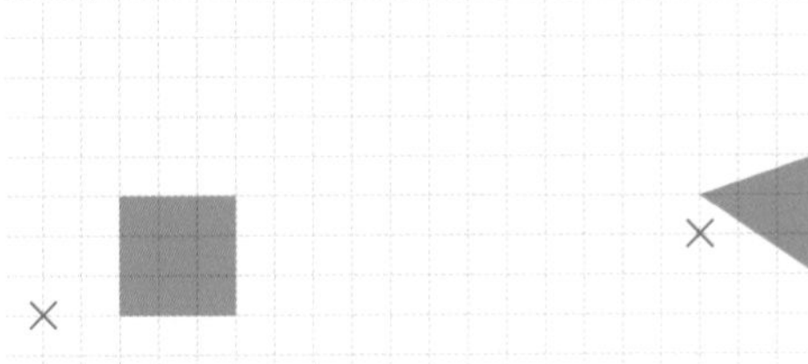

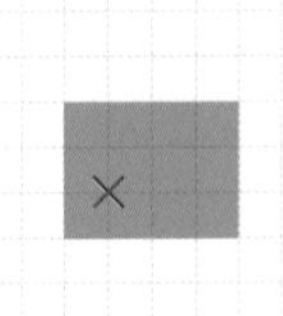

3. Shape A is enlarged by scale factor $\frac{1}{2}$ through three different centres.

   Find the centres of enlargement that map
   a) A onto B
   b) A onto C
   c) A onto D

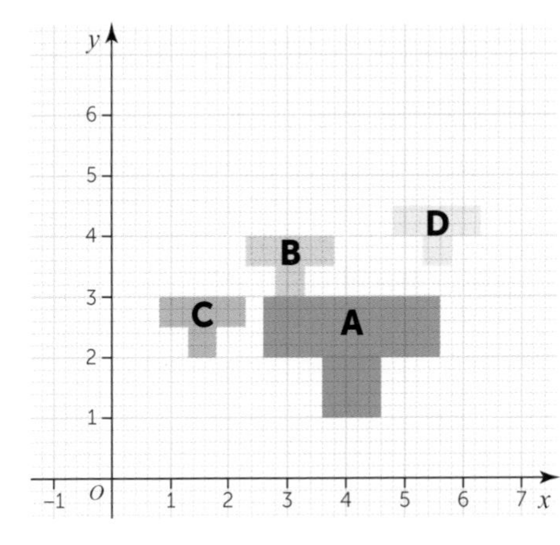

4. A triangle has coordinates (4, 3), (7, 3) and (6, 7).

   Using (3, 2) as the centre of enlargement, enlarge the triangle by scale factor 2.

   What are the coordinates of the vertices of the enlarged triangle?

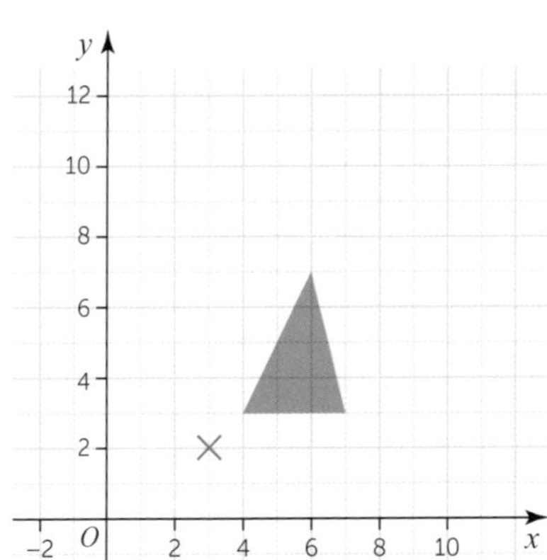


5. The diagram shows rectangle A.

   Enlarge A using (2, 2) as centre of enlargement, with scale factor 2.
   Label this as rectangle B.

   Now use (4, 12) as centre of enlargement and enlarge A by scale factor $\frac{1}{2}$ to get rectangle C.

   Write down the coordinates of the vertices of C.

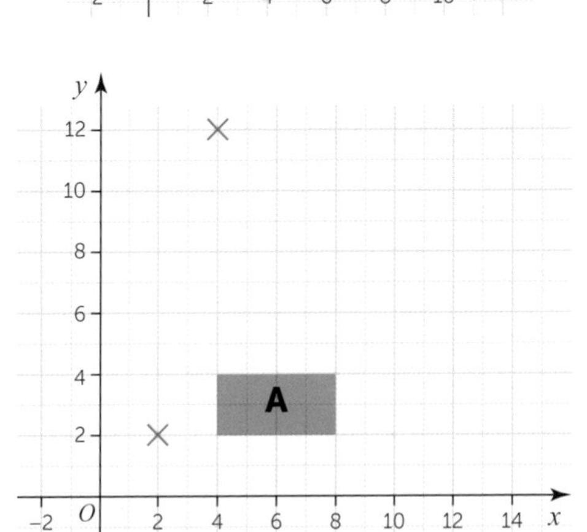

6. Pentagon D is as shown.

   Enlarge D using (−2, 1) as centre of enlargement, scale factor 3. Label the new shape E.

   Using (2, 6) as centre of enlargement, enlarge D by scale factor $\frac{1}{2}$ to get shape F.

   What are the coordinates of the vertices of F?

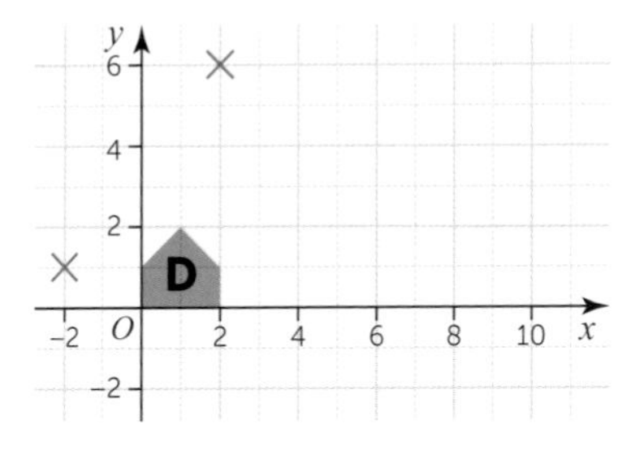

SECTION 12

# SEQUENCES

## 12.1 INTRODUCING SEQUENCES

A7 A23

### Objectives

#### Use function machines to find terms of a sequence

A **function machine** shows a set of instructions with calculations in a specific order.

**e.g.** Here is a function machine.

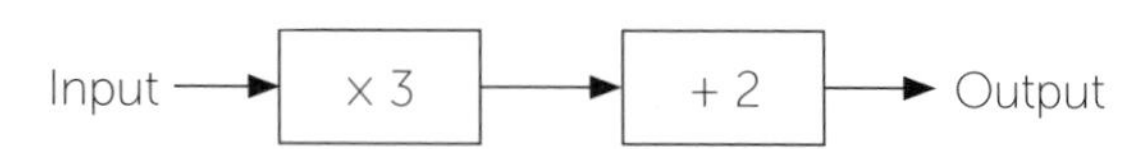


When **5** is the input, **17** is the output.

#### Find a specific term in a sequence using position-to-term or term-to-term rules

Each term in a sequence has a position.

A position-to-term rule is the rule to find the term value, given the position of the term in the sequence.

**e.g.** The position-to-term rule of a sequence is $n + 3$

To find the value of the 4th term, put $n = 4$ into the position-to-term rule

The 4th term = 4 + 3 = 7

#### Write the term-to-term definition of a sequence in words

A term-to-term rule is the rule used to work out the next term in a sequence from the current term.

**e.g.** Here is a sequence 6, 10, 14, 18, ...

The term-to-term rule is + 4

#### Find the next term in a sequence, including negative values

To continue a sequence, work out the term-to-term rule and use it to generate more terms.

**e.g.** Find the next two terms of the sequence 4, 7, 10, 13, ...

The term-to-term rule is +3, so the next three terms are **16, 19, 22**

### Practice questions

1. Here is a function machine: Input → × 2 → − 1 → Output

   a) Give the output if the input is 3
   b) Give the output if the input is 8
   c) Give the input if the output is 7
   d) Give the output if the input is −1

2. Here is a sequence **5, 8, 11, 14, ...**

   a) Write the next **two** terms in the sequence.
   b) Will says that 30 will be in the sequence. Explain how you know he is not correct.

3. Give the term-to-term rule for each of these sequences.

   a) 3, 7, 11, 15, ...
   b) 15, 17, 19, 21, ...
   c) 30, 27, 24, 21, ...
   d) −13, −8, −3, 2, ...

4. Write the next **two** terms of each of the sequences in Q3.

5. The position-to-term rule for a sequence is 'multiply by 5, add 2'.
   Write down the first 5 terms of the sequence.

6. A function machine produces a linear sequence.
The table shows some inputs and outputs from this function machine.
What is the term-to-term rule of this sequence?

| Input | Output |
|---|---|
| 1 | 7 |
| 4 | 10 |

7. Here is a sequence **8, 13, 18, 23, ...**
Anthony says that he can work out the 8th term in the sequence by multiplying 4th term in the sequence by 2.
Is Anthony correct? Give a reason for your answer.

8. Look at the sequence **1, 9, 17, 25, ...** The 50th term in the sequence is 393.
What is the 52nd term? Explain how you arrived at your answer.

9. Chloe is using this function machine.

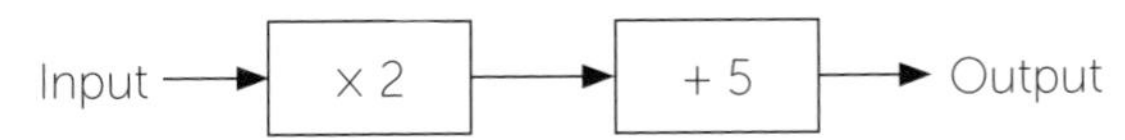


She says that if the input is a whole number, the output will be an odd number.
Is Chloe correct? Give a reason for your answer.

10. Bryan and Kate are looking at this sequence: **1, 7, 13, 19, ...** They need to find the 100th term of the sequence.
Bryan says the term-to-term rule of "+ 6" is the best way to find the 100th term.
Kate says the position-to-term rule of "multiply by 6 then subtract 5" is a better way.
Whose method would you use and why?

## 12.2 LINEAR SEQUENCES AND *n*TH TERM

A23 A24 A25

### Objectives

**Use the $n$th term of an arithmetic sequence to generate terms**

The $n$th term of a sequence describes the sequence using algebra where '$n$' represents the term number.

**e.g.** The $n$th term of a sequence is $3n + 2$

For the first term $n = 1$ — 1st term $= 3 \times 1 + 2 = 5$

For the second term $n = 2$ — 2nd term is $3 \times 2 + 2 = 8$

For the third term $n = 3$ — 3rd term is $3 \times 3 + 2 = 11$

**Deduce an expression for the $n$th term of a linear sequence**

To find the $n$th term of a linear sequence, use the term-to-term rule to show which times tables the sequence has come from. Then adjust the times table to the numbers in the sequence by adding or subtracting as required.

**e.g.** Here is a sequence — 5, 9, 13, 17, ...

The term-to-term rule is + 4

The sequence has come from the 4 times table — 4, 8, 12, 16, ...

Each term in the sequence is **+ 1** on the numbers in the times table.

The position-to-term rule is × 4 + 1

The $n$th term is $n \times 4 + 1$ or $4n + 1$

**Decide if a given number is a term in a given sequence**

**e.g.** Is 154 a term in the sequence 2, 9, 16, 23, ...?

First find the $n$th term of the sequence: **$n$th term = $7n - 5$**

If 154 is in the sequence, then an integer value of $n$ will give 154 as a term value.

So $7n - 5 = 154$

$7n = 159$

$n = 22.714...$

Since $n$ is not an integer, this means that **154 is not a term in the sequence**.

**Find the first term greater/less than a certain number**

## Practice questions

1. Use these $n$th terms, to work out the first 5 terms of each sequence.
   a) $2n + 1$  b) $3n - 1$  c) $5n + 3$  d) $4n + 7$  e) $2n - 6$

2. Harriet has been asked to find the $n$th term of the sequence 5, 9, 13, 17, ...Harriet gives the answer $n + 4$
   Explain what Harriet has done wrong and correct her answer.

3. Find the $n$th terms of these sequences.
   a) 1, 3, 5, 7, ...  b) 4, 7, 10, 13, ...  c) 8, 12, 16, 20, ...  d) 9, 18, 27, 36, ...

4. Copy and complete the table below for the sequence with $n$th term = $7n - 2$

| **Term** | 1 | 2 | 3 | 10 | 20 | 100 |
|---|---|---|---|---|---|---|
| **Sequence** | | | | | | |

5. Shaina says that 153 cannot be in the sequence with $n$th term = $3n + 1$
   Explain why Shaina is correct.

6. Look at the sequence 20, 15, 10, 5, ...
   a) What are the next two terms of the sequence?  b) What is the term-to-term rule for the sequence?
   c) Work out the 15th term of the sequence.  d) What is the $n$th term of the sequence?

7. Find the $n$th terms of these sequences.
   a) 100, 85, 70, 55, ...  b) 50, 47, 44, 42,...  c) 45, 41, 37, 33, ...

8. Jess is planning on paying money each month into a savings account for 1 year.
   In January she pays £1 into the account. In February she pays £4 into the account.
   In March she pays £7 into the account.
   The money she pays in forms a sequence.
   Work out the amount of money Jess will pay into her savings account in December.

9. Mihir has bought a television using a repayment scheme that reduces the monthly payment by £15 each month until the television is paid off. This may include a balancing payment of less than £15 in the last month.
   The first payment Mihir has to make is £200, the second is £185 and so on.
   Work out the number of months Mihir makes a payment.

10. A simple computer program is designed to generate terms of the sequence $3n + 5$ until the first output is greater than 1000.
    How many terms will be generated by the program?

# 12.3 SPECIAL SEQUENCES

## Objectives

### Generate sequences of square, cube and triangular numbers

**Square numbers** are the result of multiplying an integer by itself.

Here are the first twelve square numbers: 1, 4, 9, 16, 25, 36, 49, 64, 81, 100, 121, 144

**Cube numbers** are the result of multiplying an integer by itself and then by itself again.

Here are the first five cube numbers: 1, 8, 27, 64, 125

**Triangular numbers** are numbers from a triangular pattern like this.

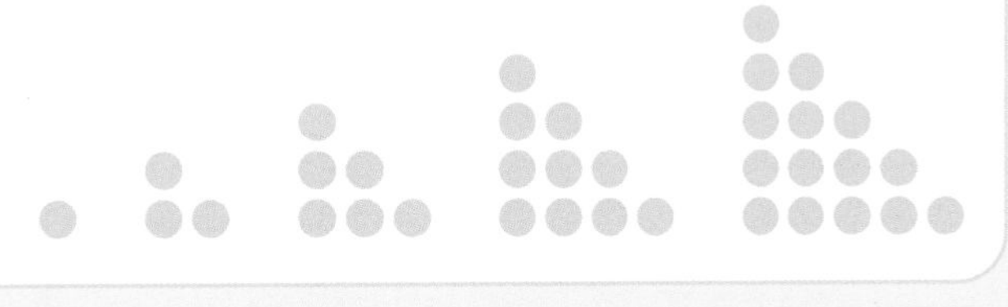

Here are the first ten triangular numbers:

1, 3, 6, 10, 15, 21, 28, 36, 45, 55

### Recognise a sequence from diagrams, draw the next term in a pattern and generate the sequence

Patterns can be shown in diagrams. Numbers from the diagrams can lead to a sequence.

e.g. Here is a sequence of patterns made from matchsticks.

The number of sticks in each pattern makes this sequence: 5, 9, 13, ...

The nth term of the sequence is $4n + 1$

The **10th pattern will have $4 \times 10 + 1 = 41$ matchsticks in it**

## Practice questions

1. Look at this sequence of patterns.
   a) Draw the next pattern.
   b) How many matchsticks will be needed for pattern 5?
   c) Will pattern 6 have double the number of matchsticks of pattern 3?

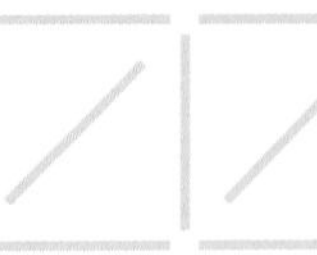

Pattern 1 Pattern 2 Pattern 3

2. Write down the next two terms of each of these sequences.
   a) 2, 5, 10, 17, ... b) 1, 3, 6, 10, ... c) 2, 8, 18, 32, ...

3. The sequence of cube numbers, 1, 8, 27, 64, .... has the $n$th term = $n^3$. This sequence 2, 9, 28, 65, ... is one more than the cube numbers; it has an $n$th term of $n^3 + 1$
   Describe in words how these sequences are linked to the $n^3$ sequence.
   a) 0, 7, 26, 63, .... b) 2, 16, 54, 128,... c) 0.5, 4, 13.5, 32, ...

4. Here is a sequence of patterns.
   a) Draw the next pattern in the sequence.
   b) How many matchsticks would be required for the 10th pattern?
   c) How many squares could be made using 50 matchsticks?

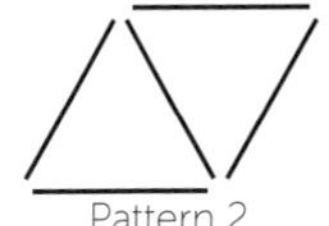
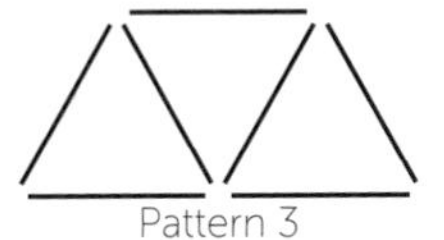

Pattern 1 Pattern 2 Pattern 3

5. The sequence of square numbers; 1, 4, 9, 16, 25, ... has the $n$th term = $n^2$. Describe how these sequences are linked to the $n^2$ sequence. (E.g. $n^2 + 1$, $3n^2$ or similar expressions involving $n^2$.)

   a) 3, 6, 11, 18, ... b) −2, 1, 6, 13, ... c) 5, 20, 45, 80, ...

6. Here are the first 4 patterns in a sequence.

   a) How many counters are required to make the next pattern?

   b) How many counters will be required for the 12th pattern?

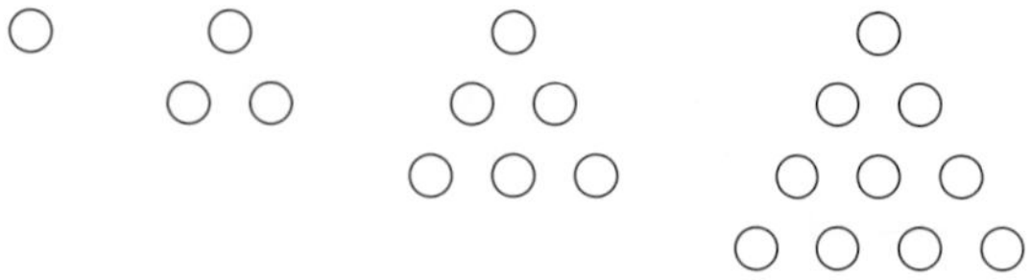

Pattern 1 Pattern 2 Pattern 3 Pattern 4

7. Here is a sequence of patterns made from squares.

   a) How many squares are needed for the next pattern?

   b) How many squares are needed for the 10th pattern?

   c) Josie says that she has used exactly 50 squares to make a pattern in the sequence. Is Josie correct? Give a reason for your answer.

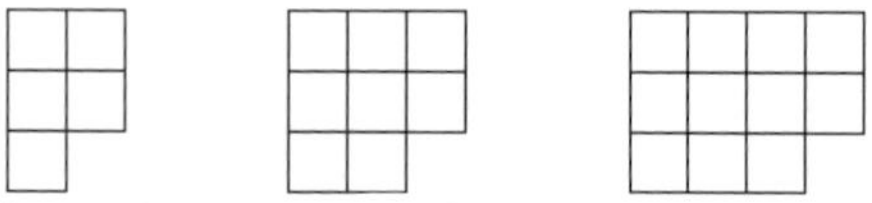

Pattern 1 Pattern 2 Pattern 3

8. Greta has made this sequence of patterns with white and black tiles.

   a) How many white and black tiles will she need for the next pattern?

   b) How many white and black tiles will she need for pattern 30?

   c) Greta has 60 black tiles and 115 white tiles.

   What is the largest pattern number she can make with these tiles?

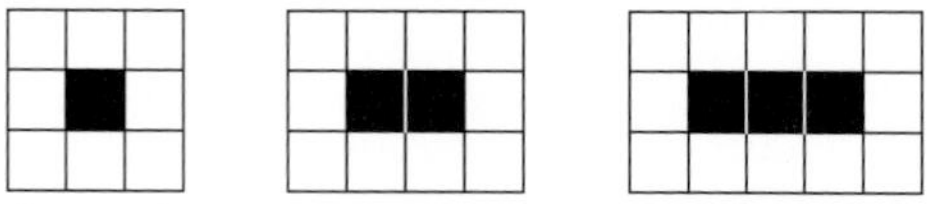

Pattern 1 Pattern 2 Pattern 3

9. Here is another sequence of patterns made from tiles.

   Simon says that he will use 32 black tiles and 40 white tiles to make one of the patterns in this sequence.

   Explain how you know that he is not correct.

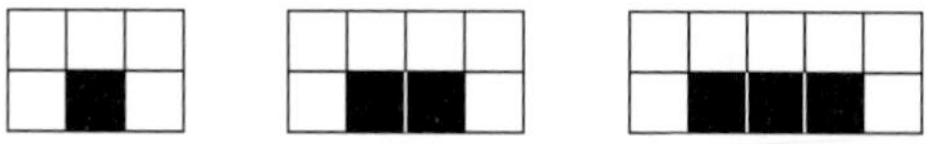

Pattern 1 Pattern 2 Pattern 3

10. Here is a sequence of patterns.

   a) How many squares make up the 4th pattern?

   b) How many squares make up the 10th pattern?

   c) Which pattern number uses 198 squares?

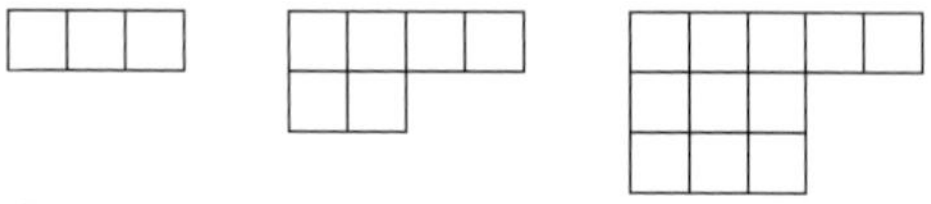

Pattern 1 Pattern 2 Pattern 3

# 12.4 GEOMETRIC SEQUENCES

## Objectives

### Continue a geometric progression and find a term-to-term rule

A **geometric sequence** has a term-to-term rule that multiplies or divides to find the next term from the current term.

**e.g.** 3, 6, 12, 24, .... is a geometric sequence.

Each term is **multiplied by 2** to get the next term.

The next term in this sequence is **48**

A **term-to-term rule** is the rule used to work out the next term in a sequence from the current term.

**e.g.** 1000, 500, 250, 125, ...
the term-to-term rule is '**divide by 2**'

3, 15, 75, 375, ...
the term-to-term rule is '**multiply by 5**'

### Recognise other types of sequence, including Fibonacci and odd/even numbers

A **Fibonacci sequence** is produced by adding two consecutive terms together to give the next term.

**e.g.** 1, 1, 2, 3, 5, 8, 13, ... $1 + 1 = 2$, $1 + 2 = 3$ etc

### Distinguish between arithmetic or geometric sequences

We can decide if a sequence is arithmetic, geometric or neither by looking at the term-to-term rule.

For **arithmetic** sequences the term-to-term rule is **add or subtract**.

For **geometric** sequences the term-to-term rule is **multiply or divide**.

**e.g.** The sequence 2, 6, 18, 54, ... is **geometric** as the term-to-term rule is $\times 3$

The sequence 4, 10, 16, 22, ... is **arithmetic** as the term-to-term rule is $+ 6$

The sequence 1, 3, 6, 10, ... is **neither** as the term-to-term rule changes

## Practice questions

1. Find the next two terms for each of the following sequences:
   a) 4, 8, 12, 16, ... b) 4, 8, 16, 32, ... c) 1, 2, 3, 5, ...
   d) 100, 50, 25, 12.5, ... e) 1, 3, 9, 27, ...

2. Find the term-to-term rule for each of the sequences in Question 1.

3. State whether the following sequences are arithmetic, geometric or neither:
   a) 1, 3, 9, 27, ... b) 6, 12, 18, 24, ... c) 2, 3, 5, 8, ...
   d) 10 000, 1000, 100, 10, ... e) 4, 6, 9, 13.5, ...

4. Henry is generating a sequence: 5, 10, 20, 40, ...
   He says that because the rule is the same between each term that the sequence is arithmetic.
   Explain why he is not correct.

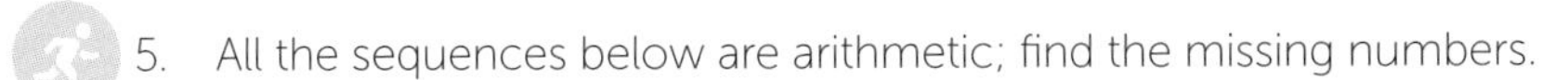

5. All the sequences below are arithmetic; find the missing numbers.
   a) 4, 9, ___, ___, 24,... b) 3, ___, ___, 15, ___, ... c) 20, ___, 14, ___, 8, ...
   d) ___, ___, 7, ___, 15, ...

6. All the sequences below are geometric; find the missing numbers.
   a) 1, 2, ___, 8, ___, ... b) 120, 60, ___, ___, 7.5, ... c) ___, 18, ___, 162, ___, ...
   d) 2, ___, ___, ___, 512, ...

7. Here are the first two terms of a geometric sequence; 4, 12, ...
   Work out the fifth term of the sequence.

8. The first term of an arithmetic sequence is 8. The fourth term of the same arithmetic sequence is 26.
   Find the 10th term of the sequence.

9. The third term of a geometric sequence is 10. The fifth term of the same geometric sequence is 40.
   What is the first term of the sequence?

10. Kate and Dan are each looking at different sequences. The first term of both their sequences is 6. The third term of both their sequences is 24
    Kate's sequence is geometric. Dan's sequence is arithmetic.
    Find the 10th term of both their sequences.

## 12.5 QUADRATIC SEQUENCES

**A23** **A24** **A25**

### Objectives

#### Recognise a quadratic sequence

A quadratic sequence has an $n$th term with $n^2$ as the highest power.

**e.g.** 2, 8, 18, 32, ...... is a quadratic sequence with the $n$th term = $2n^2$

In a quadratic sequence the difference between the terms forms an arithmetic sequence.

**e.g.**

| **Sequence** | 1 | | 4 | | 9 | | 16 | | 25 |
|---|---|---|---|---|---|---|---|---|---|
| **Difference** | | 3 | | 5 | | 7 | | 9 | |

The differences of this quadratic sequence go up in 2s.

#### Continue a quadratic sequence and use its $n$th term

A quadratic sequence is generated by substituting in values for $n$ in the $n$th term.

**e.g.** Find the first three terms of $4n^2 - 7$

First term, when $n = 1$ $\quad 4 \times 1^2 - 7 = -3$

Second term, when $n = 2$ $\quad 4 \times 2^2 - 7 = 9$

Third term, when $n = 3$ $\quad 4 \times 3^2 - 7 = 29$

### Practice questions

1. Which of the following is the $n$th term of a quadratic sequence?
   A. $n^2 + 10$ B. $2n - 3$ C. $5n^2$ D. $n^2 + n$

2. Francis wants to generate the sequence $n^2 + 3$ and has the following results:

| $n$ | 1 | 2 | 3 | 4 | 5 |
|---|---|---|---|---|---|
| **Working** | $1^2 + 3 = 2 + 3$ | $2^2 + 3 = 4 + 3$ | $3^2 + 3 = 6 + 3$ | $4^2 + 3 = 8 + 3$ | $5^2 + 3 = 10 + 3$ |
| **Term** | 5 | 7 | 9 | 11 | 13 |

   He has made a mistake. Find his mistake and correct it.

3. Find the first 5 terms of these sequences.
   a) $n^2 + 4$ b) $n^2 - 1$ c) $2n^2$ d) $n^2 - n$

4. By considering the differences, state which **two** of these sequences are quadratic.
   A. 3, 5, 8, 12, 17, ... B. 1, 5, 11, 19, 29, ... C. 3, 10, 29, 66, 127, .... D. 0, 5, 15, 30, 50, ...

5. Show that this sequence of patterns is a quadratic sequence.

Pattern 1

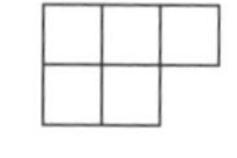

Pattern 2

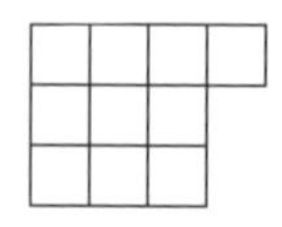

Pattern 3

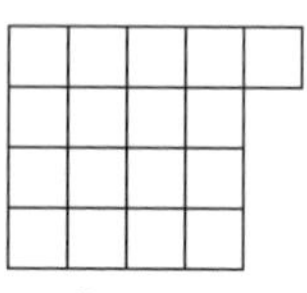

Pattern 4

6. Find the first 5 terms of these sequences.

   a) $n^2 + 2n$  b) $2n^2 + 1$  c) $2n^2 + n - 3$

7. Denise has been asked to generate the sequence $3n^2 + 2$

   This is how she works out the first term: $3 \times 1^2 + 2$

   $= 9 + 2$

   $= 11$

   She has made a mistake.

   Explain her mistake and show the correct workings.

8. Which of the following sequences are quadratic? Show your working.

   a) 2, 3, 6, 15, 42, ...  b) 5, 8, 14, 23, 35, ...  c) 2, 12, 32, 62, 102, ...

9. Graham claims that this sequence is quadratic; 2, 7, 16, 29, 46

   Is Graham correct? Give a reason for your answer.

10. Do triangular numbers form a quadratic sequence?

    Show your working.

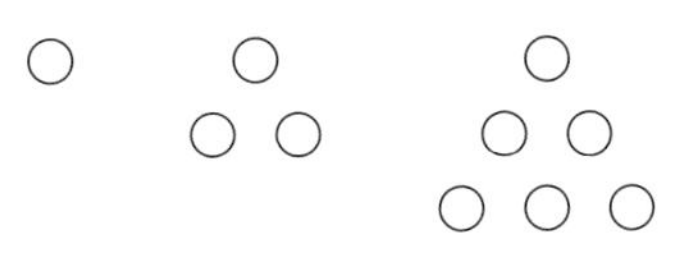

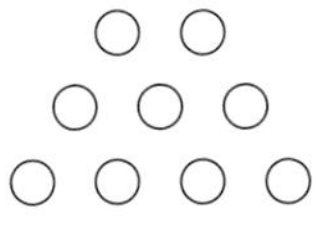

Pattern 1 Pattern 2 Pattern 3 Pattern 4

## 12.6 INEQUALITIES

A22

**What is an inequality?**

Inequalities are similar to equations but instead of giving exact answers they give a range of answers.

**e.g.** $x \geq 3$ shows that $x$ can be a value equal to or greater than 3

$x < -5$ shows that $x$ can be a value less than (not equal to) $-5$

### Objectives

**Show inequalities on number lines**

Inequalities can be shown on a number line using dots and lines or arrows.

A solid dot means that $x$ can be equal to this value.

An open dot means that $x$ cannot be equal to this value.

**e.g.** On this number line:

The range for $x > 1$ is shown in red

The range for $-2 < x \leq 3$ is shown in green

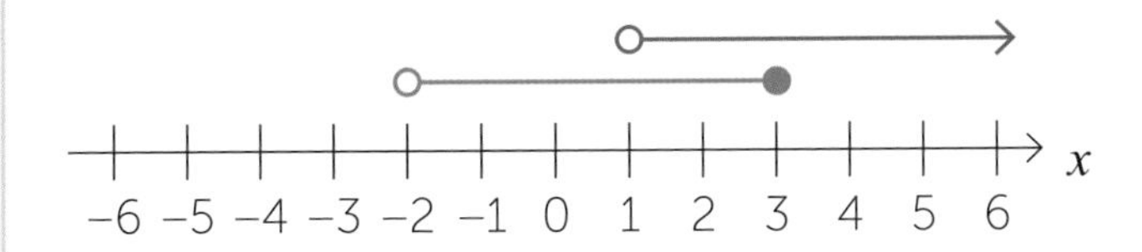


**Write down whole number values that satisfy an inequality**

Integers are whole numbers; they can be either positive, negative or zero.

Integers that satisfy inequalities are those within the range described by the inequality.

**e.g.** The integers that satisfy the inequality $-1 \leq n < 3$ are **−1, 0, 1 and 2**

**Construct inequalities to represent a set shown on a number line**

## Practice questions

1. Place one of the signs >, < or = between each pairs of numbers to make a correct statement.
   a) 9 ____ 5 b) −6 ____ −4 c) 3.04 ____ 3.4 d) $\frac{4}{5}$ ____ 0.8

2. Gerry has been asked to give the largest integer that satisfies the inequality $x < 3.5$ He gives the answer 3.4
   Explain why Gerry is not correct.

3. Which of these inequalities does this number line represent?

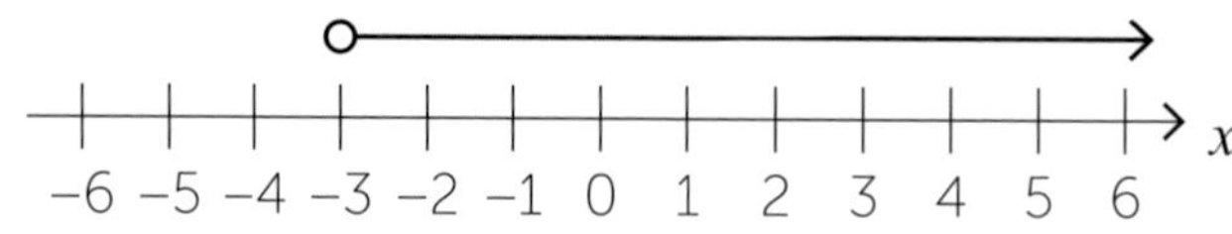


   a) $x = -3$ b) $x > -2$ c) $x < 6$ d) $x > -3$

4. Draw a number line from −4 to 4 and represent each of these inequalities on it.
   a) $x > 3$ b) $x < 1$ c) $x \geq 0$ d) $x > -4$ e) $x \leq 2$

5. Write the largest integer that satisfies each inequality.
   a) $n \leq 4$ b) $n < -3$ c) $n \leq -1.8$

6. Write the smallest integer that satisfies each inequality.
   a) $n > 0$ b) $n > -5$ c) $n \geq -3.4$

7. a) List the integers that are satisfied on this number line.
   b) Write this range as an inequality.

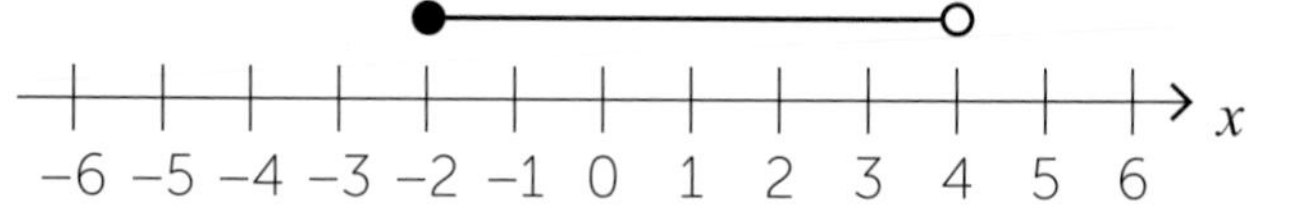

8. Toby has been asked to list the integers that satisfy the inequality $-2 \leq n < 4$.
   He gives the answer: −2, −1, 1, 2, 3, 4
   Comment on Toby's answer.

9. List the integers that satisfy each of these inequalities.
   a) $2 < n < 5$ b) $0 \leq n < 4$ c) $-1 < n \leq 3$ d) $-5 \leq n \leq -2$

10. List the integers that satisfy each of these inequalities.
   a) $2 < x + 2 \leq 5$ b) $1 \leq x - 1 \leq 3$ c) $4 \leq 2x < 10$ d) $-6 < 3x \leq 9$

# 12.7 SOLVING LINEAR INEQUALITIES

## Objectives

### Solve simple inequalities in one variable and represent the solution on a number line

Solving inequalities is like solving equations where you balance both sides, but the solution is a **range of values**.

e.g. Solve $3x + 5 \le 17$

Subtract 5 from both sides $3x \le 12$

Divide both sides by 3 $\mathbf{x \le 4}$

When solving inequalities make the variable positive.

e.g. Solve $11 - 2x > 5$

Add $2x$ to both sides $11 > 5 + 2x$

Subtract 5 from both sides $6 > 2x$

Divide both sides by 2 $\mathbf{3 > x}$

(which can also be written as $x < 3$)

### Solve two inequalities in x and find the solution sets that satisfy them

When there are two inequality sides, keep balancing all three parts of the inequality until the variable is on its own in the middle.

e.g. Solve $-7 \le x - 4 < 2$

Add 4 to all parts $\mathbf{-3 \le x < 6}$

### Show a solution set on a number line

Just like when we represented inequalities on a number line before, we can do the same after we have solved inequalities.

e.g. Solve: $3x + 5 > 17$

Subtract 5 from both sides $3x > 12$

Divide both sides by 3 $x > 4$

This can be shown on the number line.

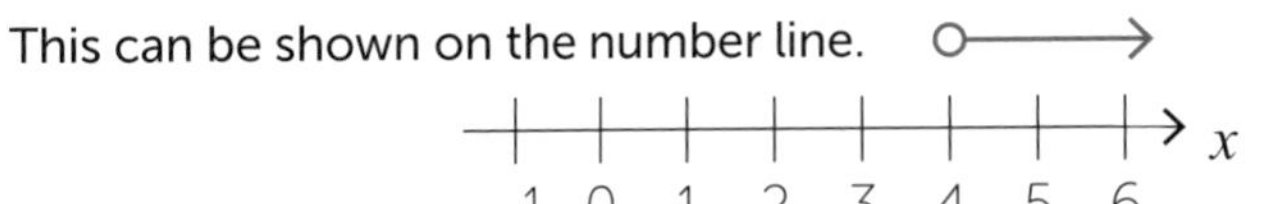


## Practice questions

1. Solve these inequalities.

   a) $x + 5 > 7$ b) $x - 3 \le 5$ c) $4x \ge 20$ d) $\frac{x}{5} < 10$ e) $-3 > x - 4$

2. Ginny has been asked to solve $y + 4 < -1$ and give the largest integer that satisfies the inequality.

   Her solution is: $y + 4 < -1$

   Add 4 to both sides: $y < 3$ so the largest integer is 3

   She has made a mistake. Find the mistake and correct it.

3. Give the smallest integer satisfied by the inequality $12 < 3n$

4. Solve these inequalities.

   a) $3x + 2 \le 17$ b) $4x - 1 > 25$ c) $7 - 2x < 0$ d) $16 > \frac{x}{3} + 5$

5. James is solving the inequality $11 - 2x \ge 5$

   His workings are: $11 - 2x \ge 5$

   Subtract 11 from both sides: $2x \ge -6$

   Divide both sides by 2 $x \ge -3$

   He has made a mistake. Find the mistake and correct it.

6. Solve these inequalities.

   a) $2 < x - 3 \le 7$ b) $-3 < y + 6 < 2$ c) $4 \le 3n + 1 < 13$ d) $-5 \le 2n - 1 \le 3$

7. a) Solve $7 - 2n \ge 3$ b) Solve $10 - 3n < 22$

   c) List all the integers that satisfy both $7 - 2n \ge 3$ and $10 - 3n < 22$

8. An office needs to buy paper for its printers.
   The budget for buying the paper is £75. A box of paper cost £2.70.
   Form an inequality to represent the situation.
   Solve the inequality and state the maximum number of boxes the office can buy.

9. Solve each inequality and represent the solution set on a number line.
   a) $-3 \leq 2x + 5 < 7$ b) $-7 < 3x + 2 < 5$ c) $-18 < 5x - 3 \leq 7$

10. A van can carry a maximum load of 500 kg.
    A delivery company uses the van to deliver packages with a mass of 30 kg each.
    The van also carries a trolley with a mass of 10 kg used to deliver the packages.
    Form an inequality to represent the situation.
    Solve the inequality and state how many packages the van is able to carry.

SECTION 13

# PROPORTION

## 13.1 CURRENCY CONVERSION

R5 R10 A14

### Objectives

**Convert between currencies**

**Exchange rate**

An exchange rate tells you how much one currency will buy in another currency.

The exchange rate can be used to convert between quantities of currency.

**e.g.** \$1 = £0.76

so \$50 = 50 × 0.76 = **£38**

and £50 = 50 ÷ 0.76 = **\$65.79 (2 d.p.)**

**Interpret information presented in a range of linear and non-linear graphs**

**Draw straight-line graphs for real-life situations, including ready reckoner graphs and conversion graphs**

**Conversion graphs**

A conversion graph can be used to convert between currencies, units and other values.

The variables on a conversion graph are in direct proportion.

The gradient of the conversion graph gives the number of $y$ units per 1 $x$ unit.

**e.g.** Here is a conversion graph between Euros and Australian dollars.

Reading off the graph we can see that €5 converts to 8 AUD\$ and that 24 AUD\$ converts to €15.

The gradient of the graph = $\frac{8}{5}$ or 1.6

This shows that **1.6 AUD\$ = 1 Euro**

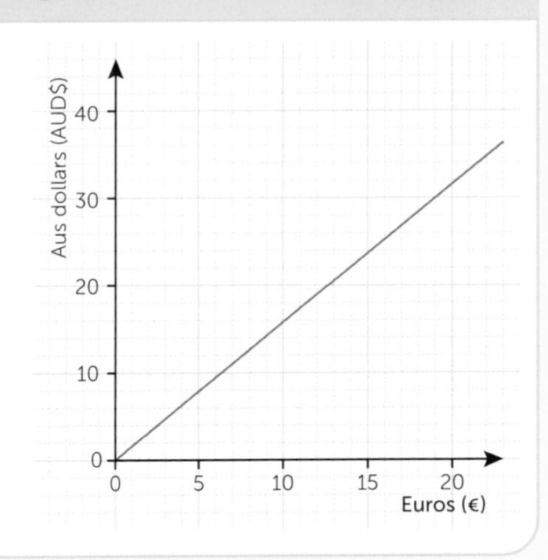


### Practice questions

1. Moira goes on holiday to Switzerland. The exchange rate is £1 = 2.10 francs.
   She changes £450 into francs. How many francs does she get?

2. The exchange rate between US dollars (\$) and South African rand is 1\$ = 13.75 rand. Jordan changes \$300 into rand.
   How many rand will he get?

3. Poppy receives 90 US dollars in exchange for £60.
   a) Work out the exchange rate for £1.
   b) Jon exchanges \$120. How many pounds (£) does he get?

4. 60 Danish krone = £5. Georgia draws a conversion graph for these currencies. Which of these graphs is correct?

A.

B.

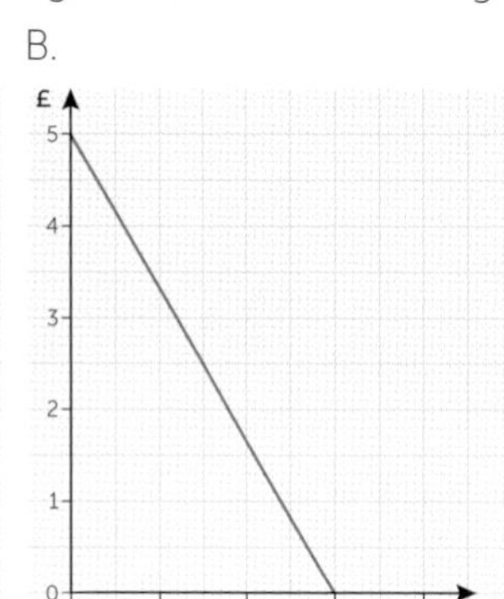


C.

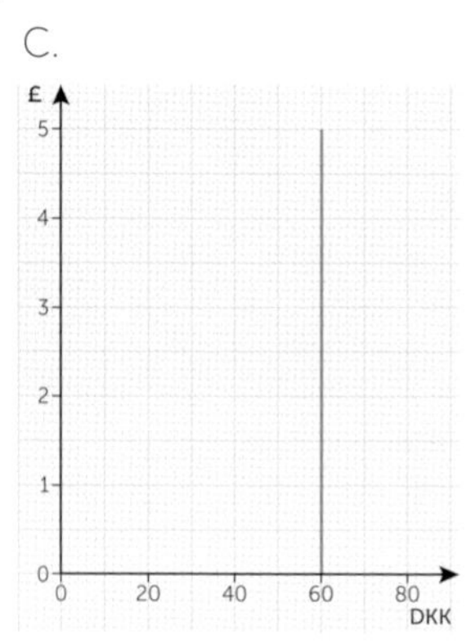


5. Carlo exchanged 13 Euros and received 500 Thai baht. Which of these is the exchange rate for Euros to Thai baht?

A. 1 Euro = 488 Thai baht

B. 1 Euro = 6500 Thai baht

C. 1 Euro = $\frac{13}{500}$ Thai baht

D. 1 Euro = 38.46 Thai baht

6. £500 = 5900 SEK Swedish krona. Which of these shows the exchange rate between the two currencies?

A. 1 SEK = £11.80

B. £1 = 11.8 SEK

C. £1 = 5900 SEK

D. 1 SEK = £500

7. The currency in Japan is the yen and the currency in India is the rupee. The table shows some amounts of rupees and the equivalent amounts in yen. Copy and complete the table.

| **rupees** | 0 | 800 | 1600 | 2000 | | |
|---|---|---|---|---|---|---|
| **yen** | 0 | | 2600 | | 5200 | 6500 |

8. £5 = €6

a) Draw a conversion graph to show this relationship.

b) Use your graph to convert £150 to Euros and to convert €120 to pounds.

c) Work out the gradient of the graph.

d) Write a formula for calculating Euros (€) from pounds (£)

9. Harry wants to exchange £560 into Euros. He has two conversions graphs, but the numbers are missing.

Is it possible for Harry to decide which bank will give him the best rate of conversion? Give a reason for your answer.

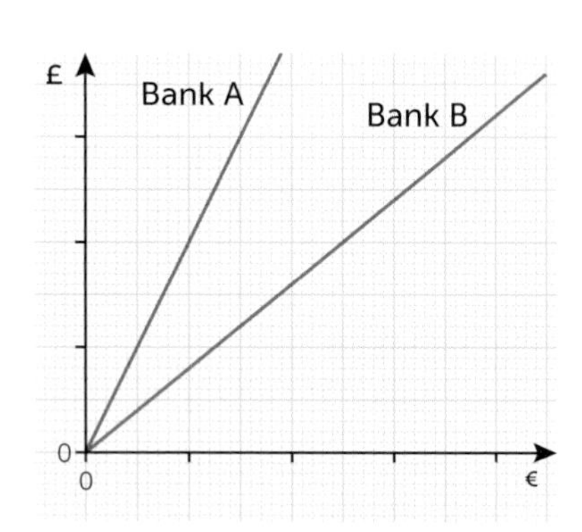

10. The currency in Japan is the yen.

Here are the exchange rates between yen and pounds.

| **Exchange rates for £1** | | |
|---|---|---|
| | Buy | Sell |
| **JPY** | 137 | 150 |

Cameron exchanges £200 to JPY. When he comes back from holiday he has 600 JPY left.

a) How many pounds did Cameron spend on his holiday?

b) Cameron changes the 600 JPY back to pounds £. How many pounds does he receive?

## 13.2 BEST BUY PROBLEMS

### Objectives

#### Convert between measures in problems involving best value

To work out which option gives the best value, compare the price per unit of amount or the quantity per unit price.

e.g. 400 g of pasta costs 88p. 700 g of pasta costs £1.47.

Either work out how much pasta per pence or the cost of 1 g of pasta for each option.

**Method 1**

400 g is 88p so 1 g is 88 ÷ 400 = **0.22p**

700 g is 147p so 1 g is 147 ÷ 700 = **0.21p**

The larger bag costs less for 1 gram, so this is the **better value**.

**Method 2**

400 g is 88p so 1p buys 400 ÷ 88 = **4.55 g (2 d.p.)**

700 g is 147p so 1p buys 700 ÷147 = **4.76 g (2 d.p.)**

The larger bag gives more grams for 1 pence, so this is the **better value**.

*Notice how the units were converted to the same (pence)*

#### Draw straight-line graphs for cost per unit

e.g. Sally is hiring a van for 1 week.

The graph shows two options.

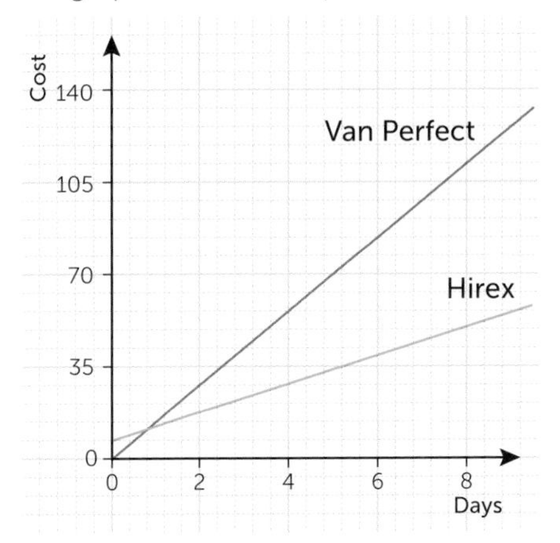


To find the option which is better value, use the graph to read off the cost for 1 week.

**Hirex has lower costs for 7 days so is the better value option.**

### Practice questions

1. Which is better value for money:
   6 bottles of orange juice for £4.80 or 5 bottles of orange juice for £4.25?

2. Place these three deals in ascending order of cost.

| Deal 1 | Deal 2 | Deal 3 |
|---|---|---|
| 8 bottles of 500 ml for £5 | 6 bottles of 500 ml for £3.50 | 2 bottles of 750 ml for £1.50 |

3. At market, one farmer offers £3.75 for 2.5 kg of apples whilst another offers £7.50 for 6 kg of apples. Which farmer offers the better deal?

4. A shop has 2 different offers on the sale of pencils.

| Offer A |
|---|
| 6 pencils for 90p |

| Offer B |
|---|
| 4 pencils for 72p |

   a) How much does one pencil cost with each offer? b) Which offer gives better value for money?

5. A 345 g bag of grapes cost £1.35. To find out how much 1 kg of grapes cost:
   Jo types in her calculator: 1.35 ÷ 345 × 1000
   Adam types in his calculator: 1.35 ÷ 0.345
   a) Who is correct? Give a reason for your answer. b) Work out the cost of 1 kg of grapes.

6. This graph shows the rates offered for 2 different mobile phone contracts.

Which company is better value for money?

Give a reason for your answer.

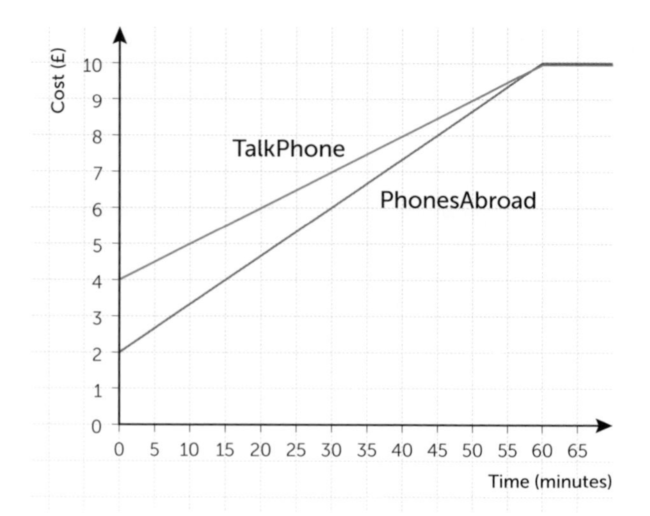


7. A baker charges one-third less per cupcake for offer B than per offer A.

He prepares a board to put outside his bakery.

| Offer A |
| --- |
| 3 cupcakes **£2.34** |

| Offer B |
| --- |
| 8 cupcakes **£.........** |

What number should he write on the board?

8. The graph shows information about the cost of two car parks.

Decide which of the sentences below are true and which are false.

Give reasons for your answers.

a) Both car parks charge the same to park for 7 hours.

b) Beaumont Road car park is more expensive than City car park if you park the car for 5 hours.

c) The longer you leave the car in the Beaumont Road car park, the more you pay.

d) City car park is cheaper if you park the car for more than 7 hours.

e) City car park charges £1/hour

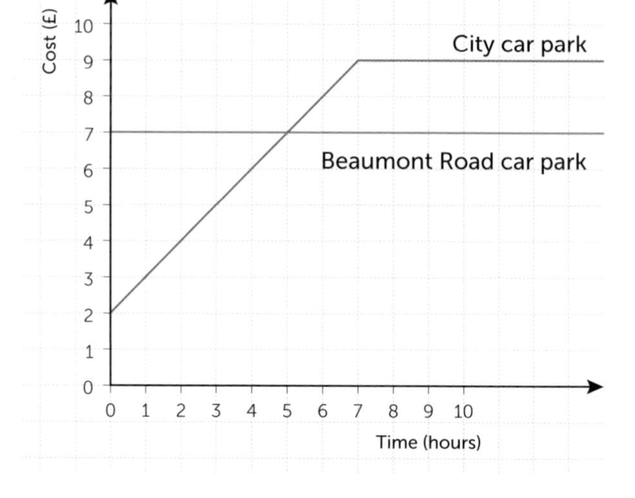


9. A shop has chocolates on offer.

What price should be charged for Offer B, so that 10 boxes is 2p less per box than Offer A?

| Offer A |
| --- |
| 6 boxes for **£2.34** |

| Offer B |
| --- |
| 10 boxes for ... |

## 13.3 STANDARD UNITS

### Objectives

**Convert between metric measures of mass, length and capacity**

Here are some of the more common metric units and their conversions.

| Mass | Length | Capacity |
|---|---|---|
| 1 tonne = 1000 kilogram<br>1 t = 1000 kg | 1 kilometre = 1000 metre<br>1 km = 1000 m | 1 litre = 1000 millilitre<br>1 l = 1000 ml |
| 1 kilogram = 1000 gram<br>1 kg = 1000 g | 1 metre = 100 centimetre<br>1 m = 100 cm | 1 litre = 100 centilitre<br>1 l = 100 cl |
| 1 gram = 1000 milligram<br>1 g = 1000 mg | 1 centimetre = 10 millimetre<br>1 cm = 10 mm | 1 $cm^3$ holds 1 ml<br>1 $m^3$ holds 1000 litres |

**Solve problems using standard units of measure**

**e.g.** A scientist has 4 kg of a metal.

She uses 1.27 kg for an experiment, 457 g is removed as a sample and 0.8 kg is given to another scientist.

Work out how much metal is left.

Metal used is 1.27 + 0.457 + 0.8 = 2.527 kg *Note how the units are the same*

Metal left is 4 − 2.527 = **1.473 kg**

### Practice questions

1. Copy and complete the missing values in each statement.
   a) 1 m = ____ cm b) 2 km = ____ m c) _____ ml = 3 litre
   d) ____ mm = 6.4 m e) 10.1 kg = ____ g f) ____ cl = 0.1 litre

2. Spot the mistakes and correct them.
   a) 1 t = 1000 mg b) 1 m = 1000 ml c) 100 cl = 1 cm d) 10 mm = 1 cm

3. In each set of units, find the odd one out.
   a) l ml mm cl b) g cm mg t
   c) mm m kg cm

4. Match together the equal values, one from each box.
   One of the measurements has two answers.

| 1 km | 1 l | 1 m | 1 g | 1 t |
|---|---|---|---|---|

| 1000 kg | 1000 mm | 1000 ml | 1000 mg | 1000 m | 100 cm |
|---|---|---|---|---|---|

5. Write each set of measurements in ascending order.
   a) 0.6 kg, 259 g, 3890 mg, 0.07 t b) 12.6 cm, 67 mm, 0.89 km, 890 mm
   c) 7.09 l, 790 ml, 0.7 l, 71 cl d) 0.01 cl, 10 ml, 0.1 l, 1 ml

6. Decide which of these statements are true and which are false.

a) $1\text{ g} = \frac{1}{1000}$ of 1 kg  b) $1\text{ kg} = \frac{1}{100}$ of 1 tonne

c) 1 l = 1000 ml  d) $1\text{ cm} = \frac{1}{10}$ of 1 m

7. Use <, > or = to complete each of these statements.

a) 2.4 m + 50 cm ____ 300 cm  b) 2250 m – 1.5 km ____ 750 m

c) 489 g + 568 g _____ 1 kg 580 g  d) 2 litre – 348 ml ____ 1 litre + 3480 ml

8. Aria has a box in the shape of cuboid.

The dimensions of the box are 120 cm, 450 mm and 0.5 m.

Which of these values is the correct capacity of the box?

a) 570.5 ml  b) 270 l

c) 2700 ml  d) 2700 cm$^3$

9. Freddie wants to post three books to his brother. He puts the books in a box.

The box has a mass of 0.12 kg.

The books have masses of 389 g, 0.43 kg and 1.08 kg.

Work out how much it costs to post the box of books.

| **Postage Charges** | |
|---|---|
| Mass of the parcel | Postage price |
| 0 – 0.5 kg | £1.25 |
| 0.51 kg – 1 kg 200 g | £3.56 |
| 1.21 kg – 2 kg | £4.99 |
| 2.01 kg – 4 kg | £7.50 |

10. Silva uses a 25 cl cup to fill a bucket that has a capacity of 12.75 litres.

How many cups does she need so that the bucket is half full?

11. One cube measures 1 cm by 1 cm by 1 cm. Another cube measures 10 mm by 10 mm by 10 mm.

Tom says that the volume of the second cube is greater than the volume of the first.

He is not correct. Give a reason why.

12 The amount (or value) in the middle is the sum of the other four measurements.

Work out the missing numbers.

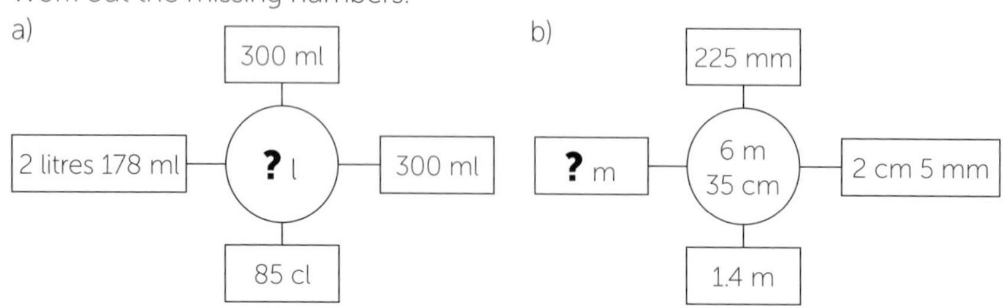

# 13.4 UNITS OF TIME

Here are the most commonly used units of time and their conversions.

| | | |
|---|---|---|
| 1000 milliseconds = 1 second | 60 seconds = 1 minute | 60 minutes = 1 hour |
| 24 hours = 1 day | 7 days = 1 week | 365 or 366 days = 1 year |
| 10 years = 1 decade | 100 years = 1 century | 1000 years = 1 millennium |

## Objectives

### Convert between units of measures of time

Remember when dealing with units of time, many of them are **not decimal**.

e.g. $1\frac{1}{2}$ hours = $1\frac{1}{2} \times 60$ = **90 minutes**

e.g. Joanne spent 336 hours sailing across the Atlantic Ocean.

Convert this time into weeks.

336 hours ÷ 24 = 14 days

14 days ÷ 7 = **2 weeks**

### Deal with remainders and decimal times

Time units can be given in decimals, but care is needed when converting to other units.

e.g. Gary ran a half marathon in 2.5 hours.

Beth ran it in 165 minutes. Who was the fastest?

Do: 2.5 hours = 2.5 × 60 minutes

= 150 minutes.

Or: 165 minutes = 165 ÷ 60 minutes

= 2.75 hours

(0.75 of an hour = 0.75 × 60 = 45 minutes)

= 2 hours 45 minutes.

### Convert between mixed units of time

## Practice questions

1. Copy and complete each statement.
   a) 120 seconds = ____ minutes
   b) 3 days = ____ hours
   c) 0.5 hours = ____ minutes
   d) ____ years = 4.3 decades

2. Convert each of the following.
   a) 5.7 minutes into seconds
   b) 3.6 hours into minutes
   c) 4.5 days into hours
   d) 3.8 centuries into years

3. Copy and complete each of these statements.
   a) 150 minutes = ____ hours ____ minutes
   b) 400 seconds = ____ minutes ____ seconds
   c) 40 hours = ____ days ____ hours
   d) 50 days = ____ weeks ____ days

4. Tanya does 40 minutes homework every weekday Monday to Friday.
   How much homework, in hours and minutes, does she do each week?

5. Use >, < = to make each of these statements true.
   a) 1000 days ____ 15 months
   b) 10 days 10 hours ____ 250 hours
   c) 2800 milliseconds ____ 28 seconds
   d) 10 millennia ____ 10 000 years

6. To convert 100 seconds into minutes, divide by 60. Copy and complete each of these statements.
   a) To convert 100 hours to days, (multiply/divide) by 7/24/60/100.
   b) To convert 100 days to weeks, (multiply/divide) by 7/24/60/100.
   c) To convert 100 centuries to years, (multiply/divide) by 7/24/60/100
   d) To convert 100 millenniums to years, (multiply/divide) by 7/24/60/100/1000

7. a) Kit earns £15/hour. One week she worked for 8 hours and 40 minutes. How much does she get paid this week?
   b) Sean earns £15/ hour too. One week he got paid £110. How many hours and minutes did he work that week?

8. Sally and Nick measure the time it takes them to finish their homework. Sally took 15 minutes and 47 seconds. Nick took 7 minutes 36 seconds longer. How long did it take Nick to complete his homework?

9. Usain Bolt ran 100 m in 9.58 seconds. Ben says: Bolt takes 9.58 ÷ 60 = 0.16 hours to run 100 m.
   Ben is not correct. Explain the mistake he has made and correct it.

10 To work out the time it takes to cook a turkey, allow 45 minutes/kg plus 20 minutes.
   How many hours and minutes will it take to cook a turkey with a mass of 4.6 kg?

11. Alice says: 1 decade = 100 years and $\frac{1}{1000}$ of 1 millennium = 1 year
    Victor says $\frac{1}{60}$ of a day = 1 hour
    Explain the mistakes they have made and correct them.

## 13.5 METRIC AND IMPERIAL CONVERSIONS

N13 R1 R10

### Objectives

**Use a given ratio to convert between metric and imperial measures**

Imperial units are 'old' units of measure. Metric units are newer units and more widely used. In the UK we use a mixture of Imperial and metric units in everyday life.

Here are some conversions from Imperial to metric units.

| Mass |
|---|
| 2.2 pounds = 1 kg |
| 7.8 stones = 50 kg |

| Length |
|---|
| 5 miles = 8 km |
| 1 inch = 2.54 cm |

| Capacity |
|---|
| 1 gallon = 4.55 litres |
| 1 pint = 568 ml |

**Solve word problems involving direct proportion**

e.g. Lo's dog needs some medicine.

This information is written on the back of the bottle.

| Dog's mass | Dose |
|---|---|
| 5 – 20 kg | 5 ml twice a day |
| 20+ kg | 7.5 ml twice a day |

The dog has a mass of 63 pounds.

Lo knows that 1 stone = 14 pounds and he has this graph to convert between kilograms and stone.

Work out the dose for the dog.

The dog's mass is 63 pounds.
63 ÷ 14 = 4.5 stone

Using the graph
4.5 stone ≈ 28 kg

**The dog should be given 7.5 ml of medicine twice a day.**

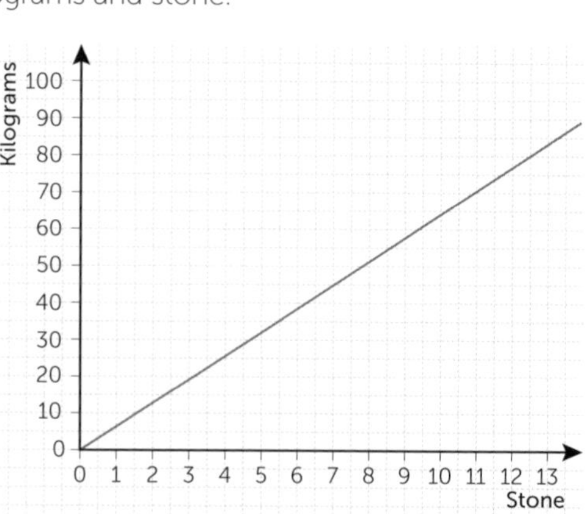

## Practice questions

1. Write these units under the correct headings in the table:

   Pints Ounces Miles Pounds Inches
   Gallons Feet Stone

| Length | Mass | Volume |
|---|---|---|
| | | |
| | | |
| | | |
| | | |

2. For each set of units, find the odd one out.
   a) kilogram pound ounce gram pint
   b) gallon litre pint ounce millilitre
   c) inch centimetre stone yard metre

3. Decide which of each of the pair is a bigger unit.
   a) 1 stone or 1 kilogram
   b) 1 inch or 1 cm
   c) 1 foot or 1 metre
   d) 1 pint or 1 litre
   e) 1 ounce or 1 g
   d) 1 mile or 1 kilometre

4. 5 miles = 8 km. Use this to convert:
   a) 20 km to miles
   b) 200 miles to km
   c) 1 mile to km

5. The table shows conversions between litres and pints.

   Copy and complete the table.

| 1 litre | 5 litres | 10 litres | 25 litres | |
|---|---|---|---|---|
| | 8.8 pints | | | 88 pints |

6. 10 metres = 32.8 feet 20 metres = 65.6 feet
   a) Draw a conversion graph to show this information.
   b) Use your graph to convert 50 feet to metres.
   c) Use your graph to convert 6 metres to feet.

7. A container holds 4.2 gallons of water. Mary used 6.5 litres of water from the container.

   How many litres of water are left in the container? (1 gallon = 4.55 litres.)

8. Jake was 1 m 45 cm tall last time he was measured. Since then, he has grown 6 inches taller.
   a) How tall is he now in metres?
   b) How tall is he now in inches? (1 inch = 2.54 cm.)

9. Lira buys 2 pounds of strawberries and 2 pints of milk. The strawberries cost £3/kg and milk cost 90p per litre.

   How much does Lira pay? (2.2 pounds = 1 kg, 1.76 pints = 1 litre)

10. Mia's car drives 45 miles per gallon. The car fuel tank holds 60 litres of petrol and is full. Mia drives to her grandma who lives 100 km away.

    5 miles = 8 km. 1 gallon = 4.55 litres.

    Does she have enough petrol in her fuel tank for the journey? Show your working.

# 13.6 DIRECT PROPORTION PROBLEMS

## Objectives

### Solve word problems involving direct proportion

Two values are in **direct proportion** when they increase or decrease in the same ratio.

**e.g.** 3 apples cost 90p, 5 apples cost £1.50, 10 apples cost £3

The ratio of number of apples to cost in pence is 1 : 30

One apple costs 30p.

Here apples and cost are in direct proportion.

However, when items are bought in bulk they may **not be in direct proportion**.

**e.g.** 3 oranges for 90p, 30 oranges for £6

The ratio of number of oranges to cost in pence in this case differs from 1 : 30 to 1 : 20

Here oranges and cost are not in direct proportion.

### Solve problems involving currency conversions and best buys, and draw conclusions

To work out which option gives the best value, compare the price per unit of amount or the quantity per unit price.

**e.g.** In the USA, 2 pounds of strawberries cost \$1.84. In the UK, 1 kg of strawberries costs £1.50.

Given that £1 = \$1.27 and 1 pound = 0.454 kg, *work out where the strawberries are cheaper; the UK or USA.*

2 pounds of strawberries is 2 × 0.454 = 0.908 kg

Convert to dollars:

In the UK, 1kg cost \$1.50 x 1.27 = \$1.905

0.908 kg cost \$1.905 x 0.908 = \$1.73

They are cheaper in the UK (\$1.73 for 2 pounds)

### Understand proportionality as equality of ratios

## Practice questions

1. The cost of a tablet is in direct proportion to the number of tablets bought. 8 tablets cost £1320. Work out the cost of 15 tablets.

2. Decide if each of these statements is true or false. Give a reason for your answer.
   a) When two quantities are in direct proportion, the ratio between them is constant.
   b) When A is directly proportional to B, if A increases by 2 units, so does B.
   c) When C is directly proportional to D, if D is halved so is C.

3. 14 balloons cost £1.12. Which of these shows the cost of 4 balloons?
   a) £1.02 b) 8p c) 28p d) 32p

4. Which one of these triangles is the odd one out? Give a reason for your answer.

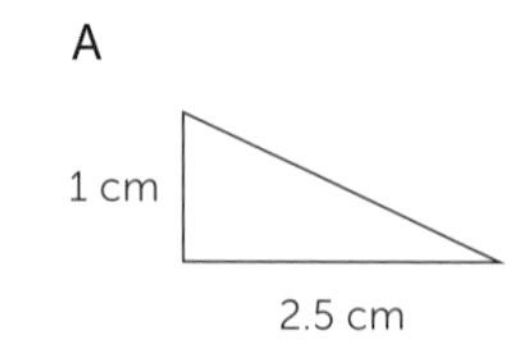


B
2.4 cm
3.6 cm

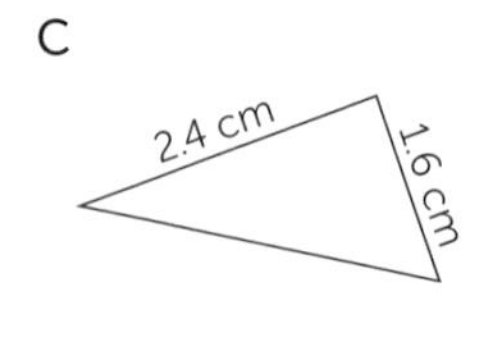


D
0.9 cm
1.35 cm

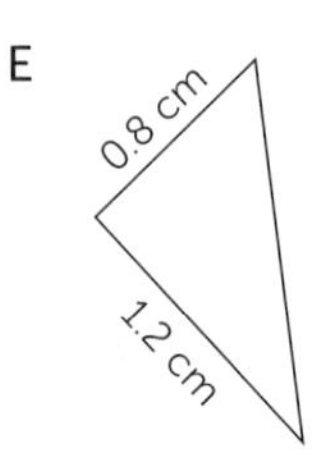


5. Harry is making 12 cupcakes. Here is part of the recipe:
   120 g butter 150 g sugar 3 eggs 132 g flour
   List the quantities of each ingredient needed to make 16 cupcakes.

6. Does this graph show variables in direct proportion?

   Give a reason for your answer.

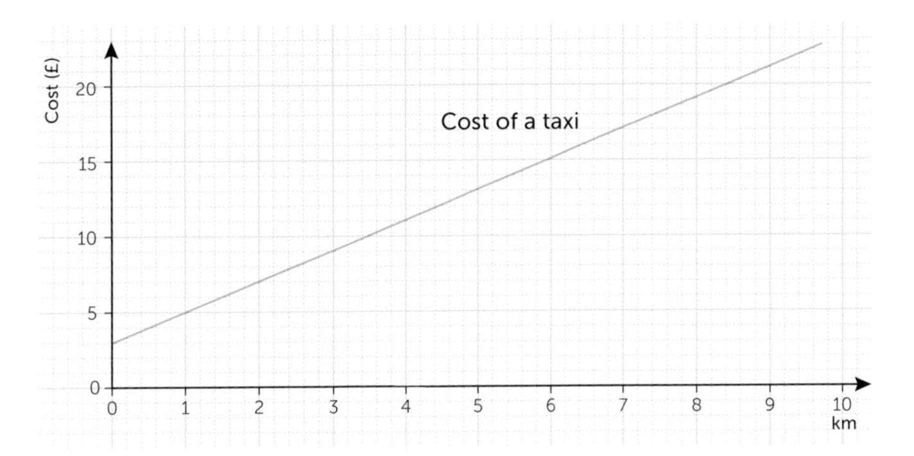


7. Helen wants to buy a new phone for her grandmother who lives in Italy.

   The phone costs €128 in Italy or £104 in the UK. Assume £5 = €6.

   Work out if it is cheaper for her to buy the phone in Italy or in the UK.

8. These two graphs show the number of students and teachers that went on a history school trip to Washington in 2019 and 2020. Use the graphs to find the ratio of students to teacher on each of the trips.

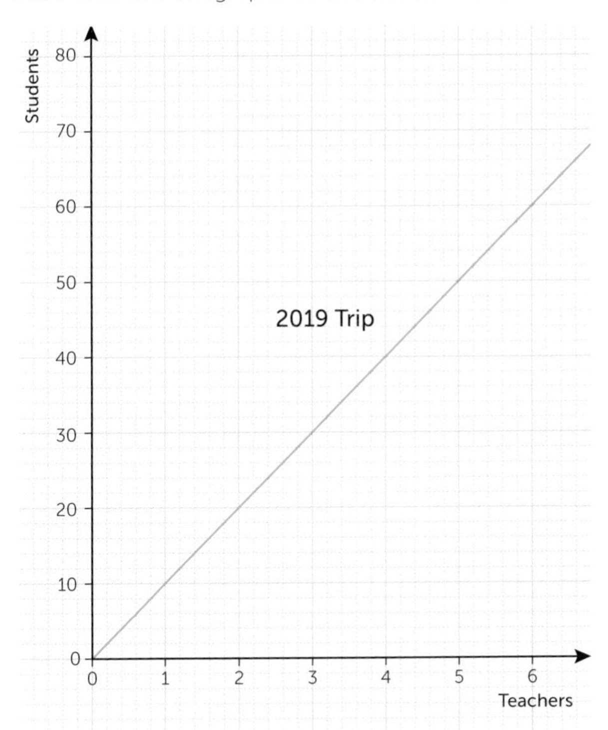


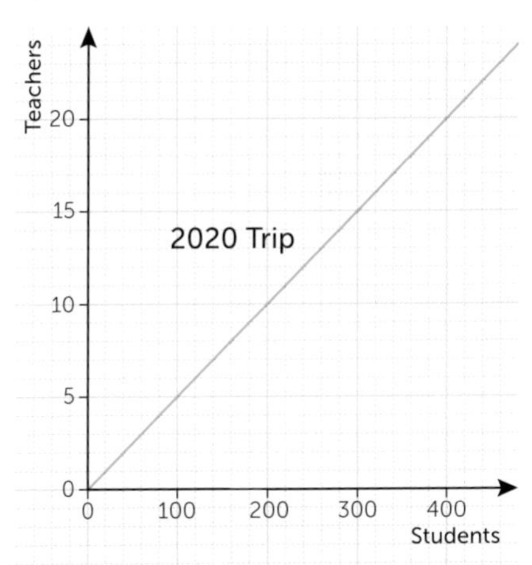


9. This is the recipe to make 6 pancakes: 2 large eggs, 120g plain flour, 250 ml milk. Jessie has 6 eggs, 300g of flour and 1 litre of milk.

   What is the greatest number of pancakes that she can make?

10. A family travelled to Barcelona where they rented a flat which cost €432.36. They hired a car for 4 days. The cost of hiring the car was €18 per day. The exchange rate is £1 = €1.14

    a) Work out the cost of the flat and car in Euros.

    b) Work out the cost of the flat and car in pounds.

11. One pint of milk in UK costs 64p. 1 litre of milk in Copenhagen costs 10 Danish krone.

    1 litre = 1.76 pints. £1 = 8.4 Danish krone. Where is the milk cheaper? Show your working.

SECTION 14
# DATA

## 14.1 CATEGORICAL AND DISCRETE DATA

S2

### Objectives

**Understand different types of data**

**Categorical data** is non-numerical and is often descriptive. For example, colours of cars.

**Discrete data** is numerical and takes distinct values. For example, number of students in a team.

**Design and use data collection sheets**

A common way of collecting data is using a tally chart.

e.g. Complete the tally chart for the following data about the number of hens' eggs stolen by a fox, over a three-week period.

Fill in the frequencies.

| | | | | | | |
|---|---|---|---|---|---|---|
| 1 | 3 | 2 | 3 | 1 | 1 | 0 |
| 2 | 2 | 0 | 1 | 1 | 2 | 2 |
| 3 | 1 | 2 | 0 | 2 | 3 | 1 |

| Eggs laid | Tally | Frequency |
|---|---|---|
| 0 | III | 3 |
| 1 | 卌 II | 7 |
| 2 | 卌 II | 7 |
| 3 | IIII | 4 |

Note that the tallies are grouped so that 卌 means five.

**Construct and interpret pictograms and bar charts.**

e.g. The following data was collected about the lunch preferences of 32 students.

| Lunch | Frequency |
|---|---|
| Sandwich | 12 |
| Chips | 6 |
| Pizza | 7 |
| Salad | 4 |
| Other | 3 |

Draw

a) A bar chart b) A pictogram

to represent this data.

**Bar chart**

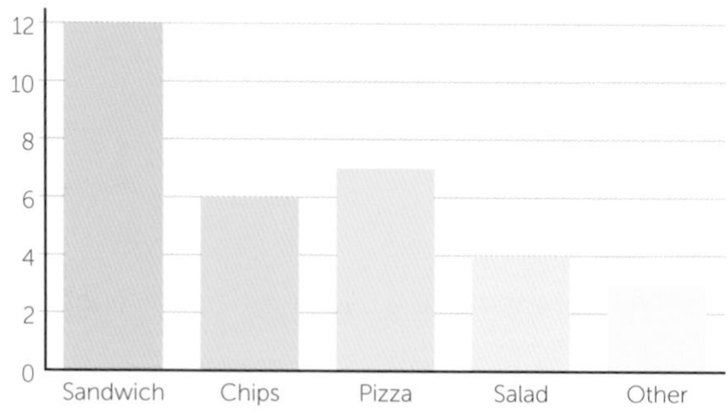


**Pictogram**

Sandwich

Chips

Pizza

Salad

Other

Key:

## Practice questions

1. The data on the right was collected about people's favourite colours:
   B = Blue, O = Orange, R = Red, P = Pink and G = Green.
   Draw and complete a tally chart for this data.

   B O R B G
   R B O R O
   O B G B R
   R G P P B

2. The data below is collected on the football team 24 people support:

   MU C A MC A MC MC MU C MU C MC
   L MC MU C MU MU MU MU L O C O

   MU = Manchester United, MC = Manchester City, C = Chelsea, A = Arsenal, L = Liverpool, O = Other
   Draw and complete a tally chart for this data. Which team was the most popular?

3. The pictogram represents the number of televisions in households:
   a) How many households had no televisions?
   b) What is the most common number of televisions in this survey?
   c) How many households were surveyed?

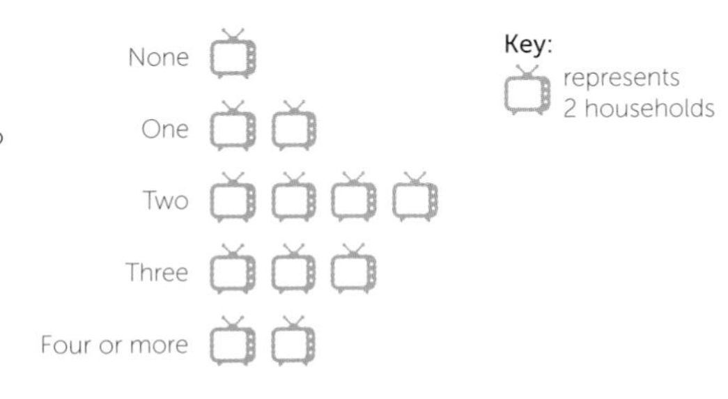


4. The bar chart shows the numbers of brothers and sisters the members of a form group have:
   a) How many members of the form have 1 brother or sister?
   b) What is the largest number of brothers or sisters that any member of the form had?
   c) How many people are in the form group?

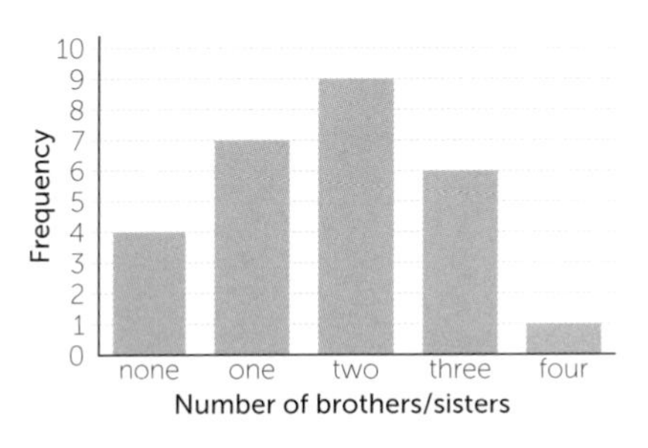


5. The tally chart represents the money to the nearest five pounds, that a group of people had in their pockets:
   Draw a pictogram of this information using the key:

| Amount (£) | Frequency |
|---|---|
| 0 | 3 |
| 5 | 6 |
| 10 | 15 |
| 15 | 12 |
| 20 | 9 |
| > 25 | 6 |

6. Jennifer is designing a sheet to collect data on pocket money received by people in her maths class; so far, she has done this:

| Name | Pocket money |
|---|---|
| | |
| | |

   Explain how Jennifer could make her sheet better.

7. The pictogram represents the hours spent shopping (to the nearest hour) by people on one day.
   a) How many people spent one hour shopping?
   b) How many people spent four or more hours shopping?
   c) How many people were surveyed about shopping?

None
One
Two
Three
Four or more

Key: represents 4 people

8. Kenny has produced and filled in the tally chart on favourite flavours of crisp:

   Suggest how Kenny could improve his tally table.

| Flavour | Tally |
|---|---|
| Plain | \|\|\|\|\|\|\| |
| Salt & Vinegar | \|\|\|\|\|\|\|\|\|\|\| |
| Cheese & Onion | \|\|\|\|\|\| |
| Smoky Bacon | \|\|\| |

9. Fifteen people donate different amounts of money to a charity.

   The bar chart below shows the amounts they put in, to the nearest pound:
   a) What was the most common amount donated by the fifteen people?
   b) What was the difference between the most donated and the least?
   c) Use the graph to calculate how much the group of fifteen people donate to the charity to the nearest pound in total.

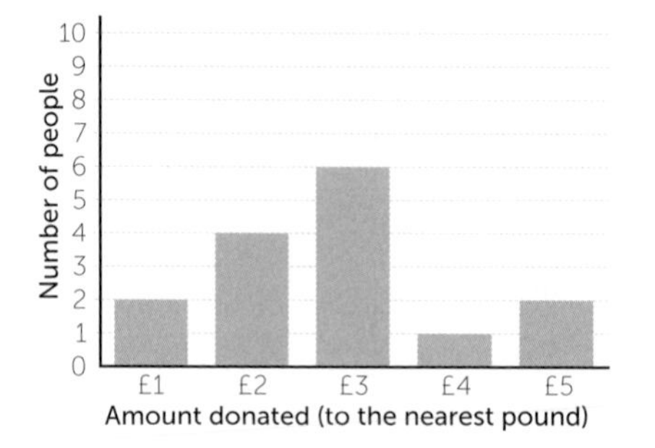


10. Gina has drawn the pictogram on the right showing how many letters all the houses in her street received on one day in a week:

    State two things that Gina needs to correct.

None
One
Two
Three
Four or more

# 14.2 DATA, CHARTS AND GRAPHS

## Objectives

### Construct and interpret pie charts

Pie charts are used to represent categorical or discrete data.

They are useful when there are just a few categories being represented.

Pie charts are best used when the data shows a clear majority in one category.

**e.g.** The frequency table shows some students' favourite sports.

a) Complete the table to find the number of degrees for each sport in the pie chart.

360° should be shared equally between

30 students: $\frac{360}{30} = 12 \Rightarrow 12°$ per student.

| Sport | Frequency | | Degrees |
|---|---|---|---|
| Football | 12 | × 12 | 144 |
| Netball | 8 | × 12 | 96 |
| Cricket | 6 | × 12 | 72 |
| Other | 4 | × 12 | 48 |
| **Total** | **30** | | **360** |

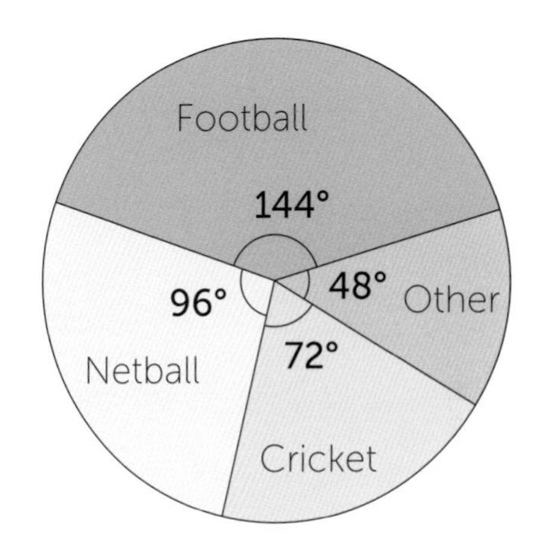


b) Draw and label the pie chart.

**e.g.** 72 teachers were asked which make of mobile phone they own.

Use the information in the pie chart to complete the frequency table below.

iPhone: The fraction of 360°: $\frac{160}{360}$ (or $\frac{4}{9}$ but it is not necessary to cancel)

This is the same fraction of 72: $\frac{160}{360} \times 72 = 32$

| Phone | Fraction of 360° | Number of people | Frequency |
|---|---|---|---|
| iPhone | 160 / 360 | $\frac{160}{360} \times 72$ | 32 |
| Samsung | 115 / 360 | $\frac{115}{360} \times 72$ | 23 |
| Nokia | 75 / 360 | $\frac{75}{360} \times 72$ | 15 |
| Other | 10 / 360 | $\frac{10}{360} \times 72$ | 2 |

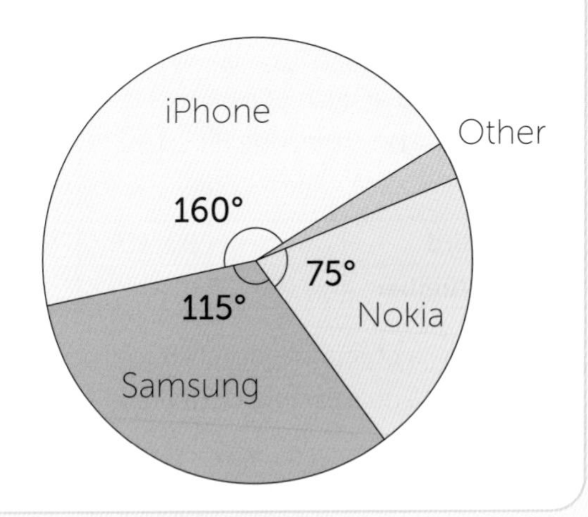


### To know which chart to use for which purpose

## To construct interpret composite and dual bar charts

These are used for comparing two (or more) sets of data.

The data sets are represented on the same axes.

**e.g.** The composite bar chart shows the social media preferences of 30 girls and 30 boys.

Use the information to draw a dual bar chart.

Reading from the composite bar chart, the following data can be found:

| | Girls | Boys |
|---|---|---|
| **WhatsApp** | 4 | 10 |
| **Instagram** | 13 | 3 |
| **Twitter** | 3 | 12 |
| **Facebook** | 10 | 5 |

The dual bar chart is represented on the bottom right.

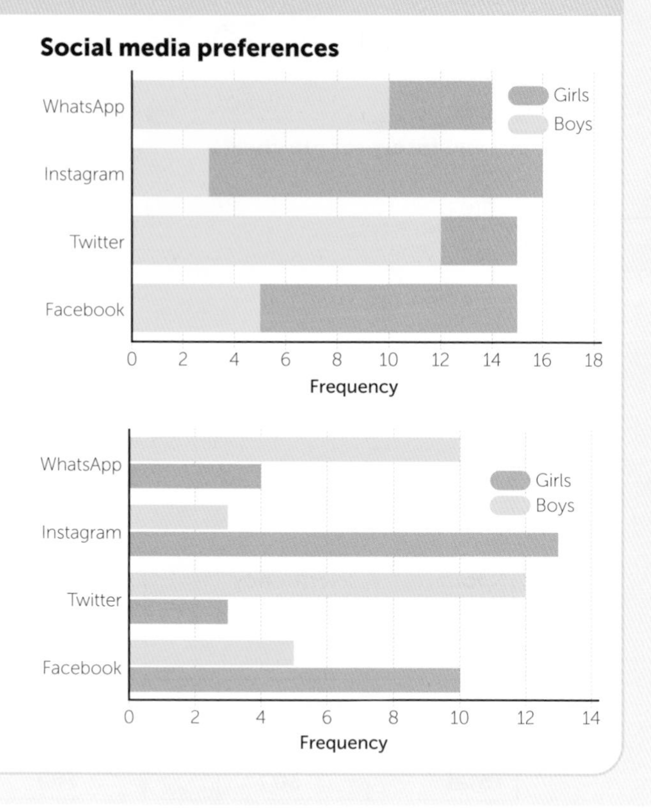


## Practice questions

1. Draw a bar chart of this information:

| **Book type** | Thriller | Romance | Biography | Historical | Fantasy |
|---|---|---|---|---|---|
| **Frequency** | 6 | 3 | 2 | 4 | 7 |

2. Gerry was asked to observe the birds that visited her bird-table in one hour; she drew a bar chart of her observations:

   a) How many birds visited the bird table altogether?

   The next day, she recorded the following:

| **Bird** | Robin | Blackbird | Starling | Blue tit | other |
|---|---|---|---|---|---|
| **Number** | 6 | 7 | 4 | 3 | 2 |

   b) Draw a composite bar chart to represent the both sets of data.

   c) Draw a dual bar chart to represent the data over two days.

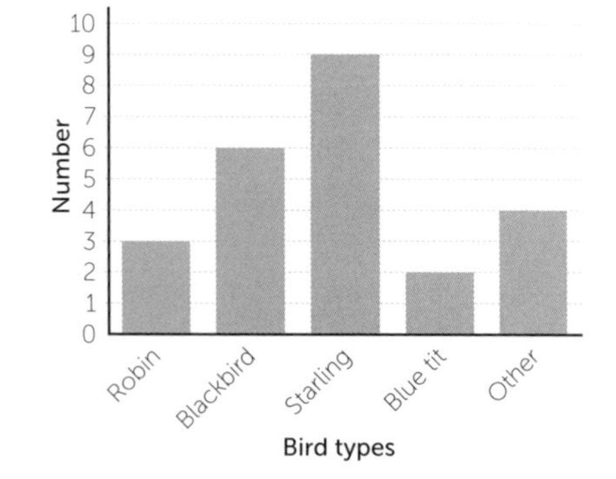


3. Louisa is drawing a pie chart to show the favourite games consoles of 40 people in her class. How many degrees will represent one person in Louisa's survey?

4. The pie chart shows how 60 students get to school:

   a) How many students came to school by bike?

   Another 12 students were interviewed and the angle for 'bike' increased to 80°

   b) How many of the 12 extra students interviewed came to school by bike?

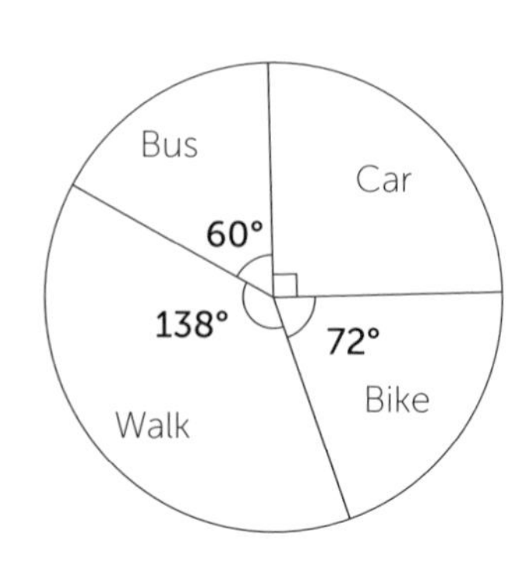

5. The dual bar chart below shows the sales of sandwiches by a kiosk one lunchtime
   a) Which sandwich filling did the kiosk sell the same amount of brown and white bread?
   b) How many more ham sandwiches on white bread were sold than on brown bread?
   c) Which filling did the kiosk sell the most of overall?
   d) How many more sandwiches on brown bread did the kiosk sell than on white bread?

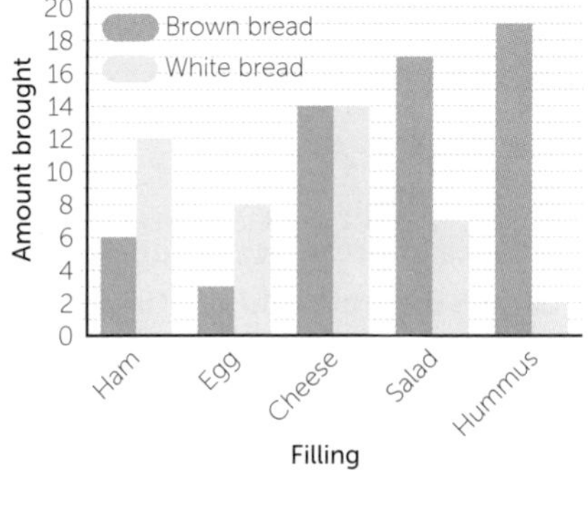

6. The pie chart opposite shows what diners in a restaurant had with the pizza they ordered:

   18 people had nothing with their pizza.

   How many people ordered pizza in the restaurant altogether?

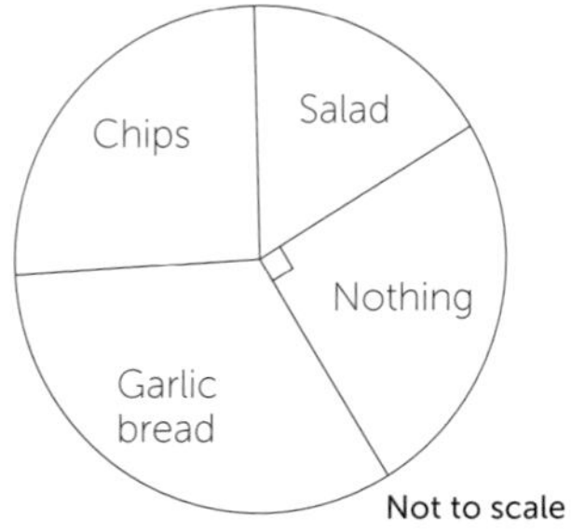

7. The pie chart opposite shows how many evenings all the members of a running club ran in one week out of a possible four sessions:

   The sector showing '4 nights' is 144° and represents 26 people.

   How many people does the pie chart represent altogether?

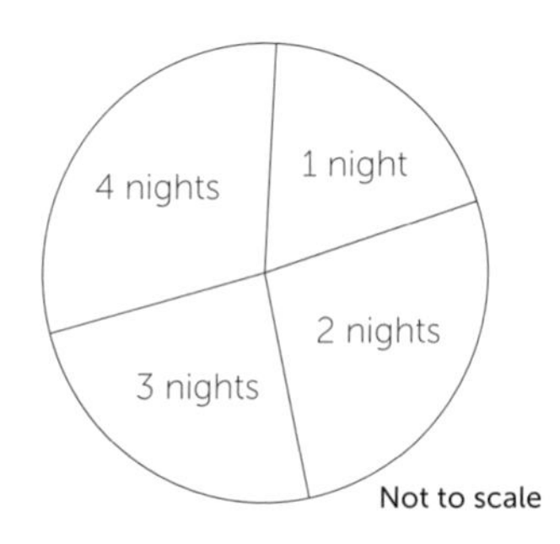

8. The pie chart opposite shows the members of a cricket club and what skill they feel they are strongest at:

   If 15 people thought that their strongest skill was bowling, how many thought their strongest skill was batting?

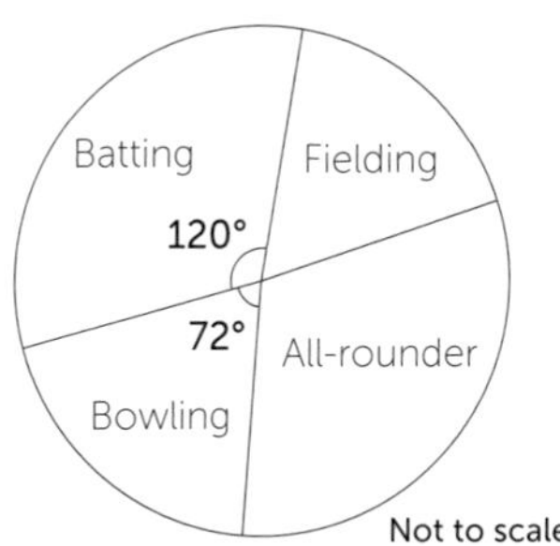


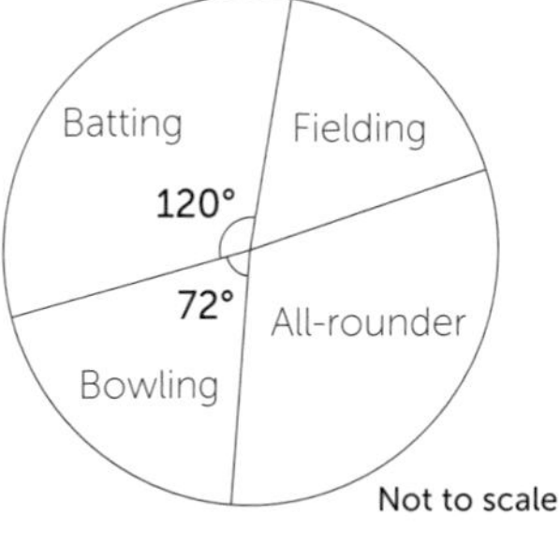

9. Two form groups were asked how they travelled to school;the results are shown in the composite bar chart.
   a) How many students walk to school in 10F?
   b) How many students from 10F and 10G ride their bike to school?
   c) How many more students get the bus in 10G compared to 10F?
   d) Which tutor group has more students?
   e) Represent this diagram as a dual bar chart.

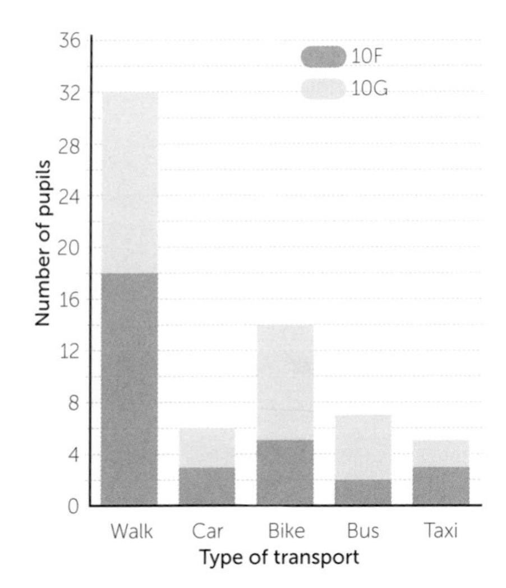


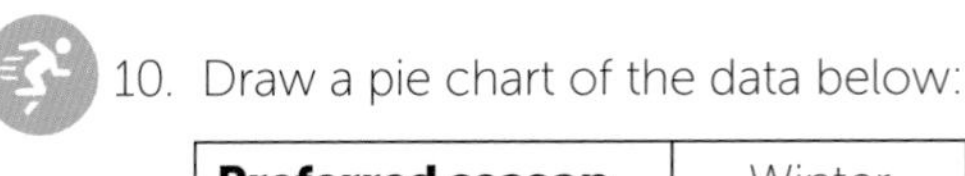

10. Draw a pie chart of the data below:

| **Preferred season** | Winter | Spring | Summer | Autumn |
|---|---|---|---|---|
| **Frequency** | 8 | 12 | 24 | 16 |

# 14.3 CONTINUOUS DATA

## Objectives

### Choose appropriate groups into which to sort discrete and continuous data

If the data values are **discrete** and thinly spread over a wide range, it is best to group them.

Ideally, groups should be of equal width, and chosen so that all data can be uniquely placed.

**e.g.** Sally records the percentages gained in a Maths test by thirty students in a grouped frequency table.
She uses the groups 0 - 20, 20 - 40, 42 - 70, 71 - 80 and 81 - 100.

State **two** things that are wrong with Sally's groupings.

Suggest a better way of grouping Sally's data.

Sally's groups are not equal in width.

There is nowhere for a score of 41 to go.

A score of 20 has two possible places to go.

A better set of groups is 0 - 20, 21 - 40, 41 - 60, 61 - 80, 81 - 100.

### Understand how to assign data values to a group

### Understand how to use inequalities for grouped continuous data

**Continuous data** values are ones that are measured, for example height, length, time.

Continuous data values are always grouped.

**e.g.** The time taken to get to school in minutes, is shown in the table.

| Time (t minutes) | Frequency |
|---|---|
| $0 < t \leq 10$ | 4 |
| $10 < t \leq 20$ | 7 |
| $20 < t \leq 30$ | 6 |
| $30 < t \leq 40$ | 3 |

Jane takes 10 minutes to get to school. Which group does Jane's data fit into? What is the maximum travelling time to school?

Jane's value fits into the $0 < t \leq 10$ group: the $0 < t \leq 10$ contains values up to, **and including** 10.
The maximum travelling time is anything more than 30, up to and including 40 minutes.

### Construct and use two-way tables for discrete and grouped data

**Two-way tables** are used to show data in an easy-to-read way.

**e.g.** The table shows the choices made by 100 students from 3 activity sessions.

| | Football | Art | Cookery | Total |
|---|---|---|---|---|
| **Male** | (15) | (18) | (13) | **46** |
| **Female** | **11** | **23** | (20) | (54) |
| **Total** | (26) | **41** | **33** | (100) |

Complete the missing values (shown here in brackets).

The key to solving this is that the total rows and columns must both add up to 100. All other values can be worked out logically.

## Practice questions

1. Create **four** groups suitable for each set of discrete data below.
   a) 17, 6, 3, 20, 5, 18, 2, 1, 3, 5, 5, 19, 19, 5
   b) Lowest value = 2, highest value = 98
   c) Lowest value = 324, highest value = 396

2. Create **four** groups suitable for each set of continuous data below.
   a) 98.3. 76.4, 22.1, 54.3, 77.3, 55. 32.1, 44.6, 88.2, 54.1, 96.4
   b) Lowest value = 7.5, highest value = 52.4
   c) Lowest value = 302.6, highest value = 319.5

3. Parveen is sorting some continuous data and has made the following groupings.

   0 - 20, 20 - 30, 30 - 40, 40 - 50, 50 - 70

   Comment on her choice. Rewrite the groups so that they are more useful.

4. Put the heights (in cm) below into the grouped frequency table:

| 155 | 161 | 173 | 148 | 149 |
|---|---|---|---|---|
| 149 | 142 | 166 | 175 | 180 |
| 170 | 155 | 167 | 140 | 152 |
| 182 | 163 | 174 | 149 | 163 |

| **Height (h centimetres)** | **Tally** | **Frequency** |
|---|---|---|
| $140 \leq h < 150$ | | |
| $150 \leq h < 160$ | | |
| $160 \leq h < 170$ | | |
| $170 \leq h < 180$ | | |
| $180 \leq h < 190$ | | |

5. Complete the two-way table below on lunches eaten by two form groups:

| | **Canteen** | **Packed Lunch** | **Total** |
|---|---|---|---|
| **10A** | 19 | | 32 |
| **10B** | | | |
| **Total** | 39 | | 62 |

   a) What fraction of 10B went to the canteen?

   b) What percentage of all students ate a packed lunch?

6. Two schools attend an art workshop, in which they can either sketch, or paint.
   School A has 35 students at the workshop.
   School B has 40 students at the workshop.
   41 students chose to do the sketching, of which 23 were from School B.
   Create and complete a two-way table for this information.
   What percentage of students from school A attended the sketching workshop?

7. 40 boys, and 40 girls attend a water sports centre and can choose one of three activities: canoeing, windsurfing, surfing. The same number of students chose windsurfing and surfing.
   26 students chose canoeing of which 12 were girls. 15 boys chose surfing.
   Create and complete a two-way table for this information.
   What percentage of those doing windsurfing are girls?

# 14.4 MEAN, MODE, MEDIAN AND RANGE

## Objectives

### Calculate and interpret the mean, mode, median and range for discrete data

There are three types of average:

**Mean**: Add up the data and divide by the number of data items.

**Median**: Order the data and find the middle item (if there are two values in the middle then find the midpoint of them).

**Mode**: Most frequently occurring data value. (If there is no data value or values that occur most frequently, then there is no mode.)

**e.g.** Find the mean, median and mode of the ages of children doing an activity:

12 15 16 14 15 12 12 16 17 14 13

**Mean**: $\frac{12+15+16+14+15+12+12+16+17+14+13}{11} = \frac{156}{11} \Rightarrow 14.2$

**Median**: Put in order: 12, 12, 12, 13, 14, **14**, 15, 15, 16, 16, 17 so the median is 14

**Mode**: 12

**Which average to use?**

The mean is used when there are no outlying values within the data set.

The mode has to be used when the values in the data set are categorical, **e.g.** eye colour.

The median is used when there are outlying (extreme) values within the data set.

- The mode is useful when the most common value or characteristic of a data set is required, **e.g.** most common shoe size or car colour.
- If there are extreme values the mean can be skewed so the median is best.

**The Range**

The range is a measure of spread, therefore shows how consistent the data is.

Find the difference between the largest and smallest pieces of data.

### Interpret and find the mean, median and range from a frequency table of discrete data

| Days food is bought | Frequency |
|---|---|
| 0 | 2 |
| 1 | 6 |
| 2 | 3 |
| 3 | 2 |
| 4 | 2 |
| 5 | 4 |

**e.g.** Above is data on how many times students bought food from the canteen in one week. For example, 2 students bought no food during the 6 days, 6 students bought food on 1 day, etc.

Calculate the mean, median, mode and range.

If the data were written out in 'raw' form:

0, 0, 1, 1, 1, 1, 1, 1, 2, 2, 2, 3, 3, 4, 4, 5, 5, 5, 5

**Mean**:

$\frac{0 \times 2 + 1 \times 6 + 2 \times 3 + 3 \times 2 + 4 \times 2 + 5 \times 4}{19}$

$= \frac{46}{19} \Rightarrow 2.4$

**Median**: There are 19 data items listed written out in 'raw' form.

The middle item is the 10th item, with a value of 2. This is the median value.

**Mode**: The group with the largest frequency is 1.

**Range**: The most days anyone bought food was 5 and the least is 0; the range is 5

## Practice questions

1. Calculate the mean to 1 decimal place, median, mode and range of each set of data below:

   a) 7, 9, 3, 6, 3, 8, 5   b 21, 20, 25, 21, 19, 19, 21, 20, 21   c) 34, 39, 41, 39, 36, 40, 35, 32, 37, 39, 34

2. Harriet is trying to calculate the mean of this set of numbers: 12, 10, 16, 13, 11

   Her calculation looks like this: 12 + 10 + 16 + 13 + 11 ÷ 5 = 53.2

   Explain what Harriet has done incorrectly and calculate the correct answer.

3. Calculate the median of each set of data below:

   a) 7, 10, 3, 5, 3, 8   b) 11, 20, 17, 15, 15, 12, 19, 21   c) 6, 9, 4, 9, 7, 10, 8, 12, 7, 9

4. Jason records the eye colours of people in his class. Which measure of average gives the best indication of a typical person's eye colour?

   Give a reason for your answer.

5. The frequency table shows the number of pieces of homework a group of students received in a week:

| **Pieces of Homework** | 3 | 4 | 5 | 6 | 7 |
|---|---|---|---|---|---|
| **Frequency** | 3 | 11 | 3 | 12 | 1 |

   Calculate the mean, median, mode and range of the number of pieces of homework received by the group.

6. Grant is hoping to average 70% in his five Maths tests this year.

   After four tests Grant is averaging 68%; what is the lowest mark Grant must achieve in his final test to finish with an average of 70%?

7. Ian, Stephanie, David and Susan collected data on how many meals out some people had in a month and collated them in the table below:

| **Number of meals out** | 0 | 1 | 2 | 3 | 4 |
|---|---|---|---|---|---|
| **Frequency** | 5 | 7 | 4 | 1 | 3 |

   Ian calculated the mean: $\frac{0 \times 5 + 1 \times 7 + 2 \times 4 + 3 \times 1 + 4 \times 3}{5} = 6$

   Stephanie said there was no mode because there was no common number.

   David said that the median was 2 because that was the group in the middle.

   Susan said that the range was $7 - 1 = 6$

   Are they correct? Explain your answer.

8. Clare and Vanessa both play cricket; their last five scores are below:

   Clare: 6, 13, 103, 142, 0 Vanessa: 38, 54, 48, 51, 39

   Both claim to have the higher average.

   They are both correct; explain how using averages.

9. A football team scored a mean of 2 goals per game in their first 4 games.

   After the 5th game their average goals per game was 3.

   How many goals did the team score in the 5th game?

10. $a$, $b$ and $c$ are all integers in ascending order.

    The median of $a$, $b$ and $c$ is 13.

    The mean of $a$, $b$ and $c$ is 15.

    The range $a$, $b$ and $c$ is 8.

    Calculate the values of $a$, $b$ and $c$.

## 14.5 ESTIMATING AVERAGES

### Objectives

**From grouped data: Estimate the mean and range and understand why these values are estimates**
**Find the modal class or group**
**Find the interval where the median lies**

The **modal class** is the group with the highest frequency. The class containing the median is the group that contains the value in the middle of the data, when it is in size order. The range can only be estimated, as the exact top and bottom values are unknown. An estimate of the range is the difference between the top of the top group and the bottom of the bottom group.

**Estimating the mean**
The mean can only be estimated because the exact values are unknown. The mid-points of each class are taken as the representative value for that group.

e.g. The table below shows the time it took 40 people to complete a puzzle:

| **Time (t mins)** | $0 < t \leq 5$ | $5 < t \leq 10$ | $10 < t \leq 15$ | $15 < t \leq 20$ | $20 < t \leq 25$ |
|---|---|---|---|---|---|
| **Frequency** | 9 | 13 | 14 | 3 | 1 |

Modal class: **$10 < t \leq 15$** (it has the highest frequency)

The median value lies between the 20th and 21st value when in numerical order.

The end of the first class interval is the 9th value. This is not far enough along the list.

The end of the second class interval is the 22nd value. This is past the median value.

The median value therefore lies in the interval **$5 < t \leq 10$**.

The estimated range: 25 – 0 = **25**

The estimated mean: $\frac{(2.5 \times 9) + (7.5 \times 13) + (12.5 \times 14) + (17.5 \times 3) + (22.5 \times 1)}{40} = \mathbf{9.25}$

### Practice questions

1. Find the midpoints of the pairs of numbers below:
   a) 10 and 20
   b) 4 and 8
   c) 0 and 30
   d) 20 and 25
   e) 15 and 30

2. The table on the right shows the times it took 20 workers to get to their office one morning:
   a) What is the modal class?
   b) In which class is the median?
   c) Find the midpoints of each group.
   d) Find an estimate for the mean.
   e) What is the maximum possible range of the data?

| **Time ($t$ minutes)** | **Frequency** |
|---|---|
| $0 < t \leq 10$ | 3 |
| $10 < t \leq 20$ | 4 |
| $20 < t \leq 30$ | 5 |
| $30 < t \leq 40$ | 7 |
| $40 < t \leq 50$ | 1 |

3. The table on the right shows the waiting times at a hospital for 30 patients:
   a) State the modal class.
   b) State the class in which the median lies.
   c) Find an estimate for the total waiting time.
   d) Find an estimate for the mean.

| **Time ($t$ minutes)** | **Frequency** |
|---|---|
| $0 < t \leq 30$ | 4 |
| $30 < t \leq 60$ | 14 |
| $60 < t \leq 90$ | 9 |
| $90 < t \leq 120$ | 3 |

4. Philip, Pascale and Matthew have been asked to look at the following set of data showing the amount of money people have in their pocket:

| **Money (£$m$)** | $0 \leq m < 10$ | $10 \leq m < 20$ | $20 \leq m < 30$ | $30 \leq m < 40$ | $40 \leq m < 50$ |
|---|---|---|---|---|---|
| **Frequency** | 6 | 7 | 4 | 1 | 2 |

a) Philip has been asked to estimate the mean. His calculations are below:

$$\frac{(10 \times 6) + (20 \times 7) + (30 \times 4) + (40 \times 1) + (50 \times 2)}{5} = 92$$

Explain why Philip must be wrong, and correct his work.

b) Pascale says that the median class is $20 \leq m < 30$ because it is in the middle. Explain what Pascale has done incorrectly.

c) Matthew says that the range of the data is estimated to be 40. Explain what Matthew could have done to get this incorrect answer.

5. A football team noted the times that they scored goals; the results are below:

| **Time ($t$ minutes)** | $0 \leq t < 15$ | $15 \leq t < 30$ | $30 \leq t < 45$ | $45 \leq t < 60$ | $60 \leq t < 75$ | $75 \leq t < 90$ |
|---|---|---|---|---|---|---|
| **Frequency** | 3 | 7 | 8 | 4 | 10 | 11 |

Estimate the mean time that the team has scored goals to 1 decimal place.

6. Kate recorded the amount she has spent at the supermarket in the last 20 weeks.

| **Money (£$m$)** | $0 \leq m < 25$ | $25 \leq m < 50$ | $50 \leq m < 75$ | $75 \leq m < 100$ | $100 \leq m < 125$ |
|---|---|---|---|---|---|
| **Frequency** | 2 | 7 | 8 | 2 | 1 |

Estimate the total amount of money Kate has spent at the supermarket in the last 20 weeks.

# 14.6 STATISTICAL DIAGRAMS AND TABLES

## Objectives

### Interpret stem-and-leaf diagrams

The stem part of the diagram is usually the left-hand group of digits, and the leaf the single, right-hand digit.
The data is distributed in numerical order

**e.g.** Represent this data in a stem and leaf diagram:
10, 17, 17, 19, 23, 23, 24, 26, 32, 33
The stem is the tens digit, and the leaves are the units (ones) digits.

| Stem | Leaves |
|---|---|
| 1 | 0 7 7 9 |
| 2 | 3 3 4 6 |
| 3 | 2 3 |

**Key**: 1 | 0 = 10

A back-to-back stem and leaf diagram displays two sets of data sharing the same stems.

**e.g.** The data shows the heights of two types of plants:
Set 1: 11, 13, 14, 20, 22, 23, 24, 24, 27, 31
Set 2: 14, 18, 21, 23, 30, 32, 33, 35, 37, 38

| Set 2 leaves | Stem | Set 1 leaves |
|---|---|---|
| 8 4 | 1 | 1 3 4 |
| 3 1 | 2 | 0 2 3 4 4 7 |
| 8 7 5 3 2 0 | 3 | 1 |

**Key**: 0 | 3 = 30 cm **Key**: 3 | 1 = 31 cm

*Note how the key is written for Set 2, for which the data is read from right to left.*

### Calculate mean, median, mode and range from bar charts

### Identify the mode from a pie chart

Averages from bar charts are calculated in the same way that the averages and range are calculated from a frequency table. The mode is the largest sector or tallest bar.

### Use stem-and-leaf diagrams to find averages and range from a set of data

**e.g.** The stem and leaf plot shows the number of people viewing an online advertisement each hour, over 15 hours.

Find the mean, median, mode and range for the data.

| Stem | Leaves |
|---|---|
| 1 | 7 9 9 |
| 2 | 5 8 8 8 [9] |
| 3 | 0 2 3 4 4 7 |
| 4 | 3 |

The mode is **28** The median is **29** (highlighted red)

The mean is $\frac{436}{15}$ = **29** The range is 43 – 27 = **26**

**Key:** 1 | 7 represents 17 people

## Practice questions

1. The bar chart shows the number of school clubs the students of 10W attend each week:
   a) Write down the modal number of clubs attended.
   b) State the greatest number of clubs attended.
   c) What is the fewest number of clubs attended?
   d) What is the range in the number of clubs attended?

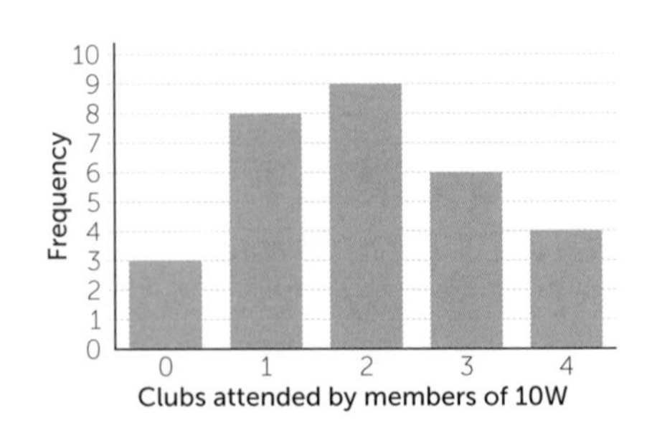


2. The data shows the ages of members of a football squad, in whole years.
   Draw a stem and leaf plot to represent the data.
   Use the key 1 | 7 to represent 17 years old

| | | | | |
|---|---|---|---|---|
| 19 | 26 | 32 | 25 | 28 |
| 20 | 17 | 36 | 41 | 29 |
| 28 | 32 | 27 | 28 | 31 |

3. The stem-and-leaf diagram shows the ages of people in a café.
   a) What is the age of the youngest person in the café?
   b) What is the age of the oldest person in the café?
   c What is the range of the ages of people?
   d) What is the modal age of people in the café?
   e) What is the median age of people in the café?

| | |
|---|---|
| 0 | 1 1 2 4 7 |
| 1 | 1 2 4 |
| 2 | 0 1 1 1 3 7 |
| 3 | 0 2 5 7 7 |
| 4 | 2 5 7 |
| 5 | 1 |

**Key:** 4 | 2 represents 42 years old

4. The bar chart shows the number of books read by twenty students:
   a) What is the modal number of books read by the students?
   b) What is the range of the number of books read?
   c) What is the median number of books read?
   d) How many books were read in total by the students?
   e) Find the mean number of books read.

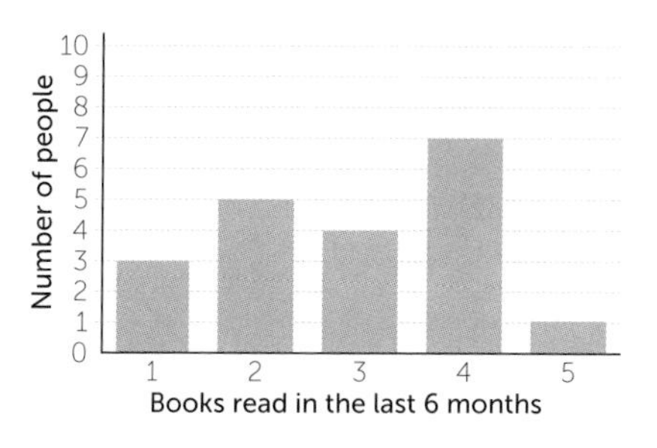


5. The back-to-back stem-and-leaf diagram shows the results of a test, out of 60 marks, for two Year 10 Maths classes, 10X and 10Y:
   a) What is the range of scores for 10X?
   b) What is the range of scores for 10Y?
   c) What was the modal score over both classes?
   d) What is the median score for 10X?
   e) What is the median score for 10Y?

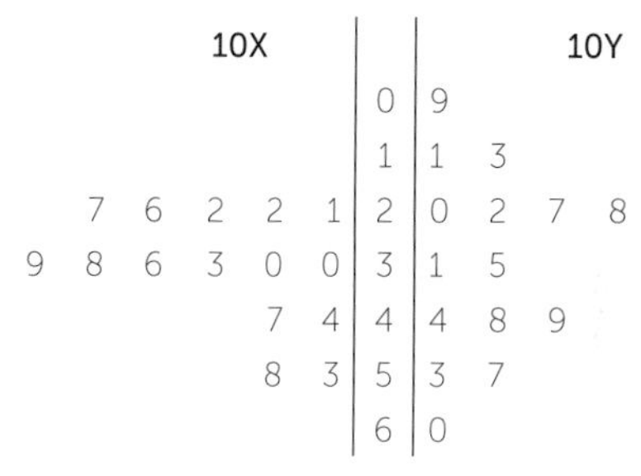

| 10X | | 10Y |
|---|---|---|
| | 0 | 9 |
| | 1 | 1 3 |
| 7 6 2 2 1 | 2 | 0 2 7 8 |
| 9 8 6 3 0 0 | 3 | 1 5 |
| 7 4 | 4 | 4 8 9 |
| 8 3 | 5 | 3 7 |
| | 6 | 0 |

Key: 1 | 2 represents 21 marks **Key:** 1 | 3 represents 13 marks

6. The bar chart shows the number of visits made by 25 people to theme parks in the last year.
   a) Find the median number of visits to theme parks.
   b) Calculate the mean number of visits to theme parks.

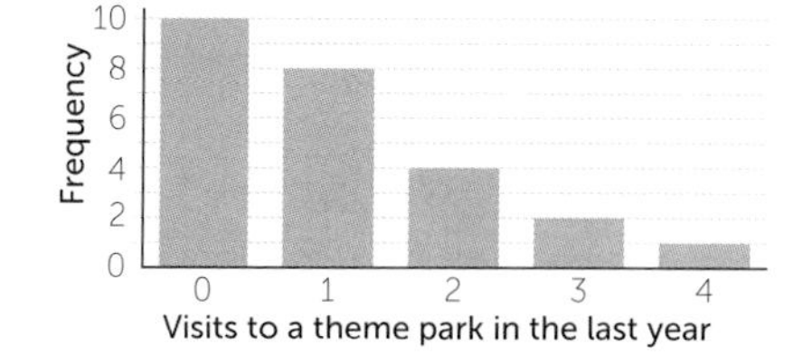

7. The stem-and-leaf diagram shows the times run in a school 100 m competition by some students, to the nearest 0.1 of a second.
   a) What was the quickest time?
   b) What was the slowest time?
   c) Find the range of the times.
   d) What was the modal time?
   e) Calculate the median time.

| | |
|---|---|
| 12 | 9 |
| 13 | 0 1 2 5 9 |
| 14 | 0 1 4 7 7 |
| 15 | 2 5 6 |
| 16 | 1 |

**Key:** 12 | 9 represents 12.9 seconds

8. The dual bar chart shows the number of brothers and sisters for two classes, 10M and 10N:
   a) Calculate the mean number of brothers and sisters for 10M to 2 decimal places.
   b) Calculate the mean number of brothers and sisters for 10N to 2 decimal places.
   c) Calculate the mean number of brothers and sisters for both classes combined to 2 decimal places

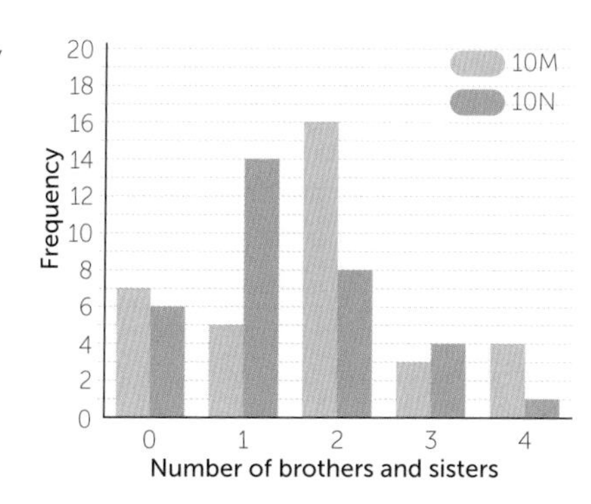

# 14.7 COMPARING POPULATIONS

## Objectives

**Compare two distributions using the mean, median, mode and range as appropriate**

In order to **compare averages**, the context of the question is key.

In general, a higher average is better but when it comes to a cost to customers or certain sports scores (such as golf), a lower one may be desirable.

When **comparing ranges**, the data set with a lower range is more reliable.

A large range shows that the data is more spread out, therefore less predictable.

**Recognise the advantages and disadvantages of different measures of average**

<table>
<tr><th></th><th>Mean</th><th>Median</th><th>Mode</th></tr>
<tr><td rowspan="3">Advantages</td><td>The most widely understood average</td><td colspan="2">Not affected by outlying data</td></tr>
<tr><td colspan="2">Has a unique value</td><td>The value always represents the data set</td></tr>
<tr><td colspan="2">Easy to find for small data sets</td><td>Can be used for categorical data</td></tr>
<tr><td rowspan="3">Disadvantages</td><td colspan="2">Does not always represent the data set</td><td>Can take more than one value</td></tr>
<tr><td colspan="2">Can only be used for numerical data</td><td>Sometimes it does not exist</td></tr>
<tr><td>Affected by outlying data</td><td colspan="2"></td></tr>
</table>

**Compare two distributions using bar charts, dual bar charts and back-to-back stem and leaf**

**Drawing conclusions**

Any conclusion must be based upon comparing averages and ranges.

**e.g.** Here is some information about the number of goals scored by two netball teams.

| | Mean | Range |
|---|---|---|
| Team A | 18.7 | 12 |
| Team B | 18.4 | 5 |

To win the league, the club has to score 15 or more goals in their next game.

Which team should the coach pick? Explain your answer.

Team B because their performance is more consistent, as their range is lower. Team A has a higher average, but their range suggests that they have more chance of scoring less than 15 goals.

## Practice questions

1. Two classes, 11A and 11B, took the same maths test. 11A had a median of 67% and a range of 32%. 11B had a median of 71% and a range of 53%.

   Make **two** comparisons of 11A and 11B's test results.

2. The frequency table shows how much it cost twenty students in 9T to equip themselves for the new school year. The minimum spent was £3 and the maximum spent was £24.00. 9U managed to equip themselves for an average cost of £12.48 with a range of £30.

   Compare 9T and 9U and how much they spent equipping themselves for the new school year.

| Amount spent by 9T (£) | Frequency |
|---|---|
| $0 < x \le 5$ | 2 |
| $5 < x \le 10$ | 6 |
| $10 < x \le 15$ | 8 |
| $15 < x \le 20$ | 3 |
| $20 < x \le 25$ | 1 |

3. The frequency tables show how much exercise two tutor groups completed in one day.

10R

| **Exercise (Mins)** | $0 \leq x < 10$ | $10 \leq x < 20$ | $20 \leq x < 30$ | $30 \leq x < 40$ | $40 \leq x < 50$ |
|---|---|---|---|---|---|
| **Frequency** | 6 | 8 | 5 | 2 | 4 |

10S

| **Exercise (Mins)** | $0 \leq x < 10$ | $10 \leq x < 20$ | $20 \leq x < 30$ | $30 \leq x < 40$ | $40 \leq x < 50$ |
|---|---|---|---|---|---|
| **Frequency** | 3 | 2 | 10 | 8 | 2 |

a) Compare the modal classes for 10R and 10S.
b) Work out the estimated means for 10R and 10S.
c) Which class, on average, does more exercise?

4. The comparative bar chart shows the pieces of homework received by members of 10F and 10G in a day.
   a) State the modal number of pieces of homework for each class.
   b) Find the median number of pieces of homework for each class.
   c) Calculate the mean number of pieces of homework for each class to 1 decimal place.
   d) Calculate the range for each class.
   e) Make two comparisons between the two classes.

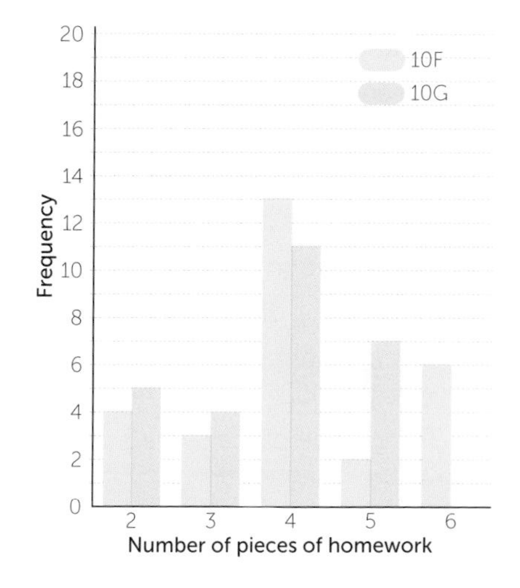

5. The back-to-back stem-and-leaf diagram shows the speeds travelled down two roads by fifteen cars:
   a) Find the median speed for each road.
   b) Find the range of the speeds on each road.
   c) The speed limit for each road is $30mph$.
   Which road should a speed camera be installed on? Explain your answer.

| A | | B |
|---|---|---|
| 8 6 | 1 | 5 6 7 7 9 |
| 9 8 5 3 1 | 2 | 0 2 3 4 4 7 |
| 9 8 6 4 2 0 | 3 | 1 4 6 |
| 4 0 | 4 | 7 |

Key: 1 | 2 represents $12mph$    Key: 1 | 6 represents 16 $mph$

SECTION 15

# PROPERTIES OF SHAPES

## 15.1 QUADRILATERALS

G4

### Objectives

**Recall and use the properties and definitions of special types of quadrilaterals**

A **quadrilateral** is a shape with four straight sides and four angles.

The four angles in any quadrilateral add to 360°, shown by drawing in one diagonal of the quadrilateral to make two triangles.

Since a triangle has angle sum 180°, the quadrilateral is 2 × 180° = 360°.

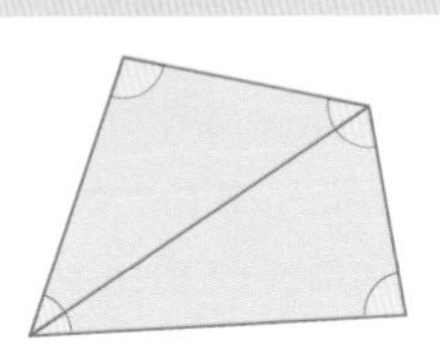

e.g. One of the smaller angles of a parallelogram is 48°.
What size is one of the larger angles?
Since the angles add to 360°, the two larger angles must be 360° – (48° + 48°) = 264°.
Hence one of the larger angles is **132°**.

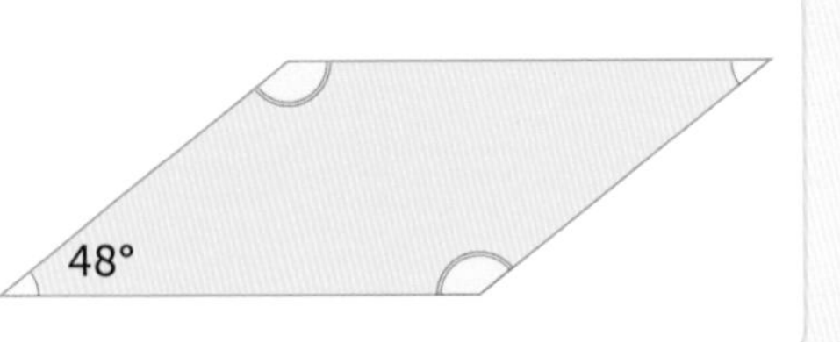


**Classify quadrilaterals by their geometric properties and name all quadrilaterals that have a specific property**

| Name | Shape | Properties |
|---|---|---|
| Square | | All four sides are equal.<br>All four angles 90°. |
| Rectangle | | Opposite sides equal.<br>All four angles 90°. |
| Rhombus | $a$, $b$, $b$, $a$ | All four sides equal.<br>Opposite sides equal.<br>Opposite sides parallel. |
| Parallelogram | $a$, $b$, $b$, $a$ | Opposite sides equal.<br>Opposite sides parallel.<br>Opposite angles equal. |
| Trapezium | | One pair of parallel sides. |
| Isosceles trapezium | | One pair of parallel sides.<br>Two pairs of equal angles. |
| Kite | $a$, $a$ | Two pairs of adjacent sides equal |

**Understand and use the angle properties of quadrilaterals including the angle sum of 360°**

Calculating areas of quadrilaterals

A rectangle of length $l$ and width $w$ has area = $l \times w$.

A parallelogram of base length $l$ and perpendicular height $h$ has area = $l \times h$.

A trapezium of parallel side lengths $a$ and $b$, and perpendicular height $h$ has area = $\frac{1}{2}(a + b)h$

A kite with diagonals of lengths $c$ and $d$ has area = $\frac{1}{2}cd$

## Practice questions

1. Write down the names of the quadrilaterals with:
   a) All side lengths the same.
   b) Opposite angles equal.

2. A rectangle has length 11 cm and width 6 cm.
   What is its perimeter and area?

3. A square has area 49 $cm^2$. What is its perimeter?

4. A parallelogram has base length 8 cm and an area of 36 $cm^2$.
   What is its perpendicular height?

5. One angle of a parallelogram is 75°.
   What sizes are the other angles?

6. A trapezium has parallel sides of 25 cm and 47 cm. Its perpendicular height is 17 cm.
   What is its area?

7. A kite has diagonal lengths of 7 cm and 14 cm.
   What is its area?

8. One of the larger angles in a rhombus is 131°.
   What is the sum of the two smaller angles?

9. A rectangle and a trapezium of the same height both have the same area. If the length of the rectangle is 24 cm, how long is the sum of the parallel sides of the trapezium?

10. In a trapezium, the two larger angles have a sum of 255°.
    What is the sum of the two smaller angles in the trapezium?

11. Ray wants to make a rectangular lawn in his garden. He wants the lawn to measure 28 m by 18 m. He buys rectangular strips of lawn which are each 9 m by 1 m.
    How many strips does he need to buy to make the lawn?

12. a) A parallelogram has sides of 8 cm and 4 cm and a perpendicular height of 3 cm. What is the area of the parallelogram?
    b) A kite has the same area as the parallelogram. One of its diagonals is 6 cm. What is the length of the other diagonal?
    c) A square is 6 times the area of the parallelogram.
    What is its side length?

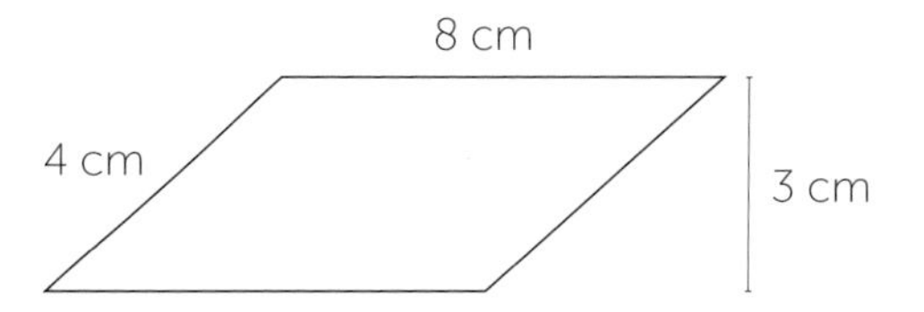

# 15.2 TESSELLATIONS

## Objectives

### Give reasons for angle calculations

Angle calculations are based around known properties.

Angles on a straight line add to 180°, angles in a triangle add to 180°, angles in a quadrilateral add to 360°, opposite/corresponding/alternate angles are equal and so on.

When solving angle problems, always give clear reasoning.

**e.g.** In the diagram, calculate the size of angles $a$ and $b$.

Give your reasoning.

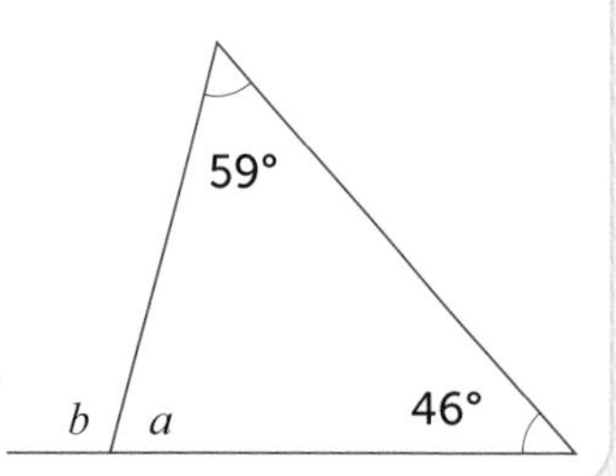


$59° + 46° = 105°$

$180° - 105° = 75°$ so angle $a$ = **75°**

Angles $a$ and $b$ are on a straight line.

$75° + b = 180°$ so angle $b$ = **105°**

### Given some information about a shape on coordinate axes, complete the shape

**e.g.** On the graph, three vertices of a parallelogram are shown.

Where could the 4th vertex be?

The 4th vertex can be at any of the following.

(1, 4), (9, 6) or (5, 0)

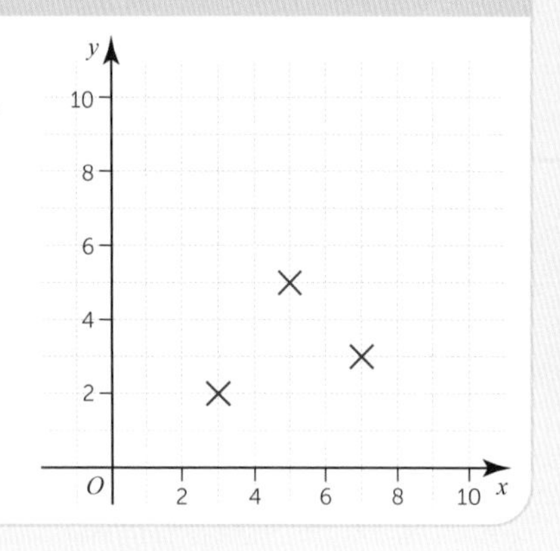


### Explain why some polygons fit together and others do not

Polygons fit together without gaps (tessellate) when the vertices that meet at a point add up to 360°.

Sum of interior angles of a polygon with $n$ sides = $(n - 2) \times 180°$

Interior angle of a regular polygon with $n$ sides = sum of interior angles ÷ $n$

**e.g.** Do regular octagons tessellate? Explain your answer.

One interior angle of a regular octagon = $\frac{6 \times 180}{8} \Rightarrow 135°$

$360 \div 135 = \frac{22}{3}$

This is not a whole number.

Regular octagons do not tessellate.

### Show step-by-step deduction when solving problems

### Use geometric language appropriately

## Practice questions

1. Do regular hexagons tessellate? How do you know?

2. Do regular pentagons tessellate? How do you know?
   Do any other polygons tessellate? What is your reasoning?

3. What are the missing angles in these diagrams?

   a)
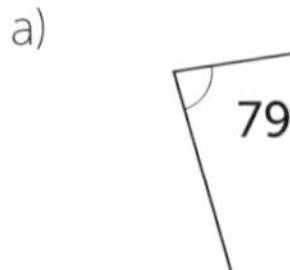
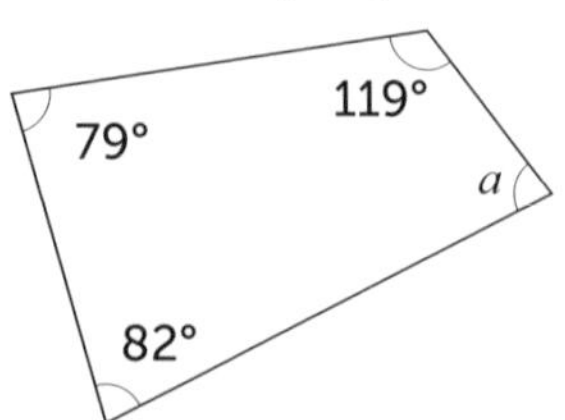


   b)
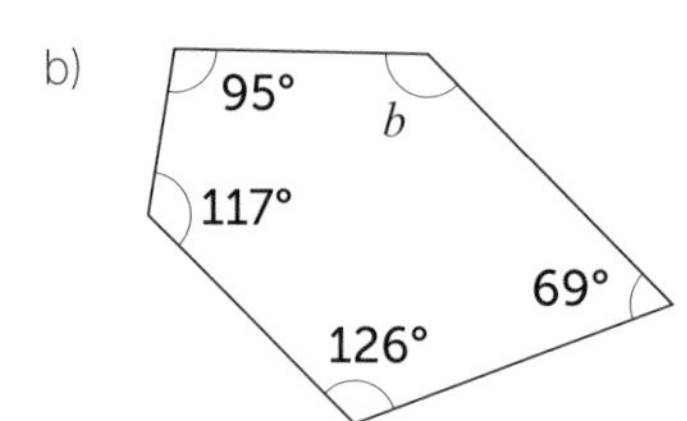

4. Susan is trying to fit some shapes together to make a floor design. She has some squares and some octagons that all have the same side lengths.
   Can Susan fit them together to make a tessellating design?

5. Ray says that any quadrilateral can form a tessellation. David says they cannot.
   Who is correct? Explain your answer.

6. On this graph, each shape needs one more point to make a polygon with rotational symmetry.
   Find where each point needs to be.

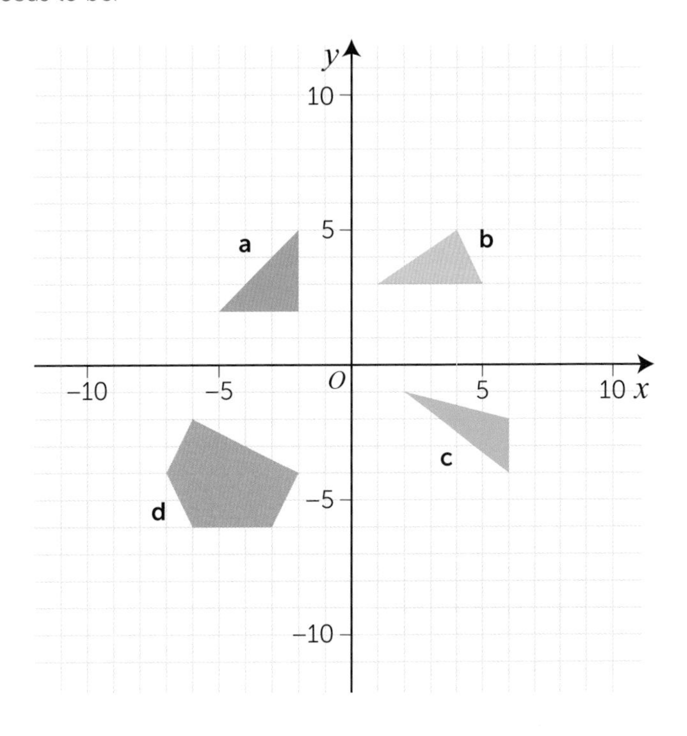


7. Add one more point to each of these shapes to make a parallelogram.

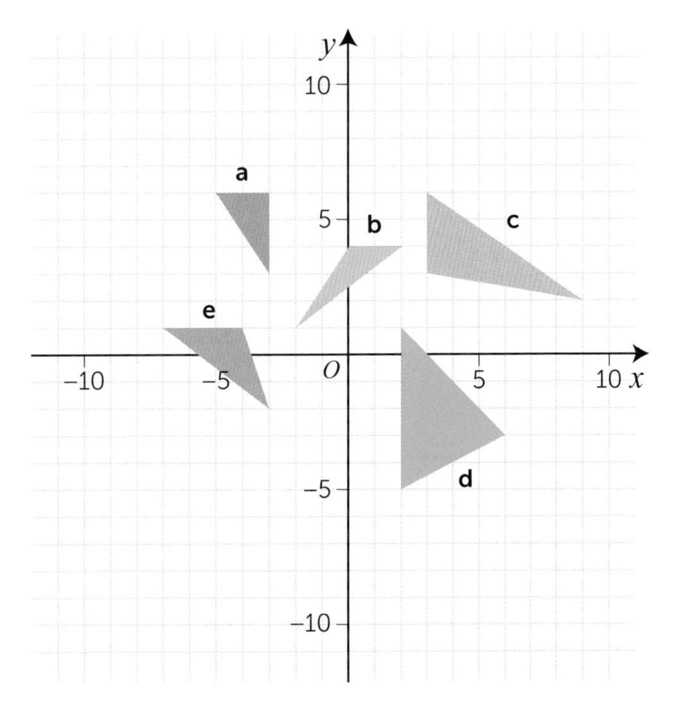

# 15.3 SOLVING GEOMETRICAL PROBLEMS

## Objectives

### Given some information about a shape on coordinate axes; complete the shape

**e.g.** The points A, B and C are three corners of a quadrilateral.
State the possible coordinates of D, the fourth corner, if the quadrilateral is:

a) a square
(3, 0)

b) a parallelogram
(7, 4), (–1, 4) or (3, 0)

c) a kite
Any coordinate on
$x = 3, y < 2$

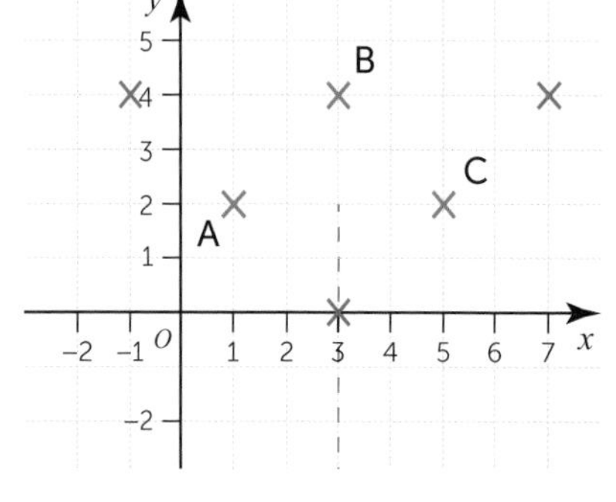


### Use geometric language appropriately

Some common examples of reasoning sentences are:

- Angles at a point on a straight line add up to 180°
- Angles around a point add up to 360°
- Corresponding angles are equal
- Alternate angles are equal
- Vertically opposite angles are equal
- Base angles of an isosceles triangle are equal
- Angles in a triangle add up to 180°

### Give reasons for angle calculations

### Show step-by-step deduction when solving problems

**e.g.** Calculate the missing angles in this triangle.

Angle $f$ = 38° (base angles of an isosceles triangle are equal)

Angle $e$ = 180° – (38° + 38°)

= 180° – 76°

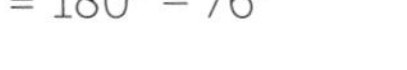

= 104° (Angles in a triangle add up to 180°)

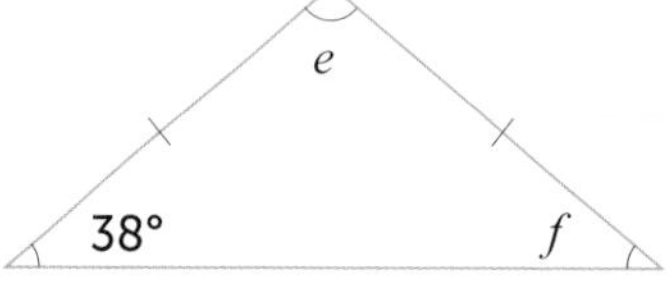


## Practice questions

1. Coordinates A, C and D show four corners of a quadrilateral.
   a) Write down a coordinate for the fourth corner, so that the quadrilateral is a square.
   b) Write down **two** possible coordinates for the fourth corner, so that the quadrilateral is a parallelogram.

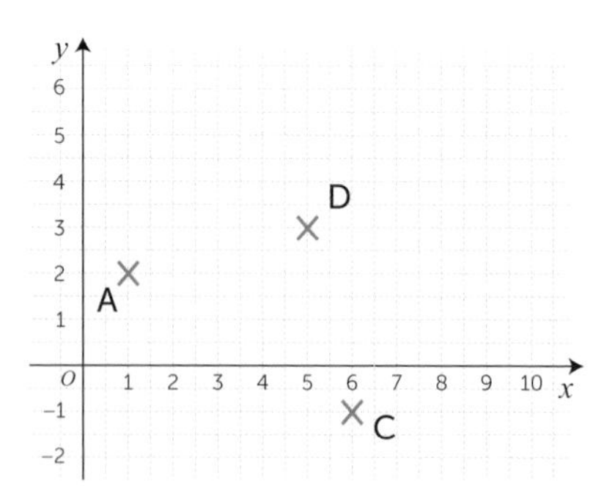


2. Find the missing angles in these triangles:

a)

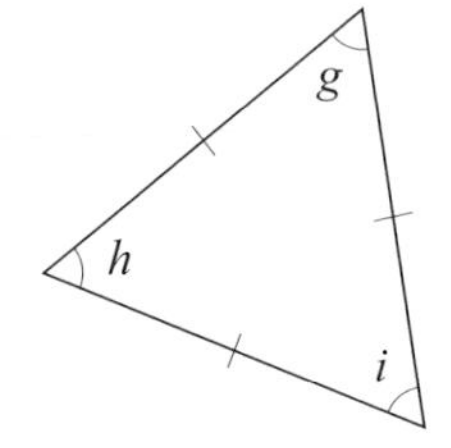


b)

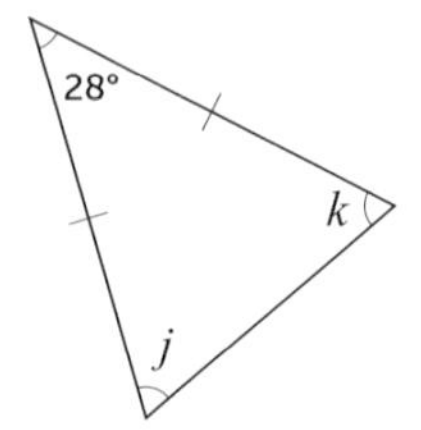

3. Calculate angles $m$ and $n$ in the two triangles shown below.

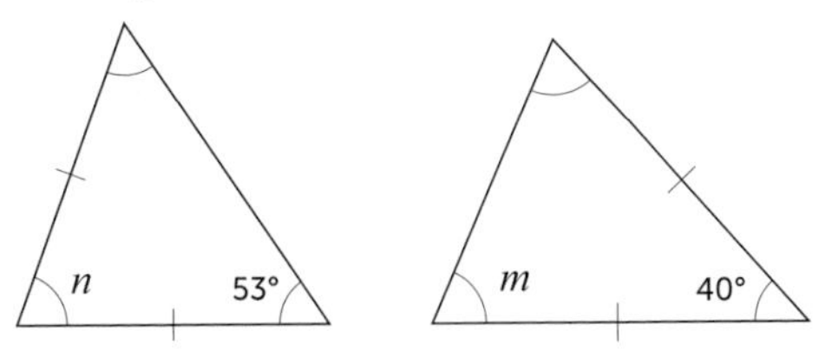

4. In the diagram, find the marked angle.
Give reasons for your answer.

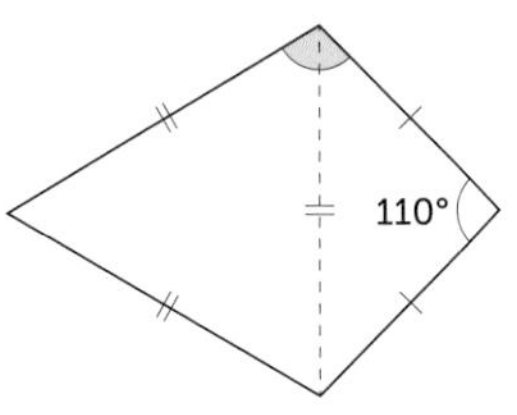

5. The diagram below shows two triangles meeting at a point on a line.

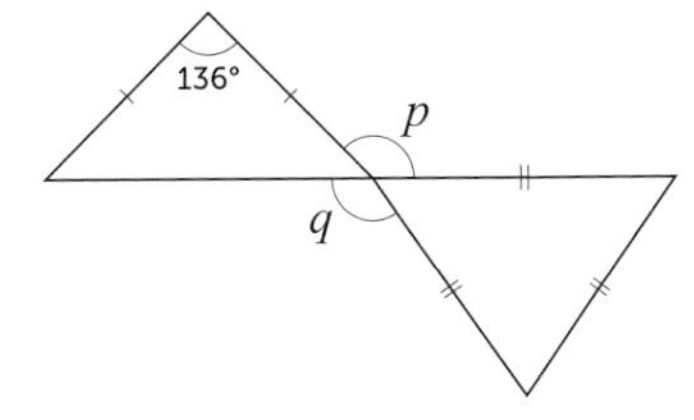


Calculate angles $p$ and $q$ from the given information.

Give reasons for your answers.

6. Calculate the angles marked with letters in these diagrams and give your reasons.

a)

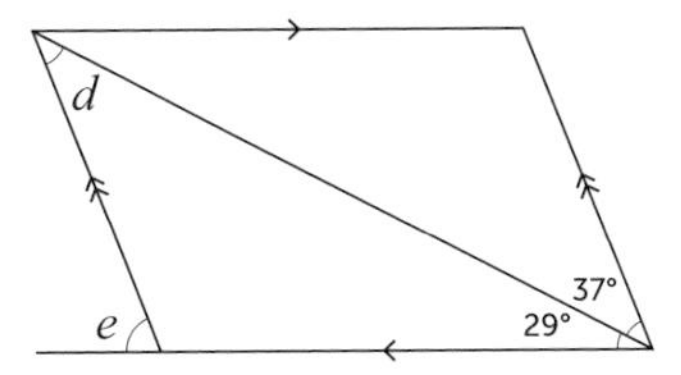

b)

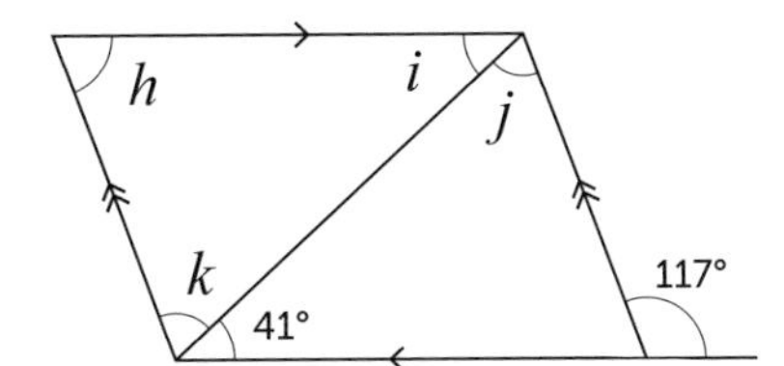

# 15.4 CIRCUMFERENCE

## Objectives

### Recall and use formulae for the circumference of a circle
### Give an answer in terms of $\pi$

A circle is a shape defined by a set of points which are all the same distance from a given point (the centre).

The circumference of a circle of radius $r = 2\pi r$, where $\pi = 3.142$ (to 3 decimal places)

The $\pi$ button on your calculator will use a value to several decimal places. $\pi$ can be left in the answer, if an exact value is required

**e.g.** A circle has a diameter of 19 cm.
What is its circumference?
Give your answer in terms of $\pi$, and then to 1 d.p.
If the diameter is 19 cm, then the radius = 9.5 cm

Circumference $= 2\pi r$
$= 2 \times \pi \times 9.5$
$= 19\pi$ cm $\Rightarrow$ 59.7 cm (1 d.p.)

**e.g.** Find the radius of a circle if its circumference is 36.3 m.

Circumference $= 2\pi r \Rightarrow r = \frac{\text{Circumference}}{2\pi}$

$r = \frac{36.3}{2\pi}$
$= 5.8$ m

### Find radius or diameter, given perimeter of a circle

### Recall the definition of a circle and identify, name and draw parts of a circle including tangent, chord, arc and segment

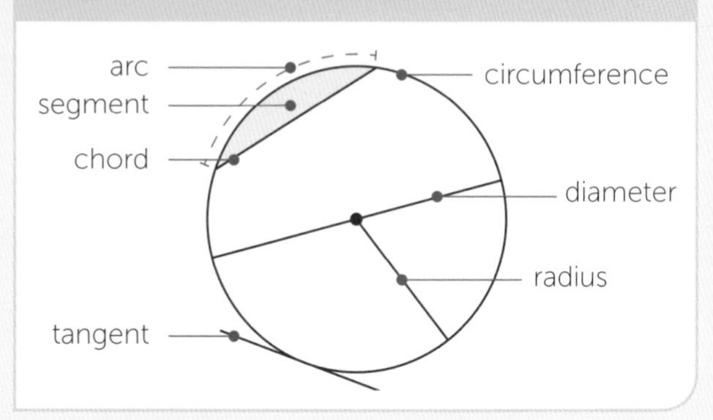


### Find the perimeters of semicircles, quarter-circles

The perimeter is the distance all the way around a shape. If part-circumferences are required, don't forget to include the other lengths that make up the perimeter.

2 cm

**e.g.** A silversmith is basing the design for a pendant on a quarter circle radius 2 cm.
Find the perimeter of the silver pendant.
Give your answer in terms of $\pi$.
If it was a full circle, circumference
$= 2 \times \pi \times 2 = 4\pi$ cm
Perimeter of pendant $= \frac{4\pi}{4} + 2 + 2$
$= (\pi + 4)$ cm

## Practice questions

*Use the $\pi$ button on your calculator in these practice questions.*

1. Calculate the circumference of circles with these radii:
   a) 7 cm  b) 25 cm  c) 16.3 m  d) 235.2 mm

2. Calculate the circumference of circles with these diameters:
   a) 12 cm  b) 45.3 m  c) 82.6 mm  d) 143.7 cm

3. If the Earth has a radius of approximately 6400 km, what is the length of the equator?

4. A tennis ball has a diameter of approximately 6.7 cm. What is its circumference?

5. What is the circumference of each of these coins?

   a) 1p coin, with diameter 2 cm

   b) 2p coin, with diameter 2.6 cm

   c) 5p coin, with diameter 1.7 cm

   d) 10p coin, with diameter 2.4 cm

6. A rope is tied 8 times around a capstan (cylindrical post). The post is 42 cm diameter.
   How long is the rope around the capstan?

7. The shooting 'circle' on a netball court is a semi-circle of radius 4.9 m.
   What is the perimeter of the semi-circle?

8. A bicycle wheel has a diameter of 630 mm including the inflated tyre.

   a) Calculate the circumference of the wheel.

   b) If the wheel turns 2400 times, what distance has it covered, to the nearest metre?

9. The radius of the stone circle at Stonehenge is approximately 15 m.
   What is the approximate perimeter of the stone circle?

10. Heather cuts a 30 cm diameter pizza into 4 quarters.
    What is the perimeter of each piece?

11. A human wrist can be taken to be roughly circular. Wrist measurements are the circumference of the arm at the wrist.

    a) What is the wrist measurement of Zach whose arm is 70 mm wide at the wrist?

    b) What is the wrist measurement of Jasmine whose arm is 55 mm wide at the wrist?

    c) If Josef has a wrist measurement of 164 cm, how wide is his arm at the wrist?

12. The shape shown is made up of a quarter circle and a semicircle.
    M is the mid-point of the radius of the larger circle.
    Find the perimeter of the shape.
    Leave $\pi$ in your answer.

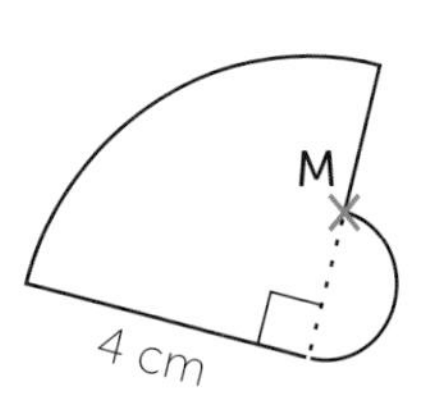

# 15.5 CIRCLE AREAS

## Objectives

### Recall and use formulae for the area enclosed by a circle
### Give an answer in terms of $\pi$

The area A of a circle = $\pi r^2$ ($r$ = radius)

The $\pi$ button on your calculator will use a value to several decimal places, but you can leave your answers in terms of $\pi$.

e.g. A circle has diameter 28 cm.

Calculate its area in terms of $\pi$

If the diameter is 28 cm, then $r$ = 14 cm

Area: $\pi r^2 = \pi \times 14 \times 14$

$= $ **$196\pi$ cm$^2$**

### Find radius or diameter, given area of a circle

e.g. A circle has area 34.2 mm$^2$. Calculate its diameter.

Area of circle $= \pi r^2$

$\Rightarrow \quad r = \sqrt{\frac{\text{Area}}{\pi}}$

$r = \sqrt{\frac{34.2}{\pi}} \quad \Rightarrow$ 3.3 mm

diameter = 2 × 3.3 $\Rightarrow$ **6.6 mm (1 d.p.)**

### Find the areas of semicircles, quarter-circles

e.g. A semicircle fits into a rectangle measuring 8 m x 4 m.

a) What is the area of the rectangle?

The rectangle has area 8 × 4 = **32 m$^2$**

b) What is the area of the semicircle?

The semicircle has area $\pi \times 4 \times 4 \div 2 =$ **$8\pi$ m$^2$**

c) What is the area of the rectangle not covered by the semicircle?

Area = **$(32 - 8\pi)$ m$^2$**

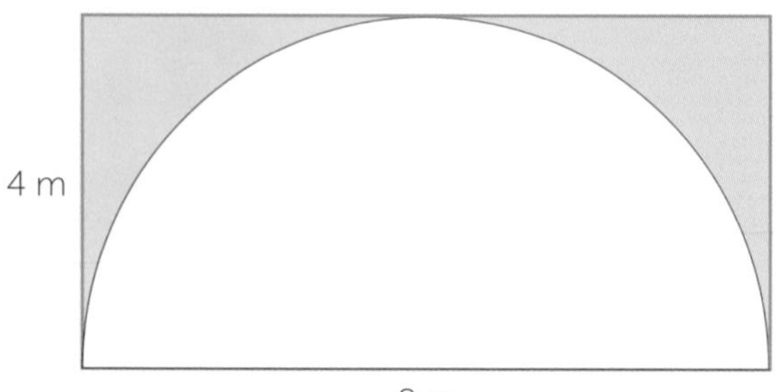


## Practice questions

*Use the $\pi$ button on your calculator in these practice questions.*

1. Calculate the area of circles with these radii, to 1 decimal place:
   a) 5.3 cm b) 23.8 mm c) 173 cm d) 35.8 m

2. Calculate the area of circles with these diameters, in terms of $\pi$:
   a) 64 cm b) 16 m c) 38 mm d) 642 cm

3. Calculate the radius of circles with these areas:
   a) 48.3 cm$^2$ b) 162.84 m$^2$ c) $81\pi$ mm$^2$ d) $729\pi$ m$^2$

4. Calculate the diameters of circles with these areas:
   a) 81.9 m$^2$ b) 327.14 mm$^2$ c) $169\pi$ cm$^2$ d) $1024\pi$ m$^2$

5. What is the area of each of these coins, to 1 decimal place?
   a) 1p coin, with diameter 2 cm
   b) 2p coin, with diameter 2.6 cm
   c) 5p coin, with diameter 1.7 cm
   d) 10p coin, with diameter 2.4 cm

6. What is the area of two joined semicircles, each of diameter 48 cm?

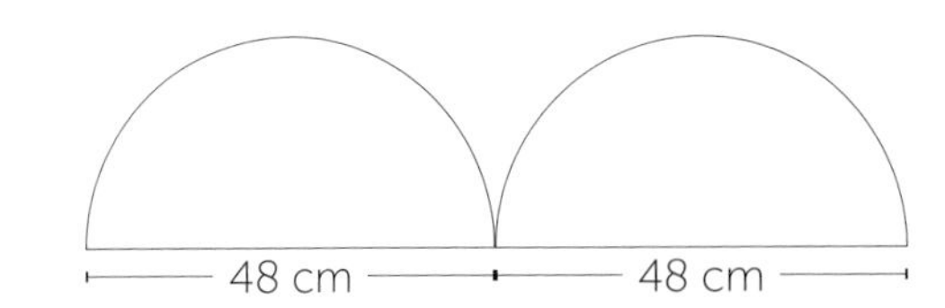

7. A circular fishpond has a circumference of 25 m. There is a circular island in the middle of the pond of radius 1 m.
Calculate the area of the surface of the water.

8. A protractor is made from a semi-circle of diameter 10 cm and a thin rectangular strip 10 cm by 5 mm.
What is the area of the protractor?
Give your answer in terms of $\pi$.

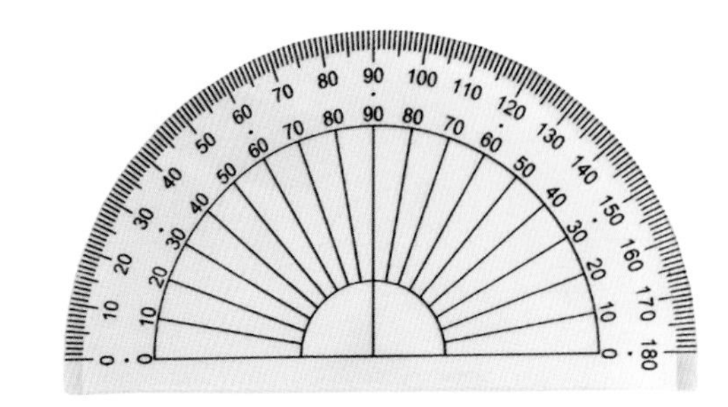

9. A circle is drawn on a square of side 8 cm so it touches the sides.
What area of the square is not covered by the circle?
Give your answer in terms of $\pi$ and in figures to 1 d.p.

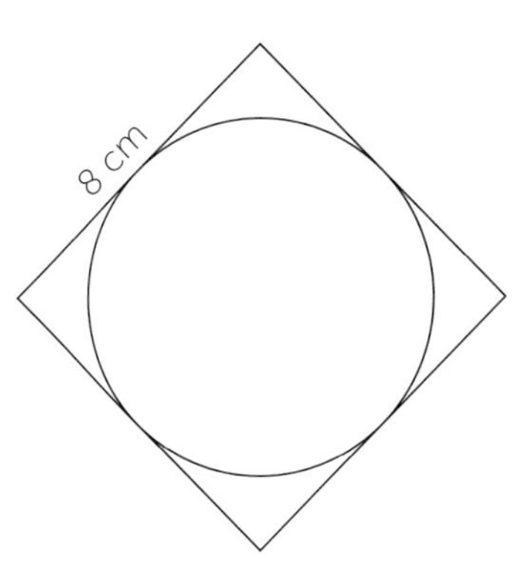

10. A rectangle 9 cm x 12 cm fits inside a circle as shown.
What area of the circle is not covered by the rectangle?
Give your answer in terms of $\pi$.

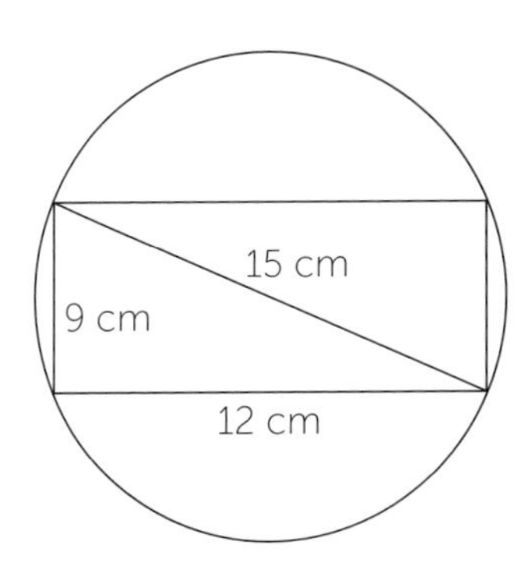

# 15.6 3D SHAPES

## Objectives

**Draw sketches of 3D solids and know the terms face, edge and vertex**

**Identify, name and know the properties of common solids: cube, cuboid, cylinder, prism, pyramid, sphere and cone**

All 3D shapes have length, breadth and height.

They are solid shapes, not flat like squares and triangles, for example.

The flat surfaces of a 3D shape are called **faces**.

Any two faces that are joined meet along an **edge**.

Where three or more edges meet at a point, it is called a **vertex** (plural vertices).

Not all 3D shapes have all flat faces; some have curved surfaces, for example cones, cylinders or spheres.

**e.g.** How many faces, edges and vertices has a cube?

Describe them.

A cube has **6** faces, all of which are squares and all the same size.

It has **12** edges, all of which are equal in length and joined at right-angles.

It has **8** vertices, with 3 edges meeting at each vertex.

## Practice questions

1. Which 3D shape has 4 triangular faces and one square face?
   Where might you see one in real life?

2. A prism has the same cross-section along its entire length, which matches the face at each end of the prism.
   Here is a pentagonal prism. It has 2 pentagonal faces and 5 rectangular faces.
   It has 15 edges and 10 vertices.
   a) Sketch a triangular prism and state how many faces, edges and vertices it has.
   b) Sketch a hexagonal prism and state how many faces, edges and vertices it has.

3. Sketch a cylinder. How many faces does it have?
   What shape are they?
   What other surfaces does a cylinder have?

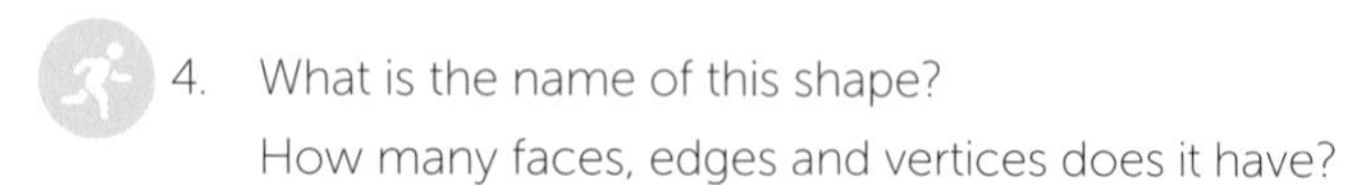

4. What is the name of this shape?
   How many faces, edges and vertices does it have?

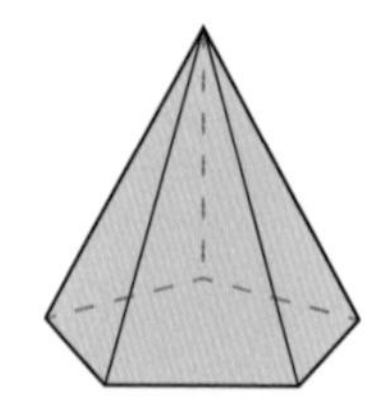

5. Four balls each of diameter 6 cm fit tightly into a box, as shown.
   a) Work out the volume of the box.
   b) Work out the surface area of the box.

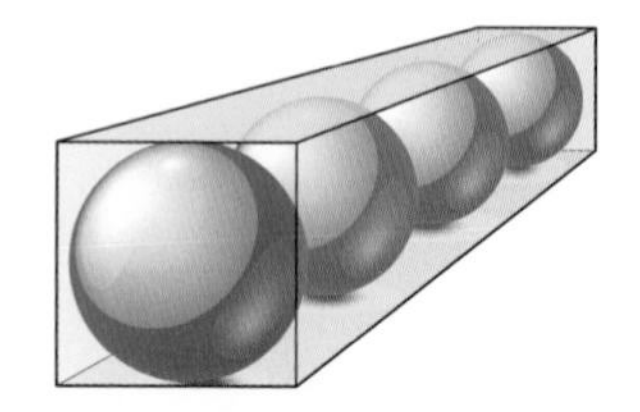

6. Complete the table below for the number of faces, edges and vertices for each shape.

| Shape | Name | Faces | Edges | Vertices |
|---|---|---|---|---|
| | Cuboid | | | |
| | Tetrahedron (Triangular-based pyramid) | | | |
| | Octagonal prism | | | |

7. Compare a cylinder, a cone and a sphere. What do they have in common? How do they differ? How would you describe their properties in terms of faces, edges and vertices?

8. a) Name a 3D shape that has more faces than vertices.
   b) Name a 3D shape with the same number of faces as a cube, but which is not a cube.
   c) Name two different 3D shapes each with an odd number of faces.

## 15.7 PLANES OF SYMMETRY AND NETS

G1 G12

### Objectives

#### Identify and sketch planes of symmetry of 3D solids

If a 3D shape can be cut into 2 symmetric pieces, it is said to have a plane of symmetry.

Some 3D shapes have several planes of symmetry.

For example, a cuboid:

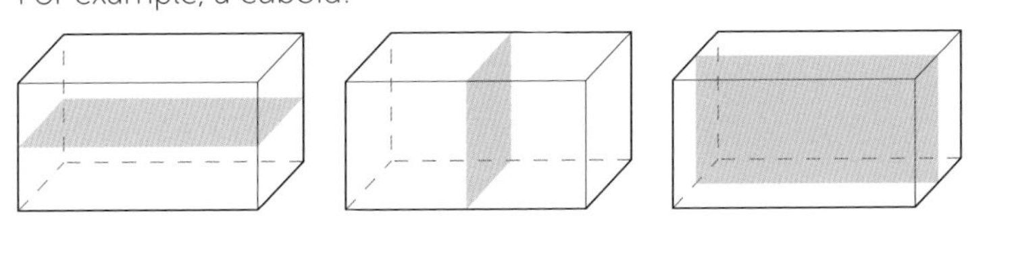

#### Sketch nets of cuboids and prisms

Many 3D shapes in everyday use can be made from a net. A net is a flat (2D) shape that can be folded into a 3D shape.

For example, here is one possible net for a cube:

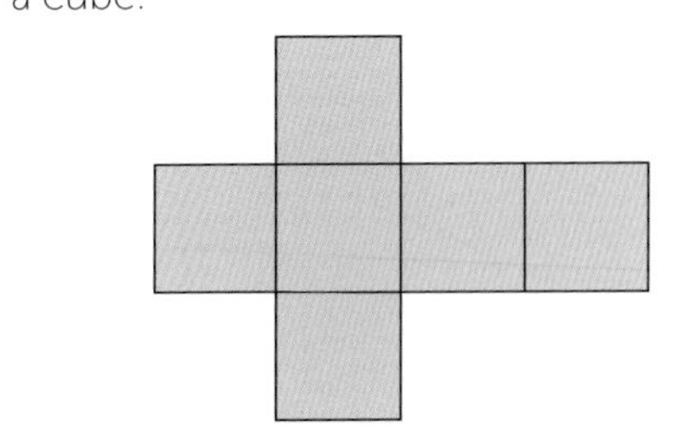

## Practice questions

1. How many planes of symmetry does each of these 3D shapes have?
   a) a cube
   b) an equilateral triangular prism?
   c) a regular hexagonal prism?
   d) a cylinder?

2. What shapes are the planes of symmetry for each of the shapes in question 1?

3. What shape is each plane of symmetry in a sphere?

4. Sketch each shape shown below and add one plane of symmetry on each shape.
   a)
   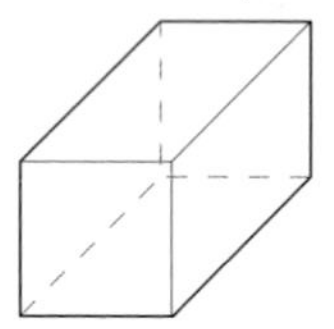
   b)
   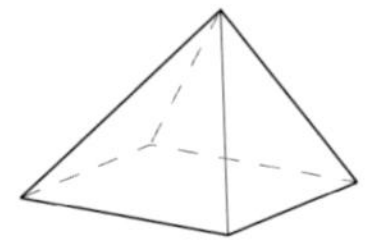
   c)
   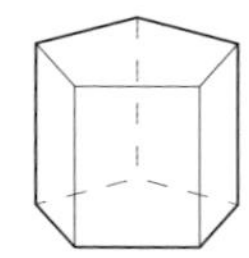
   d)

5. Sketch four different nets for a cube with side lengths 6 cm.

6. Draw an accurate net for each of these cuboids:
   a)
   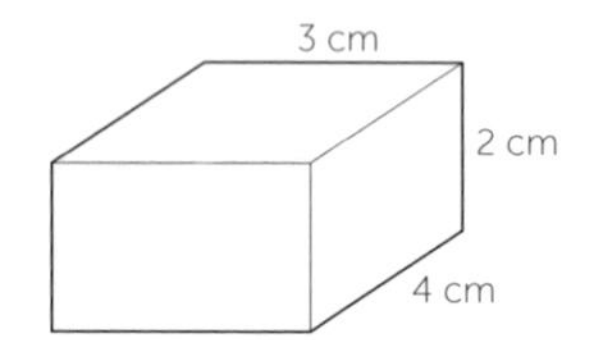

   b)
   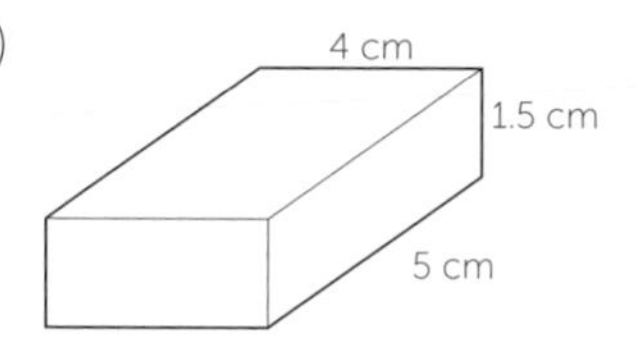


7. A square-based pyramid has base length 7cm and slanting edge length 5 cm.
   Draw an accurate sketch of a net for the pyramid.

8. A tetrahedron is made from four equilateral triangles of length 3.5 cm.
   Draw an accurate sketch of the net for the tetrahedron.

9. Draw accurate nets for these prisms:
   a)
   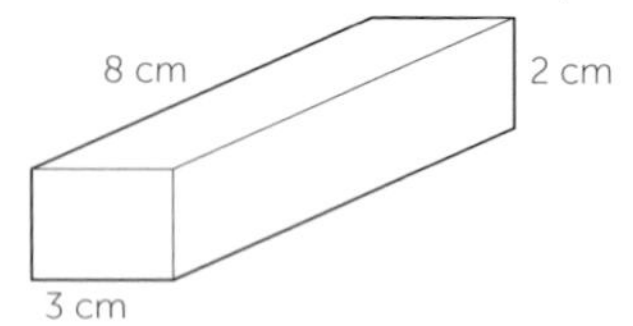

   b)
   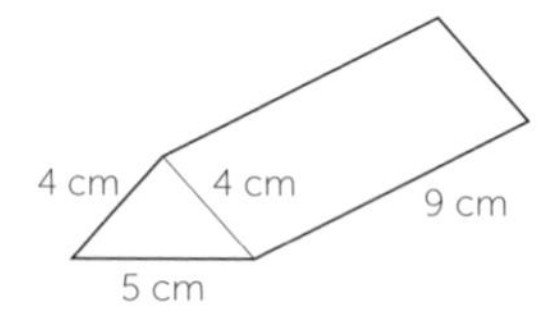

   c)
   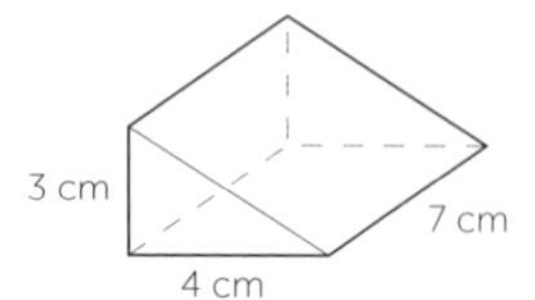

   d)
   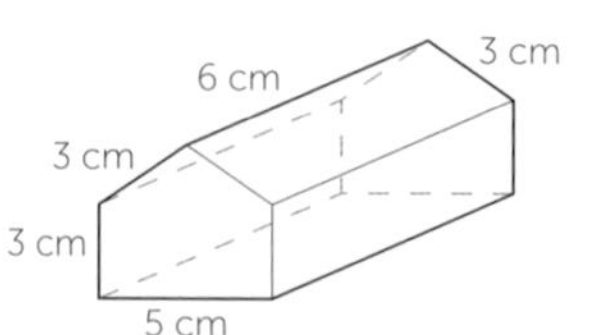

10. A triangular prism has dimensions as shown:
    a) Sketch an accurate net for the prism
    b) Measure the height of the prism
    c) What are the dimensions of the planes of symmetry of the prism?

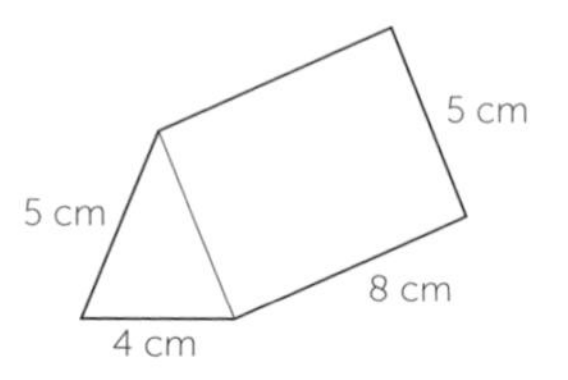

SECTION 16

# APPLICATIONS OF NUMBER

## 16.1 DEALING WITH MONEY

R9 N12

### Objectives

**Know and understand terms used in household finance: profit, loss, cost price, selling price, interest and interest rates**

**Profit** = selling price – cost price

Loss = cost price – selling price

Percentage profit = $\frac{\text{profit}}{\text{cost price}} \times 100$

Percentage loss = $\frac{\text{loss}}{\text{cost price}} \times 100$

**e.g.** Adam buys a bike for £160. He sells it a year later for £120.

Loss = £160 – £120

= £40

Percentage loss = $\frac{40}{160} \times 100 = 25$ %

**Interest** is often paid by banks on money which is deposited. Interest is usually charged by banks on money which is loaned. The **interest rate** is the percentage of the amount of money which is to be paid or charged.

**e.g.** Liam deposits £2000 in a bank account.

The bank pays 3% interest per year on deposits.

Liam leaves the money in the bank for 2 years.

Interest gained in 1 year is 3% × £2000 = **£60**

Interest gained in 2 years is 2 × £60 = **£120**

At the end of 2 years Liam has **£2120 in his account**.

This type of interest is called **simple interest**. Simple interest is only calculated on the original amount deposited.

**Solve problems set in the context of financial mathematics**

### Practice questions

1. A shirt cost £28. It is reduced in a sale by 20%. Work out the sale price of the shirt.
2. Lucy gets a 3.5% pay rise. Her salary was £38 000. Work out her new salary.
3. Stuart buys £300 worth of shares. The value of the shares increases by 22%. Work out the value of the shares now.
4. Valentina buys a house for £300 000 and sells it later for £330 000.
   Work out the percentage profit she makes on the house.

5. Ben sells some items at a car boot sale. Work out the percentage profit or loss he makes on each item.

| **Item** | Packet of cards | Book | Collectable football cards |
|---|---|---|---|
| **Buying price** | £1 | £6 | £3 |
| **Selling price** | 50p | £4 | £3.50 |
| **Percentage profit or loss** | | | |

6. Catriona has £6. She wants to buy a ribbon that originally costs £8 but the shop offers 20% off.
   Does she have enough money to buy the ribbon? Show your working.

7. Beth's account is £9.99 in debit. Ali's account is £9.99 in credit.
   Who has the most money in the account? Give a reason for your answer.

8. Mr Smith deposits £580 in the bank. The bank pays 4.2% interest per year.
   a) Work out the interest that Mr Smith receives in 1 year.
   b) Work out the total amount in Mr Smith's account at the end of the year.

9. Mrs Grey deposits £650 in the bank for one year. At the end of the year, there is £679.25 in total in the account.
   a) Calculate how much interest Mrs Grey got that year.
   b) Work out the interest rate which the bank pays on the account.

10. The school uniform costs £80 when new. The price of the second-hand uniform is £20. Catherine thinks that the percentage reduction in the uniform price is 25% Catherine is wrong. Explain why.

11. Kate bought 12 bottles of perfume for £185 in total and sold them all for £16.50 each.
    Work out her percentage profit or loss.

12. Ralph runs an Italian restaurant. He calculates his expenses and sales for the last two months.
    Part of his notes have been smudged.
    Give an example of what the expenses and sales could have been for December and January.

| | December | January |
|---|---|---|
| Expenses | £[illegible] | £[illegible] |
| Sales | £[illegible] | £[illegible] |
| | 32% profit | 8% loss |

13. Tom makes 100 cupcakes for the school fair. He spent £20 buying the ingredients. He sold 60% of the cupcakes for 90p each. He then reduced the price by 30% and sold the rest of them. He paid £12.50 for hiring the stall.
    Work out the percentage profit or loss that Tom made.

# 16.2 FINANCIAL MATHEMATICS

**Objectives** **Solve problems set in the context of financial mathematics**

**Know and understand terms used in household finance: interest and interest rates, simple interest, income tax and VAT**

**Simple interest** is interest paid by a bank on money deposited in a savings account. The interest is worked out as a percentage of the amount deposited into the account. The interest rate is set by the bank.

**e.g.** A bank gives a simple interest rate of 2.2% per year on a saving account.

A different bank gives a simple interest rate of 0.25% interest per month.

Joseph has £2500 savings to save over 3 years.

Bank 1: 2.2% on £2500 is 0.022 × £2500
= **£55 per year**

Bank 2: 0.25% on £2500 is 0.0025 × £2500
= **£6.25 per month** or £6.25 × 12
= **£75 per year**

**Bank 2 offers the most interest per year.**

**Income tax** is a tax paid on your earnings each year. The first £12 500 you earn is not taxed. All earnings between £12 501 and £50 000 are taxed at 20%

Earnings of £50 001 and above are taxed at 40%

**e.g.** Tony earns £58 000 in a year. How much tax does he pay?

£12 500 is not taxed.

20% tax rate:
£50 000 – £12 500 = £37 500
20% of £37 500 = £7500

40% tax rate:
£58 000 – £50 000 = £8000
40% of £8000 = £3200

Total tax paid is £7500 + £3200 = **£10 700**

**VAT** or **Value Added Tax** is a tax payable on products or services. The government fixes the rate of VAT. Assume the current rate of VAT is set at 20%.

**e.g.** A washing machine costs £350 + VAT. What is the price of the washing machine?

VAT is 20% of £350 = **£70**

Selling price of the washing machine is £350 + £70 = **£420**

## Practice questions

1. Martin invests £4500 into a savings account. The simple interest rate is 3.4% per year.
   Work out the amount of interest earned in 1 year.

2. Dania invests £250 for 4 years. The simple interest rate is 4% per year.
   How much money will she have in her account at the end of 4 years?
   A. £260 B. £40 C. £160 D. £290

3. Colin invests £1850 for 5 years. He receives £203.50 interest after 5 years.
   a) How much interest does he receive each year? b) What is the interest rate for this account?

4. Maria buys a new washing machine. The price of a washing machine is £520 plus 20% VAT.
   Which of these calculations show how to work out the price of the washing machine?

| A. 520 × 20% | B. 520 × 1.2 | C. 520 × 1.02 | D. 520 + 520 × 0.2 | E. 520 × 0.2 |
|---|---|---|---|---|

5. In a shop, the price of a camera is £80 plus VAT. Online, the same camera costs £90 plus £3.99 delivery.
   Which is the cheaper way to buy the camera?

6. Mr Kilpatrick deposits £850 in the bank. The bank pays 3.8% interest.
   a) Work out the interest that he receives in 1 year.
   b) Work out the total amount in his account at the end of 1 year.

7. Maeve earns £28 000 per year. The first £12 500 she earns is not taxed. Earnings between £12 501 and £50 000 are taxed at 20%. Tax is deducted from her pay each month.
   a) How much tax does she pay in one year? b) How much is her pay each month after tax has been deducted?

8. Javed invests some money in a bank account. Simple interest is paid at rate of 4% per year. After 2 years he earns £80 interest.
   How much money did he invest?

9. Lucy earns £62 500 per year. These are Lucy's calculations to work out how much tax she pays.
   £12 500 is not taxed.
   20% tax rate: £62 500 − £12 500 = £50 000 20% of £50 000 = £10 000
   40% tax rate: £62 500 − £50 000 = £12 500 40% of £12 500 = £5000
   Tax paid is £12 500 + £10 000 + £5000 = **£27 500 tax**
   She has made mistakes in her calculations. Find her mistakes and correct them.

10. Tom puts £1500 into an investment that pays 3.65% simple interest per year. He takes the money out of the account after 5 years and 3 months.
    How much money does he take out altogether? Give your answer to the nearest pound.
    Hint – *What fraction of the year is 3 months?*

## 16.3 SPEED

R1

R11

N13

### Objectives

**Understand and use compound measure of speed**

**Speed** is a measure of rate of change of distance over time

Speed = $\frac{\text{distance}}{\text{time}}$

The formula can also be shown in a formula triangle.

**e.g.** Lisa travelled 12 km between 11.00 and 11.30

Find her speed.

Speed = $\frac{\text{distance}}{\text{time}}$

Speed = $\frac{12}{0.5}$ = **24 km/h**

Notice how the time is put into the formula in hours

**Calculate average speed, distance and time**

**Average speed** = $\frac{\text{total distance travelled}}{\text{total time taken}}$

**e.g.** A cyclist took 4 minutes to cycle 1 km and a further 8 mins to cycle 1100 m.

Work out the average speed of the cyclist in km/h.

Total distance is 1 km + 1100 m = 2.1 km

Total time is 4 minutes + 8 minutes = 12 minutes

Convert time into hours gives: Total time is $\frac{12}{60}$ = 0.2 hour

Average speed = 2.1/0.2 = **10.5 km/h**

**Convert between metric speed units**

Units of speed can be converted using standard conversions.

**Remember**: 10 mm = 1 cm 100 cm = 1 m 1000 m = 1 km
60 seconds = 1 minute 60 minutes = 1 hour

**e.g.** Convert 15 m/s into km/h

15 m/s = 15 ÷ 1000 km/s
= 15 ÷ 1000 × 60 km/minute
= 15 ÷ 1000 × 60 × 60 km/h
= **54 km/h**

## Practice questions

1. A motorist travels 132 km in 2 hours. Work out her average speed.

2. A duck swims a distance of 21 m in 3 minutes 30 seconds.
   a) Work out the average speed of the duck in m/minute. b) Give the average speed of the duck in m/s.

3. A train travels at an average speed of 156 km/h for 2 hours 20 minutes.
   Work out the total distance travelled.

4. Convert:
   a) 15 m/s to m/minute b) 50 m/minute to m/h c) 80 m/minute to km/h
   d) 12 km/h to km/minute e) 10 km/h to m/minute

5. Which **one** is the odd one out? Give a reason for your answer.
   A. 300 m/minute B. 5 m/s C. 18 km/h D. 3 km/minute

6. Lucy and Anthon walk 200 m in 3 minutes. To find their speed in km/h they do the following calculations.

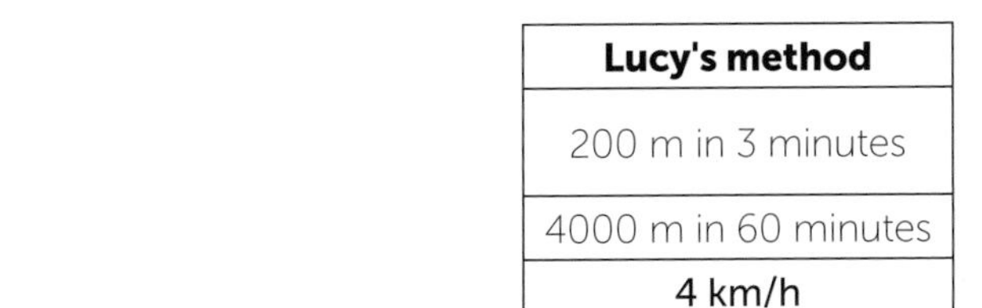

| Lucy's method |
|---|
| 200 m in 3 minutes |
| 4000 m in 60 minutes |
| 4 km/h |

| Anthon's method |
|---|
| 0.2 km $\div \frac{3}{60}$ hours |
| |
| 4 km/h |

   Explain their methods

7. Jane walks at an average speed of 4.2 km/h to her dentist 2 km away. She leaves home at 9.30 am for an appointment at the dentist at 10 am. Will she be at the dentist in time? Show your working.

8. These are the descriptions of four short journeys.

| Amal | Bob | Charlie | Dianne |
|---|---|---|---|
| 600 m in 50 seconds | 1.2 km in 2 minutes | 1 km 600 m in 2 minutes | 3.5 km in 2 minutes 50 seconds |

   Write them in order of average speed, slowest first.

9. Jake goes for a car journey. He drives at a speed of 30 km/h for the first 10 km and at 50 km/h for the next 20 km.
   Jake says that his average speed for the whole journey = (30 + 50) ÷ 2 = 40 km/h.
   Is he correct? Show your working.

10. Farah took 40 minutes to run from the library to her grandmother's house.
    Without stopping she then ran a further 5 km to the park at a speed of 3.5 m/s.
    If she reached the park at 10.50 am, what time, to the nearest minute, did she leave the library?

# 16.4 COMPOUND MEASURES

## Objectives

### Understand and use standard compound measures

A **compound measure** is **one** measure which is found from two different measures.

**e.g.** Speed, density and pressure are examples of compound measures.

**e.g.** A car uses 12 litres of petrol to drive 150 km.

Fuel consumption of the car, or km per litre is a compound measure.

The fuel consumption of the car is
150 km per 12 litres = 150 ÷ 12 = 12.5 km/litre

### Change freely between related compound units

Another compound measure is **rate of pay**.

Rate of pay $= \frac{\text{pay}}{\text{hours worked}}$

**e.g.** Tom works a total of 22 hours and gets paid £138.60

His rate of pay is $\frac{138.60}{22}$ = £6.30/hour

**Population density** is the number of people per km$^2$. This is another compound measure.

Population density $= \frac{\text{population}}{\text{area}}$

**e.g.** Manila has one of the largest population densities in the world at 41 515 people/km$^2$. Greenland has one of the smallest population densities at 0.03 people/km$^2$.

## Practice questions

1. In the supermarket, eggs are sold in a range of quantities.
   a) Work out the price per egg for each box size.
   b) Which of the quantity offers the best value for money?

| Box | Price |
|---|---|
| 6 eggs | £1.50 |
| 10 eggs | £2.30 |
| 12 eggs | £2.64 |

2. These are the amounts of text messages that four friends sent.

| Connor | Laurie | Owen | Annie |
|---|---|---|---|
| 12 text messages in 3 minutes | 11 text messages in 2.5 minutes | 4 messages in 50 seconds | 25 messages in 5 minutes |

True or false?

Owen has the fastest rate of sending texts - Laurie has the slowest.

3. The table shows the carbon dioxide emissions of three different countries.

| Country | Carbon dioxide emissions (millions of tonnes) | Population (millions) |
|---|---|---|
| A | 1230 | 110 |
| B | 680 | 79 |
| C | 639 | 70 |

Amal thinks that Country C has the lowest emission per person.

Is she correct? Show your working.

4. Crisps are sold in bags of two sizes. To work out which of the bags is the better value for money, Tia finds the price per gram of each bag of crisps. Michelle finds the mass of crisps she can buy with £1.

Use either method to work out which of the bags is better value for money.

Small bag

50 g
costs 55p

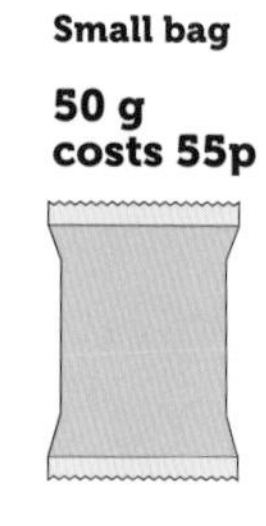

5. The table shows the population and areas of three cities.

| City | Population (millions) | Area (km²) | Number of people/km² |
|---|---|---|---|
| Tokyo | 14.1 | 2191 | |
| Berlin | 3.56 | 891.8 | |
| Rome | 4.2 | 1285 | |

a) Copy and complete the table. Round the number of people/km² to the nearest whole number.

b) Which of the cities has the lowest population density?

6. Bertie drives 234 km and uses 18 litres of fuel.

a) What is the average fuel consumption of his car in km/litre?

b) Bertie has 30 litres of petrol in his car to drive a journey of 300 km.
Based on the average fuel consumption, does he have enough fuel in his car?

7. Paul works at the local café for 18 hours per week and gets paid £117. He also works at the supermarket for 22 hours per week and gets paid £148.50. Paul wants to keep just one job; the one that pays the better rate.
Which one should he keep? Show your working.

8. Lora and Michael plant two trees in their garden. Lora's tree grows 125 cm in 4 years.
Michael's tree grows 100 cm in 3 years.

a) How many centimetres does each tree grow in 1 year?

b) How long does it take for each tree to grow 25 cm?

c) Whose tree is the fastest growing? Show your working.

9. Last week Ellie worked 5 hours each day from Monday to Friday and earned £175. This week she worked 26 hours in the week and 8 hours at the weekend. The rate of pay at the weekend is 50% more than in the week.
How much did she get paid in total this week?

10. Lucy's car can travel 230 miles on 6.7 gallons of petrol. Bob's car can travel 420 km on 25 litres of petrol.
Using the conversions: 1 mile = 1.6 km, 1 gallon = 4.5 litres, work out which car is more economical.
Show your working.

## 16.5 STANDARD FORM

N9

### Objectives

**Convert large and small numbers into standard form**

A number in **standard form** is written in the form $a \times 10^n$

$a$ is a number between 1 and 10 (but not 10) $n$ is an integer

**e.g.** $2.3 \times 10^5$ $7 \times 10^1$ $3.4 \times 10^{-2}$ are all numbers in standard form.

To convert a number into standard form, rewrite as the product of powers of 10 and a, which lies between 1 and 10

Remember: multiplying by $10^{-2}$ is the same as dividing by $10^2$

**e.g.** $4000 = 4 \times 10 \times 10 \times 10 = \mathbf{4 \times 10^3}$

$0.094 = 9.4 \div 10 \div 10 = \mathbf{9.4 \times 10^{-2}}$

**Convert numbers from standard form into ordinary numbers**

To convert from standard form to ordinary numbers, multiply or divide by the power of 10 as specified.

**e.g.** $7 \times 10^4 = 7 \times 10 \times 10 \times 10 \times 10 =$ **70 000**

$3.9 \times 10^2 =$ **390**

$6 \times 10^{-4} =$ **0.0006**

$9.4 \times 10^{-1} =$ **0.94**

**Order numbers written in standard form**

## Practice questions

1. Convert to ordinary numbers:

   a) $10^4$  b) $10^1$  c) $10^3$  d) $10^{10}$

2. Match up the numbers which are equal to each other.

| $10^{-3}$ | $10^{-1}$ | $10^{-2}$ | $10^{-4}$ |
|---|---|---|---|

| 0.0001 | 0.001 | 1 | 0.1 | 0.00001 | 0.01 |
|---|---|---|---|---|---|

3. Write each of these numbers in standard form.

   a) 500  b) 9000  c) 7500
   d) 43 000  e) 870  f) 653

4. Copy and complete each statement using numbers in the form $10^n$.

   a) $0.08 = 8 \times$ ____  b) $0.000\,9 = 9 \times$ ______  c) $0.004\,3 =$ ____ $\times 10^{-3}$
   d) $0.018\,4 = 1.84 \times$ _____  e) $0.007\,53 = 7.53 \times$ ____  f) $0.000\,030\,2 =$ ____ $\times$ _____

5. Write each of these standard form numbers as ordinary numbers.

   a) $3 \times 10^5$  b) $2.9 \times 10^4$  c) $8.07 \times 10^2$
   d) $6.873 \times 10^6$  e) $4 \times 10^{-2}$  f) $6.1 \times 10^{-3}$
   g) $5.23 \times 10^{-4}$  h) $1.105 \times 10^{-5}$

6. Which **one** in each set is the odd one out? Give a reason for your answer.

   a) $10 \times 10^2$  $4.5 \times 10^0$  $3 \times 10^7$  $2.8 \times 10^2$
   b) $3.7 \times 10^{-3}$  $2 \times 10^{11}$  $11 \times 10^{-2}$  $4.9 \times 10^{-5}$
   c) $8 \times 10^{0.5}$  $4.567 \times 10^{-3}$  $3.03 \times 10^5$  $1 \times 10^{-4}$
   d) $1.6 \times 10^4$  $3.7 \times 10^8$  $2.01 \times 10^{-3}$  $0.7 \times 10^{-2}$

7. Use >, < and = to make each of these number statements true.

   a) $0.000\,7$ ___ $7 \times 10^{-4}$  b) 360 ___ $36 \times 10^2$
   c) 52 000 ___ $5.2 \times 10^4$  d) $4 \times 10^0$ ___ $3.8 \times 10^0$

8. Place each of these sets of numbers in ascending order.

   a) $7.3 \times 10^2$  $7.5 \times 10^4$  $7 \times 10^7$  $7.1 \times 10^2$
   b) $2.2 \times 10^{-3}$  $2.05 \times 10^{-1}$  $2.6 \times 10^{-3}$  $2.8 \times 10^{-2}$
   c) $8.9 \times 10^1$  $8.7 \times 10^{-1}$  $8.79 \times 10^2$  $8.8 \times 10^{-2}$
   d) $5 \times 10^0$  $5 \times 10^{-2}$  $5.2 \times 10^3$  $5.4 \times 10^{-2}$

9. The average distance from Venus to the Sun is 108 000 000 km. The average distance from Mercury to the Sun is $5.79 \times 10^7$ km. Alice thinks that Mercury is further away from the Sun than Venus.
   Is she correct? Give a reason for your answer.

10. The table shows the dimensions of some small organisms.

    a) Which organism has the greatest length?
    b) Which organism has the smallest length?

| Organism | Length (m) |
|---|---|
| Bacteria | 0.000 002 1 |
| Virus | 0.000 000 3 |
| Atom | $1 \times 10^{-9}$ |

11. There are mistakes in each of the statements below. Identify the mistakes and correct them.

    a) $8 \times 10^{-1} = -8$  b) $3200 = 32 \times 10^3$  c) $5.3 \times 10^{-2} = 0.53$

# 16.6 CALCULATING WITH STANDARD FORM

## Objectives

### Add, subtract, multiply and divide numbers in standard form

When adding or subtracting numbers in standard form, change the numbers to ordinary numbers, add or subtract them, and then change the answer back to standard form.

**e.g.** $3 \times 10^{-2} + 4 \times 10^{-3} = 0.03 + 0.004$
$= 0.034$
$= 3.4 \times 10^{-3}$

When multiplying or dividing numbers in standard form, multiply or divide the $a$ values.
**e.g.** $a \times 10^n$
Use the laws of indices to multiply or divide the powers of 10 parts.
Check the final answer is still expressed in standard form.

**e.g.** $(4 \times 10^8) \times (5 \times 10^2) = 4 \times 5 \times 10^8 \times 10^2$
$= 20 \times 10^{10}$
$= 2 \times 10^{11}$

$\frac{6 \times 10^2}{12 \times 10^9} = (6 \div 12) \times (10^2 \div 10^9)$
$= 0.5 \times 10^{-7}$
$= 5 \times 10^{-8}$

### Know how to enter numbers in standard form into a calculator

**e.g.** To input the standard form number $7.98 \times 10^7$ into a calculator use these keys:

7 . 9 8 × $10^x$ 7 =

The number 79 800 000 will be displayed.
Alternatively, use these keys:

7 . 9 8 × 1 0 $x^{\blacksquare}$ 7

### Interpret a calculator display using standard form

## Practice questions

1. Work out each calculation **without** a calculator. Give your answers in standard form.
   a) $(4 \times 10^6) \times (2 \times 10^3)$ b) $(3 \times 10^7) \times (1.5 \times 10^2)$ c) $(4 \times 10^4) \times (3 \times 10^6)$ d) $(2.5 \times 10^3) \times (4 \times 10^6)$

2. Work out each calculation **without** a calculator. Give your answers in standard form.
   a) $(8 \times 10^9) \div (2 \times 10^2)$ b) $(5 \times 10^7) \div (2.5 \times 10^2)$ c) $\frac{8 \times 10^9}{8 \times 10^7}$ d) $\frac{1.5 \times 10^9}{6 \times 10^7}$

3. Use a calculator to work out each of these. Give your answers in standard form.
   a) $(8.3 \times 10^5) \times (1.2 \times 10^3)$ b) $(3.06 \times 10^7) \times (9.8 \times 10)$
   c) $(8.4 \times 10^{11}) \div (12 \times 10^6)$ d) $(17.375 \times 10^8) \times (1.25 \times 10^3)$

4. Spot the mistakes in these calculations and correct them.
   a) $(3 \times 10^7)^2 = 9 \times 10^9$ b) $(6 \times 10^5)^2 = 3.6 \times 10^{10}$ c) $(4 \times 10^6)^2 = 8 \times 10^{12}$ d) $(5 \times 10^3)^2 = 2.5 \times 10^5$

5. Work out each calculation without a calculator.
   Give your answers in standard form.
   a) $(6 \times 10^7) + (3 \times 10^7)$ b) $(3 \times 10^8) - (1.5 \times 10^8)$ c) $(11.3 \times 10^9) - (1.3 \times 10^9)$
   d) $(4 \times 10^4) + (3 \times 10^3)$ e) $(4 \times 10^3) - (4 \times 10^2)$

6. Match each calculation with one of the answers in the circles.

A. $(2.1 \times 10^2) + (6.9 \times 10^2)$ B. $(2.2 \times 10^4) - (1.3 \times 10^4)$ C. $(1.5 \times 10^3) \times (0.6 \times 10^1)$ D. $(2.7 \times 10^5) \div (3 \times 10)$

$9 \times 10^3$

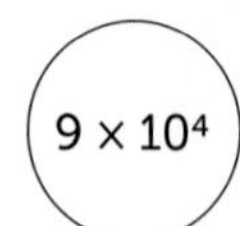


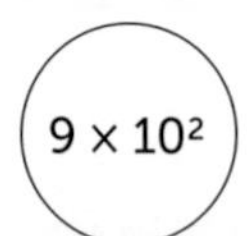


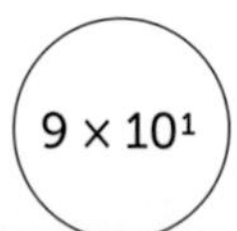


7. A 3D electron microscope magnifies objects 1 000 000 times. A virus has the length of $2.8 \times 10^{-6}$ cm. How long will the virus appear in the microscope? Give your answer in mm.

8. The base of a microchip is in the shape of a square with length of $3.55 \times 10^{-3}$ mm. Find the area of the base. Give your answer in mm$^2$.

9. This is part of Lily's work: $(6 \times 10^7) \div (2 \times 10^2) = 3 \times 10^9$
Lily has made a mistake. Find her mistake and correct it.

10. A plane travels at a speed of $1.2 \times 10^3$ km/h. The distance around the earth is 40 000 km.
How long will it take for the plane to go around the earth 50 times?
Give your answer in standard form.

11. Light travels at approximately $3 \times 10^9$ metres per second.
The speed of light is $8.8 \times 10^5$ times faster than the speed of sound.
How fast is the speed of sound? Give your answer in standard form.

SECTION 17

# FURTHER GRAPHS

## 17.1 REAL LIFE GRAPHS

R11 R14 A10 A12 A14

### Objectives

**Draw and interpret distance-time graphs**

A **distance-time** graph measures distance on the $y$-axis and time on the $x$-axis

**Find and interpret gradient as the rate of change in distance-time graphs**

**Calculate speed, distance and time from a distance-time graph**

**e.g.** This distance-time graph shows a journey in a car.

The **gradient** of the line on a distance-time graph represents the **speed**.

$$\text{Speed} = \frac{\text{Distance}}{\text{Time}}$$

**e.g.** The speed of the first part of the journey shown is

$$\frac{40}{1} = 40 \text{ km/h}$$

The speed between 1 and 1.5 hours is zero, since the line is flat

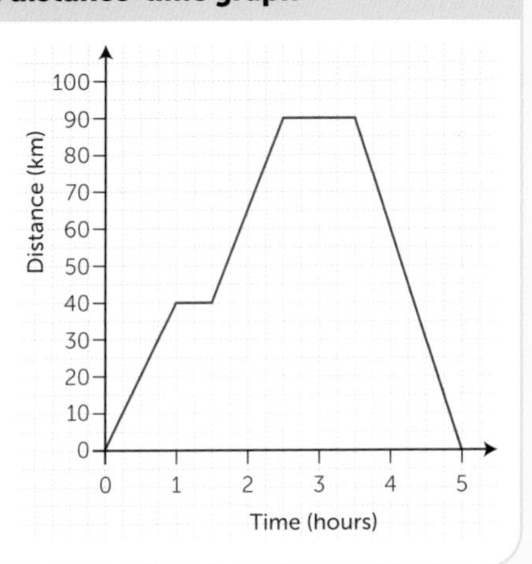


**Draw straight line graphs for real-life situations**

Some graphs show other situations from real-life.

Here are two examples.

**e.g.** This flask is filled with liquid at a constant rate.

A graph of the depth of liquid over time looks like this.

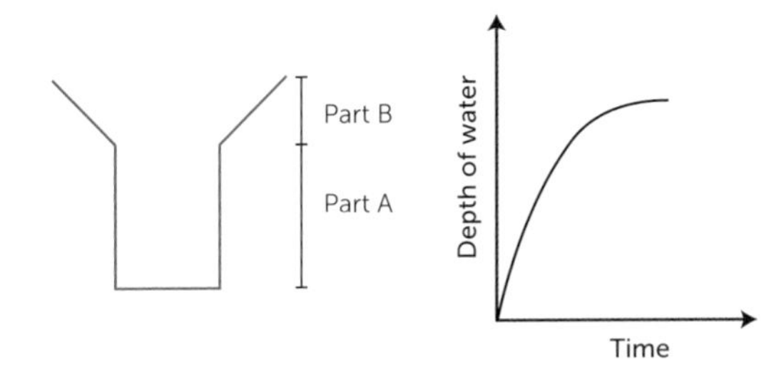


**e.g.** A delivery firm uses this graph to work out the cost of a delivery depending upon the distance travelled.

A delivery at a distance of 20 km costs £50

A delivery which costs £65 is 30 km away

The gradient of the graph gives the costs per km travelled.

The $y$-intercept of the graph gives the call-out cost (£20)

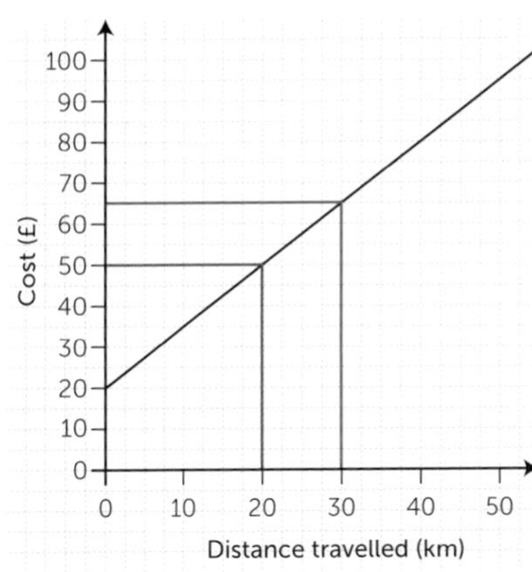

## Practice questions

1. This graph is used by a gardener to work out the cost of a job.
   a) How much does a job lasting 4 hrs cost?
   b) How much does a job lasting 7.5 hrs cost?
   c) How long should the gardener work for £130?
   d) How much does the gardener charge per hour?

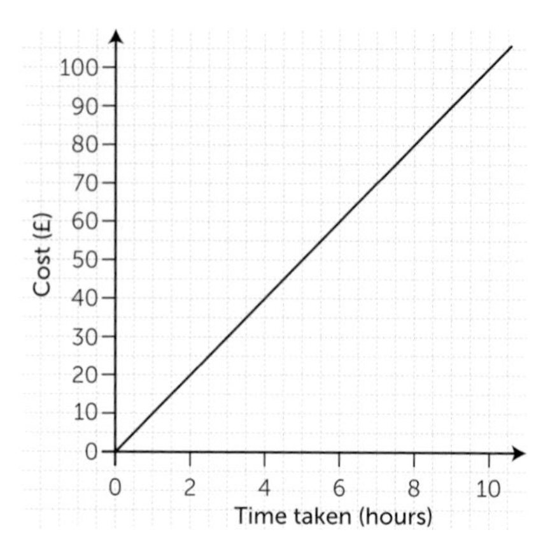


2. This distance-time graph shows a journey made by Scarlett.
   First she went to the shops, and then she went to her friend's house.
   a) What time did Scarlett set off from home?
   b) How long did it take her to get to the shop?
   c) How long did the shopping take?
   d) How far away from Scarlett's home does her friend live?
   e) What was Scarlett's speed from her home to the shop?
   f) What was Scarlett's speed from the shop to her friend's house?

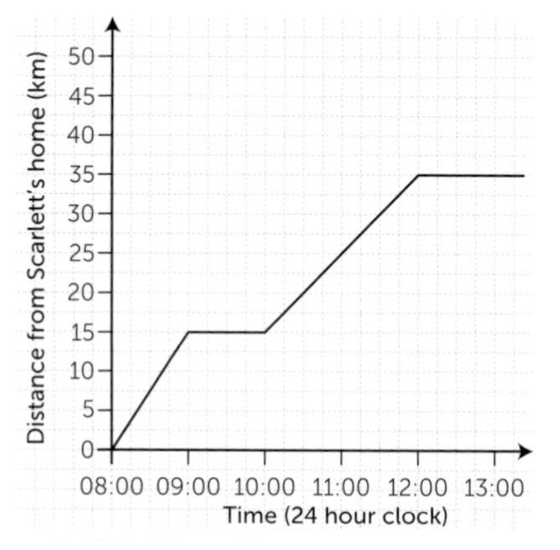


3. The cup below is filled at a constant rate with water. Three graphs are plotted to show the depth of the water over time as the cup is filled.

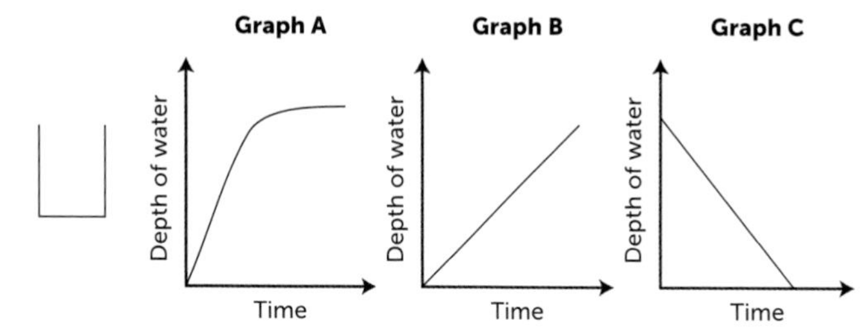


   Which is the correct graph for the cup? Give a reason for your answer.

4. This graph shows how a maintenance firm works out its bills.
   a) What is the cost of a job lasting 6 hours?
   b) A job cost a customer £45; how long did the job take?
   c) The firm has a call-out charge; how much is the call-out charge?
   d) Frank says that a job lasting 16 hours will cost double the amount a job lasting 8 hours costs.
      Explain why he is not correct.

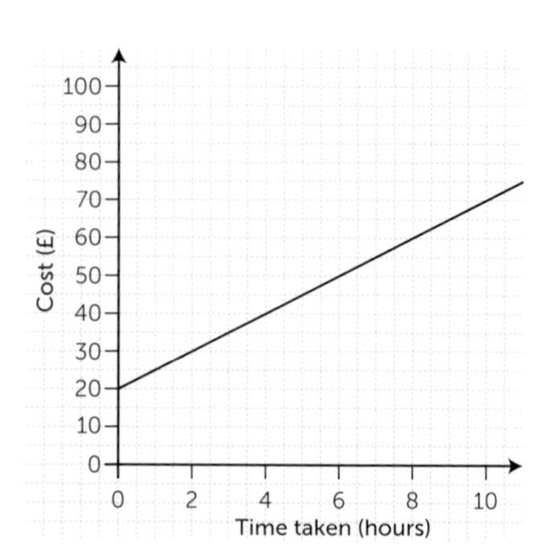

5. Ursula is filling this bowl with water at a constant rate.

   She draws a graph to show the depth of water against time as the bowl fills.

   Is her graph correct? Give a reason for your answer.

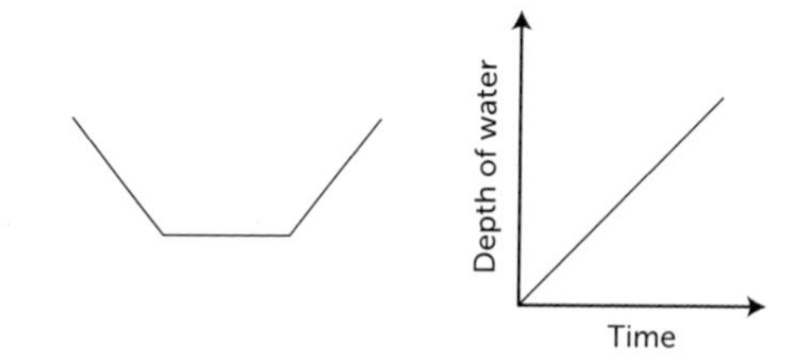


6. Gemma wants two parcels delivered. One parcel needs to travel 10 km and the other needs to travel 30 km.

   She has a choice of using two delivery firms, whose charges are shown on the graph.

   Which company should she use to send each parcel? Give reasons for your answers.

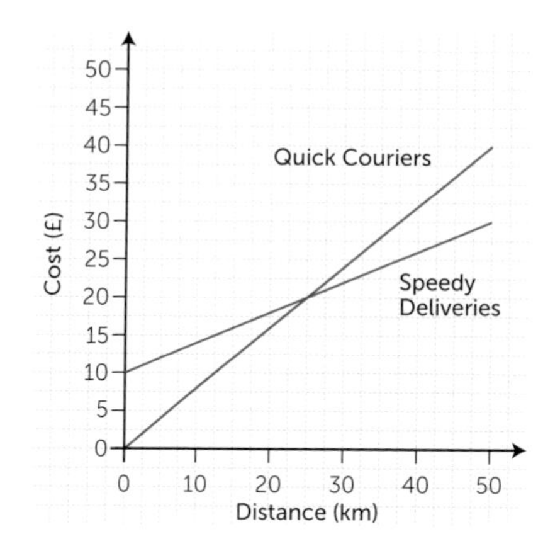

7. This distance-time graph shows a trip made by Kez to call at the shops, go to the gym and then go home.

   a) How far away are the shops from Kez's house?
   b) How long did it take Kez to get to the shops?
   c) What was his average speed to the shops?
   d) How long did he spend at the gym?
   e) What was his speed when returning home?

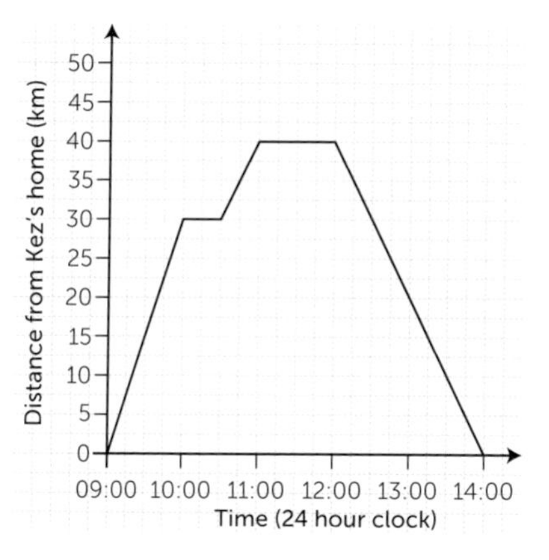

8. A company wants to draw a graph to show its customers how it calculates its bills.

   The company charges a £10 call-out fee plus £30 per hour. They produced this graph.

   Comment on the company's graph.

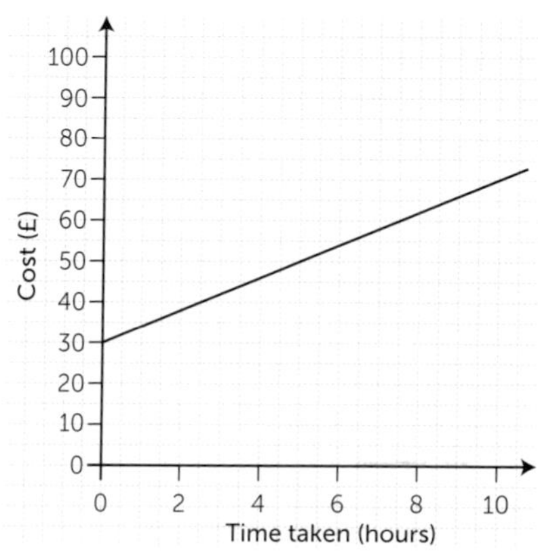

9. A paddling pool is filled with water.

   There are two hosepipes which can be used.

   Jenifer got into the pool for a while.

   Here is a graph showing the depth of the water in the paddling pool over time.

   Describe which hosepipes are used when.

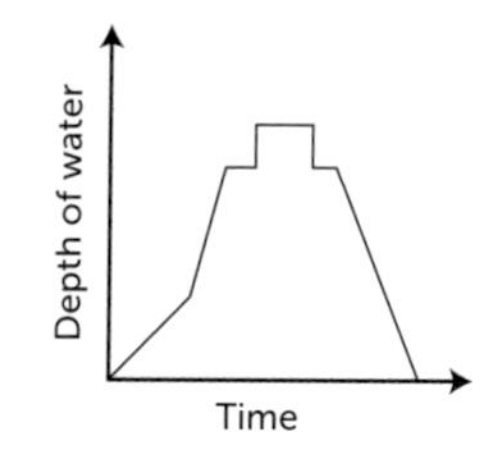

10. This distance-time graph shows part of Jo's trip out to the cinema.
    a) At what speed did Jo travel to the cinema?
    b) How long did the film last?
    c) Jo travelled back home at a constant speed of 30 km/h.
    Copy this graph and draw the final part of the journey on your graph.

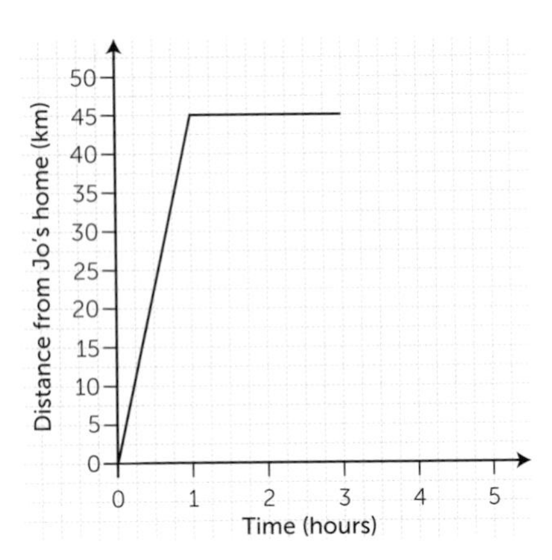


## 17.2 QUADRATIC FUNCTIONS

A11 A12 A14

### Objectives

#### Identify quadratic expressions

A quadratic expression has the **highest power** of the variable (typically $x$) squared; so, $x^2$

e.g. These are all **quadratic** expressions:

$x^2$, $3x^2 + 1$, $-2x^2$, $4x + x^2$

These are **not** quadratic expressions:

$2x + 3$, $x^3 + x^2$, $\frac{1}{x^2} + 5$

#### Generate and plot points on a quadratic curve given an equation and table

To plot a graph of a quadratic function, **substitute** values for $x$ into the function to find values of $y$.

The graph of a quadratic function is a curve called a parabola.

Take **care when substituting** into an equation; remember that squaring a negative produces a positive answer. **Brackets** can be very helpful.

e.g. Plot a graph of $y = x^2 - 5$

| $x$ | −3 | −2 | −1 | 0 | 1 | 2 | 3 |
|---|---|---|---|---|---|---|---|
| $y$ | 4 | −1 | −4 | −5 | −4 | **−1** | 4 |

For $x = -3$; $y = (-3)^2 - 5 = 9 - 5 = 4$

For $x = 2$; $y = (2)^2 - 5 = 4 - 5 = -1$

Plot the $x$ and $y$ coordinates and join them up to give the curve, which is called a parabola.

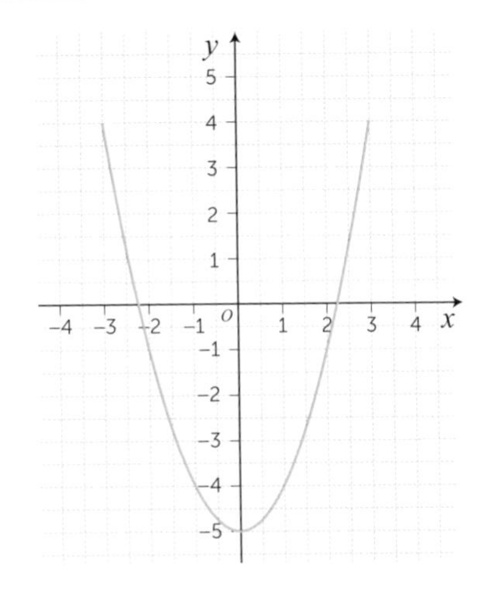


### Practice questions

1. Which of these functions are a quadratic function?

| Function A | Function B | Function C | Function D |
|---|---|---|---|
| | | | |

2. a) Copy and complete the table of values for the function $y = x^2 + 3$

| $x$ | −3 | −2 | −1 | 0 | 1 | 2 | 3 |
|---|---|---|---|---|---|---|---|
| $y$ | | | | | | | |

b) Plot a graph of this function on axes from $x = -3$ to 3 and $y = 0$ to 12

3. Match each graph with the correct quadratic function.

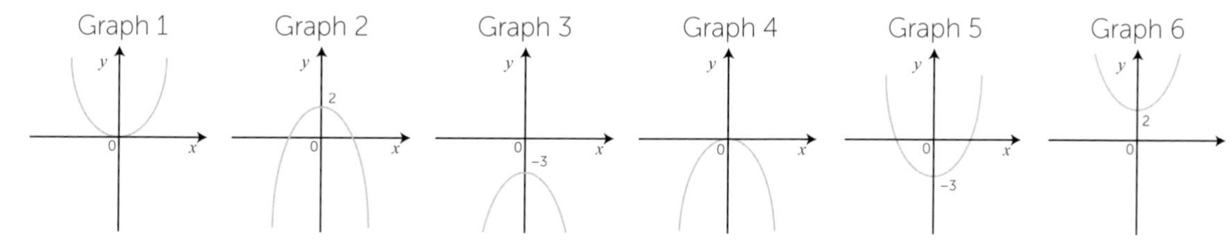


a) $y = x^2$  b) $y = -x^2$  c) $y = x^2 + 2$

d) $y = 2 - x^2$  e) $y = x^2 - 3$  f) $y = -x^2 - 3$

4. Declan wants to plot the function $y = x^2 - 4x + 1$ but has made some errors.

Here are some of Declan's workings: When $x = -2$

$$y = -2^2 - 4x -2 + 1$$
$$= -4 - 8 + 1$$
$$= 5$$

Find his errors and correct them.

5. a) Copy and complete the table of values for the function $y = 5 - x^2$

| $x$ | −3 | −2 | −1 | 0 | 1 | 2 | 3 |
|---|---|---|---|---|---|---|---|
| $y$ | | | | | | | |

b) Plot a graph of this function on axes from $x = -3$ to 3 and $y = -5$ to 5

6. a) Copy and complete the table of values for the function $y = x^2 + 2x$

| $x$ | −5 | −4 | −3 | −2 | −1 | 0 | 1 |
|---|---|---|---|---|---|---|---|
| $y$ | | | | | | | |

b) Plot the graph of this function on axes from $x = -5$ to 1 and $y = -5$ to 15

7. a) Copy and complete the table of values for the function $y = x^2 - 3x$

| $x$ | −2 | −1 | 0 | 1 | 2 | 3 | 4 |
|---|---|---|---|---|---|---|---|
| $y$ | | | | | | | |

b) Plot the graph of this function on axes from $x = -2$ to 4 and $y = -5$ to 10

8. a) Copy and complete the table of values for the function $y = x^2 + x - 2$

| $x$ | −3 | −2 | −1 | 0 | 1 | 2 | 3 |
|---|---|---|---|---|---|---|---|
| $y$ | | | | | | | |

b) Plot the graph of this function on axes from $-3 \leq x \leq 3$ and $-5 \leq y \leq 10$

# 17.3 QUADRATIC GRAPHS

## Objectives

### Identify and interpret roots, intercepts and turning points of a quadratic graph

The **roots** of a quadratic function are the points where the curve crosses the x-axis.

These can be found by drawing a graph of the function.

**e.g.** The roots of $y = x^2 - 6x + 8$ are $x = 2$ and $x = 4$

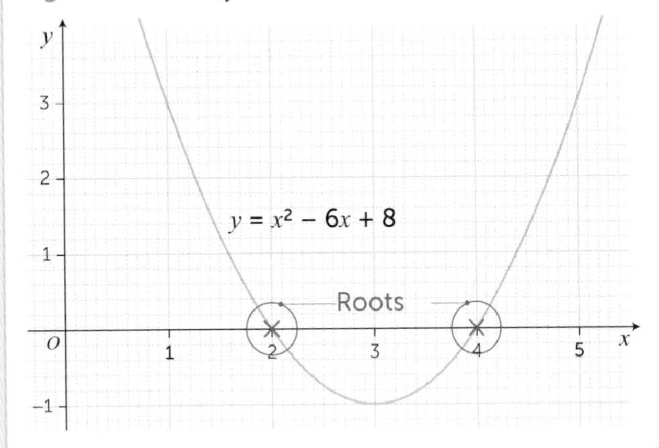


### Find approximate solutions using a quadratic graph

The **solutions** to equations are where the graphs of the functions intersect.

**e.g.** The solution to $x^2 - 6x + 8 = 3$ is shown where graphs of $y = x^2 - 6x + 8$ and $y = 3$ meet

They meet at (1, 3) and (5, 3)

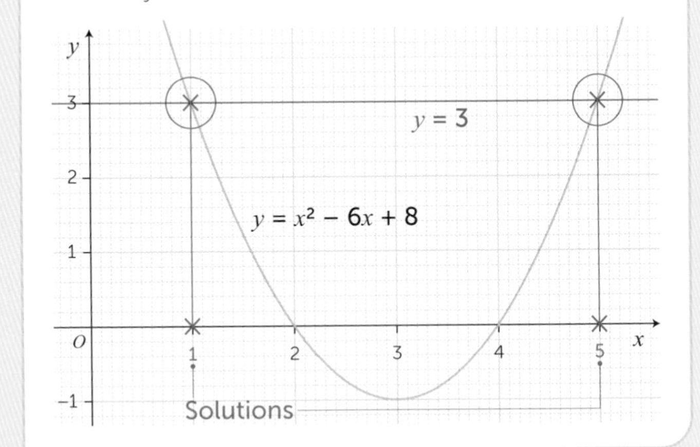


### Identify the line of symmetry of a quadratic graph

Graphs of quadratic functions have a vertical **line of symmetry**.

They also have a **turning point**; the line of symmetry passes through the turning point.

**e.g.** Here is a graph of the function $y = x^2 - 4x + 3$

The equation of the line of symmetry is $x = 2$

The $x$-intercepts (roots) and the $y$-intercept are shown.

The turning point is at $x = 2$, $y = -1$

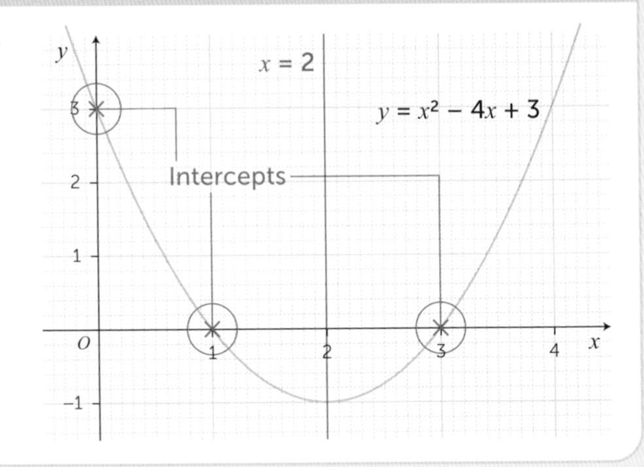


## Practice questions

1. a) Complete the table of values for the function $y = x^2 - 4$

| $x$ | −3 | −2 | −1 | 0 | 1 | 2 | 3 |
|---|---|---|---|---|---|---|---|
| $y$ | | | | | | | |

b) Plot a graph of the function on axes like these.

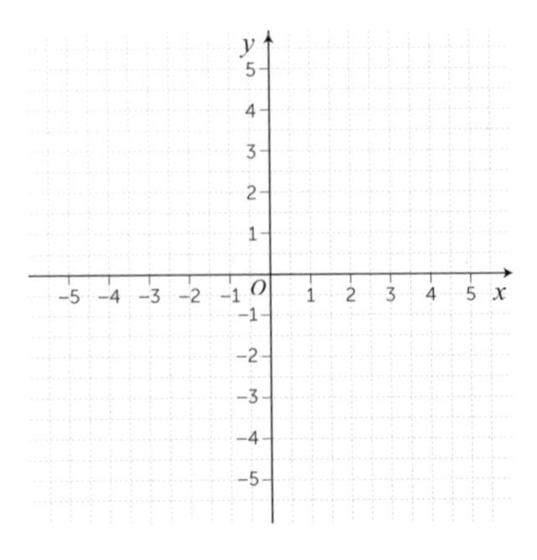

2. Use your graph from Q1 to find:
   a) The solutions to the equation $x^2 - 4 = 0$
   b) The line of symmetry for the function $y = x^2 - 4$
   c) State the coordinates of the turning point.

3. a) Complete the table for the function $y = x^2 - 1$

| $x$ | −3 | −2 | −1 | 0 | 1 | 2 | 3 |
|---|---|---|---|---|---|---|---|
| $y$ | | | | | | | |

   b) Plot a graph of the function on axes like these.

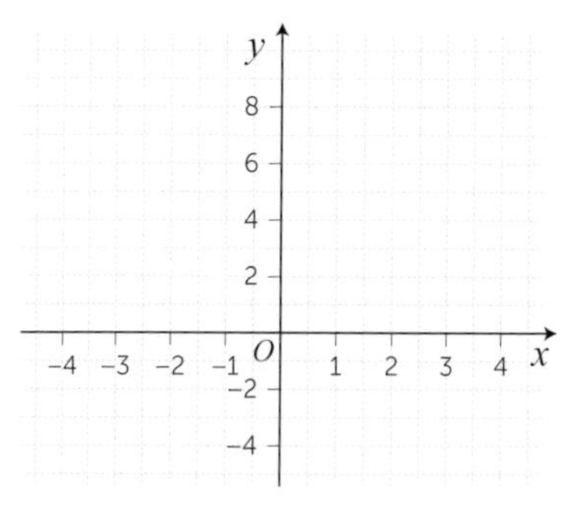


4. Use your graph from Q3 to find:
   a) The roots of the function $y = x^2 - 1$
   b) The line of symmetry for the function $y = x^2 - 1$
   c) Use your graph to solve the equation $x^2 - 1 = 3$

5. a) Copy and complete the table for the function $y = x^2 - 2x$

| $x$ | −2 | −1 | 0 | 1 | 2 | 3 | 4 |
|---|---|---|---|---|---|---|---|
| $y$ | | | | | | | |

   b) Plot a graph of the function on axes $-2 \le x \le 4$ and $-5 \le y \le 10$
   c) Write down the roots of the function $y = x^2 - 2x$
   d) State the coordinates of the turning point of $y = x^2 - 2x$
   e) What is the line of symmetry for the function $y = x^2 - 2x$ ?

6. a) Copy and complete the table for the function $y = x^2 - 3x - 4$

| $x$ | −2 | −1 | 0 | 1 | 2 | 3 | 4 | 5 |
|---|---|---|---|---|---|---|---|---|
| $y$ | | | | | | | | |

   b) Plot a graph of the function on axes $-2 \le x \le 5$ and $-10 \le y \le 10$
   c) Explain how you could find the $y$-intercept without plotting the function.
   d) What are the roots of the function $y = x^2 - 3x - 4$ ?
   e) State the equation of the line of symmetry for the function $y = x^2 - 3x - 4$

7. Plot the graph of $y = x^2 + 2x + 1$ for $-4 \le x \le 2$ on graph paper. Use your graph to:
   a) Find the roots of $y = x^2 + 2x + 1$
   b) Solve $x^2 + 2x + 1 = 4$
   c) Solve $x^2 + 2x + 1 = 2$

8. Plot the graph of $y = 5 + 4x - x^2$ for $-2 \le x \le 6$ on graph paper. Use your graph to:
   a) State the coordinates of the $x$-intercepts and the $y$-intercepts on the graph.
   b) Solve $5 + 4x - x^2 = 0$
   c) Solve $5 + 4x - x^2 = 3$

## 17.4 RECIPROCAL AND CUBIC GRAPHS

### Objectives

**Recognise, sketch and interpret graphs of the reciprocal function $y = \frac{1}{x}$ with $x \neq 0$**

A reciprocal function has a variable, usually $x$, as part of the denominator.

**e.g.** Each of these is a reciprocal function

$y = \frac{1}{x}$ $y = \frac{3}{x}$ $y=\frac{1}{x+2}$ $y = \frac{1}{2x}$

**Use graphical representations of inverse proportions to solve problems in context**

To draw a graph of the reciprocal function $y = \frac{1}{x}$, first complete a table of values for $x$ and $y$.

| $x$ | 1 | 2 | 4 | 8 | 10 |
|---|---|---|---|---|---|
| $y$ | 1 | 0.5 | 0.25 | 0.125 | 0.1 |

Plot the coordinates onto a set of axes and join them up to draw the graph.

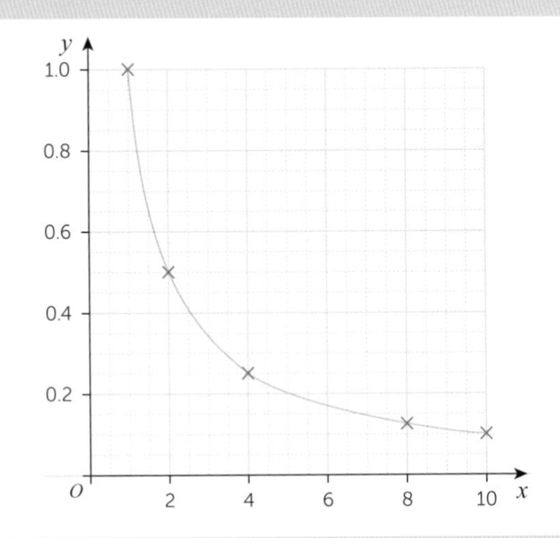


**Recognise and sketch a cubic function**

A cubic function has $x^3$ as the largest power of $x$

**e.g.** These are all cubic functions.

$y = x^3$ $y = x^3 - 5$ $y = 2x^3$ $y = x^3 - x^2 - 1$

These are not cubic functions.

$y = x^3 - 2x^4$ $y = \frac{1}{x^3}$

**Generate points and plot a cubic function using a table**

To plot a graph of a cubic function, substitute values of $x$ into the function to find values of $y$.

Plot the $x$ and $y$ coordinates and join them up.

**e.g.** Here is the cubic function $y = x^3 - 3x + 2$

Complete the table of values and plot the graph.

| $x$ | −2 | −1 | 0 | 1 | 2 |
|---|---|---|---|---|---|
| $y$ | 0 | 4 | 2 | 0 | 4 |

Cubic graphs often have two turning points like this one.

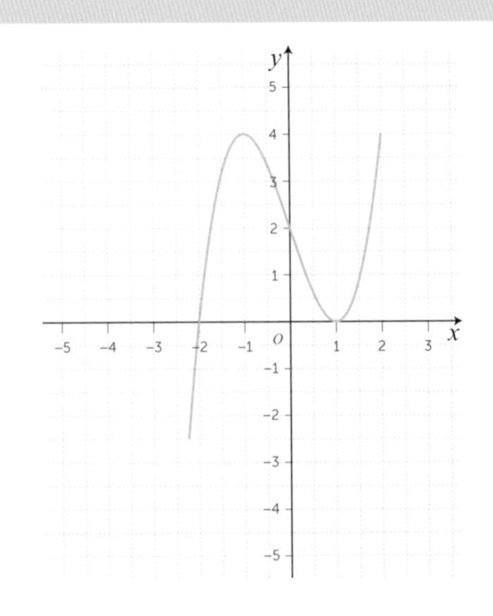

## Practice questions

1. State which type of function; linear, quadratic, cubic or reciprocal, each of these is.

   a) $y = \frac{1}{2}x$  b) $y = 3x^2$  c) $y = \frac{5}{x}$  d) $y = x^3 - 4$

2. For each of these graphs state whether they are linear, quadratic, cubic or reciprocal.

| Graph A | Graph B | Graph C | Graph D |
|---|---|---|---|
| | | | |

3. a) Copy and complete the table of values for the reciprocal function $y = \frac{1}{x}$

| $x$ | 0.1 | 0.25 | 0.5 | 1 | 2 | 3 | 4 |
|---|---|---|---|---|---|---|---|
| $y$ | | | | | | | |

   b) Draw a graph of the function on graph paper.

4. a) Copy and complete the table of values for the function $y = x^3$

| $x$ | −3 | −2 | −1 | 0 | 1 | 2 | 3 |
|---|---|---|---|---|---|---|---|
| $y$ | | | | | | | |

   b) Draw a graph of the function for $-3 \leq x \leq 3$

5. a) Copy and complete the table of values for the function $y = x^3 + 3$

| $x$ | −3 | −2 | −1 | 0 | 1 | 2 | 3 |
|---|---|---|---|---|---|---|---|
| $y$ | | | | | | | |

   b) Draw a graph of the function for $-3 \leq x \leq 3$

6. Match each graph to its equation.

a)

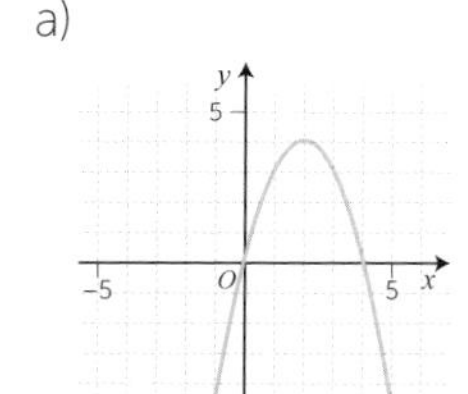

b) 

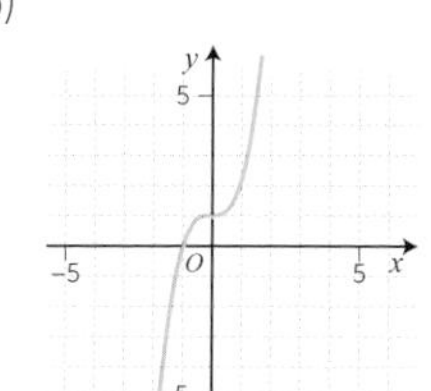

c)

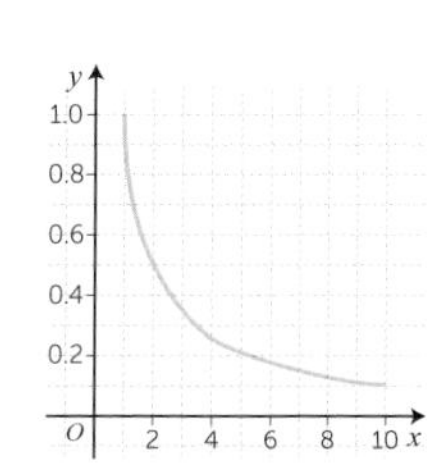

d)

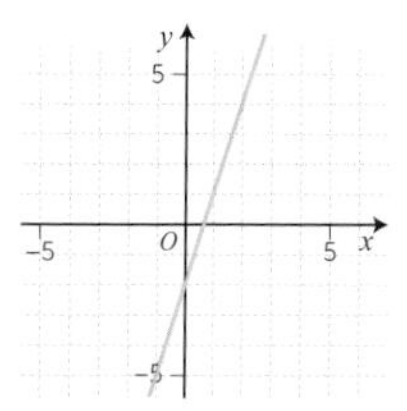

i) $y = 3x - 2$  ii) $y = \frac{1}{x}$  iii) $y = 4x - x^2$  iv) $y = x^3 + 1$

7. Mia wants to use a graph to show the time it takes to travel a distance of 30 miles, with speeds varying from 10 mph to 30 mph. She uses the formula Time = $\frac{\text{Distance}}{\text{Speed}}$.

   a) Complete the table below.

| Speed (mph) | 10 | 20 | 30 | 40 | 50 | 60 |
|---|---|---|---|---|---|---|
| Time (hours) | 3 | 1.5 | | | | |

   b) Will the graph be a straight line or a curve? Explain your answer.

   c) Draw the graph of Time against Speed to verify your answer. (Speed is the horizontal axis).

# 17.5 SOLVING SIMULTANEOUS EQUATIONS GRAPHICALLY

## Objectives

### Solve two linear simultaneous equations graphically

Solving two equations together is called solving equations simultaneously or solving simultaneous equations.

The point where the graphs of two equations meet is the solution to both equations.

The solution is the x and y coordinates.

**e.g.** Solve these simultaneous equations from the graph.

$y = 2x + 1$

$x + y = 4$

The solution is where the lines meet at **(1, 3)**, i.e. at $\mathbf{x = 1, y = 3}$

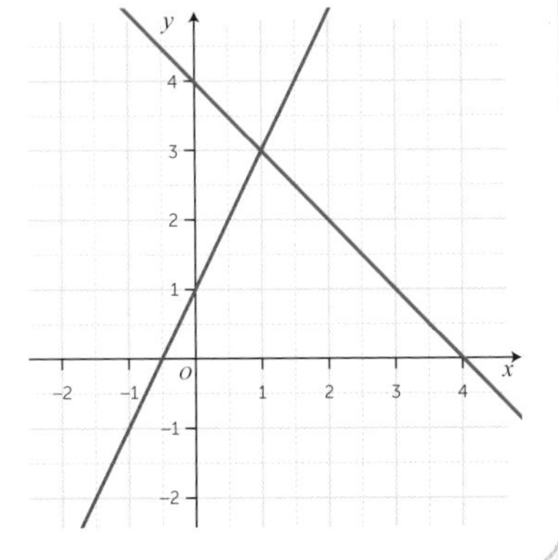


### Solve simultaneous equations representing a real-life situation, interpreting the solution in the context of the question

When solving simultaneous equations in context, think about what the variables ($x$ and $y$) represent.

## Practice questions

1 Here are the graphs of $y = x$ and $y = 4 - x$

Use the graphs to solve the simultaneous equations:

$y = x$

$y = 4 - x$

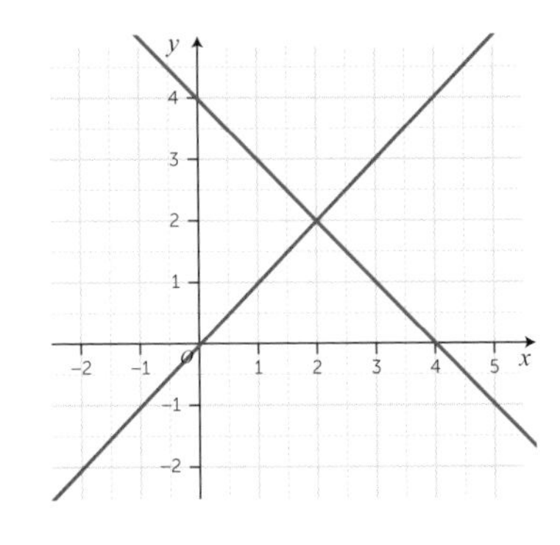


2. Here is the graph of $y = x + 2$

a) Copy this graph and on the same axes plot the graph of $y = 3x$

b) Use your graphs to solve the simultaneous equations:

$y = x + 2$ **and** $y = 3x$

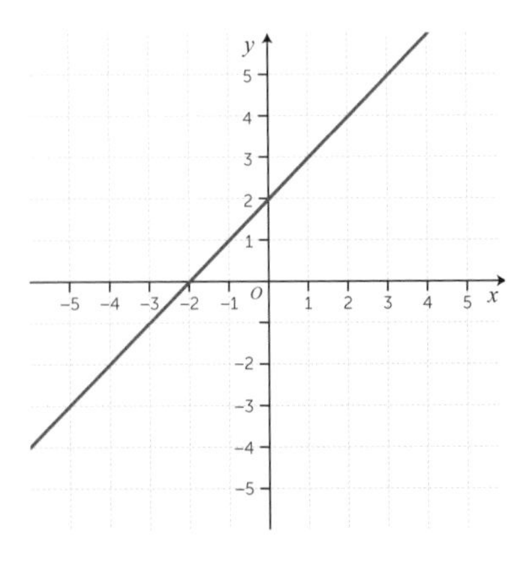

3. a) Copy and complete the table for $y = 3x - 1$

| $x$ | 0 | 1 | 2 | 3 |
|---|---|---|---|---|
| $y$ | | | | |

b) Plot the coordinates on a grid with the $x$-axis and $y$-axis between –5 and 10

c) On the same axes plot the graph of $y = x + 1$

d) Use your graphs to solve the simultaneous equations: $y = 3x - 1$ **and** $y = x + 1$

4. Here is the graph of $x + y = 4$

Use your graph to solve the simultaneous equations:

$x + y = 4$

$y = 2x - 5$

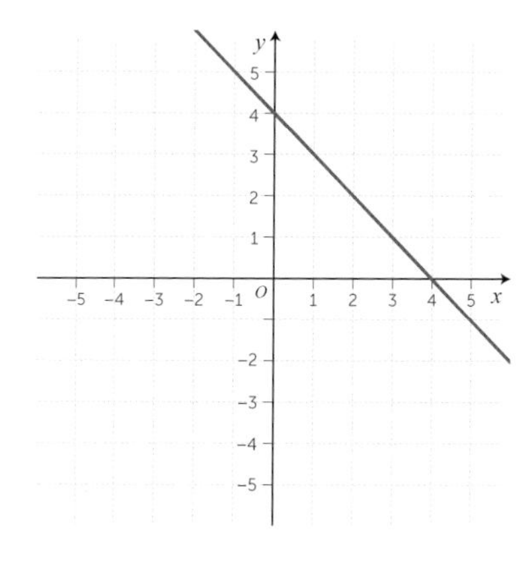

5. a) Draw a set of axes $-10 \leq x \leq 10$ and $-10 \leq y \leq 10$

b) Plot the graph of $y = 7 - 2x$

c) Use your graph to solve the simultaneous equations: $y = 7 - 2x$ **and** $y = x - 5$

6. Scientists are tracking the paths of two particles. For a short period of time:

**Particle A** is travelling along a path of $y = 3x - 5$

**Particle B** is travelling along a path of $y = \frac{1}{2}x$

a) Plot these paths on a single set of axes with $x$ and $y$ between – 10 and 10.

b) Find the point where the particles meet.

7. Copy and complete the table of values for the function $y = x^2 - 3$

| $x$ | –3 | –2 | –1 | 0 | 1 | 2 | 3 |
|---|---|---|---|---|---|---|---|
| $y$ | | | | | | | |

a) Plot the coordinates on a set of axes with x and y between – 5 and 10

b) On the same axes plot the graph of $y = x - 1$

c) Use your graphs to solve the simultaneous equations: $y = x^2 - 3$ **and** $y = x - 1$

SECTION 18

# GEOMETRY

## 18.1 CONSTRUCTIONS

G1 G2

### Objectives

**Make accurate drawings of triangles and other 2D shapes using a ruler and a protractor**

**Constructions**

Many diagrams featuring straight lines, angles and 2D shapes can be drawn using only straight edges and compasses. They involve no direct measurement and rely on geometric properties.

**e.g.** Construct an equilateral triangle on base AB.

a) Draw arc, centre A, radius equal to AB.

b) Draw arc, centre B, same radius to cross at C.

c) Complete the triangle ABC.

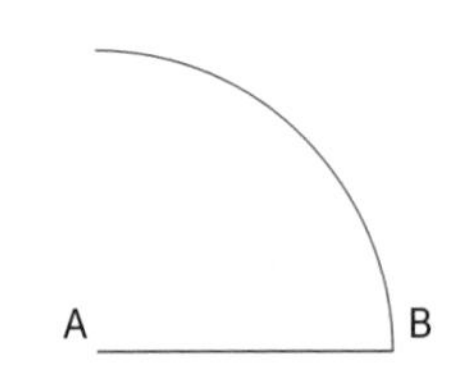


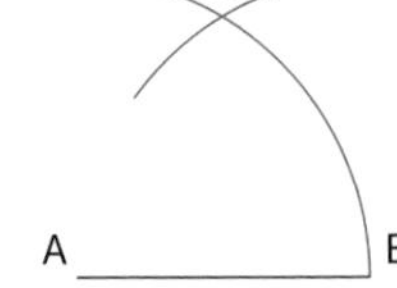


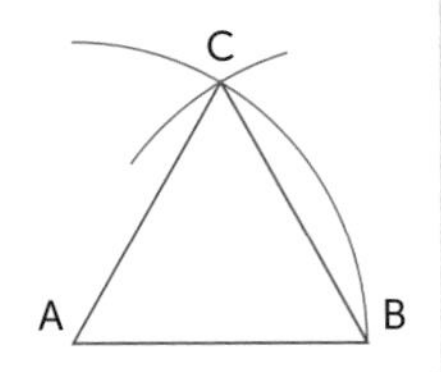


**Use straight edge and a pair of compasses to do standard constructions: perpendicular bisector of a given line; perpendicular from a point to a line; bisector of a given angle; angles of 90°, 45°**

**Constructions**

**e.g.** Construct the perpendicular bisector on the line CD.

a) Draw arc, centre C, radius more than half of CD.

b) Draw arc, centre D, same radius to cross at E and F.

c) Join EF which cuts CD at its midpoint.

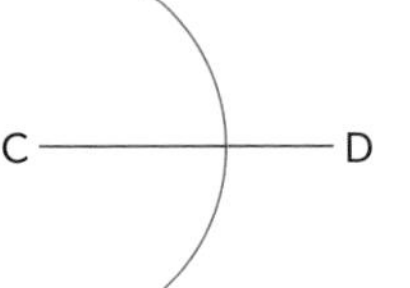


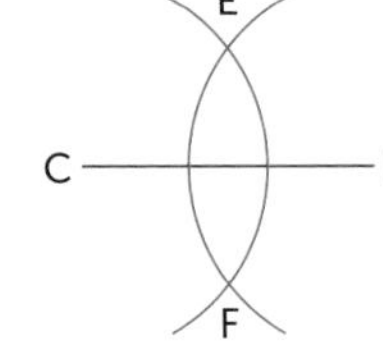


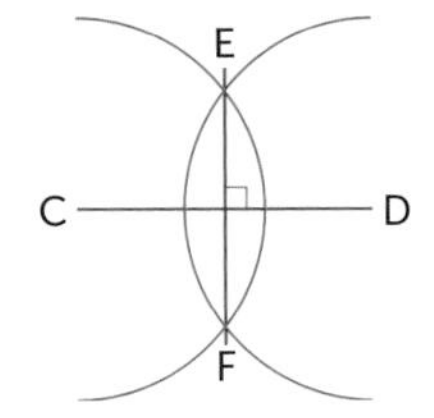


**e.g.** Construct the perpendicular at X on the line AB.

a) Draw 2 arcs, centre X, to cross AB at E and F.

b) Draw arc, centre E, radius longer than EX.

c) Draw arc, centre F, with the same radius, to cross at point C.

d) Join CX. This is the perpendicular at X.

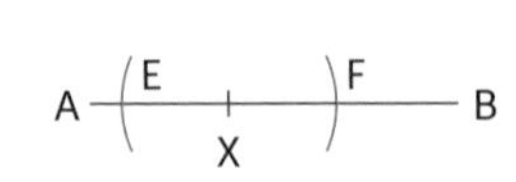

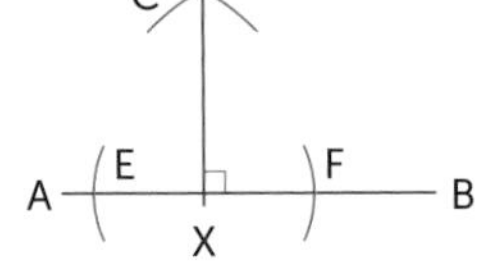


**e.g.** Construct the bisector of angle GHJ.

a) Draw arc, centre H to cross HJ at K and HG at L.

b) Draw arc, centre K, and draw arc, centre L, same radius to cross at M.

c) Join HM, the angle bisector of GHJ.

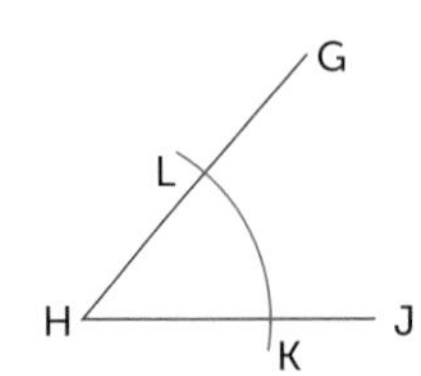


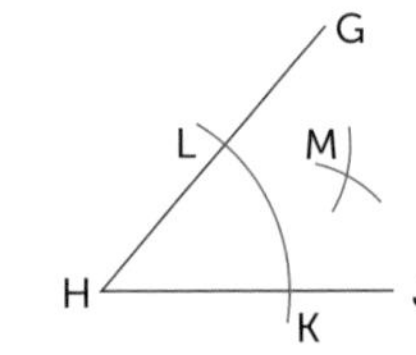


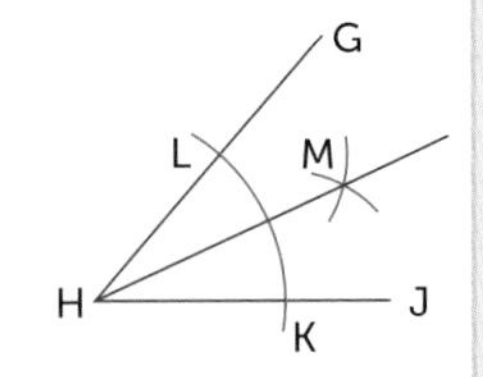

## Practice questions

1. Draw a straight line and label it LM.
   Using only compasses, construct the perpendicular bisector of LM.

2. Draw two straight lines that meet at a point and label the resulting angle PQR.
   Using only compasses, construct the bisector for angle PQR.

3. Construct the perpendicular at X on the line VW.

4. Draw a triangle and label the vertices X, Y and Z.
   Construct the angle bisectors for each angle in triangle XYZ.

5. Draw a circle, radius 4 cm and construct a regular hexagon inside it.

6. Construct a perpendicular from point H to line JK.

H •

J ——————————————— K

7. Copy the triangle ABC.
   Construct the perpendicular bisector for each side of triangle ABC.

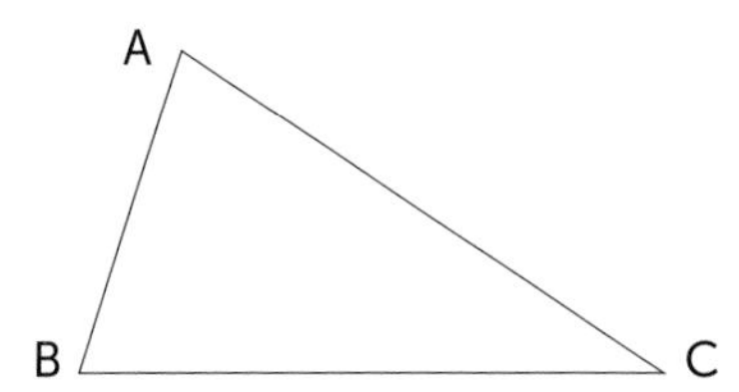

8. Construct the perpendicular from each vertex of triangle DEF to its opposite side.

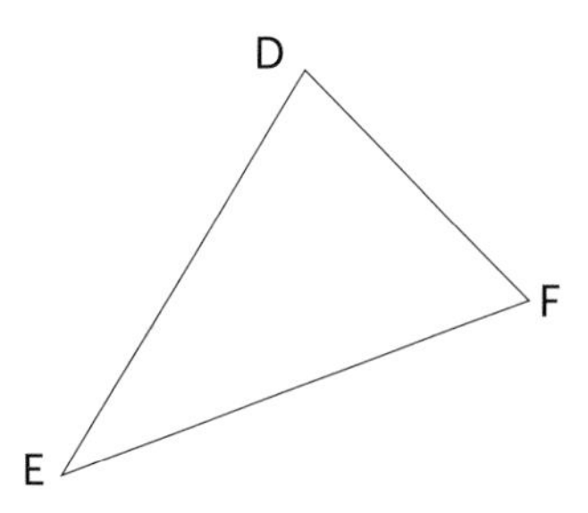


9. Draw a straight line and label it JK. Construct an angle of 60° on the line JK.
   Then bisect this angle to make two angles of 30° each.

10. Construct the perpendicular bisectors on each side of quadrilateral LMNO.

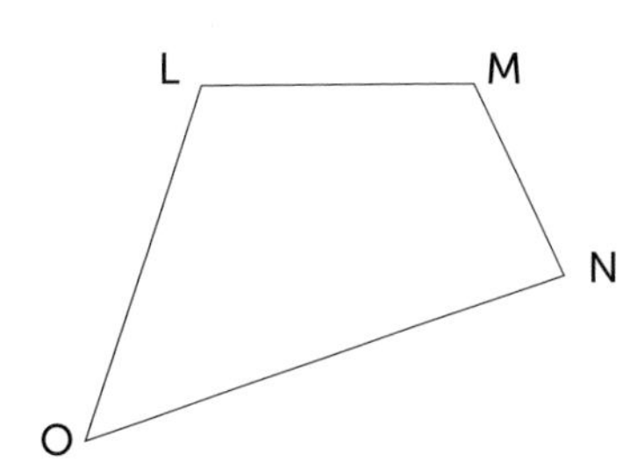

# 18.2 LOCI

## Objectives

### Draw and construct diagrams from given instructions

When a point is positioned according to a set of conditions, it marks out a path or locus.

Loci is the plural of locus.

Two particular loci are the basis for many others – see the worked examples.

**e.g.** Draw the locus of a point that is always 3 cm from fixed point A.

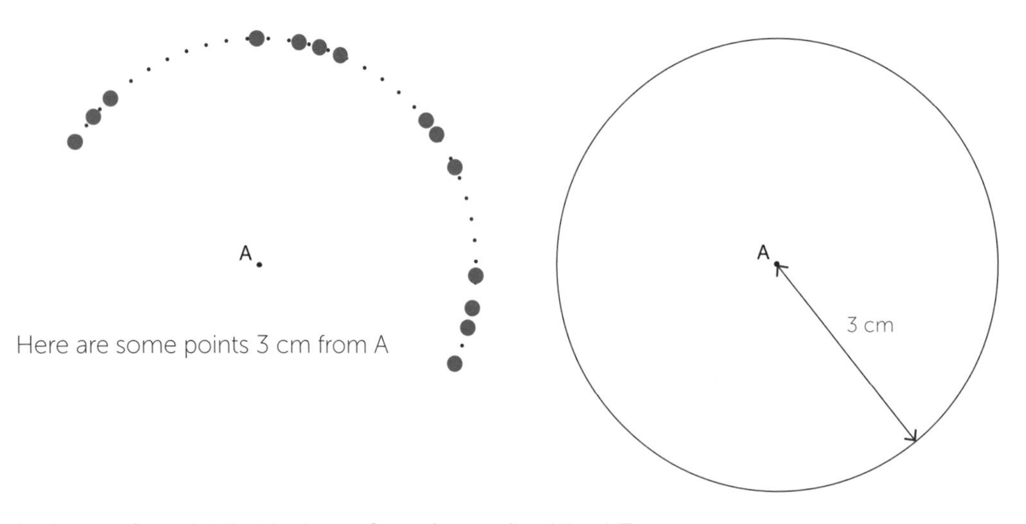


**e.g.** Draw the locus of a point that is always 2 cm from a fixed line YZ.

The dotted line shows where any point is 2 cm from the fixed line YZ.

The locus consists of 2 straight sections parallel to YZ and 2 semi-circular arcs around the endpoints Y and Z.

### Solve 2D locus problems using constructions: including bearings and regions satisfying a combination of loci

### Know that the perpendicular distance from a point to a line is the shortest distance to the line

## Practice questions

1. A is a fixed point. Sketch the locus of the point P for each of the following:
   a) AP = 3 cm  b) AP = 5 cm  c) AP = 6.3 cm  d) Shade the area represented by AP < 4 cm

2. C and D are two fixed points 7 cm apart. Sketch the locus of the point P for the following situations:
   a) CP = DP  b) CP = 5 cm and DP = 5 cm
   c) Shade the area P such that P is always within 6 cm of the line CD.

3. a) A goat is tethered to a stake in a field by a rope 6 m long. Describe and sketch the area which the goat can graze.
   b) The farmer puts a straight fence across the field which is 3 m from the stake at its closest point. Describe and sketch the area the goat can now graze.

4. EFGH is a square of side 8 cm. For each of the following loci the point P only moves within the square. Sketch the locus in each case.

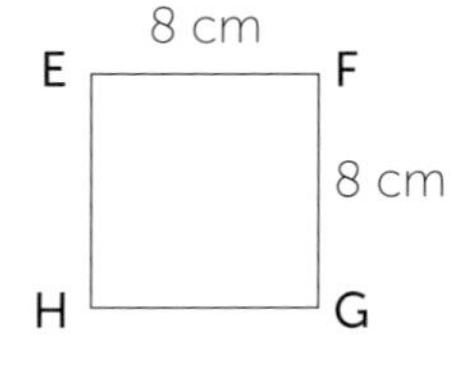


   a) EP = FP  b) FP < GP
   c) EP = GP  d) GP < 8 cm
   e) HP > 5 cm  f) GP > 9 cm

5. A bicycle moves along a straight flat road.
   a) What is the locus of the centre of the front wheel of the bicycle?
   b) What is the locus of a point on the rim of the front tyre?

6. A dog is tethered to the corner of a shed 5 m × 3 m, by a rope of length 4 m.
   Sketch the area that the dog can guard.

7. A phone company wishes to put up a mast between towns Ashbridge (**A**) and Bedlow (**B**). The towns are 10 km apart.
   The mast must be within 6 km of A and 7 km of B. Sketch the area where the mast could be placed.
   (Use a scale of 1 cm = 1 km)

   A • ——— 10 km ——— • B

8. The map shows a field that is 80 m square. There are 3 large trees in the field and a power line that cuts across the field.
   a) Draw an accurate scale drawing of the field using a scale of 1 cm = 10 m.
   b) Becky wants to fly a kite in the field.
      She can't fly the kite within 30 m of the power line. She can't fly the kite within 20 m of any tree. Sketch where she can safely fly the kite.

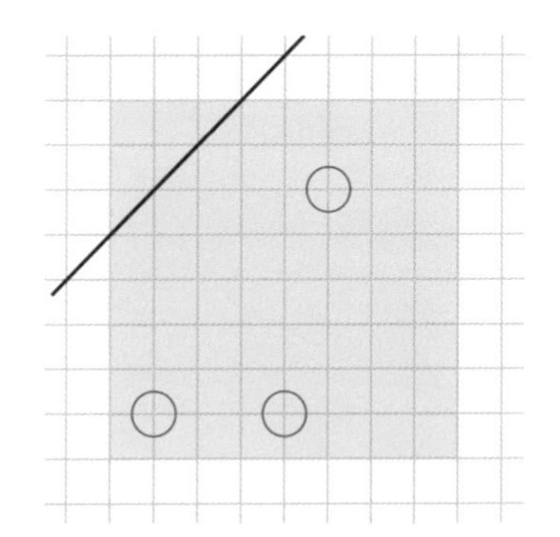

9. A (1, 4) is a fixed point. A point P moves so that it is always less than 3 units from A and always nearer the $x$-axis than the $y$-axis.
   Sketch the region that satisfies this condition.

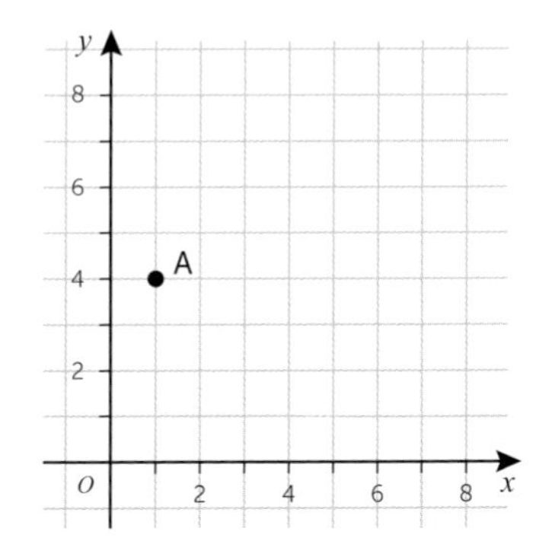

## 18.3 CONGRUENCE

### Objectives

**Use the basic congruence criteria for triangles (SSS, SAS, ASA and RHS)**

If two shapes are congruent, they are the same size and shape.

For two triangles to be congruent they need to satisfy one of these four conditions:

- All 3 sides are equal (SSS)

- Two sides and the angle between them are equal (SAS)
- Two angles and the corresponding side are equal (ASA)
- A right angle, hypotenuse and one other side are equal (RHS)

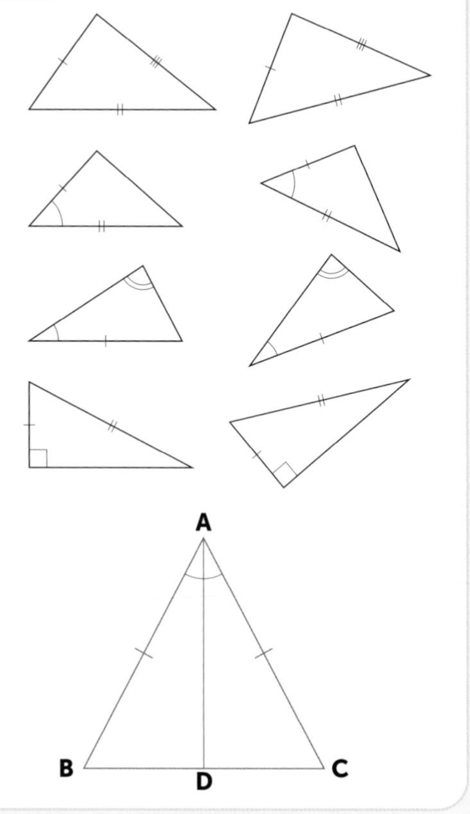


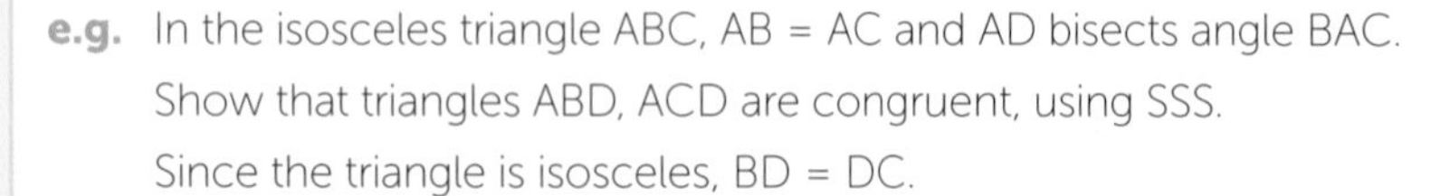

e.g. In the isosceles triangle ABC, AB = AC and AD bisects angle BAC.
Show that triangles ABD, ACD are congruent, using SSS.
Since the triangle is isosceles, BD = DC.
AB = BC is given.
AD is in both triangles.
All 3 sides are equal, so the triangles are congruent (SSS).

**Solve angle problems involving congruence.**

### Practice questions

1. The triangle EFG is divided into two congruent triangles EFM and EGM as shown
   Angle EFM = 55°
   Use congruency to:
   a) Write down the size of ∠EGM
   b) Work out the size of ∠FEG

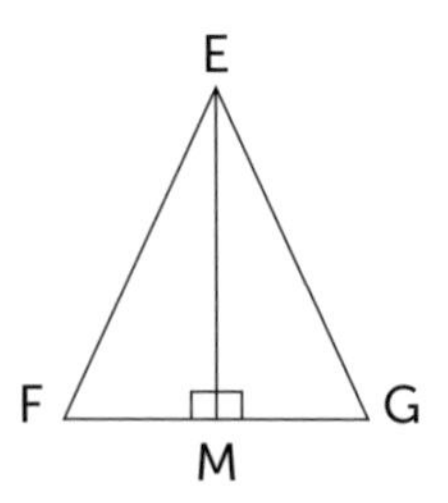


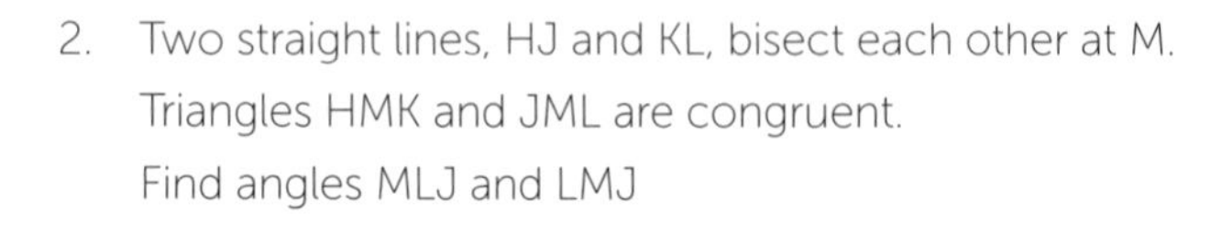

2. Two straight lines, HJ and KL, bisect each other at M.
   Triangles HMK and JML are congruent.
   Find angles MLJ and LMJ

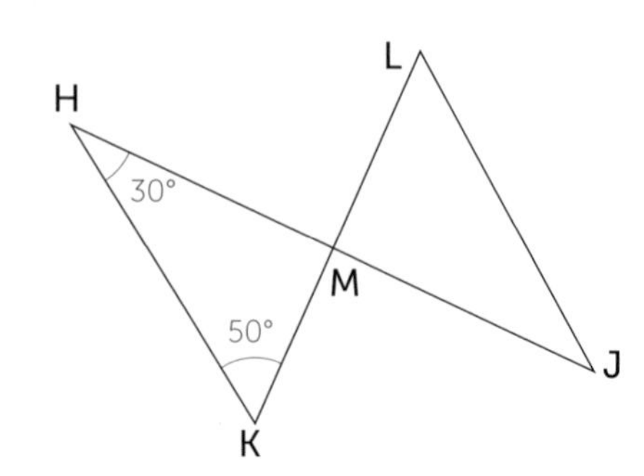

3. In quadrilateral PQRS, QR = RS and S and Q are right-angles.
   Triangles PQR and PSR are congruent.
   a) State which of SSS, SAS, ASA or RHS applies.
   b) Does RP bisect angle QPS? Give a reason for your answer.

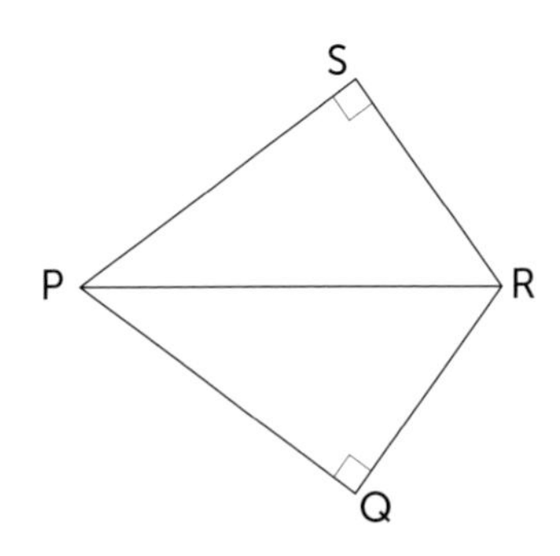


4. Match each of the congruence criteria for triangles (SSS, SAS, ASA, RHS) with a pair of triangles given below:

a)

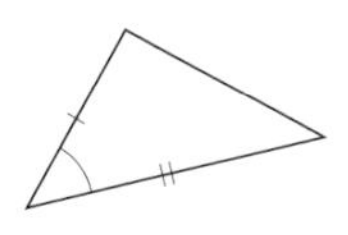

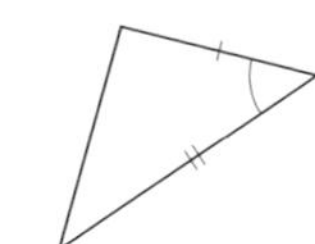

b)

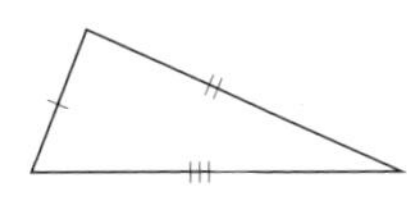

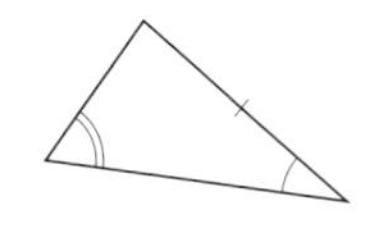

c)

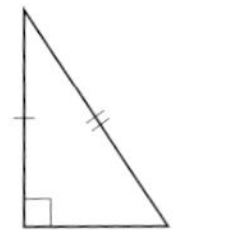

d)

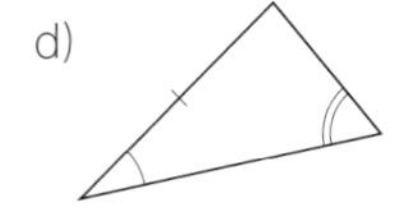

5. In a quadrilateral ABCD, AD = CD and BD bisects angle ADC.
   a) Triangles BCD and BAD are congruent. Which of the congruency criteria (SSS, SAS, ASA, RHS) applies?
   b) Does BD bisect angle ABC? Give a reason for your answer.

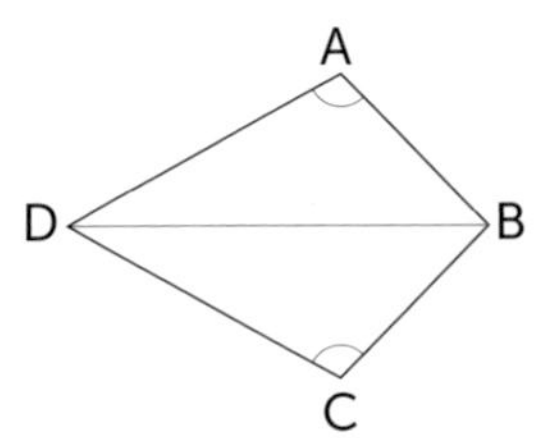

6. E and F are points on a circle, centre O.
   EF is a chord and G is on EF so that OG is perpendicular to EF.
   OG = 3 cm, EF = 8 cm.
   Which of these statements are true?
   (i) Area of triangle OEG = Area of triangle OFG
   (ii) OE = OF
   (iii) ∠OEG = 2 × ∠EOF

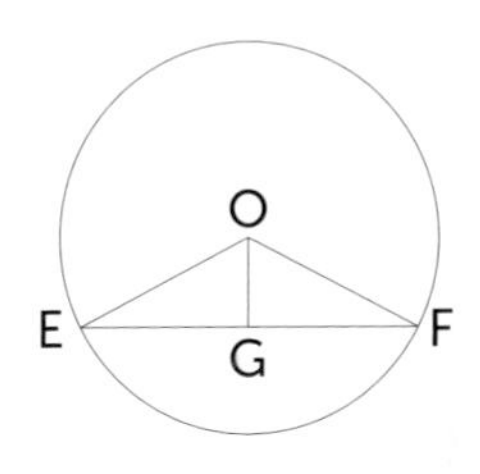

7. a) Are these triangles congruent? Explain your answer.
   b) Find the length of XZ.

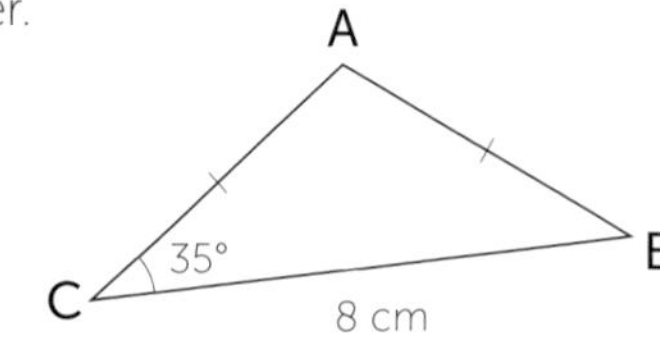


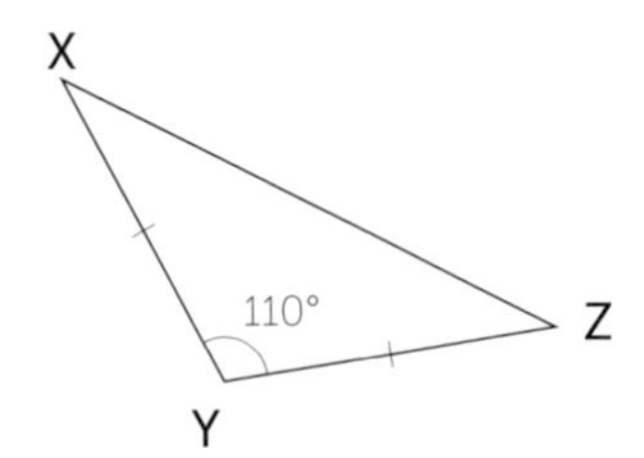

8. HML and KMJ are straight lines with HM = KM and JM = LM.
   Are triangles HJM and KLM congruent?
   Explain your answer.

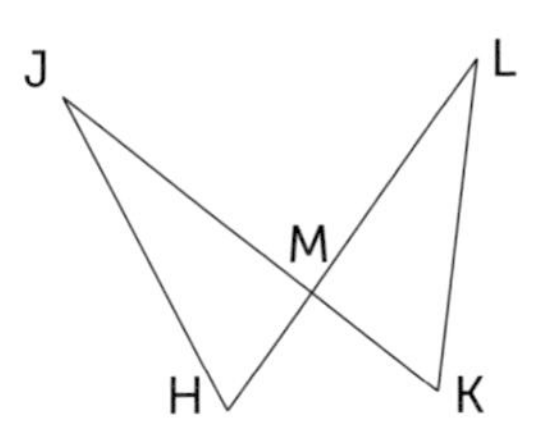

# 18.4 SIMILARITY

## Objectives

### Solve problems to find missing lengths in similar shapes

Two shapes are mathematically similar if corresponding angles are all equal and corresponding sides are in the same ratio.

One shape may be a scaled up, scaled down, rotated or reflected version of the other.

If two shapes are similar, then the scale factor of the enlargement of corresponding sides on the object and image is the same.

**e.g.** Triangles XYZ and ABC are similar.

XY = 5 cm, YZ = 8 cm, BC = 15 cm and AC = 27 cm.

Calculate the length of sides AB and XZ.

Scale factor of the enlargement: $\frac{BC}{YZ}$

$= \frac{24}{8} \quad \Rightarrow 3;$

All sides of XYZ are multiplied by 3: AB $= 5 \times 3 \quad \Rightarrow x = 15$ cm

$y \times 3 = 27 \quad \Rightarrow y = 9$ cm

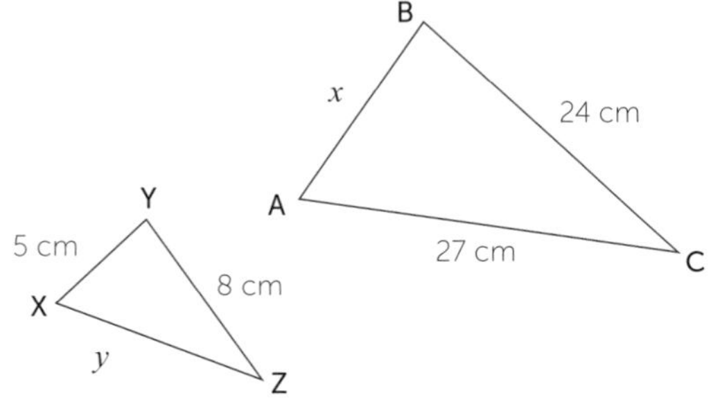


### Understand similarity of triangles and of other plane shapes, including all circles or all regular polygons with equal number of sides

### Solve angle and length problems using similarity

## Practice questions

1. Triangles ABC and DEF are similar.
   a) What is the size of angle DEF? Explain your answer.
   b) Work out the length of BC.
   c) Work out the length of DE.

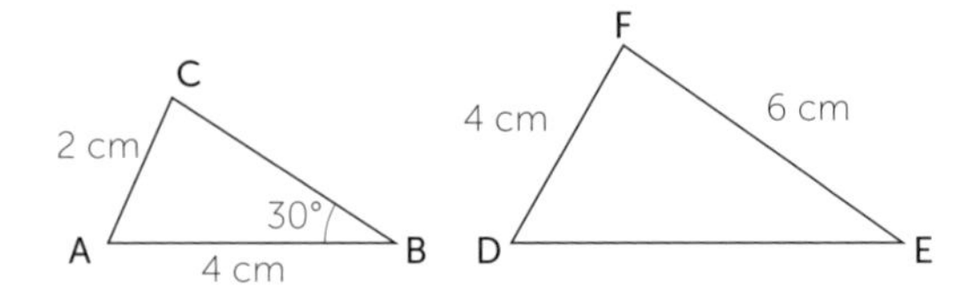


2. One rectangle measures 7 m × 3 m, and another measures 14 m × 7 m.
   Are the two rectangles similar? Explain your reasoning.

3. Rectangles G and H are similar.
   Work out the value of length y.

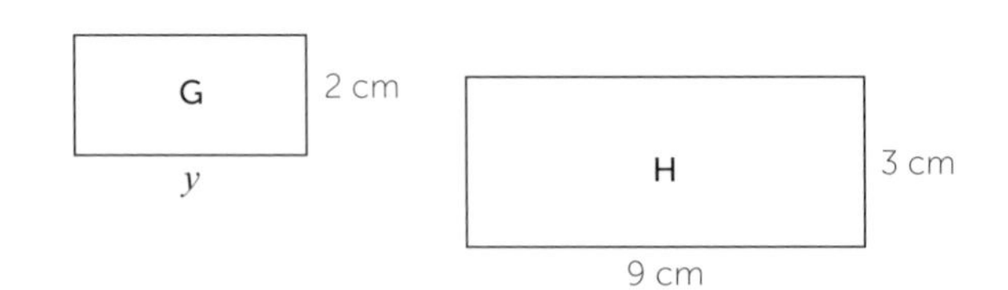


4. Pentagons J and K are similar.
   a) What is length $m$?
   b) What is length $n$?

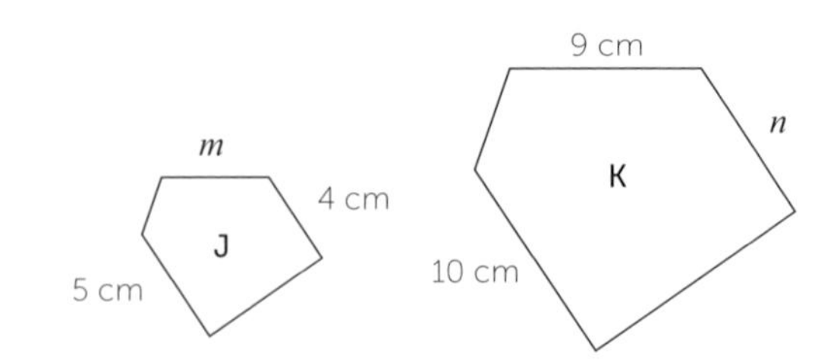

5. These two quadrilaterals are similar.
   Find the angles $a$, $b$, $c$, $d$ and $e$.
   Find lengths $x$ and $y$.

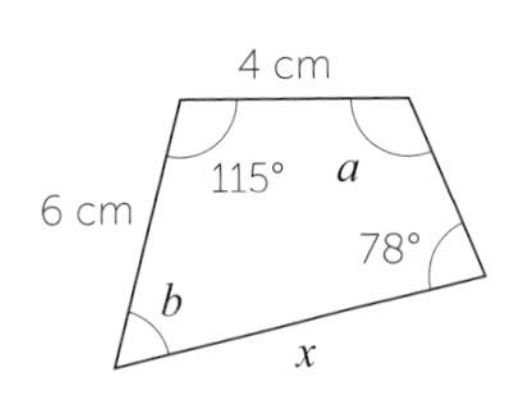


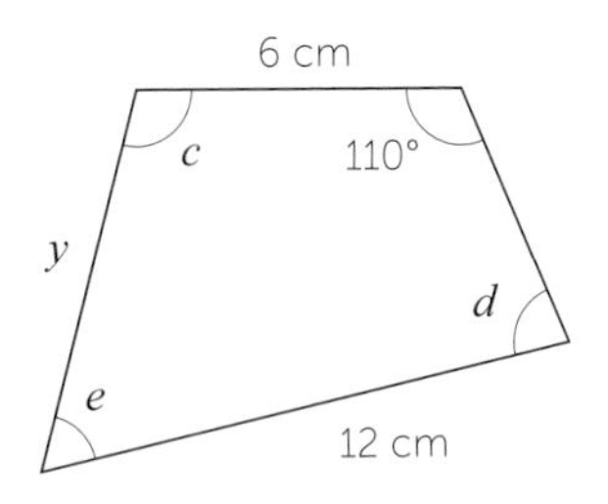

6. In these two similar triangles, the scale factor of enlargement is 3.5
   Find the values of $p$, $q$ and $r$.
   What type of triangles are these?

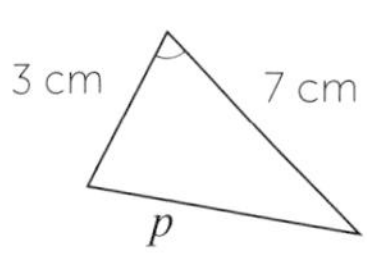


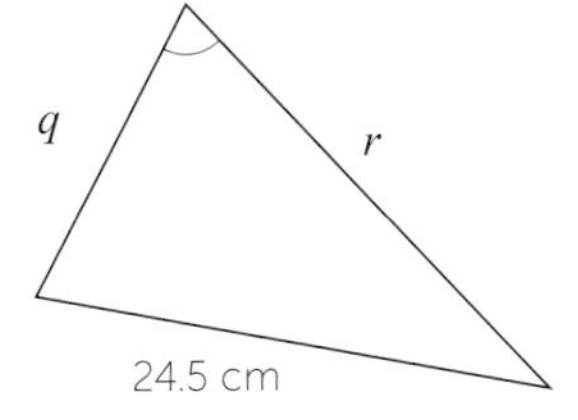

7. Four quadrilaterals (not drawn to scale) are shown below.

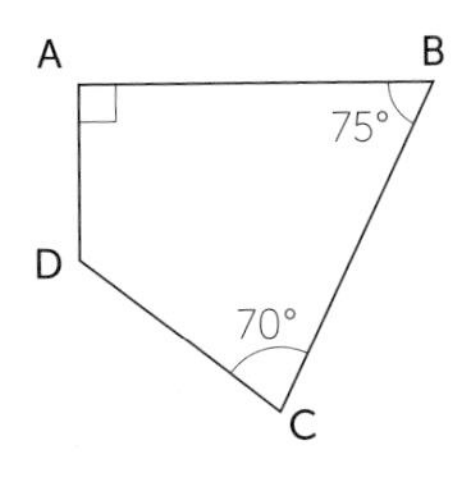


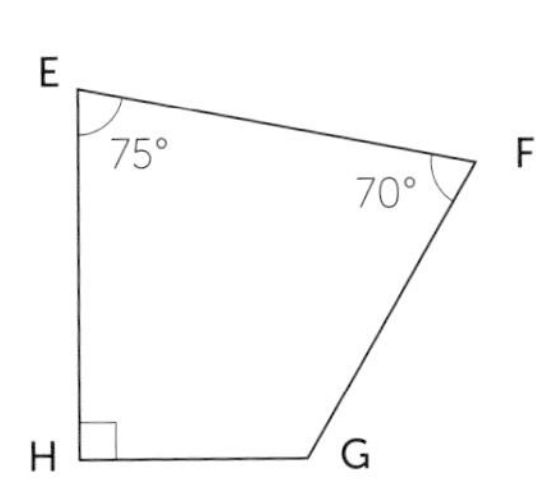


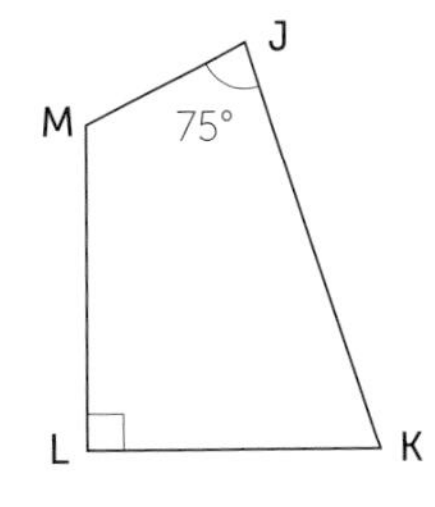


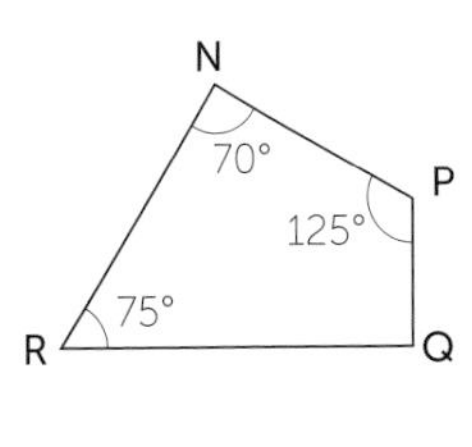


   a) Calculate ∠ADC
   b) Which of the quadrilaterals is not similar to the other three?

8. In triangle RST, UV is parallel to RS.
   Draw, and label triangles RST and UVT as separate diagrams.
   a) Calculate the length of SV.
   b) Calculate the length of TU.

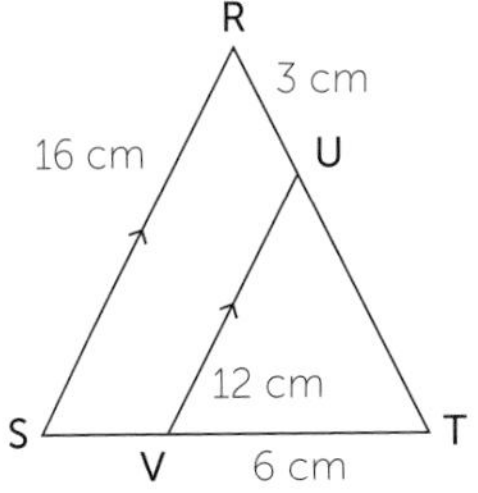
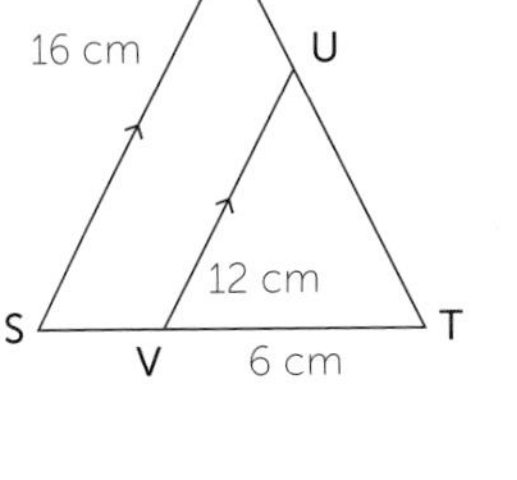

9. In triangle LMN, PQ is parallel to MN.
   Draw and label triangles LPQ and LMN as separate diagrams.
   a) What is the length of PM?
   b) What is the length of NM?

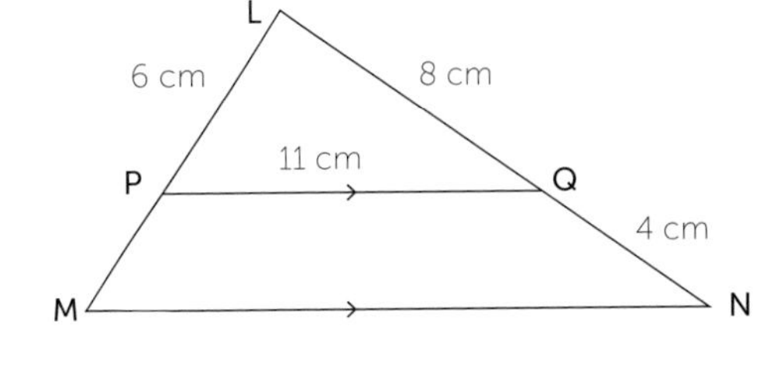

10. Six shapes are shown below.
   a) Two of the shapes are congruent.
      Write down the letters of these shapes
   b) One of the shapes is similar to A.
      Write down the letter of this shape

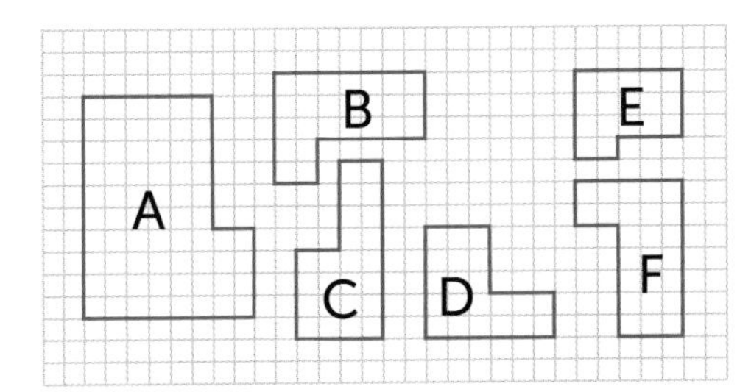

# 18.5 PYTHAGORAS' THEOREM

## Objectives

**Understand, recall and use Pythagoras' theorem in 2D to justify if a triangle is right-angled or not**

Triangle ABC has sides of lengths $a$, $b$ and $c$ units.

On each side, a square has been added.

Pythagoras' theorem states that, in any right-angled triangle with side lengths $a$, $b$ and $c$, where $c$ is the longest side (hypotenuse), it is always true that $a^2 + b^2 = c^2$.

Equally, if $a^2 + b^2 = c^2$ for any triangle, then it must be right-angled.

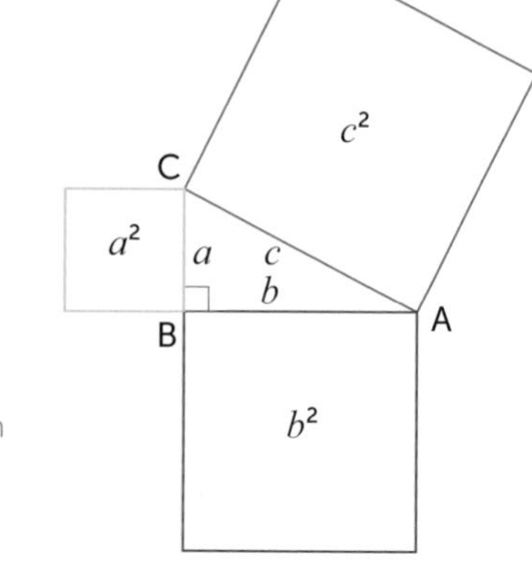


**e.g.** Calculate the length of the hypotenuse on this right-angled triangle:

$a^2 + b^2 = c^2 \Rightarrow 4^2 + 11^2 = c^2$

$16 + 121 = c^2$

so $c^2 = 137$

and $c = \sqrt{137}$ cm

The answer can be left exactly, as a surd, or to a given degree of accuracy: $c$ = **11.7 cm** (1 d.p.)

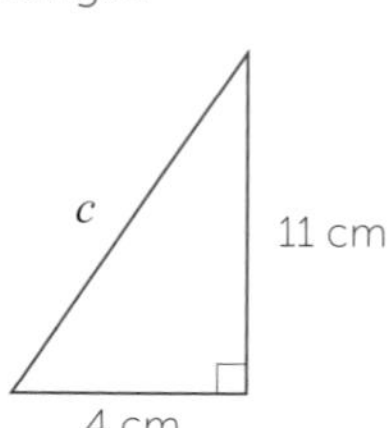


**e.g.** Find the missing side on this right-angled triangle:

$a^2 + 15^2 = 25^2 \Rightarrow a^2 = 25^2 - 15^2$

$a^2 = 625 - 225$

$a^2 = 400$

$a = \sqrt{400}$

$a$ = **20 cm**

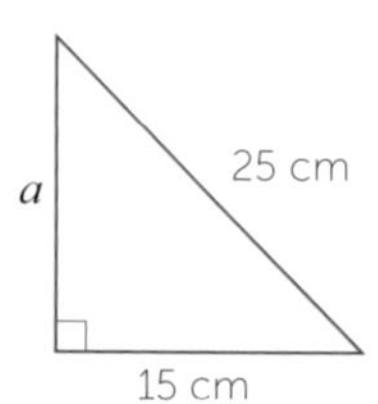


**Calculate length of a line segment AB given pairs of points; leave answers in surd form**

**Calculate the length of the hypotenuse and of a shorter side in a right-angled triangle, including decimal lengths and a range of units**

## Practice questions

1. In each of these right-angled triangles, calculate the length of the unknown side.

a) b) c)

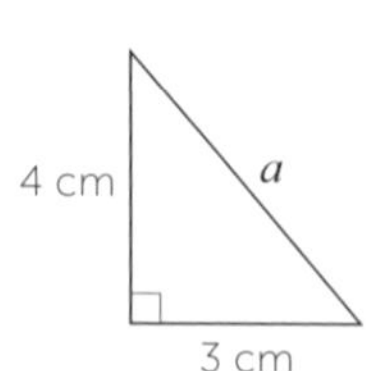


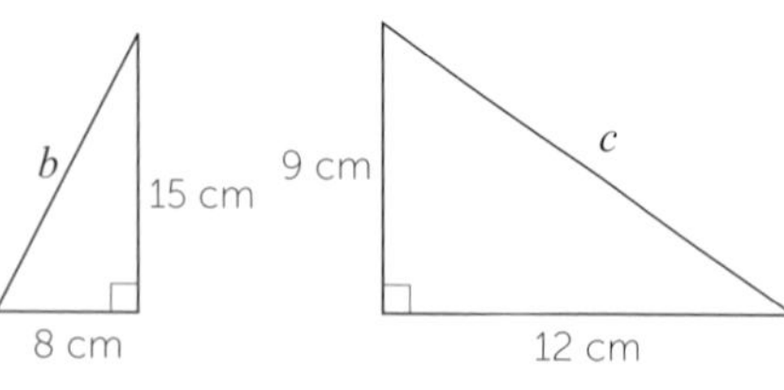


2. Work out the missing lengths in these right-angled triangles:

a) b) c)

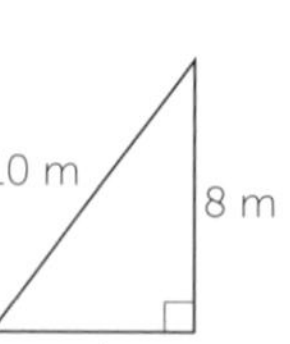


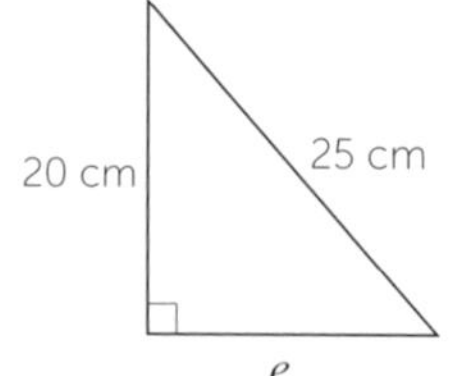


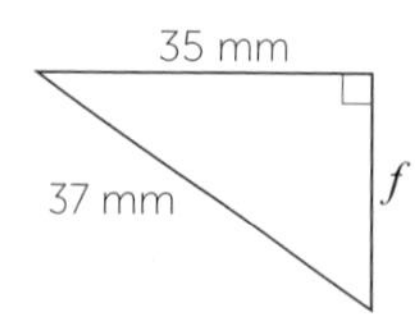

3. By using Pythagoras' Theorem, decide which of these triangles is right-angled.

a) b) c)

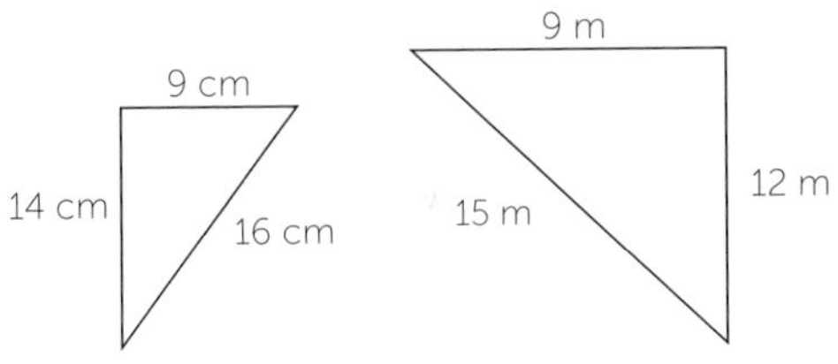


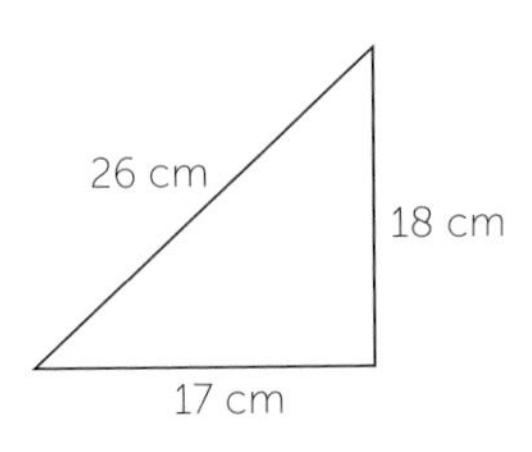


4. What is the length of the diagonal of this rectangle?
   Give your answer in cm, to 2 decimal places.

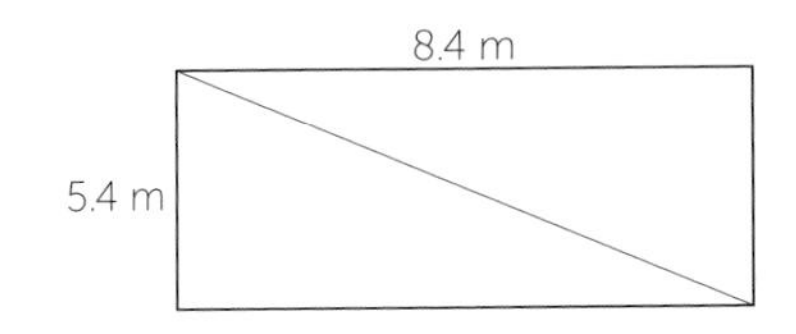


5. A ladder leans against a wall.
   The foot of the ladder is 1.3 m from the wall and the ladder is 6 m in length.
   How far up the wall does the ladder reach?

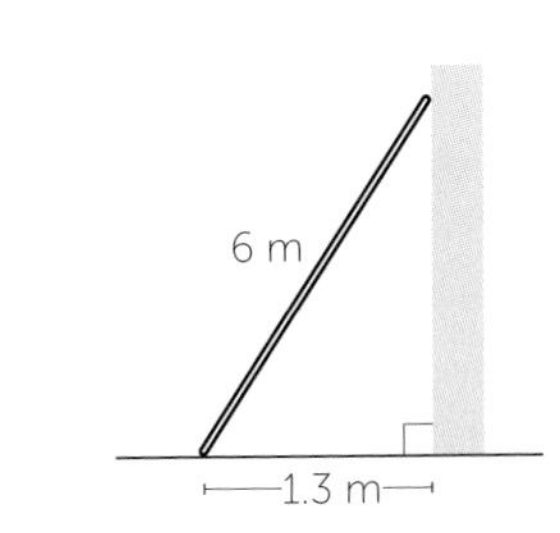


6. Find the lengths of the missing sides in this isosceles triangle.

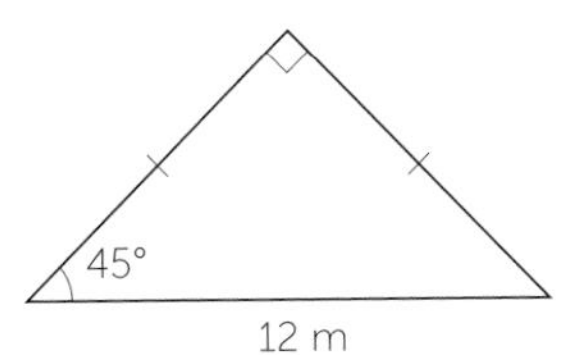


7. What are the side lengths for this square?

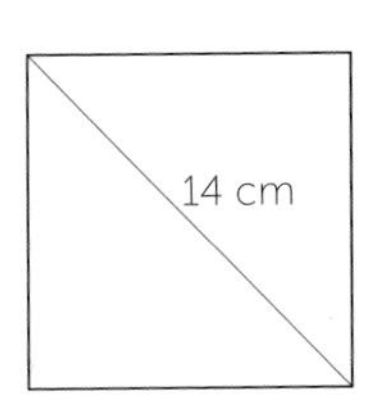


8. Find the length of the line joining (3, 1) and (11, 7), which is the hypotenuse of this right-angled triangle.

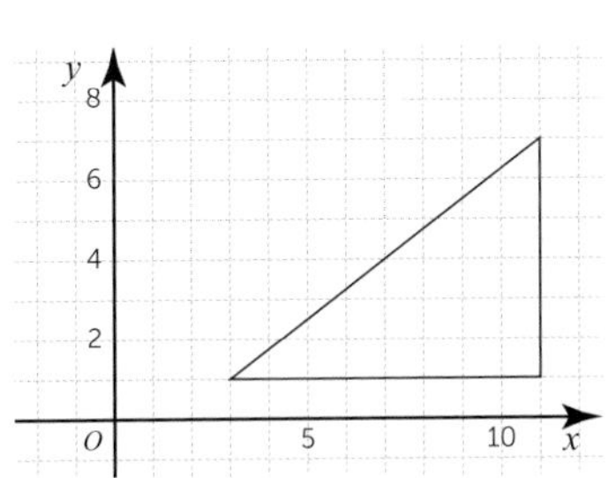


9. Use Pythagoras' theorem to find the lengths of each hypotenuse in these two triangles.

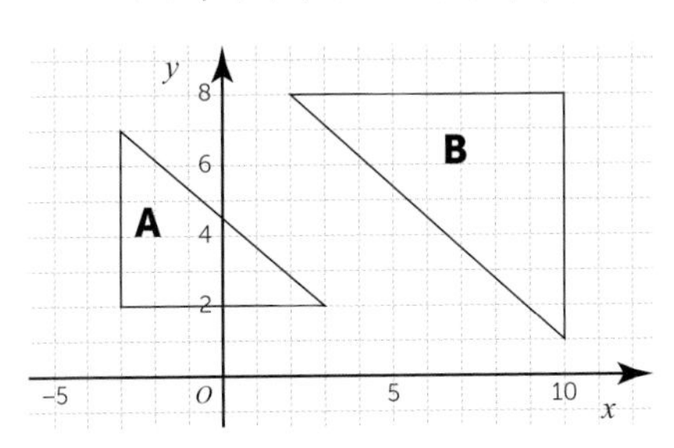

10. A triangle is formed by joining the points A (2, 6), B (2,1) and C (14, 1).
    a) Work out the lengths of each side of the triangle, AB, BC and AC.
    b) Use the answers to part a) to show that the triangle is right-angled.

# 18.6 GEOMETRIC PROOFS

## Objectives

### Understand a proof that the exterior angle of a triangle is equal to the sum of the interior angles at the other two vertices

Prove that the exterior angle of a triangle is equal to the sum of the opposite interior angles.

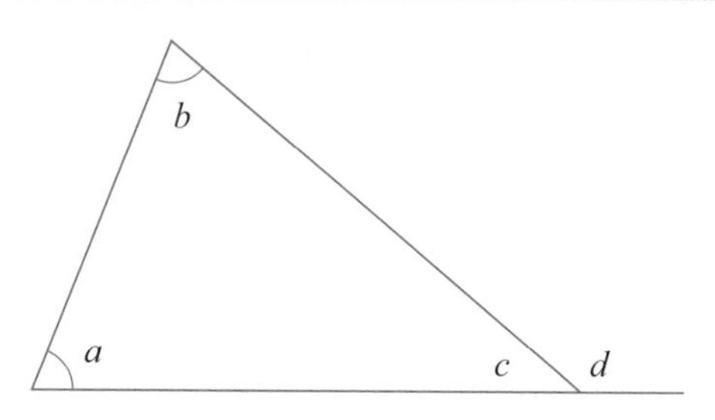


Proof

| | | |
|---|---|---|
| In the triangle, | $a + b + c$ | $= 180°$. |
| On the straight line, | $c + d$ | $= 180°$. |
| Hence | $a + b + c$ | $= c + d$ |
| and so | $a + b$ | $= d$ |

### Use the symmetry property of an isosceles triangle to show that base angles are equal

Some geometric properties of triangles can be deduced and proved from these known facts:

- The angles in a triangle always add to 180°
- Angles on a straight line add to 180°
- An isosceles triangle always has two equal angles and two sides of equal length and is hence always symmetrical
- Equilateral triangles have three equal sides and three angles of 60°.

### Use the side/angle properties of isosceles and equilateral triangles to prove results

## Practice questions

1. In the questions below, use the symmetry of the isosceles triangles to help you find the missing angles.

a)

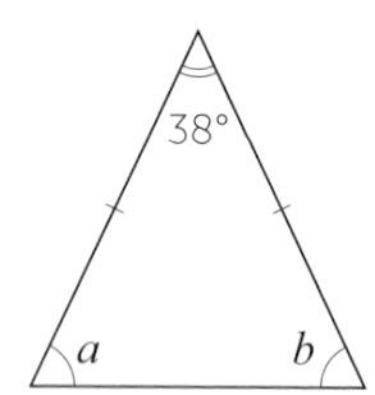


b)

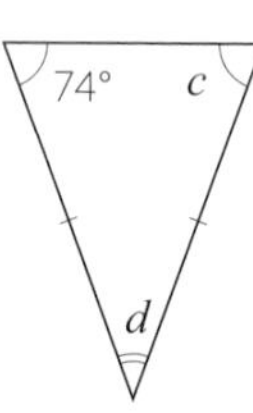


c)

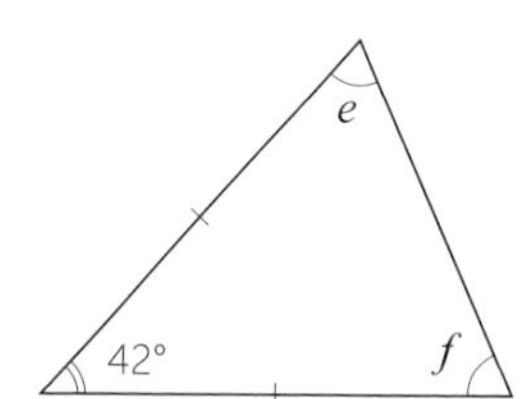


2. Find the missing angles in these triangles.

a)

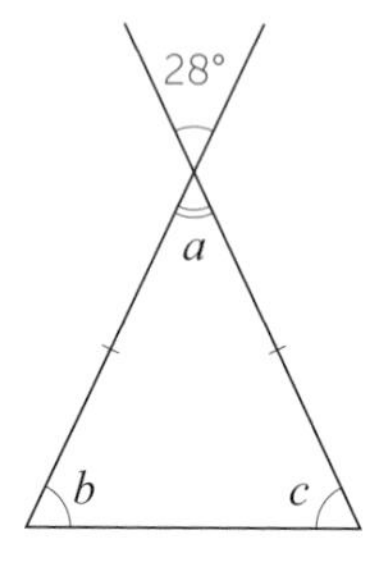


b)

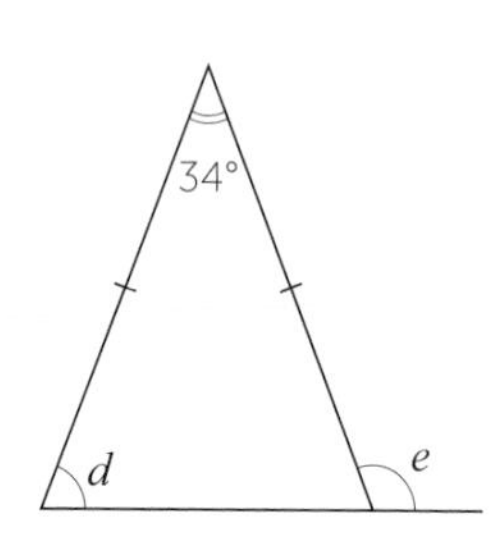


c)

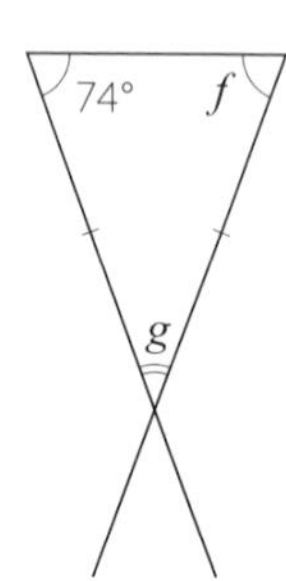

3. What size are the interior and exterior angles of an equilateral triangle?

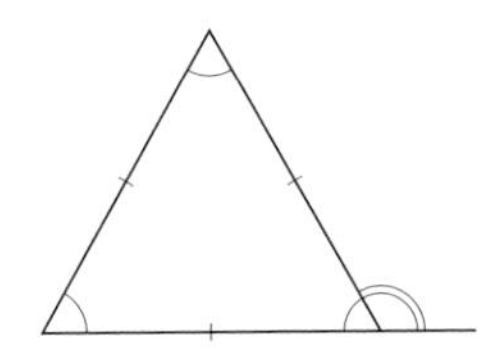

4. Find the size of angles $\boldsymbol{a}$, $\boldsymbol{b}$ and $\boldsymbol{c}$ in this shape.
   Explain your answers.

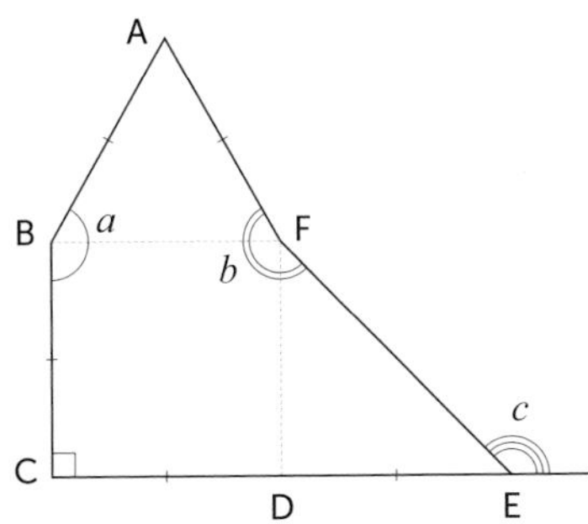


5. What is the missing angle in this triangle?
   Give reasons for your answer.

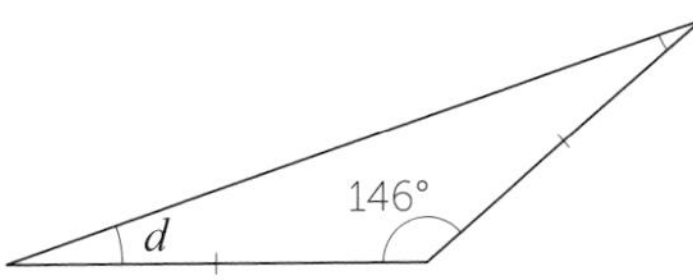


6. In each case, find the missing exterior angle.

a)

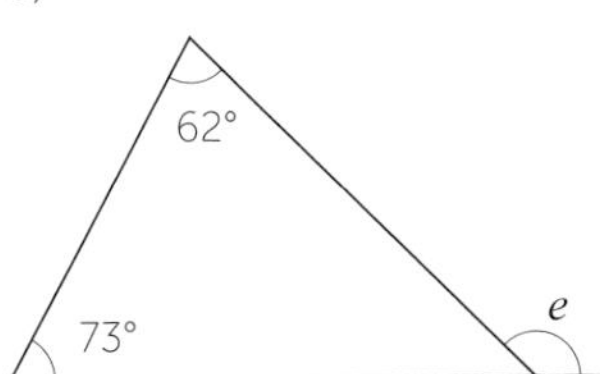


b)

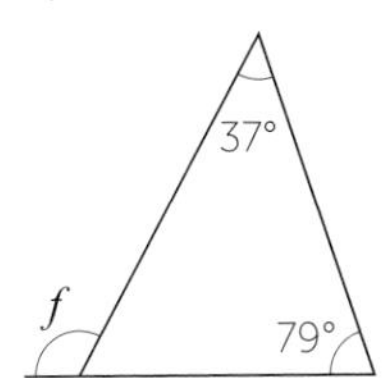


c)

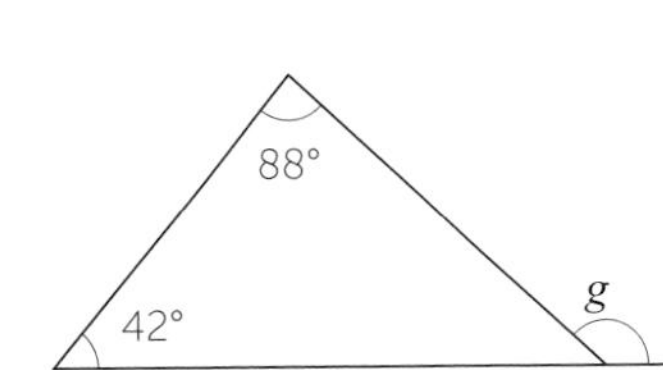


d)

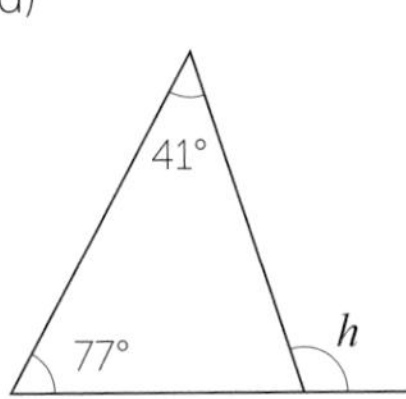

7. What are the missing angles in these diagrams?
   Give reasons for your answers.

a)

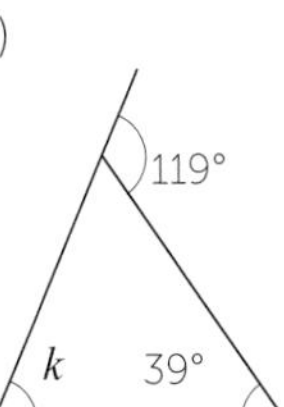


b)

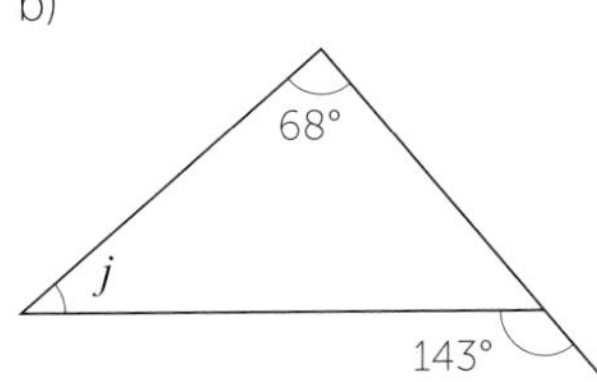

SECTION 19

# EQUATIONS AND IDENTITIES

A4 A17 A21

## 19.1 KNOWLEDGE CHECK

### Objectives

**Expand a single bracket and simplify expressions**

Multiply all terms inside the bracket by the term on the outside of the bracket.

Using a grid can help.

**e.g.** Expand $3(2x + 5) = 6x + 15$

| × | $2x$ | $+ 5$ |
|---|---|---|
| 3 | $6x$ | $+ 15$ |

**Factorise into a single bracket**

To factorise terms, find the common factor in each term and place this outside a bracket.

Put the remaining factors inside the bracket. Using a grid can help.

**e.g.** Factorise $12x - 18 = 6(2x - 3)$

| × | $2x$ | $- 3$ |
|---|---|---|
| 6 | $12x$ | $-18$ |

**e.g.** Factorise fully $15y^2 + 25y = 5y(3y + 5)$

| × | $3y$ | $+ 5$ |
|---|---|---|
| $5y$ | $15y^2$ | $25y$ |

**Form and solve an equation from a worded description**

**Solve linear equations involving brackets with the variable on one side**

Solve equations involving brackets by expanding the brackets before balancing both sides of the equation.

**e.g.** Solve $3(2x + 1) = 18$

$6x + 3 = 18$ expand the brackets

$6x = 15$ subtract 3 from both sides

$\mathbf{x = 2.5}$ divide both sides by 6

### Practice questions

1. Expand the brackets in each of these expressions.
   a) $2(x + 7)$ b) $3(2y - 5)$ c) $n(n + 2)$ d) $6(4 - 3p)$

2. Factorise fully these expressions.
   a) $4x + 12$ b) $15a - 10$ c) $p^2 - 8p$ d) $6d - 4e + 10f$

3. Solve these equations.
   a) $5x + 7 = 22$ b) $3(y - 4) = 33$ c) $15 - n = 7$ d) $4(5 - g) = 16$

4. Kelly has expanded the bracket $7x(2x - 5)$ to give $9x - 12x$
   Find the mistakes and correct them.

5. Ben is 7 years older than Matthew. The sum of their ages is 39.
   Form an equation from this information and solve the equation to find the ages of Ben and Matthew.

6. Here is a square.
   Work out the value of $x$

   16

   $2x + 5$

7. Tilda has been asked to factorise $y^2 - 7y$
   She writes this answer $= y(y^2 - 7)$
   Is she correct? Give a reason for your answer.

8. Gerry has 5 more sweets than Frieda.
   Henri has twice the number of sweets as Frieda.
   Between them, Frieda, Gerry and Henri have 57 sweets.
   Form and solve an equation to find how many sweets Frieda has.

9. The perimeter of this right-angled triangle is 30 cm
   Find the area of the triangle.

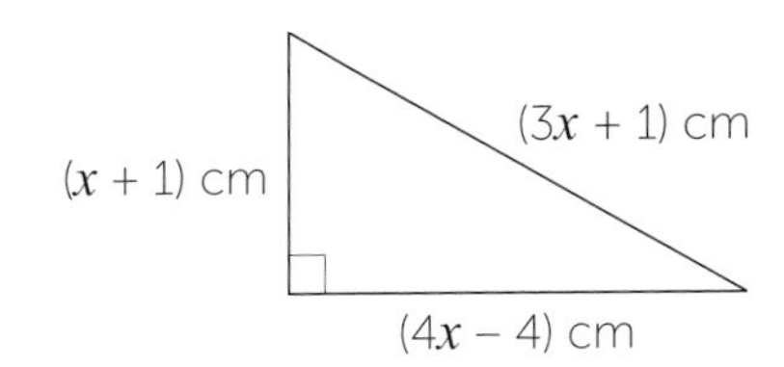


10. The area of this rectangle is 84 cm$^2$
    Find the value of $x$.

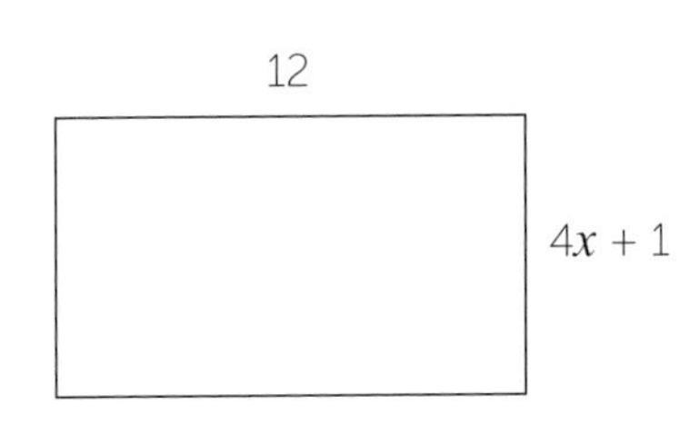

## 19.2 HARDER LINEAR EQUATIONS

### Objectives

**Solve equations where the unknown appears on both sides, with or without brackets**

Remember that the letters work exactly like numbers (they are numbers) so you can add or subtract them to balance the equation.

**e.g.** To solve $5x - 1 = 2x + 11$

$3x - 1 = 11$ subtract $2x$ from both sides

$3x = 12$ add 1 to both sides

$x = 4$ divide both sides by 3

**Solve equations where the variable and/or solution is negative**

Try to keep the $x$ term positive if possible.

**e.g.** To solve $11 - 3x = 2x - 4$

$11 = 5x - 4$ add $3x$ to both sides

$15 = 5x$ add 4 to both sides

$x = 3$ divide both sides by 5

**Solve equations with fractional coefficients and/or solutions**

When dealing with fractions in an equation or expression you can choose to:

- Add or subtract as fractions, or
- Multiply through by the denominator(s) and solve as you have done before

**e.g.** To solve $\frac{5 - 3x}{4} - \frac{2x + 7}{5} = 1$

$\frac{5(5 - 3x) - 4(2x + 7)}{20} = 1$ put over a common denominator

$\frac{25 - 15x - 8x - 28}{20} = 1$ expand the brackets

$\frac{-23x - 3}{20} = 1$ collect like terms

$-23x - 3 = 20$ multiply both sides by 20

$-23x = 23$ add 3 to both sides

$x = -1$ divide both sides by −23

### Practice questions

1. Solve these equations.

   a) $2y - 2 = y + 5$ b) $3n + 7 = 2n - 1$ c) $3x + 2 = 5x - 4$

   d) $2a + 3 = 6a - 7$ e) $5g + 3 = g - 5$

2. Here is a square.

   Find the value of $x$.

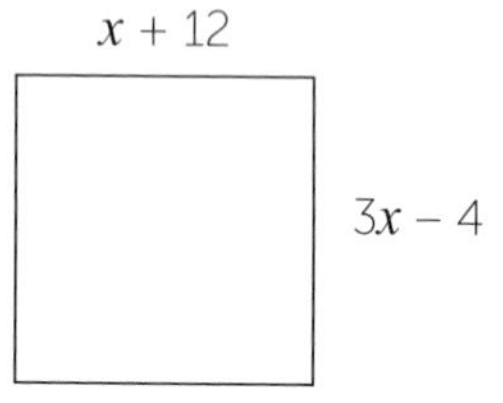


3. Jamie is solving the equation $18 - 3n = 2n + 3$.

   He does this:

   $3n = 2n + 21$ Add 18 to both sides

   $n = 21$ Subtract $2n$ from both sides

   Jamie has made a mistake. Find his mistake and correct it.

4. Solve these equations.

   a) $x + 4 = 19 - 2x$
   b) $\frac{2x + 3}{5} = 4$
   c) $3(2x + 1) = 4x - 5$
   d) $5 = \frac{7 - x}{2}$
   e) $2x + 3 = \frac{3x - 1}{2}$

5. Portia has tried solving the equation $\frac{x + 2}{3} = \frac{2x + 1}{5}$

   She started by doing this: $3(x + 2) = 5(2x + 1)$

   She has made a mistake.

   Find her mistake and correct it. Solve the equation.

6. Chloe and Dima are aged $2x - 1$ and $3x + 5$ respectively. Edgar and Fran are aged $x + 8$ and $5x - 12$ respectively.

   The sum of Chloe and Dima's ages is equal to the sum of Edgar and Fran's ages.

   Who is the oldest? Show your working.

7. Loretta has five numbers: $n$, $n + 3$, $2n - 1$, $2n + 5$ and $3n + 4$

   The mean of these five numbers is 13.

   Find the range of the five numbers.

8. Here is an isosceles triangle.

   a) Work out the value of $x$.

   b) Work out the perimeter of the triangle.

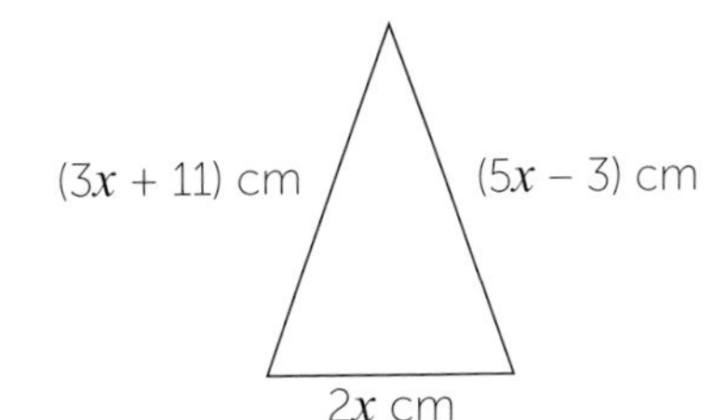

9. Solve $\frac{6x - 1}{2} - \frac{4x + 5}{2} = 2$

10. The area of the square is double that of the triangle. Find the value of $y$.

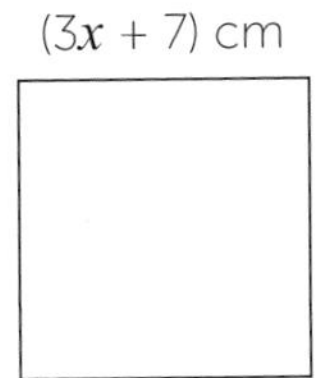


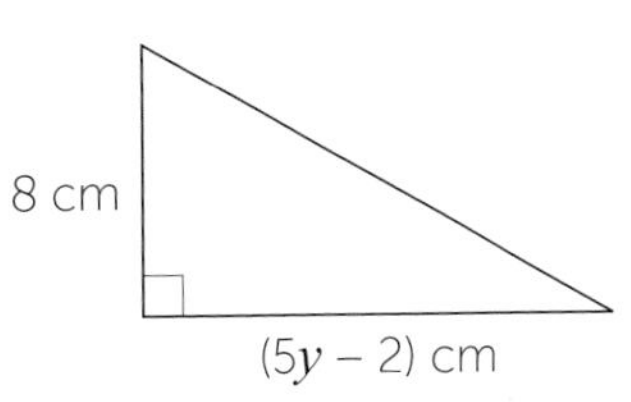

# 19.3 PRODUCT OF TWO BINOMIALS

## Objectives

### Expand the product of two algebraic expressions with brackets

A binomial is an expression with two parts such as $x + 3$

When multiplying two binomials together, each term is multiplied by every other term.

Using a table can be helpful.

**e.g.** Expand and simplify $(x + 3)(x - 2) = x^2 + 3x - 2x + 6$

$= x^2 + x + 6$

| $\times$ | $x$ | $+3$ |
|---|---|---|
| $x$ | $x^2$ | $+3x$ |
| $-2$ | $-2x$ | $-6$ |

### Square a linear expression

When you square a number, you multiply it by itself.

Squaring a linear expression means to multiply a linear expression by itself.

Write the product out as two brackets to avoid forgetting to multiply terms.

**e.g.** Expand and simplify $(2x - 3)^2 = (2x - 3)(2x - 3)$

$= 4x^2 - 6x - 6x + 9$

$= 4x^2 - 12x + 9$

| $\times$ | $2x$ | $-3$ |
|---|---|---|
| $2x$ | $4x^2$ | $-6x$ |
| $-3$ | $-6x$ | $+9$ |

## Practice questions

1. Expand and simplify each of these expressions.
   a) $(y + 1)(y + 4)$ b) $(n + 2)(n + 7)$ c) $(x - 2)(x - 6)$
   d) $(p - 3)(p - 1)$ e) $(t + 5)(t + 2)$

2. A shape is made from a square and two rectangles.
   Write down an expression for the area of this shape.
   Expand and simplify your expression.

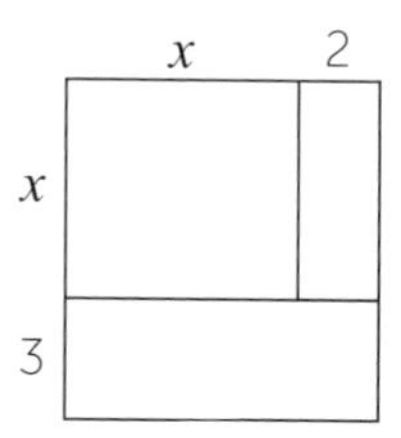


3. Expand and simplify each of these expressions.
   a) $(w + 1)(w - 2)$ b) $(x - 5)(x + 2)$ c) $(d - 7)(d - 3)$
   d) $(y + 6)(y - 3)$ e) $(t + 2)(t - 9)$

4. Duncan has been asked to expand and simplify $(x + 4)^2$
   He gives the answer $x^2 + 16$
   Explain what Duncan has done incorrectly. Give the correct answer.

5. Melissa has 4 less than a number.
   Freya has 3 more than the same number.
   Write down an expression for the product of their two numbers.

6. A square has sides of length $n - 5$

   Write down an expression for the area of the square. Expand and simplify your expression.

7. Expand and simplify each of these expressions.

   a) $(2x + 1)(x + 2)$ b) $(3y - 2)(y - 5)$ c) $(2n + 5)^2$

   d) $(3d - 1)(d + 4)$ e) $(2v + 5)(3v - 1)$

8. Write down an expression for the area of each shape. Expand and simplify your expressions.

   a) Rectangle b) Square c) Triangle

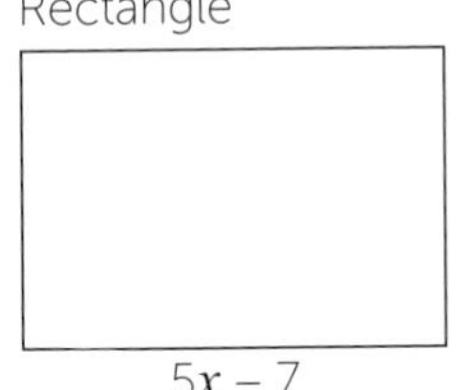


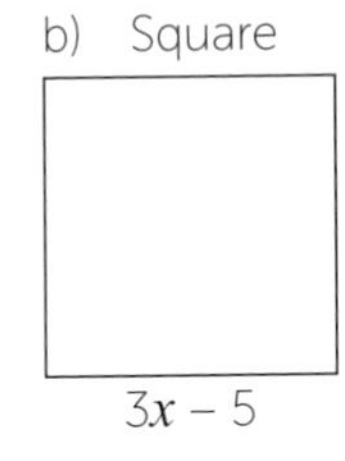


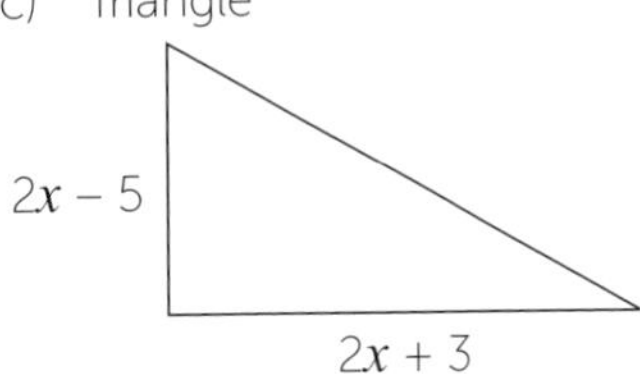

9. Expand and simplify the expression: $(n + 4)(n - 7) - (n - 2)^2$

10. Find an expression for the shaded area in this diagram.

    Expand and simplify your expression.

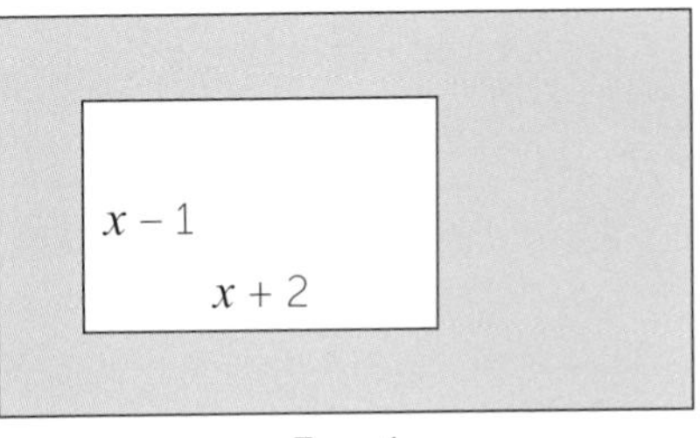


## 19.4 FACTORISING QUADRATICS

A4

### Objectives

#### Factorise a quadratic expression into two brackets

When two binomials are multiplied they make a quadratic expression.

To factorise a quadratic expression, means to express it as the product of two binomials.

Since this is the opposite of expanding two binomials; the table can be used in reverse.

| × | | | |
|---|---|---|---|
| | $x^2$ | | |
| | | $-6x$ | |
| | | | $+8$ |

**e.g.** Factorise $x^2 - 6x + 8$

Write down pairs of numbers that multiply to give $+8$

$+1, +8$ $-1, -8$ $+2, +4$ $-2, -4$

$-2$ and $-4$ add to give $-6$, so these are the numbers in the binomials

Complete the rest of the table:

$x^2 - 6x + 8 = (x - 2)(x - 4)$

| × | $x$ | | $-4$ |
|---|---|---|---|
| $x$ | $x^2$ | | $-4x$ |
| | | $-6x$ | |
| $-2$ | $-2x$ | | $+8$ |

#### Factorise $x^2 - a^2$ using the difference of two squares

When the product of two **certain** binomials is found; the answer given is the difference of two squares.

**e.g.** $(x + 3)(x - 3) = x^2 - 3x + 3x - 9$

$= x^2 - 9$

$x^2$ is a square and 9 is a square; so, this expression is the **difference of two squares**.

This only happens when the $x$ terms **cancel** each other out.

When factorising the difference of two squares; think which binomials would give these terms when multiplied.

**e.g.** Factorise $x^2 - 4 = (x + 2)(x - 2)$

Factorise $x^2 - 100 = (x + 10)(x - 10)$

## Practice questions

1. Find all the pairs of integers that multiply to make the following integers.
   a) 5 b) 8 c) 20 d) 12 e) 25

2. Factorise the following expressions.
   a) $a^2 + 6a + 8$ b) $k^2 - 5k + 4$ c) $n^2 - 9n + 20$
   d) $y^2 + 7y + 12$ e) $x^2 - 10x + 16$

3. The area of this rectangle is given by the expression $n^2 + 5n + 4$
   The length of the rectangle is given by the expression $(n + 4)$
   Write down an expression for the width of the rectangle.

   $n^2 + 5n + 4$

4. Find all the pairs of integers that multiply to make the following integers.
   a) −3 b) −10 c) −4 d) −18 e) − 24

5. Factorise the following expressions.
   a) $t^2 + 4t - 5$ b) $q^2 + 5q - 6$ c) $a^2 - a - 6$
   d) $x^2 + 2x - 15$ e) $y^2 - 8y - 20$

6. Factorise each of these difference of two squares.
   a) $x^2 - 64$ b) $y^2 - 81$ c) $r^2 - 121$

7. Simon is factorising $p^2 - 5p - 6$ and gets $(p - 2)(p - 3)$
   He has made a mistake.
   Explain what he has done wrong and correct his answer.

8. Georgia and Henry are factorising $n^2 - 4n - 5$

| **Georgia's solution:** | **Henry's solution:** |
|---|---|
| Pairs that multiply to −4 are: | Pairs that multiply to −5 are: |
| 1, −4    −1, 4    −2, 2 | −1, −5 |
| None add to make −5 | They add to make −4 |
| Answer: It doesn't factorise | Answer: (x − 1)(x − 5) |

They have both made mistakes. Find their mistakes and correct them.

9. Write down expressions for the height and base of this triangle.
   Find another pair of expressions which also work.

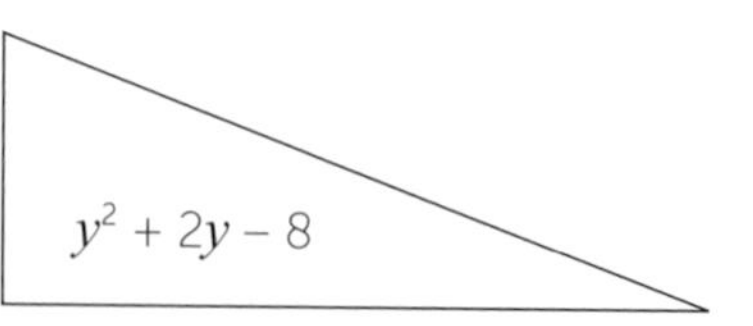


10. The expression for the area of this rectangle is $2x^2 - 5x - 3$
    Find expressions for the length and width of the rectangle.

    $2x^2 - 5x - 3$

# 19.5 IDENTITIES

## Objectives

**Understand the symbols =, ≠ and ≡**

= is an 'equals sign'; it shows that both sides have the same value. **e.g.** $3x + 2 = 14$

≠ is a 'not equals sign'; it shows that both sides do not have the same value. **e.g.** $5 + 4 \neq 7$

≡ is an 'identity symbol'; it shows that both sides are always equal. **e.g.** $3x + 2x \equiv 5x$

**Understand the difference between equations and identities**

**Equations** only work for certain values of the variable; an equation can be solved to find the value of that variable for which it works.

**Identities** are two algebraic expressions that are **always** equal for any value of the variable.

**Argue mathematically to show that algebraic expressions are equivalent**

## Practice questions

1. Insert the symbol ≠ or = between each of these number statements.
   a) $3 \times -4$ ____ 12
   b) $5 + (4 - 9)$ ____ 0
   c) $(-4)^2 - 2 \times 9$ ____ 2

2. State whether each of these is an identity or not.
   a) $a + a + a + a = 4a$
   b) $2n + 3n - n = 12$
   c) $x \times 5y = 5xy$
   d) $3p^2 \times 4p = 7p + 8$
   e) $5a + 2b - 3a + 7b = 2a + 9b$

3. Piers has an identity, but his pen leaked, and ink has covered some of it.
   $6(2x + ■) \equiv ■x + 30$
   What has the ink covered? Show your working.

4. State whether each of these is an equation or an identity.
   a) $3x + 2 = x - 6$
   b) $2n + 6 = n(n + 6)$
   c) $5(y - 3) = 5y - 15$
   d) $4t + 5 - t - 2 = 3(t + 1)$
   e) $2(w + 3) - (w - 2) = w + 8$

5. Clara has been asked to write an identity using the expression $4(x - 3) - 2(x + 5)$
   Her workings are:
   Expand the brackets: $4x - 12 - 2x + 10$
   Simplify: $2x - 2$
   Write the identity: $4(x - 3) - 2(x + 5) \equiv 2x - 2$
   She has made some mistakes. Find the mistakes and correct them.

6. Show that $6(2x - 5) - 5(x - 3) \equiv 7x - 15$ is correct. You must show each stage of your working.

7. Loh says he has an identity, but someone has spilled coffee on part of it.
   $(x - 8)(x - 3) \equiv x^2$ ■
   Complete the identity.

8. Which of these is an identity?
   a) $(y - 4)(y - 6) = y^2 - 6y - 4y + 24$
   b) $(x + 5)(x - 7) = x^2 + 2x - 35$
   c) $a^2 + 3a - 10 = (a - 2)(a + 5)$
   d) $(2h + 1)(h + 3) = 2h^2 + 7h + 3$

9. $a$, $b$, $c$ are integers. Find the value of $a$, $b$ and $c$ in each of these identities.
   a) $(x - 3)(x + 5) \equiv ax^2 + bx + c$
   b) $(ay - 4)(y + b) \equiv 2y^2 - cy + 8$
   c) $(ag + 1)(g + 3) \equiv 6g^2 + 19g + c$
   d) $(2p + 1)(3p + b) \equiv cp^2 + 7p + 2$

# 19.6 PROVING IDENTITIES

## Objectives

### Express even, odd, multiples, squares and consecutive integers using algebra

**Even numbers** are all divisible by 2; they are all multiples of 2

This means in algebra an even number can be described as $n$ (any integer) times 2

**e.g.** An even number is $\mathbf{2n}$ or $2n + 2$, or $2n - 2$ or $2n + 4$ and so on, where $n$ is any integer.

**Odd numbers** are one more or one less than even numbers.

They can be described using algebra by comparing them to even numbers.

**e.g.** An odd number is $2n + 1$ or $2n - 1$ or $2n + 7$ and so on, where $n$ is any integer.

**Multiples of a number**

The multiple of a number can be written in algebra using the same method.

**e.g.** A multiple of 5 is $5n$ or $5n + 5$ or $5n - 10$ and so on, where $n$ is any integer.

### Answer 'show that' problems using algebra

When asked to 'show that' something is the case you must use algebra to show that it works for every value that fits the description.

**e.g.** Show that two consecutive odd numbers always sum to an even number.

Consecutive odd numbers means two odd numbers which are next to each other.

Let one odd number be $2n + 1$
So the next odd number will be $2n + 3$

The sum of these two odd numbers
$= 2n + 1 + 2n + 3$
$= 4n + 4$
$= 4(n + 1)$

This must be an even number since 4 is a factor of the number.

### Prove identities

## Practice questions

1. $n$ is any integer. Decide for each expression whether it is describing an odd or an even number.
   Give a reason for your answer.
   a) $2n$  b) $2n - 4$  c) $2n + 1$  d) $4n + 6$  e) $6n - 3$

2. John has the following question:
   $x$ is any integer. Is the expression $4x + 3 - 2x - 4$ an odd or even number?
   He says that $4x + 3 - 2x - 4 \equiv 2x - 1$ which is even because it has $2x$ in it.
   Explain why John's answer is not correct.

3. Show that $7y + 6 - 5y - 3 + y$ produces a multiple of 3 when $y$ is an integer.
   Show each step of your working.

4. Each expression below gives a multiple of an amount when $n$ is an integer.
   Find the multiple for each expression.
   a) $6n - 7 + 4n - 8$  b) $4(3n + 1) + 3(n - 3)$  c) $5(2n - 1) - 3(2n - 7)$

5. Show, using algebra, that two consecutive even numbers sum to an even number.
   Write a sentence at the end to explain your answer.

6. Gerry has been asked to prove that the sum of three consecutive integers is a multiple of three.
   He has done this: $3n + (3n + 3) + (3n + 6) \equiv 9n + 9$
   $\equiv 3(3n + 3)$ which is a multiple of 3
   He has made a mistake. Find his mistake and correct it.

7. Nina says: "If you square an even number and add 3 you always get a prime number."
   Show that she is wrong by finding a square number for which this does not work.

8. Show that the sum of $a(a + 1)$ and $a(5 - a)$ is an even number when $a$ is an integer.
   Show your working.

9. Lizzie has to prove that the sum of three consecutive even numbers is a multiple of 6.
   Here are her workings: $n + (n + 2) + (n + 4) \equiv 3n - 6$
   $\equiv 3(n - 2)$ which is a multiple of 3 (not 6)

   Find the mistake she has made and correct it.

10. Show, using algebra, that the square of an odd number is always an odd number.
    Show all your working. Write a sentence to explain your answer.

## 19.7 SOLVING ALGEBRAIC PROBLEMS

A21

### Objectives

**Form an expression or equation from a given geometric situation**

Remember that the letters in algebra represent numbers and work like numbers.

When set an algebraic problem think about how you would do it if you had numbers.

**Solve geometric problems algebraically**

When asked to solve algebraically, you may need to know these mathematical terms.

Sum (add) Difference (subtract) Product (multiply)

**e.g.** There are three bags of marbles.

Bag B contains seven more marbles than Bag A.

Bag C contains three times as many marbles as Bag A.

The sum of the marbles in the three bags is 52.

Work out how many marbles there are in each bag.

Let Bag A = $n$ marbles; then Bag B = $n + 7$ marbles and Bag C = $3n$ marbles

The sum of marbles in the three bags is $n + n + 7 + 3n$ and this equals 52

So, $n + n + 7 + 3n = 52$ solving this equation gives $n = 9$

**So Bag A = 9 marbles, Bag B = 16 marbles and Bag C = 27 marbles**

### Practice questions

1. The perimeter of this rectangle is 44 centimetres.
   Find the value of $a$.

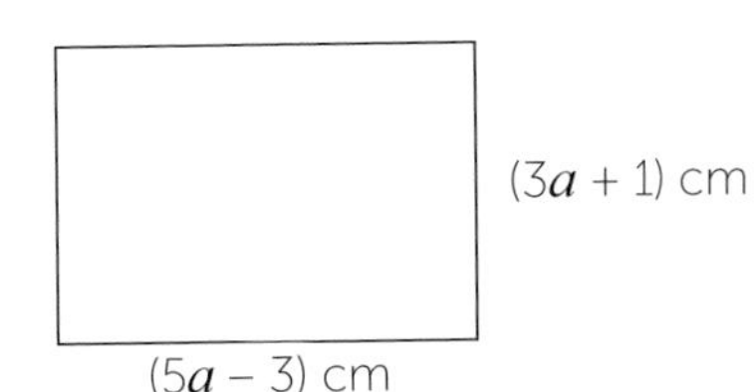


2. Ariana is $y$ years old. Beatrice is 5 years older than Ariana. Chloe is three times as old as Beatrice.
   The sum of their ages is 50 years.
   Work out their ages.

3. Find the value of $x$ in this triangle.

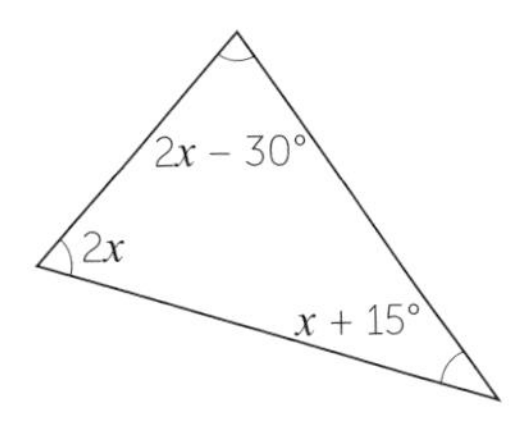

4. Col is $n$ years old; Dea is three times older than Col.
   The difference between their ages is 18 years.
   Work out the ages of Col and Dea.

5. Tea costs $x$ pence at a café. Coffee costs twice the price of tea.
   Fran orders 2 teas and 3 coffees and his bill totals £7.60
   Form and solve an equation to work out the cost of tea and coffee at the café.

6. Rob, Steve and Tom have some sweets.
   Steve has 5 more sweets than Rob.
   Tom has 1 less sweet than Steve.
   The three friends want to share their combined sweets exactly, without any left over.
   Can they do this? Show your working.

7. Work out the size of angle $y$ in this isosceles triangle.

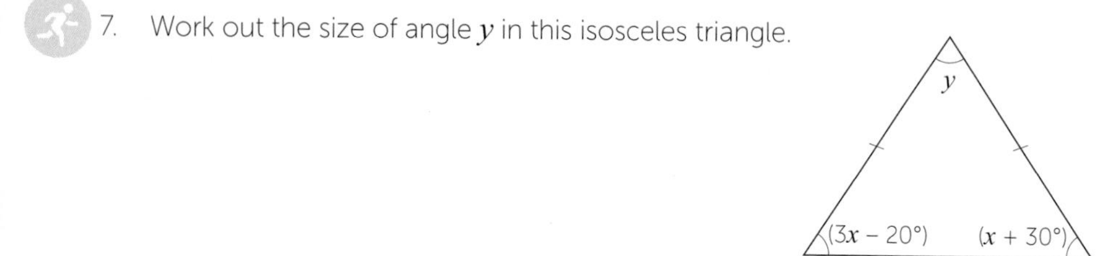


8. Four numbers $n$, $n + 4$, $2n - 1$, $4n - 7$ are listed in order from smallest to largest.
   The mean of the four numbers is 15
   Calculate the range and median of the four numbers.

9. This rectangle has an area of 14 $cm^2$
   Show that $x^2 - 5x - 28 = 0$

   $(x - 7)$ cm

   $(x + 2)$ cm

10. Expand and simplify $(x + 3)(x - 3)$
    Use your answer to work out $103 \times 97$ without a calculator.

SECTION 20

# TRIGONOMETRY

## 20.1 RATIOS IN SIMILAR SHAPES

G20 R12

### Objectives

**Understand and use similarity of 2-D shapes**

**Investigate and use the ratio of side lengths of right-angled triangles**

Two triangles are similar if:

- all the angles are the same, or
- all the sides are in the same ratio, or
- two pairs of sides are in the same ratio and the angles between them are equal

e.g. Here are two similar right-angled triangles. Find the values of $a$, $b$, $c$ and $x$.

Explain your reasoning.

Corresponding angles are the same in similar shapes.

Therefore $a = 25°$, $b = 65°$ and $c = 65°$

The ratios of corresponding sides are the same for similar shapes.

4 cm is half of 8 cm

So $x$ is half of 11 cm

So $x$ = **5.5 cm**

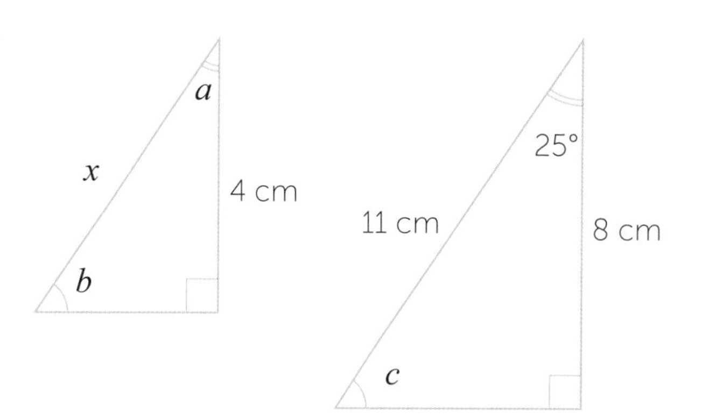


Using ratios:

4 cm : 8 cm = 1 : 2

So $x$ : 11 = 1 : 2

$\frac{x}{11} = \frac{1}{2}$

$x = 11 \times \frac{1}{2}$

$x$ = **5.5 cm**

### Practice questions

1. Which of the following statements are true?
   A Two isosceles triangles are always similar
   B Two equilateral triangles are always similar
   C Two right-angled triangles are always similar

2. State whether each statement below is true or false. Give a reason in each case.
   A All regular pentagons are similar
   B. All hexagons are similar
   C. All squares are similar
   D. All rectangles are similar

3. a) Work out the sizes of angles $s$ and $t$.
   b) Work out the lengths $j$ and $k$.
   c) What is the scale factor of enlargement?

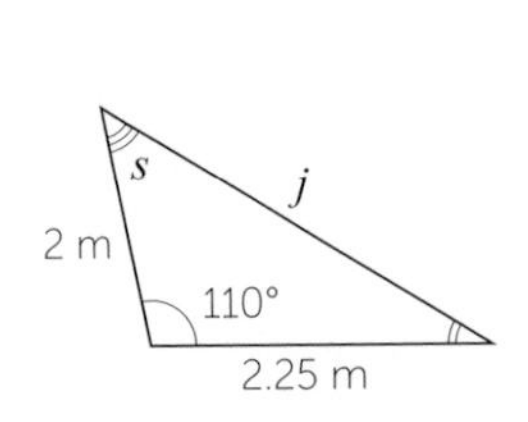


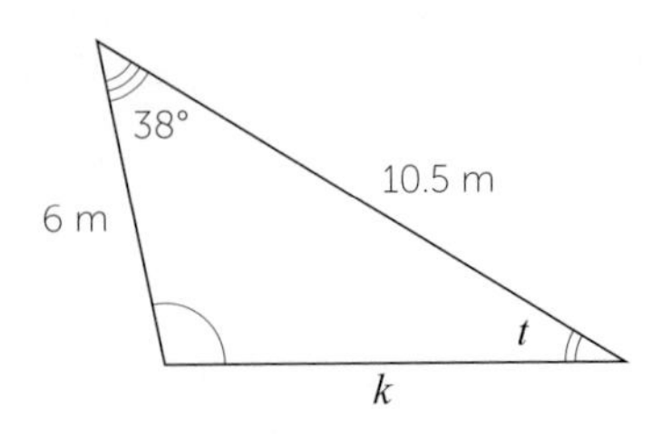


4. Here are two similar right-angled triangles.
   a) Work out the size of angles $m$, $n$ and $p$.
   b) Work out the length $d$.

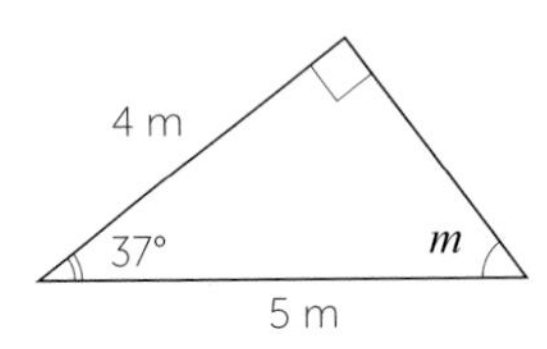


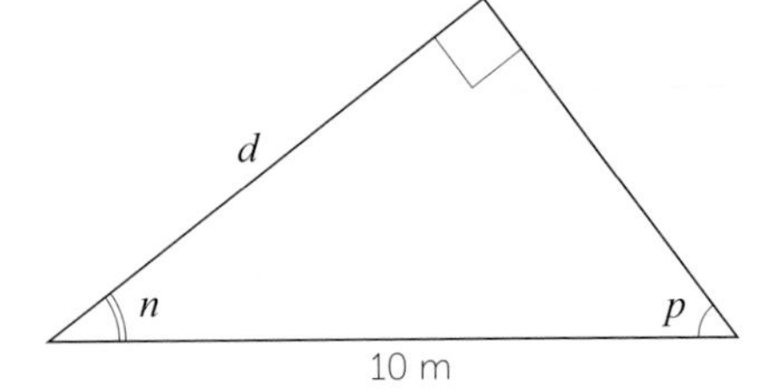


5. The right-angled triangles shown on the right are similar.
   The ratio of sides BC : AC = 0.5.
   Calculate the value of $x$.

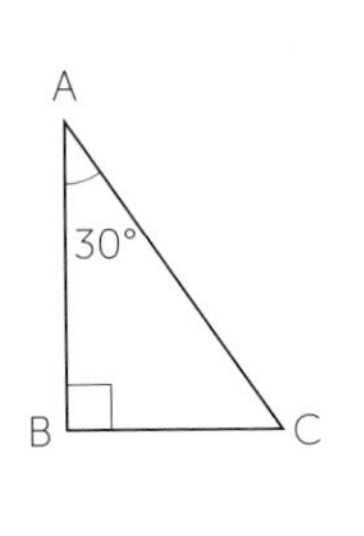


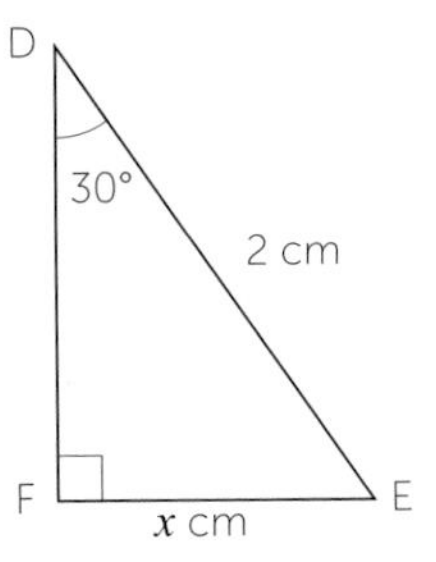


6. Here are two similar rectangles.
   a) Calculate the value of $x$.
   b) Calculate B, the area of the larger rectangle.

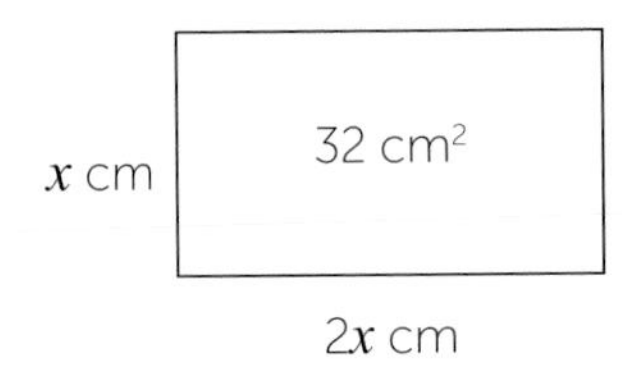


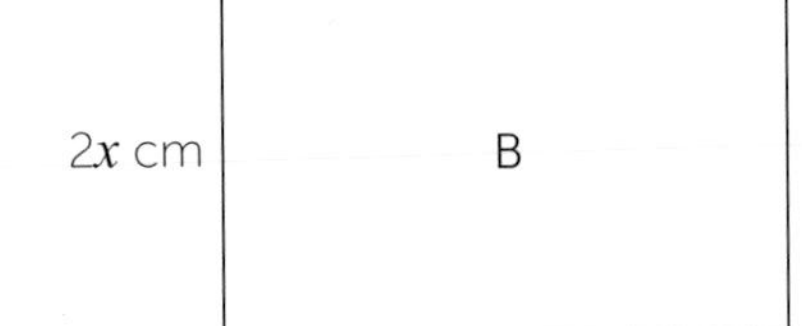


7. Here are two triangles.
   Bernie says that since ED = 3 x AB and DF = 3 x AC, the two triangles are similar.
   Is she correct? Explain your answer.

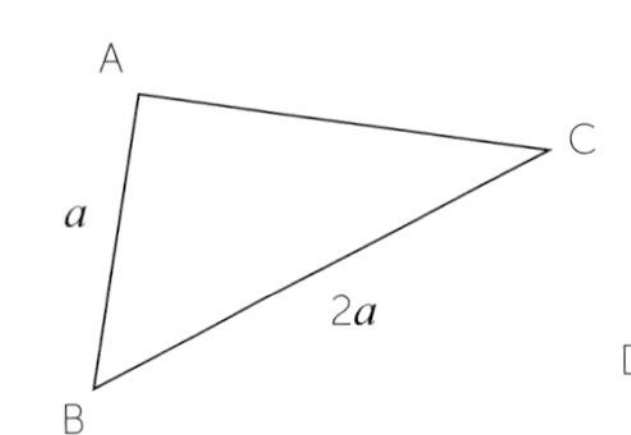


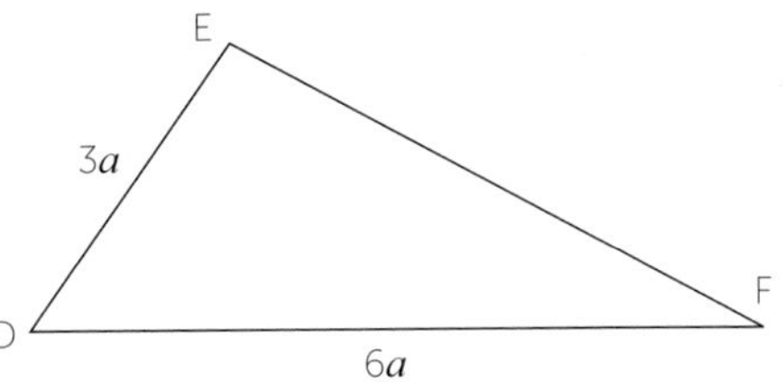

8. The triangles on the right are all similar.
   The ratio of sides XY : YZ = 3 : 2
   a) Work out the value of YZ.
   b) Work out the ratio of $p$ to $q$. Give your answer in its simplest form.

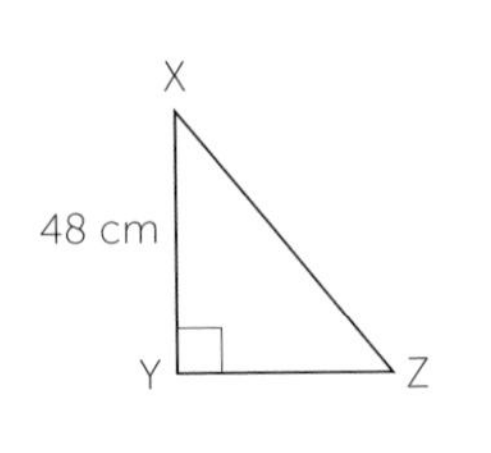


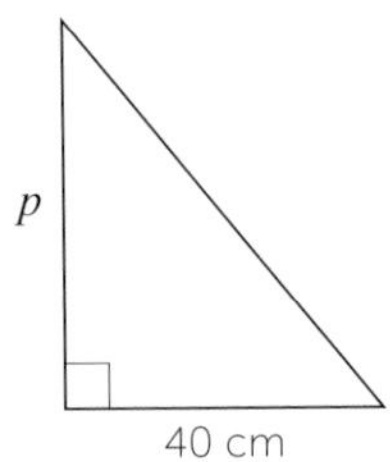


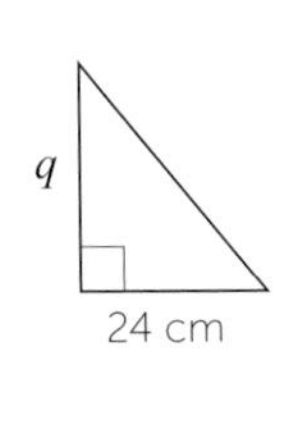

9. In the diagram shown:
   DE is parallel to AC. DE = 1.6 cm. BC = 7.2 cm, BE = 2.4 cm.
   a) Work out the length of AC.
   b) BD is 2 cm. Work out the length of BA.

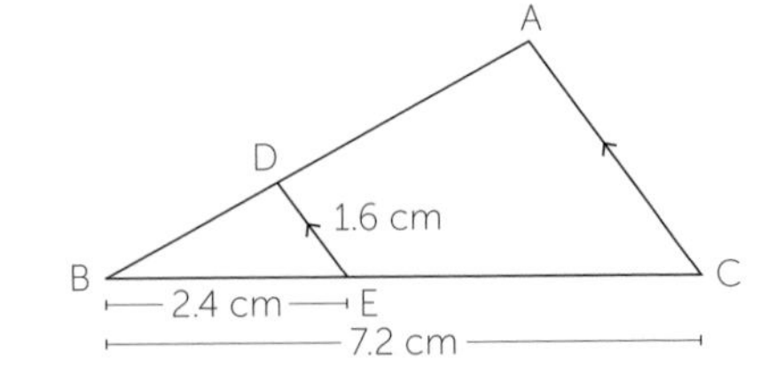

# 20.2 TRIGONOMETRIC RATIOS

## Objectives

### Know and use the names of sides in right-angled triangles

In any right-angled triangle, the longest side is called the hypotenuse.

The other two sides are called the opposite and adjacent sides, depending on the angle given or needed.

### Know the trigonometric ratios

There are three ratios that apply to right-angled triangles, labelled according to angle $x$, as shown:

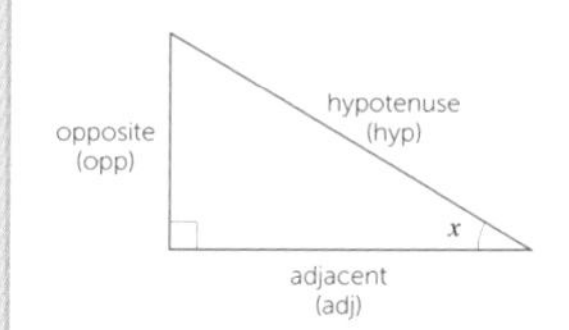


opp : hyp sine $x$ (sin $x$) $= \frac{\text{opposite}}{\text{hypotenuse}}$

adj : hyp cosine $x$ (cos $x$) $= \frac{\text{adjacent}}{\text{hypotenuse}}$

opp : adj tangent $x$ (tan $x$) $= \frac{\text{opposite}}{\text{adjacent}}$

### Know which ratio to apply to find missing values in a right-angled triangle

**e.g.** State which ratio (sin, cos or tan) is being shown in these diagrams.

a)

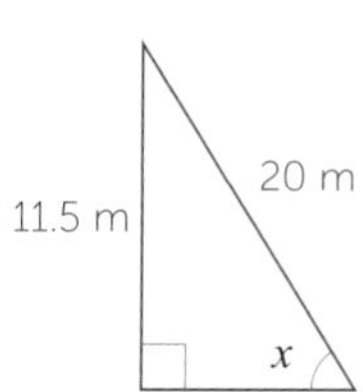


b)

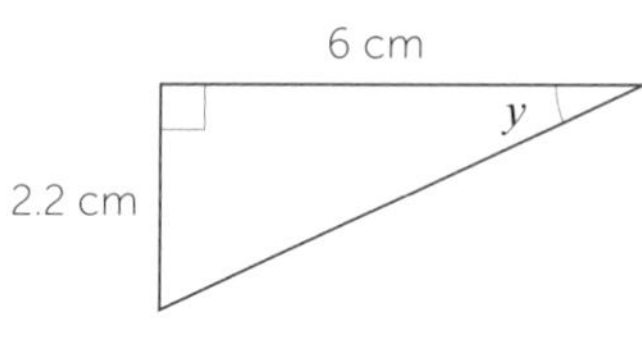


c)

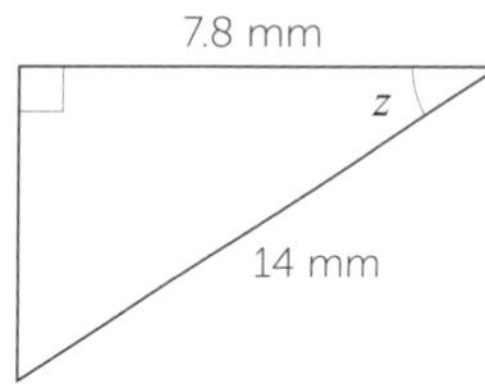


sin $x$ = 11.5 ÷ 20

$\frac{11.5}{20} = 0.576$

tan $y$ = 2.2 ÷ 6

$\frac{2.2}{6} = 0.367$

cos $z$ = 7.8 ÷ 14

$\frac{7.8}{14} = 0.557$

You can use your calculator to find the sine, cosine or tangent of an angle.

**e.g.** Press the **sin** button and then the number buttons **3** and **0** to find sin 30°. Press = to display the result.

a)

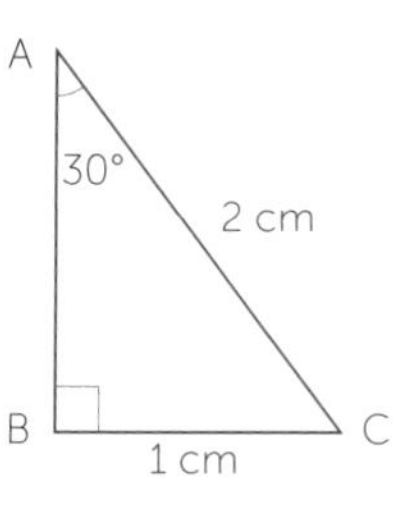


b)

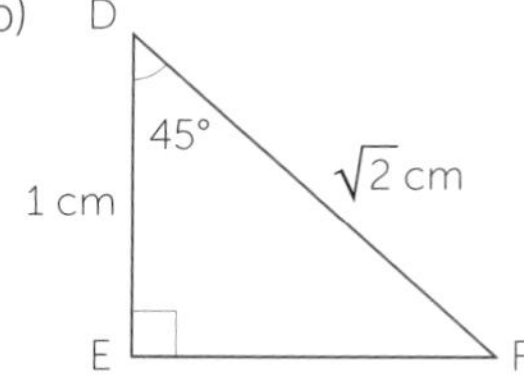


c)

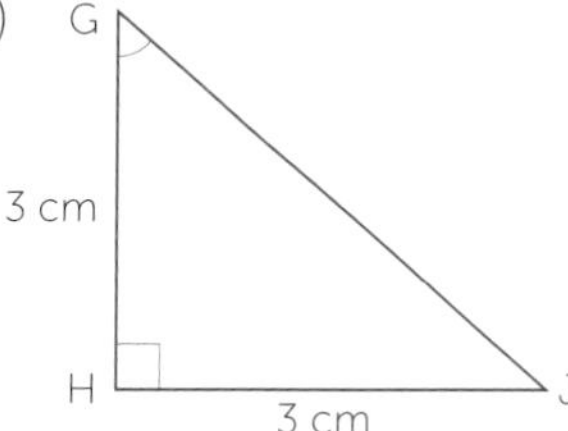


**e.g.** Use your calculator to verify these values for sin 30°, cos 45° and tan 45°.

a) sin 30° $= \frac{1}{2} = 0.5$ b) cos 45° $= \frac{3}{\sqrt{18}} = 0.707$ c) tan 45° $= \frac{3}{3} = 1$

## Practice questions

1. Which side in each triangle is the hypotenuse, the opposite side and the adjacent side for the marked angle? (Example: $a$ is the opposite side for angle $\theta$)

a)

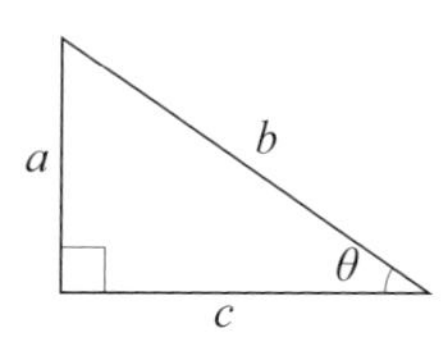


b)

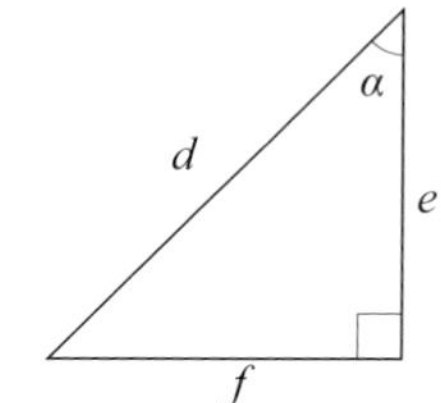


c)

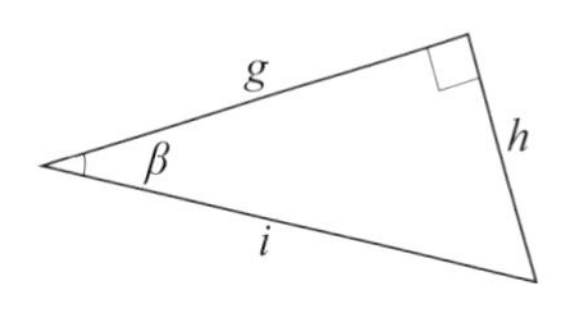

2. In each of these triangles, which pair of sides form the ratio for sine, cosine and tangent of the marked angle? (Example: $\sin\theta = \frac{b}{a}$)

a)

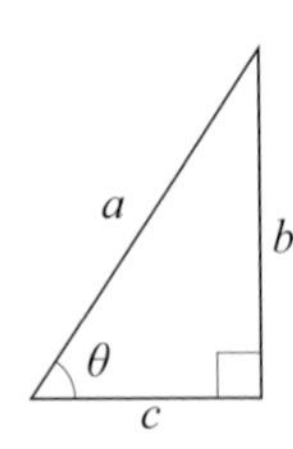


b)

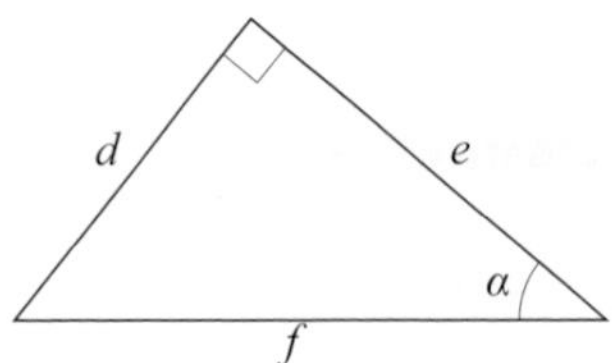


c)

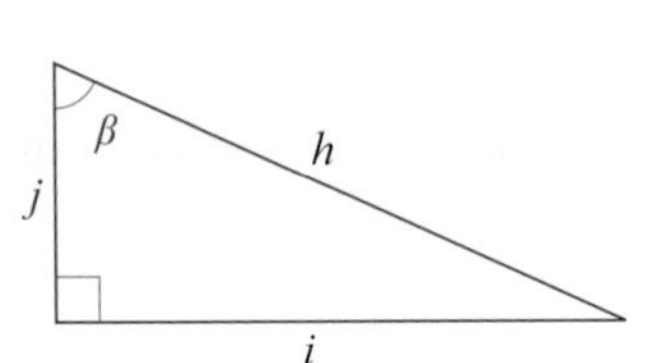


3. Use a calculator to find the following values to 3 decimal places:

a) sin 32°  b) cos 78°  c) tan 14°

4. Which of sin, cos or tan can be used with the lengths and angles given, to find x, y and z, respectively?

a)

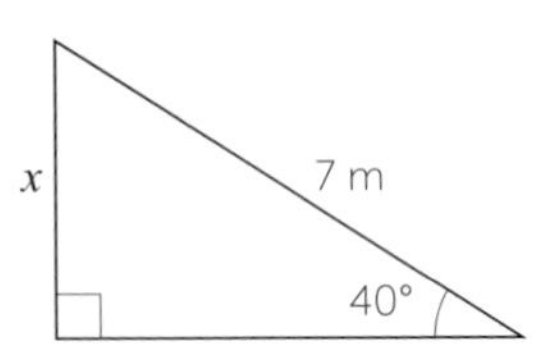


b)

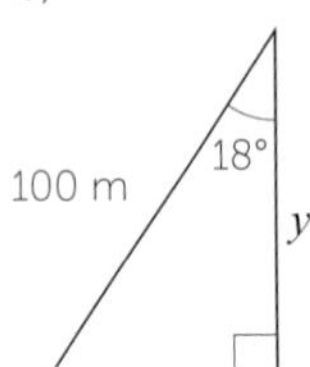


c)

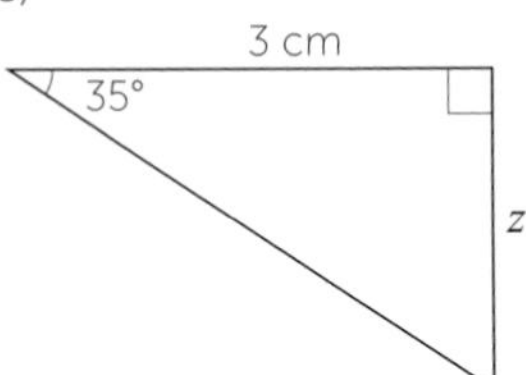


5. Use the information in the right-angled triangle shown (no calculators) to decide whether the following statements are true or false:

a) $\cos 60° = \frac{3}{6}$  b) $\sin 30° = \frac{3}{6}$  c) $\sin 60° = \frac{6}{3}$  d) $\cos 30° = \frac{3}{6}$

6. For each triangle, choose the appropriate trigonometric ratio and work out the unknown side. Give your answer to 2 decimal places.

a)

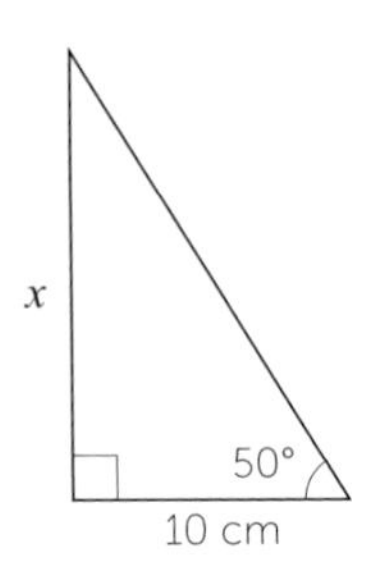


b)

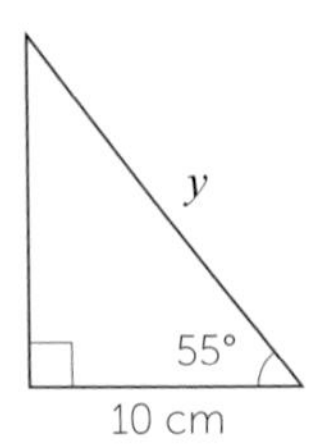


c

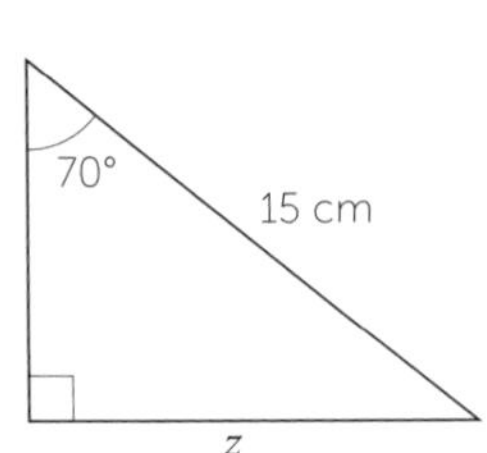


7. Find the value of $x$ in the following triangles.

a)

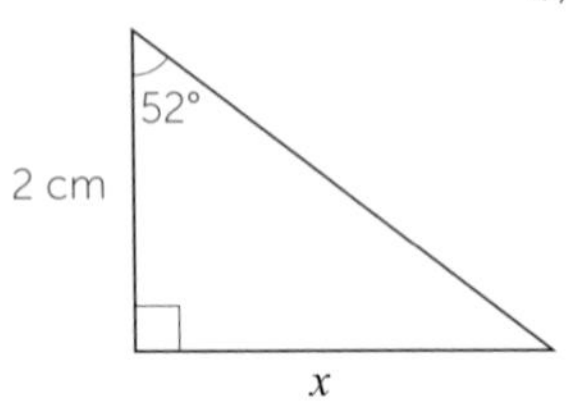


b)

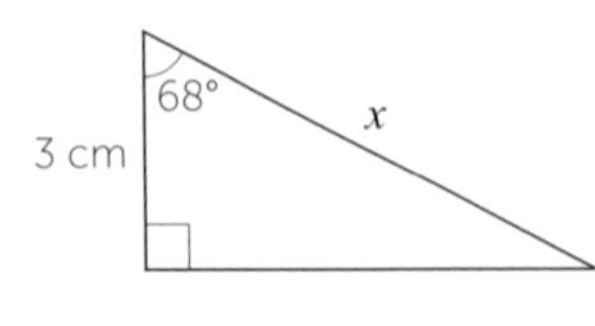


c

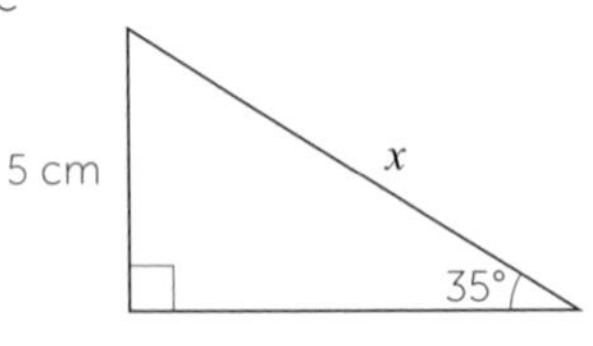


8. Which of the following angles has a tan of 0.625, calculated to 3 decimal places?

A 19°  B 26°  C 32°  D 43°

# 20.3 USING TRIGONOMETRIC RATIOS

## Objectives

### Use trigonometric ratios to find missing lengths in right-angled triangles

In the diagram, **O** is the side **Opposite** to angle $x$, **A** is the side **Adjacent** to angle $x$ and **H** is the **Hypotenuse**.

The Trigonometric ratios associated with these sides are:

$\sin x = \frac{O}{H}$ $\cos x = \frac{A}{H}$ $\tan x = \frac{O}{A}$

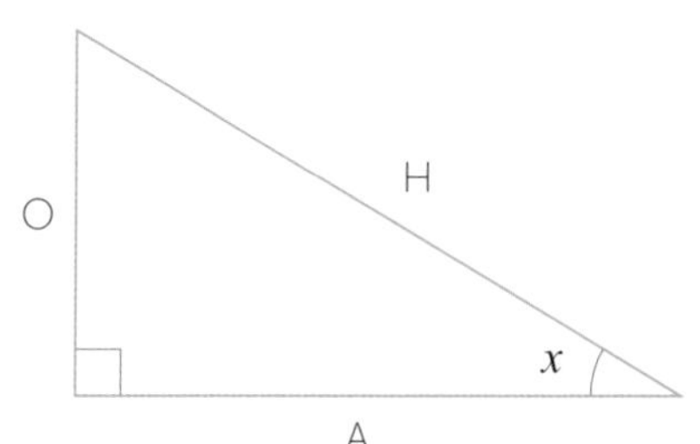


To find an unknown length or angle:

- Label the sides according to the angle to be found, or that is given
- Cross out the side that is not needed, and not known
- Use the remaining sides to determine which ratio to use
- Set up an equation and rearrange it

**e.g.** Find the length of side $y$ in this right-angled triangle:

The ratio required connects O and H, therefore use sine.

$\sin 35° = \frac{y}{42}$

$y = 42 \times \sin 35°$

$y =$ **24.1 cm (1 d.p.)**

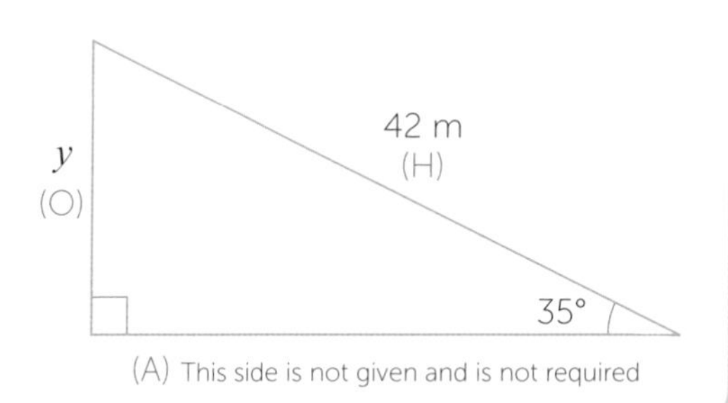


### Use trigonometric ratios to find missing angles in right-angled triangles

Given the cosine, sine or tangent of an angle, you can use a calculator to find the angle.

**e.g.** $\cos^{-1}$ (0.5) means 'The angle whose cosine is 0.5'. Press the Shift button or equivalent on your calculator, then the cos button, to access the $\cos^{-1}$ function. Enter 0.5, =, and the answer, 60, should be displayed.

**e.g.** Find the angle, $x$, that goes with each ratio, to the nearest degree.

a) $\cos x = 0.62$ — **$\cos^{-1}$ (0.62) = 52°**

b) $\sin x = 0.86$ — **$\sin^{-1}$ (0.86) = 59°**

c) $\tan x = 0.2$ — **$\tan^{-1}$ (0.2) = 11°**

**e.g.** Calculate the value of angle $p$ in this right-angled triangle:

The ratio required connects O and A, therefore use tangent.

$\tan p = \frac{2.9}{1.3}$

$p = \tan^{-1}\left(\frac{2.9}{1.3}\right)$

$p =$ **65.9° (1 d.p.)**

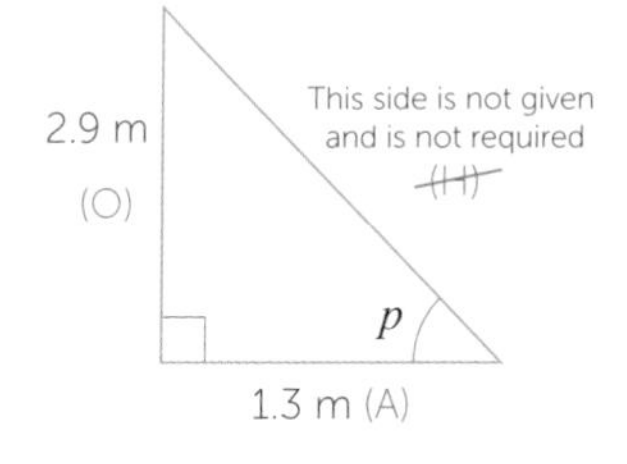


## Practice questions

1. Stu is standing 5 m from the base of a tree. He measures the angle to the top of the tree to be 47°.

   Work out the height of the tree. Give your answer to 1 decimal place.

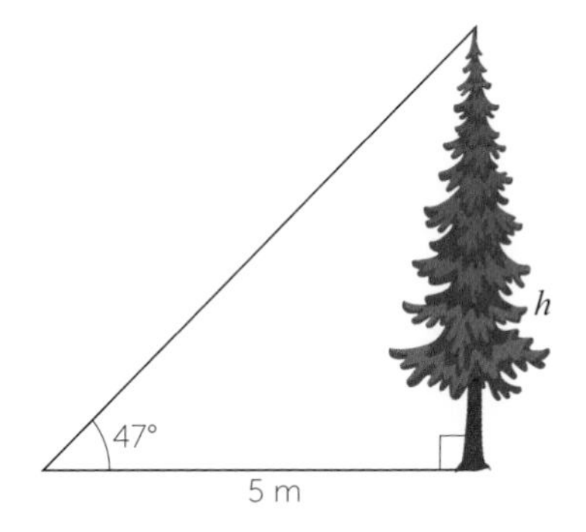

2. Find the angles marked with a letter:

a)

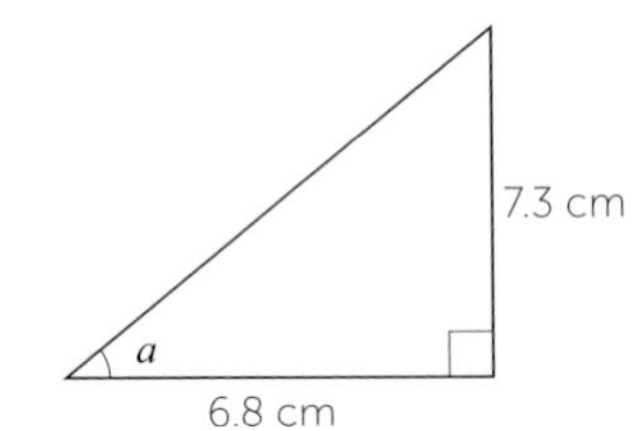


b)

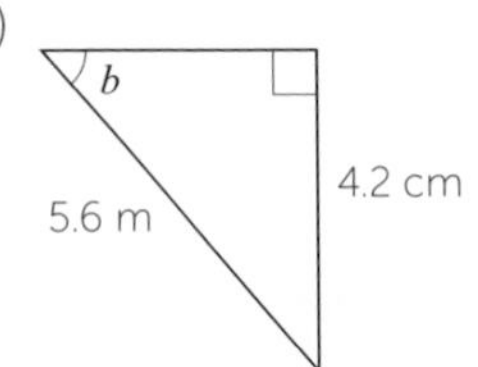


c)

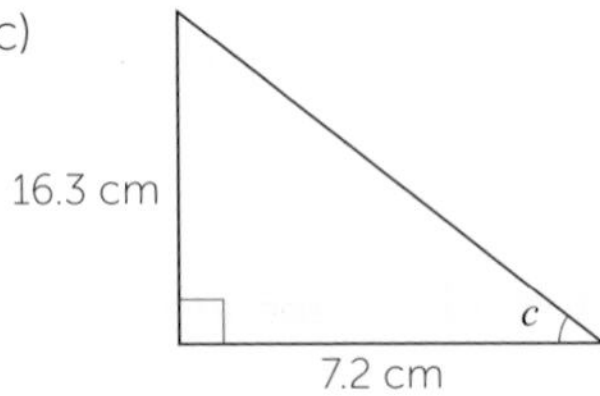


3. Find the angles marked with a letter:

a)

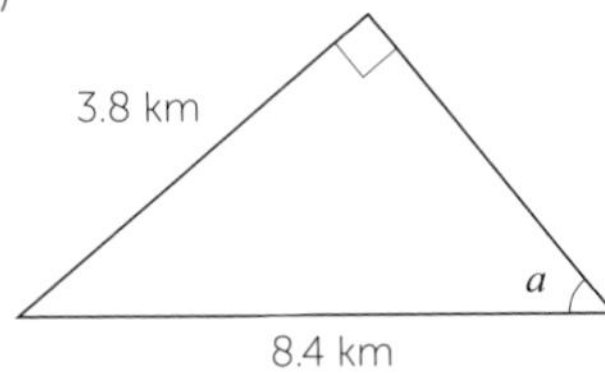


b)

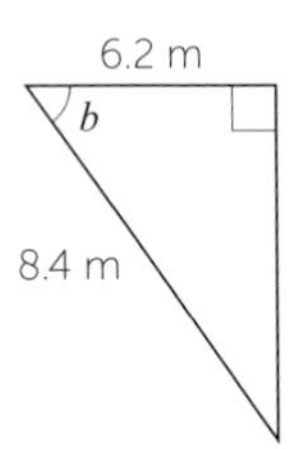


c)

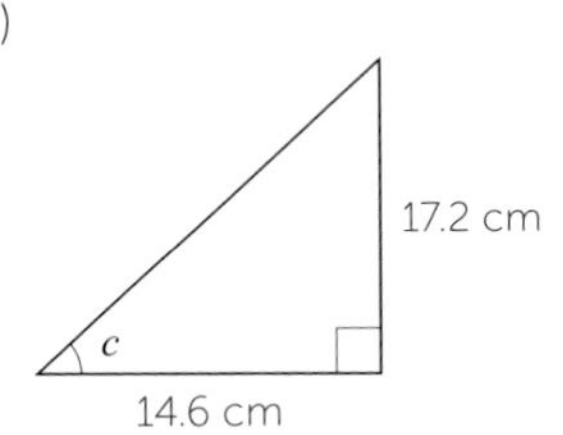


4. Find the lengths marked with a letter:

a)

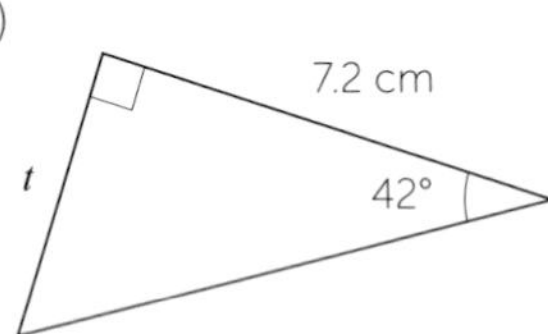


b)

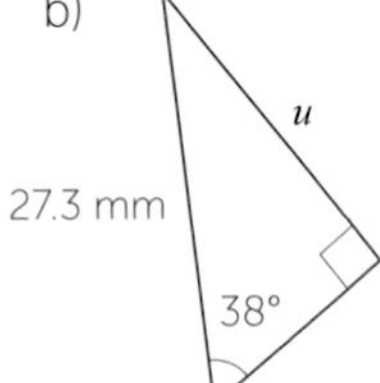


c)

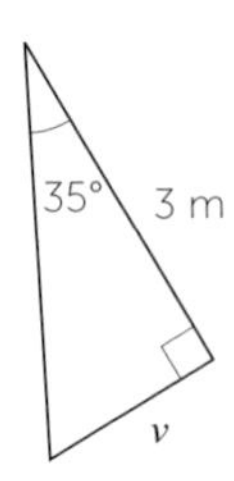


d)

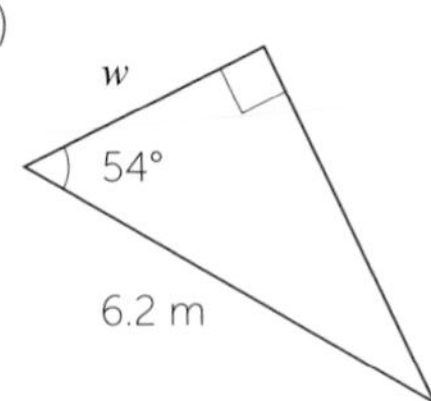


e)

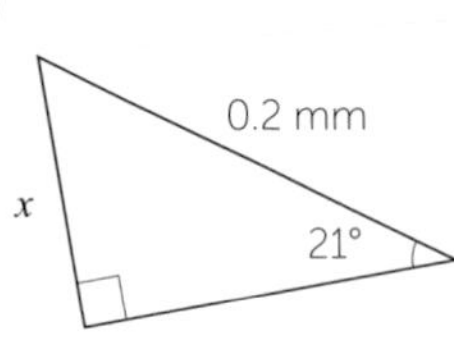


f)

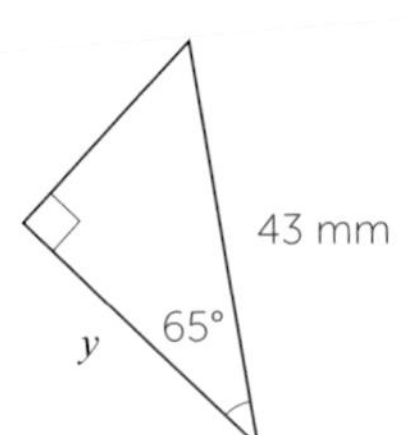


5. Find the length or angle marked with a letter:

a)

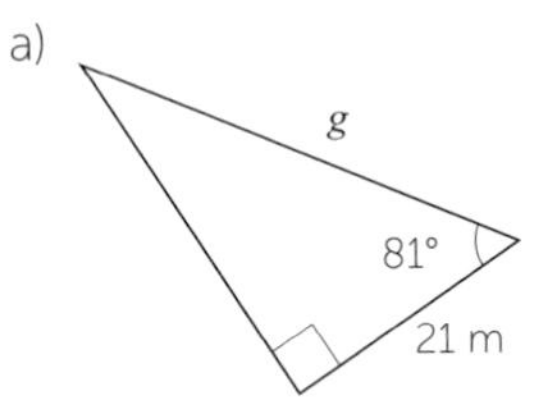


b)

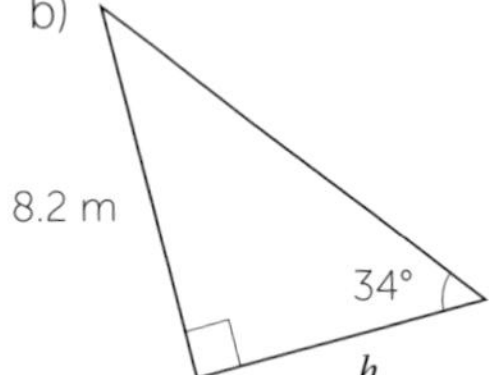


c)

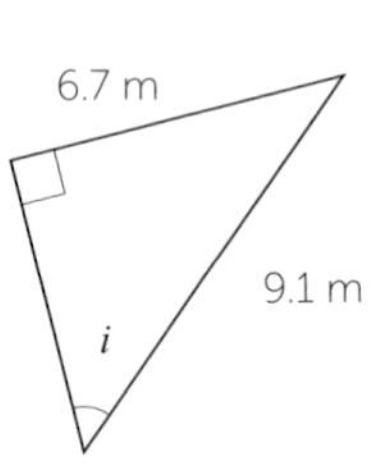

6. In triangle ABC, the point D lies on AC so that BD is perpendicular to AC.

a) Use triangle ABD to work out:

(i) length BD (ii) length AD

b) Use triangle BCD to work out length CD.

c) Use your answers to parts a) and b) to work out length AC.

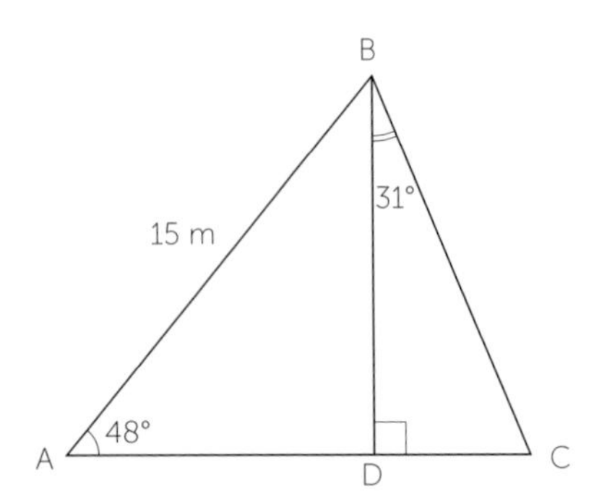

# 20.4 TRIGONOMETRY IN CONTEXT

## Objectives

### Find and use angles of elevation and depression

The **angle of elevation** of an object is the angle between the horizontal line of sight and the line from the observer's eye to the object.

If the object is below the level of the observer, then the angle between the horizontal line of sight and the object is called the **angle of depression**.

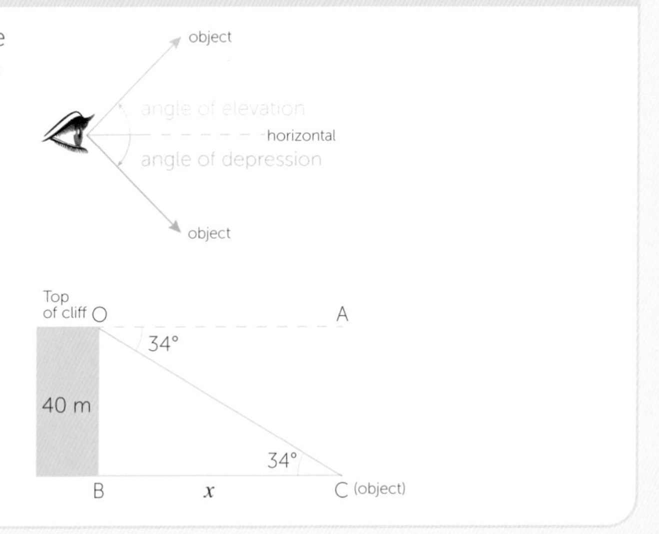


e.g. From the top of a vertical cliff 40 m high, the angle of depression of an object that is level with the base of the cliff is 34°.

Angle of depression AOC = 34°

angle BOC = 90 – 34 ⇒ BOC = 56°

### Use the trigonometric ratios to solve problems in right-angled triangles

How far, in metres, is the object at C from the base of the cliff, B?

Using trigonometry:

$$\tan 56° = \frac{x}{40}$$

$$x = 40 \times \tan 56°$$

$$x = 59.3 \text{ m}$$

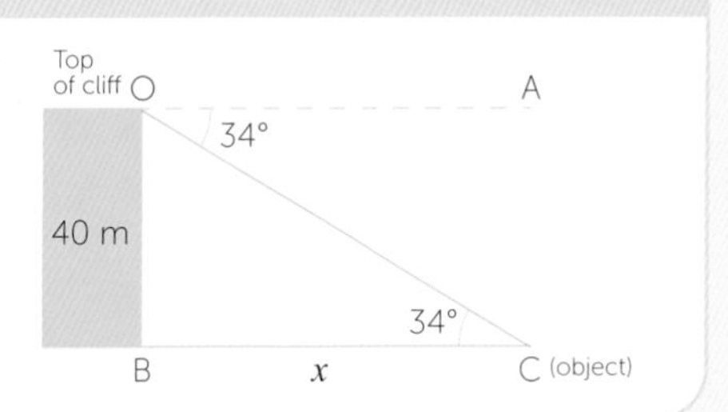


## Practice questions

1. Jo flies a kite above a field at the end of 85 m of string. The angle of elevation to the kite measures 78°.

   How high is the kite above Jo's head?

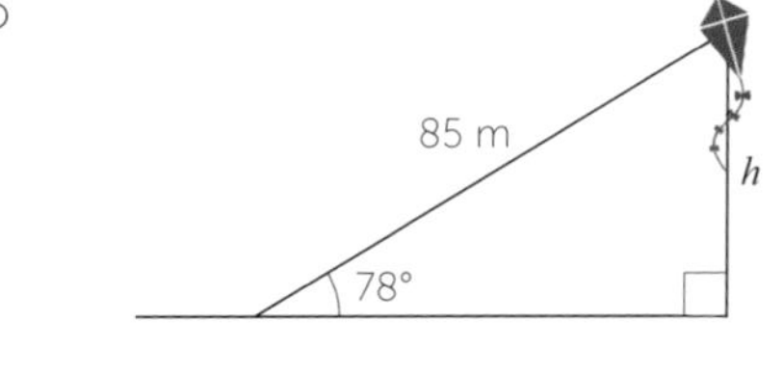


2. From an airplane at an altitude of 4500 m, the angle of depression to a building on the ground measures 64°.

   Find the distance from the plane to the building.

3. From a point on the ground 15 m from the base of a flagpole, the angle of elevation of the top of the pole measures 47°.

   How tall is the flagpole?

4. From the top of a vertical cliff 54 m high, the angle of depression of a buoy level with the base of the cliff is 32°.
   How far is the buoy from the cliff base?

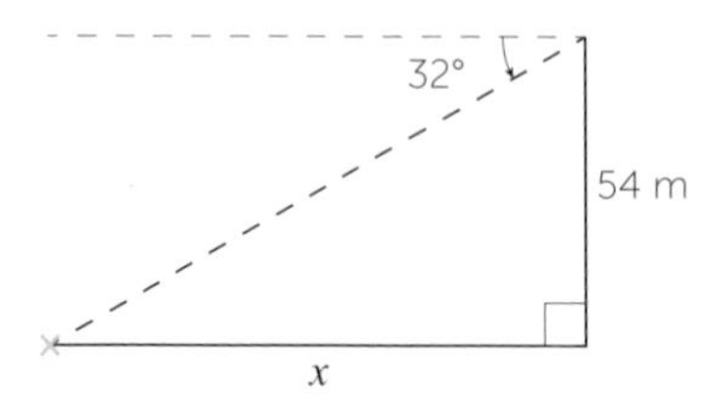


For questions 5 to 10, draw a diagram, then solve the problem.

5. Findlay flies a drone which travels diagonally upwards.
   After flying a distance of 74 m on the diagonal, it hovers above a puddle on the ground.
   The angle of elevation of the drone is 67°.
   a) How high off the ground is the drone?
   b) How far away is the puddle, directly under the drone?

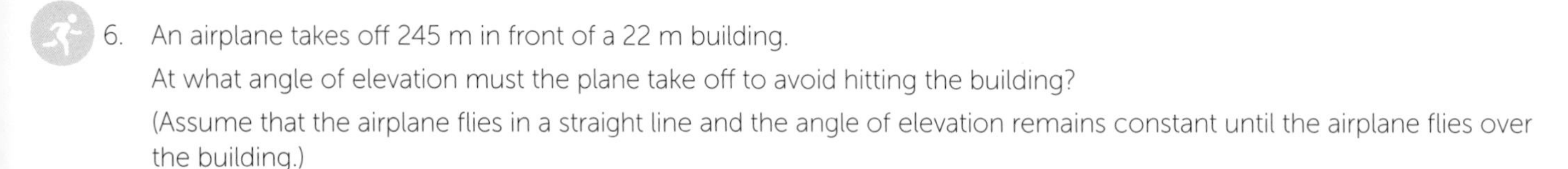

6. An airplane takes off 245 m in front of a 22 m building.
   At what angle of elevation must the plane take off to avoid hitting the building?
   (Assume that the airplane flies in a straight line and the angle of elevation remains constant until the airplane flies over the building.)

7. A 3.5 m ladder is used to climb over a 3 m high wall.
   At what angle of elevation must the ladder be positioned to reach the top of the wall?

8. Standing on a cliff 142 m above the sea, Zach sees an approaching ferry and measures its angle of depression as 13°.
   a) How far from shore is the ferry?
   b) Now Zach sees a second ship beyond the first.
      The angle of depression of the second ship is 7°.
      How far apart are the ships? Give your answer the nearest metre.

9. From a plane flying due east at 548 m above sea level, the instantaneous angles of depression of two ships sailing due east measure 36° and 23°.
   How far apart are the ships? Give your answer to the nearest metre.

10. From standing on the ground, Brodie looks up at a tall office block. He is standing 75 m from the base of the building and the angle of elevation to halfway up the block is 52°.
    a) Calculate the height of the office block.
    b) What is the angle of elevation from Brodie to the top of the block?

## 20.5 SPECIAL ANGLES

### Objectives

**Know the exact values of sin $x$ and cos $x$ for $x$ = 0°, 30°, 45°, 60°, 90°**

**Know the exact value of tan $x$ for $x$ = 0°, 30°, 45° and 60°**

In the equilateral triangle ABC, the length of the perpendicular bisector AD is calculated using Pythagoras' theorem.

In the right-angled isosceles triangle EFG, Pythagoras' theorem is used to calculate the hypotenuse EF.

From these special right-angled triangles, the following relationships can be found:

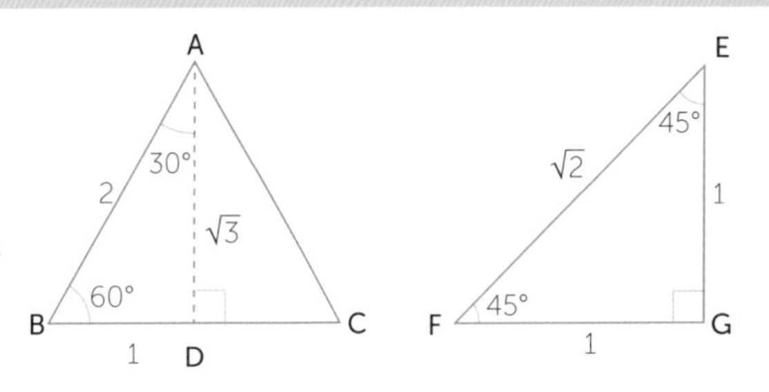


$\sin 30° = \frac{1}{2}$ $\quad$ $\sin 45° = \frac{1}{\sqrt{2}}$ $\quad$ $\sin 60° = \frac{\sqrt{3}}{2}$

$\cos 30° = \frac{\sqrt{3}}{2}$ $\quad$ $\cos 45° = \frac{1}{\sqrt{2}}$ $\quad$ $\cos 60° = \frac{1}{2}$

$\tan 30° = \frac{1}{\sqrt{3}}$ $\quad$ $\tan 45° = 1$ $\quad$ $\tan 60° = \sqrt{3}$

Also,

$\sin 0° = 0$ $\quad$ $\cos 0° = 1$ $\quad$ $\tan 0° = 0$

$\sin 90° = 1$ $\quad$ $\cos 90° = 0$

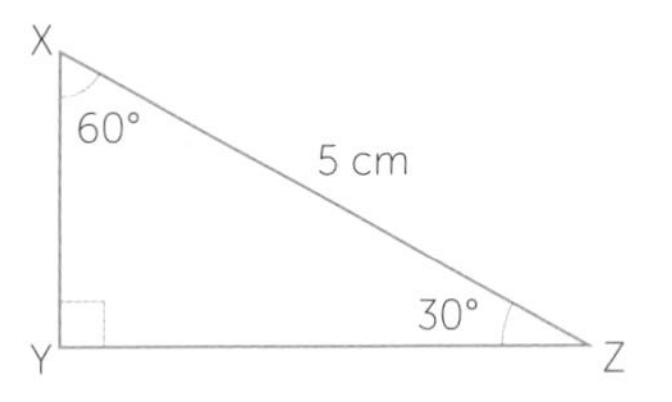


**e.g.** For the triangle shown work out the lengths of XY and YZ:

$\sin 30° = \frac{XY}{5}$ $\qquad$ $\cos 30° = \frac{YZ}{5}$

$XY = 5 \times \sin 30°$ $\qquad$ $YZ = 5 \times \cos 30°$

$= 5 \times \frac{1}{2}$ $\qquad$ $= 5 \times \frac{\sqrt{3}}{2}$

**XY = 2.5 cm** $\qquad$ **YZ = 4.33 cm**

### Practice questions

Do not use a calculator for questions 1 - 6. Leave your answers as surds in terms of $\sqrt{2}$ or $\sqrt{3}$.

1. Work out the lengths of:
   a) BC $\qquad$ b) AC

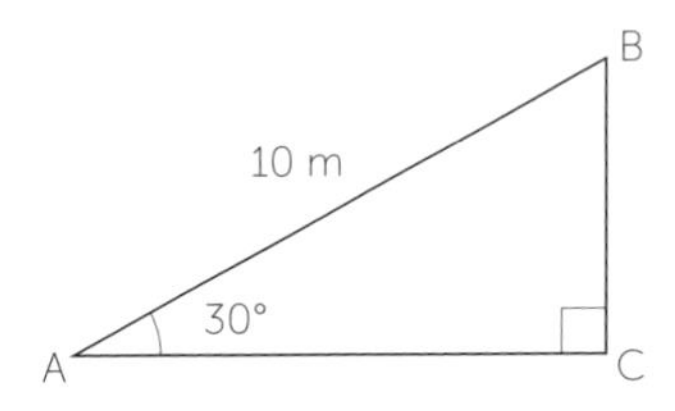


2. Work out the lengths of:
   a) DE $\qquad$ b) EF

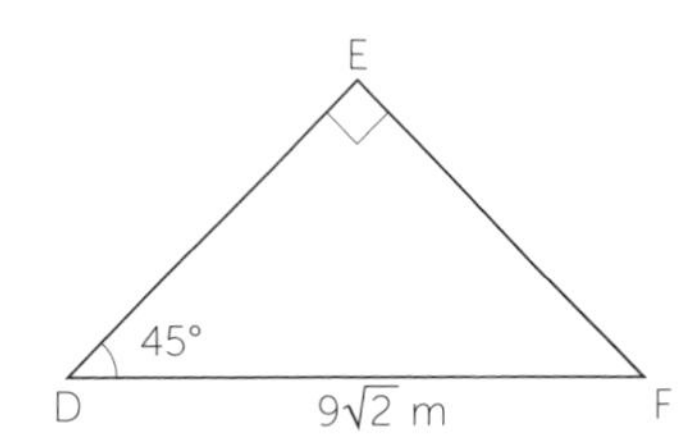


3. Write down the values of:
   a) sin 45° $\qquad$ b) cos 60° $\qquad$ c) tan 30°
   d) cos 0° $\qquad$ e) tan 45° $\qquad$ f) sin 60°

4. Write down the size of the angle given by:

   a) $\cos^{-1}\frac{1}{\sqrt{2}}$ b) $\sin^{-1}\frac{1}{\sqrt{2}}$ c) $\tan^{-1}\sqrt{3}$

5. Work out the lengths of the missing sides in these triangles:

   a)

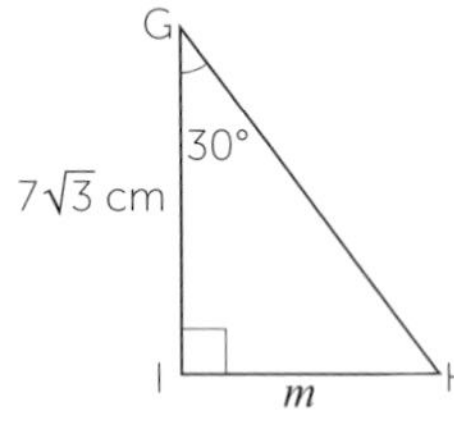


   b)

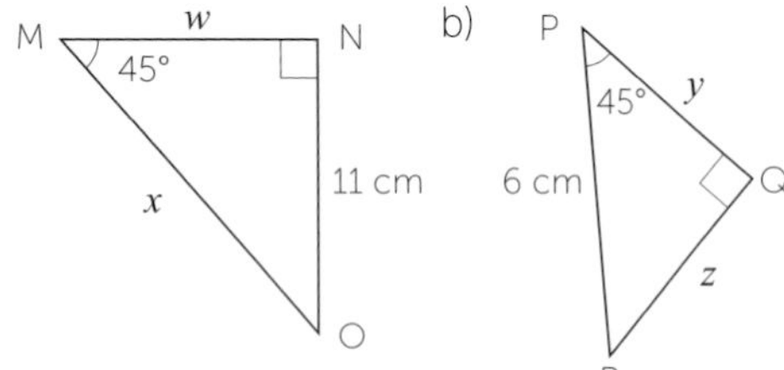

6. Work out the lengths of the missing sides in these triangles:

   a) M w N 45° x 11 cm O

   b) P 45° y 6 cm Q z R

## 20.6 TRIGONOMETRY PROBLEMS

G20 G21

### Objectives

**Use the trigonometric ratios to solve 2D problems in right-angled triangles**

In a right-angled triangle, the following identities are true:

$\sin x = \frac{\text{opposite}}{\text{hypotenuse}}$

$\cos x = \frac{\text{adjacent}}{\text{hypotenuse}}$

$\tan x = \frac{\text{opposite}}{\text{adjacent}}$

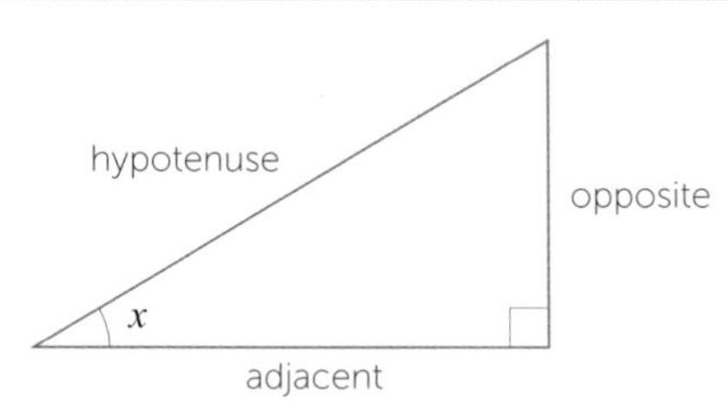


**Worked example**

A ladder, 4.4 m long, stands against a wall.

The ladder makes an angle of 74° with the ground.

a) How high up the wall does the ladder reach?

b) How far is the bottom of the ladder from the wall?

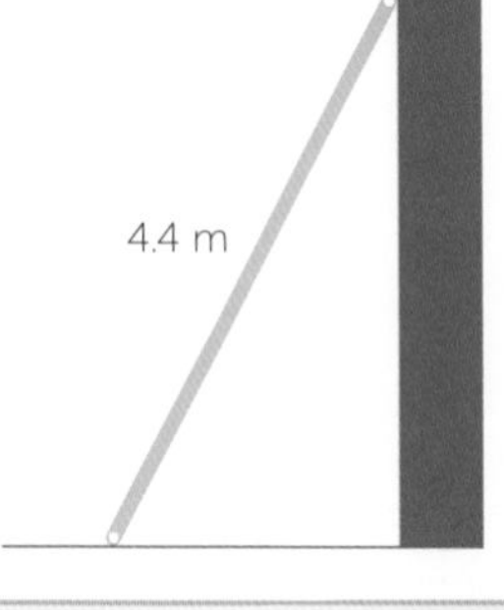


Solution:

a) Height = 4.4 × sin 74° = 4.4 × 0.96 = **4.23 m**

b) Bottom of ladder to wall = 4.4 × cos 74°
= 4.4 × 0.28 = **1.21 m**

## Practice questions

1. A cat is lying on the ground, 5 m from the base of a tree, looking up at a bird in the tree. The angle of elevation of the bird from the cat is 42°. What is the straight line distance from the cat to the bird?

2. An aircraft takes off and climbs at an angle of 8°.
   How high is the aircraft when it has travelled 24 km from its take-off point?

3. A ship sails from Plymouth on a bearing of 146° for 39 km to a buoy.
   a) How far is the buoy east of Plymouth?
   b) How far is the buoy south of Plymouth?

4. A plank is standing against a building, making an angle of 12° with the building.
   If the bottom of the plank is 1 m from the building, how long is the plank?

5. Kieron is flying a kite. The string is 90 m long. The angle between the string and the horizontal is 52°. If Kieron is holding the string 1.3 m above the ground, how high is the kite above the ground?

6. A garden shed has a sloping roof. The end panel of the shed is shown.
   If the slope of the roof is 24°, how high is the tallest part of the shed?

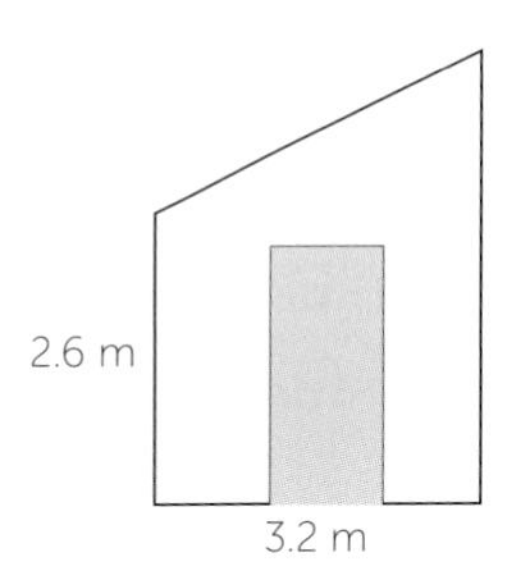


7. A footbridge spans a railway cutting. The footbridge is 28 m long.
   Use the dimensions shown in the diagram to calculate the angle the footbridge makes with the horizontal.

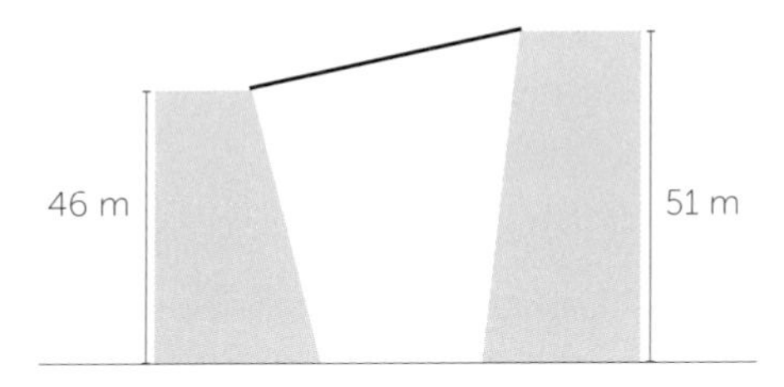


8. A ramp is to be fitted at a local community centre to allow disabled access.
   The entrance is 0.65 m above ground level and the ramp must make an angle of 16° with the ground.
   Calculate the length of the ramp correct to 2 decimal places.

9. The leaning Tower of Pisa is 54.56 m tall.
   It leans 3.80 m away from the vertical at its highest point.
   a) What is the angle of tilt as measured from the vertical?
   b) If the angle of tilt was 5.5°, how far from the vertical would the top be?
   c) If the tower tilts more than 5.5°, it will fall. If the distance from the vertical increases at 1 cm per year from now, how many years will it take until the tower falls?

SECTION 21

# STATISTICS

## 21.1 KNOWLEDGE CHECK

S2

### Objectives

**Recap categorical, discrete and continuous data**

**Categorical data** values are not numerical, for example eye colour.

**Discrete data** values are distinct quantities, for example number of pets.

**Continuous data** values are measured and can be written to a given number of decimal places, for example, height, weight or time.

**Recap interpreting statistical diagrams and tables**

**Recap finding averages from frequency tables**

**Mean** – add up and divide by the number of pieces of data

**Median** – order the data and find the middle value

**Mode** – the most common data value

**Range** – the difference between the highest and lowest value in the data set.

Discrete data is grouped when it is spread thinly over a large range. For example, the test results, as percentages, of a class of 20 students.

Continuous data is always grouped. Every value possible over a given range must fit in the table.

For example, $10 \leq x < 30$ takes any value from 10 up to, but not including, 30.

### Practice questions

1. State whether the following data is discrete or continuous:
   a) The heights of the members of a basketball squad.
   b) The number of pets owned by students in a maths class.
   c) The time taken by competitors to complete a race.
   d) The length of jumps in a long jump competition.
   e) The number of people competing in each event on a sports day.

2. Gerry has drawn the following bar chart on data for pets owned by members of his maths class:

   Comment on Gerry's bar chart.

   Gerry says that the median pet is 'cat'. Do you agree with Gerry?

   Explain your answer.

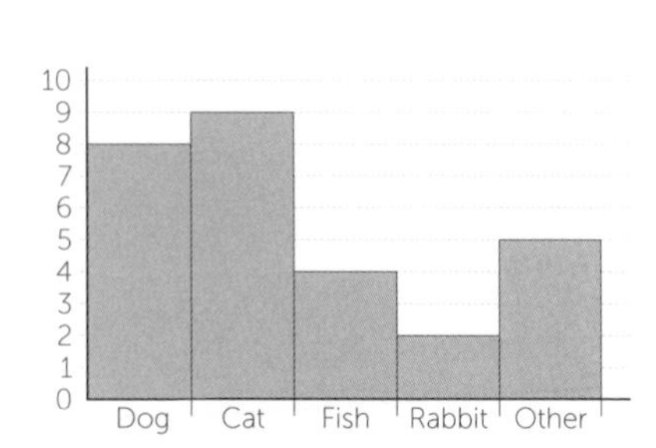


3. A five-a-side football team has players with a mean age of 27. One week a team member aged 25 is ill and a substitute aged 35 plays instead.

   How did this substitute player change the mean age of the team?

4. The following data was collected on the numbers of sandwiches bought by the workers in an office in one week:
   a) What is the modal number of sandwiches bought by the office workers?
   b) What was the median number bought?
   c) What was the mean number bought?

| Number of sandwiches | Frequency $f$ |
|---|---|
| 0 | 3 |
| 1 | 2 |
| 2 | 3 |
| 3 | 1 |
| 4 | 2 |
| 5 | 9 |

5. James is drawing a pie chart of the following data from a restaurant regarding the choices made by its customers in a week.
   Explain what James has done wrong.

| Main Meal | Frequency |
|---|---|
| Steak | 55 |
| Chicken | 70 |
| Fish | 20 |
| Vegetarian | 35 |

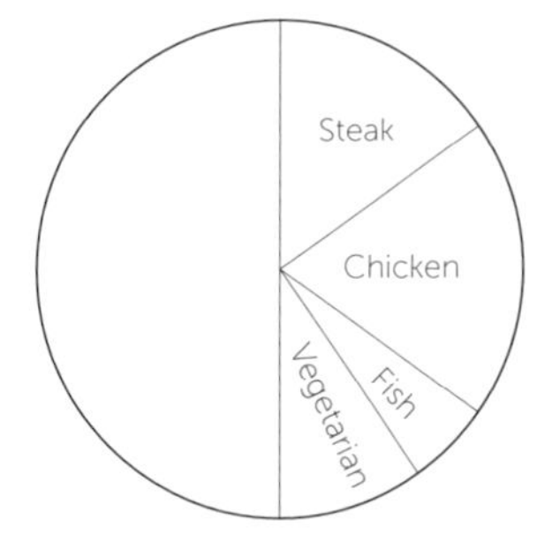


6. Jonty needs to achieve a mean average of 60% over four exams to get a place at a new school. He has a mean average of 57% in English, Science and Maths. What percentage does he need in a fourth subject, to get a place at his new school?

7. Twenty people were asked how much money they had in savings.
   a) Show this data in a bar chart.
   b) Estimate the mean amount of savings of these twenty people.
   c) Gavin says this data is continuous.
      Do you agree with Gavin?
      Explain your answer.

| Savings (£$x$) | Frequency |
|---|---|
| 1 – 200 | 4 |
| 201 – 400 | 7 |
| 401 – 600 | 6 |
| 601 – 800 | 2 |
| 801 – 1000 | 1 |

8. Fleur plays cricket. After 9 innings her batting average is 47.
   What is the minimum Fleur needs to score in her 10th innings to have a batting average of at least 50?

9. The speeds of 15 cars were measured on two different roads and put in the stem-and-leaf diagram below:
   a) Find the median speed of the cars on each road.
   b) Find the range of speeds on each road.
   c) One of the roads has speed bumps.
      Identify which one?
      Explain your answer.

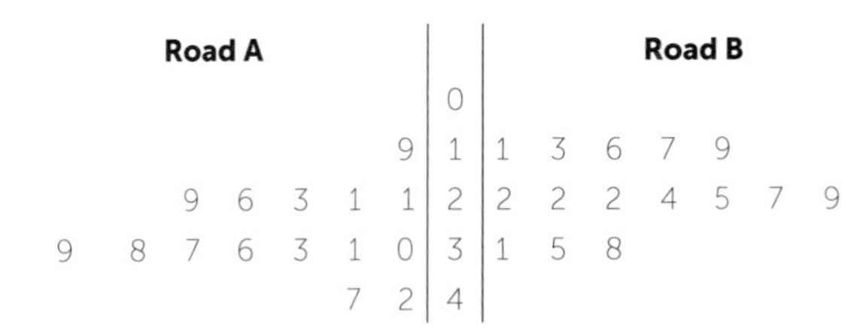

| Road A | | Road B |
|---|---|---|
| | 0 | |
| 9 | 1 | 1 3 6 7 9 |
| 9 6 3 1 1 | 2 | 2 2 2 4 5 7 9 |
| 9 8 7 6 3 1 0 | 3 | 1 5 8 |
| 7 2 | 4 | |

Key: 6 | 2 represents 26 km/h    Key: 2 | 6 represents 26 km/h

10. Data has been collected on the ages of people visiting two different libraries in a 1-hour period.
    The results are shown on the dual bar chart.
    a) How many people visited Library A?
    b) Estimate the mean age of visitors to Library A and Library B.
    c) Calculate the minimum possible age range of visitors to either of the libraries.

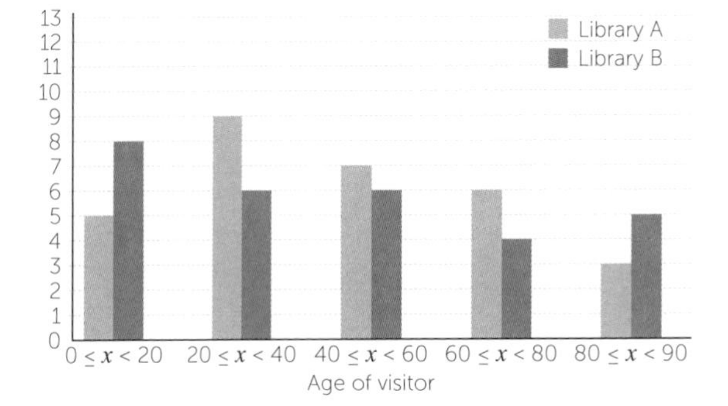

# 21.2 SAMPLING

## Objectives

### Understand when it is appropriate to use either a sample or a whole population

A 'population' includes all possible people, or items relevant to the study.

**e.g.** Ben wants to test the hypothesis: "Most Year 7 pupils bring sandwiches for lunch".

Suggest a suitable population from which Ben can get his results.

Ben could use the population of Year 7 pupils at his own school, or he could use a larger population of Year 7 pupils by looking at a number of schools.

A 'sample' is a subset of the whole population, chosen to represent the population from which it is taken.

In the above example, if the population is only Ben's school, it may be small enough that he can question the whole population about whether they bring sandwiches for lunch.

If Ben chose a larger population, for example all Year 7 pupils in his county, he would need to take a sample of students to ask. Using the whole population would be too expensive and time-consuming.

### Understand when a set of data might be biased

Bias may occur when the elements chosen for the sample do not represent the whole population.

**e.g.** Santosh Marketing are testing the hypothesis: "Most people prefer to shop in a supermarket than online".

They set up a stall outside a local supermarket and ask people their preferences.

Explain why their sample is biased. How can they improve their results?

Santosh are only asking people who use the supermarket. Online shoppers would not be available to ask.

They could knock on doors in the area, selecting random addresses.

### Explain why a sample might not represent the entire population

Using the whole population in the study would ensure that all measures are tested, thus avoiding bias.

However, if the test involves destruction - **e.g.** the life-span of a battery, this would destroy the whole population.

Here are some other advantages and disadvantages of using populations or samples:

| Whole Population | |
|---|---|
| **For** | **Against** |
| Reliable results | Time consuming |
| All opinions collected | Expensive |

| Sample | |
|---|---|
| **For** | **Against** |
| Quicker to do | Potentially biased |
| Relatively cheap | Some opinions may be missed out |

## Practice questions

1. For the following topics, state in each case whether the whole population or a sample should be surveyed.
   a) Voting intentions at a general election.
   b) School canteen choices of a tutor group.
   c) Activity choices for a year group at a school.
   d) Favourite snack type of 10-year-olds.
   e) Most popular TV channel at 9pm on a Saturday evening.

2. Gretel is doing a survey to find out who are the most popular actors in the country. She has decided to ask her friends who their favourite actor is.
   Explain why Gretel's survey is likely to produce biased results.

3. In a school there are 800 boys and 1200 girls; a sample of 100 students will be asked how they travel to school.
   The sample has the same proportion of boys as the population.
   How many boys and how many girls are in the sample?

4. Jimmy intends to take a sample from all the pupils who use the school canteen. He suggests picking the first 10 people from his year group he sees at the canteen.
   This might not give him a random sample; suggest how Jimmy could make his sample random.

5. Ebony wants to find out how many people buy music in her local town. She goes to her local entertainment store to ask people the following question:
   How often do you buy music each month?
   Explain why Ebony's results might be biased.

6. A school has the following numbers of students in each year group:
   Year 7: 220
   Year 8: 160
   Year 9: 200
   Year 10: 180
   Year 11: 240
   A sample of 50 students is to be selected for a survey.
   The sample should contain the same proportion of each year group as the whole population of students.
   How many of each group should be selected?

## 21.3 TIME SERIES

### Objectives

#### Work out time intervals for graph scales

A time series graph shows data points collected over a given period of time.

The data often follows a pattern or trend. Look for a repeating pattern in sales.

**e.g.** The data below shows the quarterly sales made by a swimwear company over a period of five years:

| **Year** | **2014** | | | | **2015** | | | | **2016** | |
|---|---|---|---|---|---|---|---|---|---|---|
| **Quarter** | 1 | 2 | 3 | 4 | 1 | 2 | 3 | 4 | 1 | 2 |
| **Sales (£1000s)** | 15 | 28 | 60 | 37 | 18 | 31 | 59 | 27 | 22 | 42 |

| **Year** | **2016** | | **2017** | | | | **2018** | | | |
|---|---|---|---|---|---|---|---|---|---|---|
| **Quarter** | 3 | 4 | 1 | 2 | 3 | 4 | 1 | 2 | 3 | 4 |
| **Sales (£1000s)** | 50 | 41 | 30 | 47 | 65 | 34 | 35 | 52 | 69 | 40 |

A time series graph is plotted for this data.

Comment on the trend.

The general trend for this data is that sales are increasing, although there was a dip in sales in 2016.

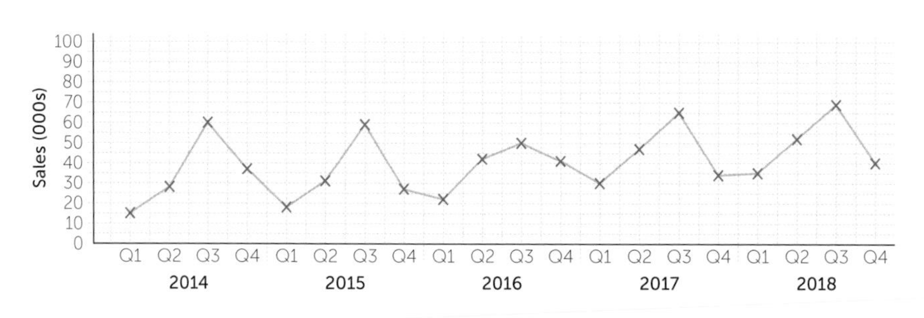


#### Construct tables for time series data

#### Produce and interpret line graphs for time series

#### Recognise simple patterns and relationships in line graphs

### Practice questions

1. A scarf manufacturing company has the following sales figures:

| **Year** | **2014** | | | | **2015** | | | | **2016** | |
|---|---|---|---|---|---|---|---|---|---|---|
| **Quarter** | 1 | 2 | 3 | 4 | 1 | 2 | 3 | 4 | 1 | 2 |
| **Sales (£1000s)** | 35 | 15 | 10 | 45 | 40 | 10 | 10 | 40 | 45 | 20 |

| **Year** | **2016** | | **2017** | | | | **2018** | | | |
|---|---|---|---|---|---|---|---|---|---|---|
| **Quarter** | 3 | 4 | 1 | 2 | 3 | 4 | 1 | 2 | 3 | 4 |
| **Sales (£1000s)** | 25 | 45 | 45 | 25 | 30 | 45 | 55 | 25 | 20 | 60 |

Work out a suitable scale and plot the sales data.

Make a comment about the trend in sales.

2. The time series shows visitor numbers over a year at a tourist attraction.
   a) How many visitors did the tourist attraction have in February?
   b) How many visitors did the tourist attraction have in August?
   c) Explain why the tourist attraction is busiest in July and August.
   d) The attraction's management are thinking of closing for a quarter of the year; which months should they choose?
   e) The owners were thinking of closing whenever the visitor numbers drop below 30,000. What would be the problem with doing this, according to the time series?

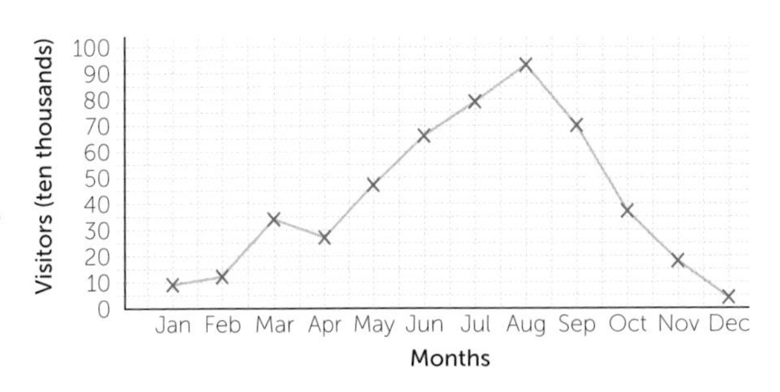

3. The time series graph shows sales for a firewood supplier:
   a) Which is the most popular quarter for the company?
   b) Why do you think this is?
   c) Which quarters are least popular?
   d) Why do you think this is?
   e) Describe the trend in the company's sales over the time period shown in this time series.

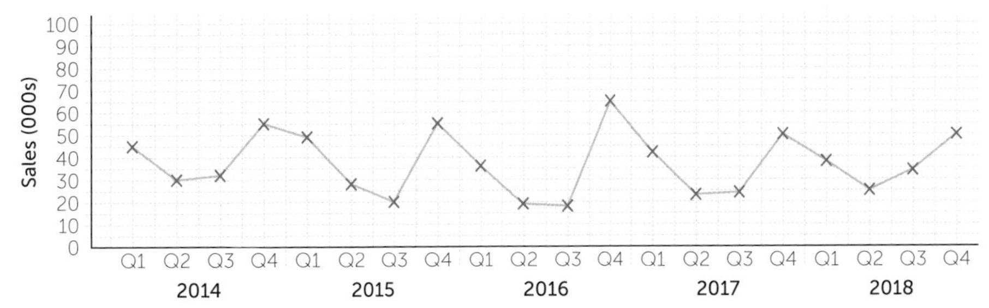


4. The time series graph shows the number of battery-powered electric cars sold worldwide each year since 2013.
   a) How many electric vehicles were sold in 2019?
   b) Comment on the underlying trend in the sales of electric vehicles since 2010.
   c) Justin estimates that more than 7 million electric cars will be sold in 2020. Do you agree? Justify your answer.

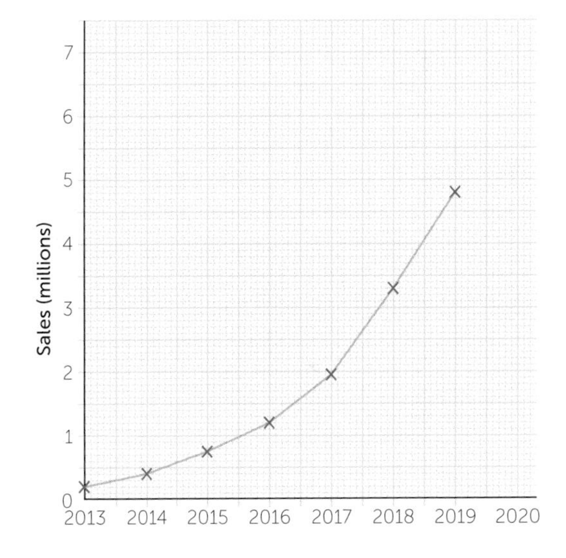


5. A baby panda weighed 90 g at birth. It was weighed every week and at 8 weeks, it weighed 1.4 kg. The table shows how much it weighed at the end of each week.
   a) In which week did the mass of the panda increase the most? Explain how you can get the answer from the graph.
   b) The panda was ill during this period and did not drink as much milk as usual. How old was the panda when it fell ill? Explain your answer.

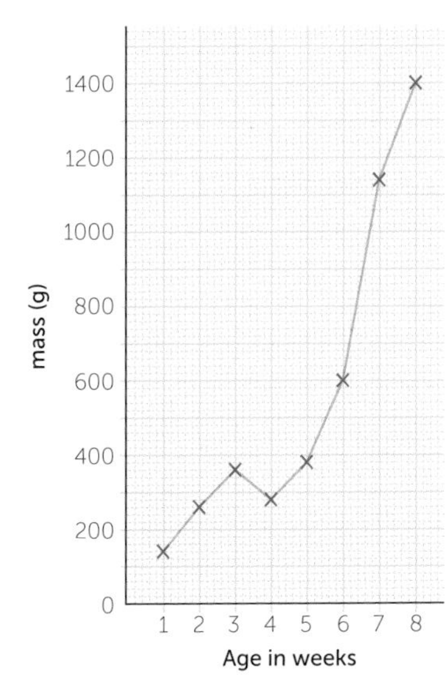

# 21.4 SCATTER DIAGRAMS

## Objectives

### Draw scatter graphs and interpret points including identifying outliers

Plotting points on a scatter graph is exactly like plotting coordinates.

| **Student** | A | B | C | D |
|---|---|---|---|---|
| **Maths** | 32 | 53 | 69 | 83 |
| **Science** | 25 | 62 | 50 | 25 |

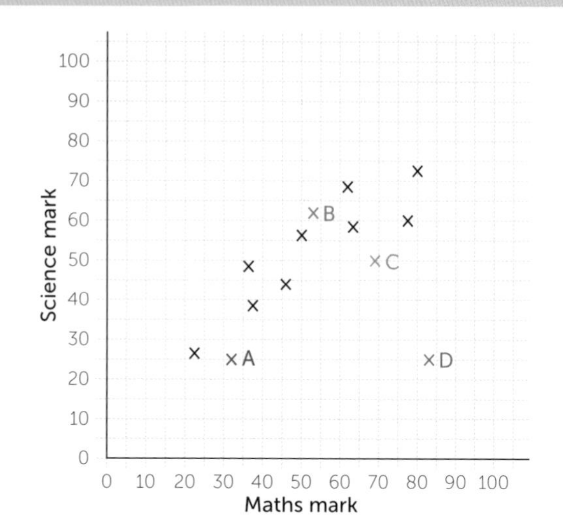


e.g. The scatter graph shows the maths and science marks obtained by a class.

A scatter graph is a good way of displaying two sets of data to see if there is a **correlation** or connection between them.

**Positive correlation** means that as one variable increases, so does the other one, or as one variable decreases, so does the other.

**Negative correlation** means that as one variable increases, the other one decreases, and vice versa – for example average temperature and sales of warm clothing.

**No correlation** means the two variables are not connected.

Points A, B, C and D have been added to the graph.

Point D (83, 25) is an 'outlier' as it doesn't fit the pattern of the rest of the points.

It represents the person who scored 83% in Maths and only 25% in Science.

### Draw by eye a line of best fit on a scatter graph
### Understands what a line of best fit represents

A 'line of best fit' is a straight line that follows the trend.

It is usually drawn 'by eye' and splits the points roughly into two groups of equal size.

The line of best fit does not go beyond the points plotted.

The line of best fit allows us to make an estimate of one variable, based on another.

For example, the mark in Science likely to be achieved by a student who got 70% in Maths is around 65%. This process is called **interpolation**.

It is not safe to extend the line of best fit in either direction (i.e. '**extrapolate** the data'). You cannot deduce that a student who achieves 10% in Science will get 0% for Maths.

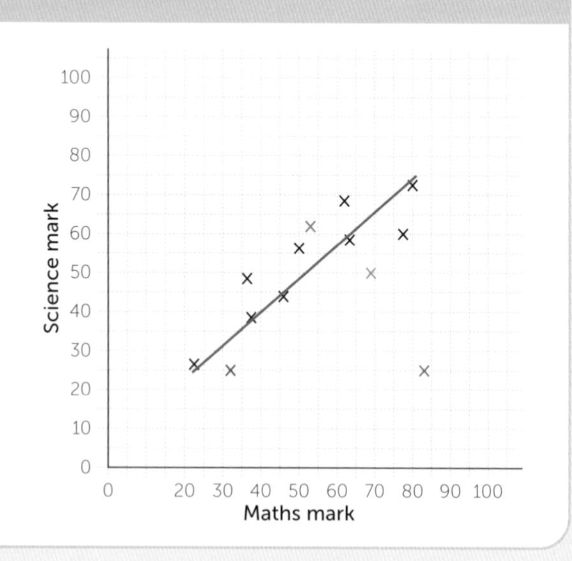

## Practice questions

1. The following data was collected regarding the temperature and the hours of sunlight:

| **Temp °C** | 15 | 8 | 11 | 3 | 9 | 18 | 7 | 9 | 12 |
|---|---|---|---|---|---|---|---|---|---|
| **Hours of sunshine** | 8 | 3 | 6 | 1.5 | 5 | 9.5 | 4.5 | 7 | 6.5 |

   a) Plot a scatter graph of this data and use it to describe a relationship between the hours of sunshine, and temperature.

   b) One day there was 3.5 hours of sunshine and the temperature was 14°C. What can you say about this reading?

2. A café recorded the sales of hot chocolate during a ten-day period:

| **Temp °C** | 4 | 7 | 8 | 13 | 19 | 18 | 15 | 9 | 2 |
|---|---|---|---|---|---|---|---|---|---|
| **Sales** | 88 | 85 | 79 | 60 | 15 | 22 | 44 | 51 | 32 |

   a) Plot this data as a scatter diagram.

   b) Comment on the relationship between temperature and sales of hot chocolate.

   c) Identify any outliers in the data.

3. The scatter graph shows the values of some mid-range cars against their age in years:

   A 9-year-old car has been valued at £7500.

   a) Plot this as a point on the scatter diagram.

   b) Comment on its value compared to the other cars represented.

   c) Draw some conclusions about this car, from the graph.

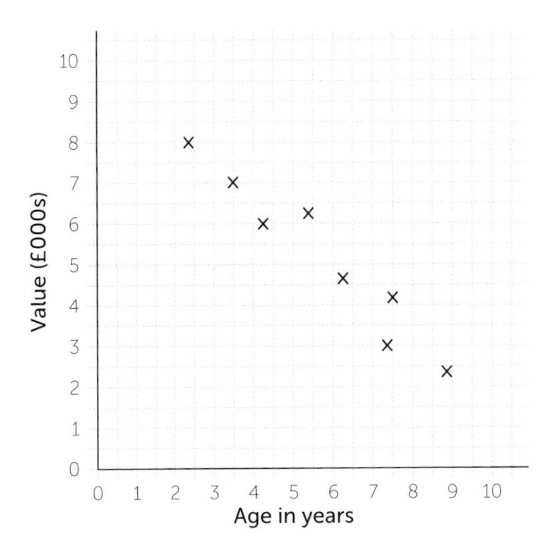

4. Hattie has drawn the following scatter graph with a line of best fit:

   Write two things that Hattie has done incorrectly.

   Explain why they are incorrect.

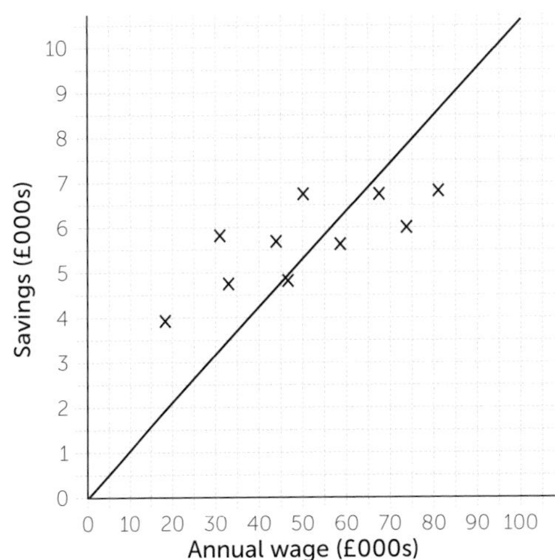

# 21.5 CORRELATION

## Objectives

### Interpret scatter graphs and correlation in terms of the relationship between two sets of variables

When describing a relationship, it must be written in the context of the values on the scatter diagram.

The value on the $x$-axis is an independent variable.

e.g. Describe the relationship between outside temperature and ice-cream sales, shown on this scatter diagram.

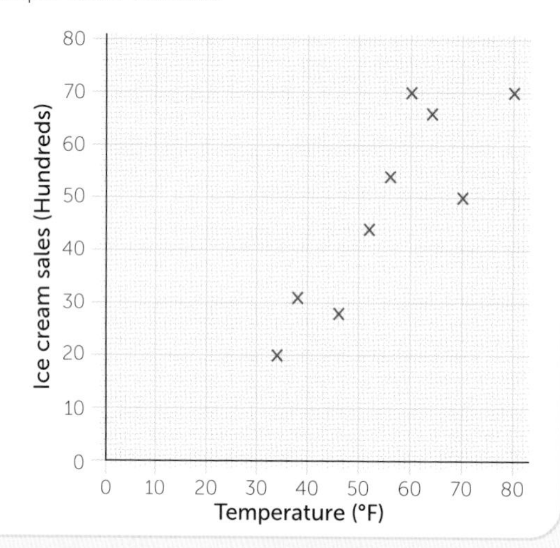


The higher the outside temperature, the more ice-creams are sold.

Note that the sentence should read 'The higher the $x$-axis label, the higher (or lower) the $y$-axis label'.

### Understand that correlation does not imply causation

A positive, or negative correlation does not prove that the change in the $x$ variable causes the change in the $y$ variable.

There could be a third variable that affects both variables.

e.g. A scatter diagram shows a strong positive correlation between teachers' earnings and house prices.

Guy says that teachers' salaries affects house prices.

Explain why Guy is wrong.

Teachers' salaries and house prices are both affected by the strength of the economy and are not directly related.

### Distinguish between positive, negative and no correlation

Correlation can be **positive**, **negative**, **weak** or **strong**. Strong correlation shows points close to the line of best fit. Weak correlation shows points that are far away from the line of best fit. Where the points are randomly placed, there is no correlation, and no line of best fit.

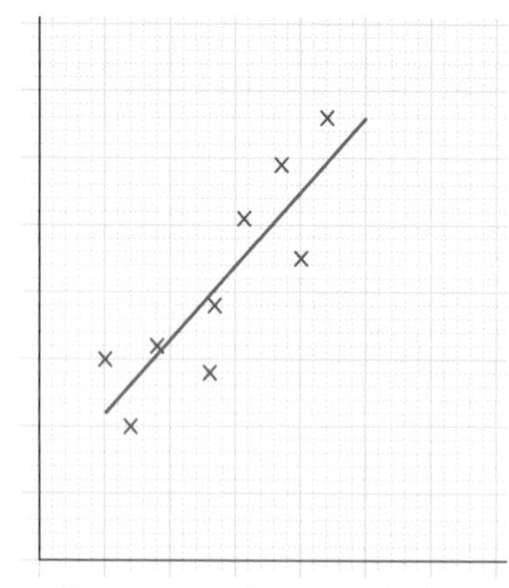

Strong, positive correlation

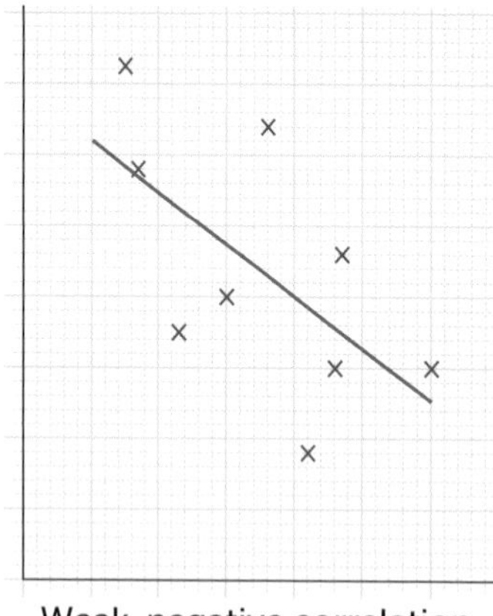

Weak, negative correlation

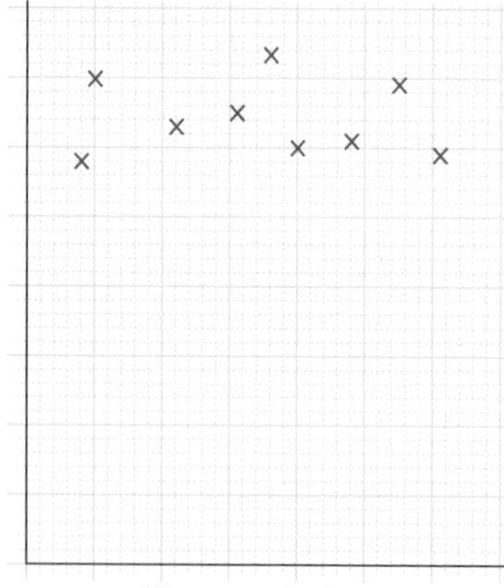

No correlation

**Predict values using a line of best fit, understanding the dangers of doing so**

Draw the line of best fit by eye and read off the values from the graph, to make predictions.

Do not use the line of best fit to make predictions beyond the range of values represented.

e.g. The scatter graph below shows the results in Maths and Science tests for twelve students.

Another student takes the Maths test and scores 60.

What is his expected Science score?

**58 marks**. This is within the range of the data, so valid.

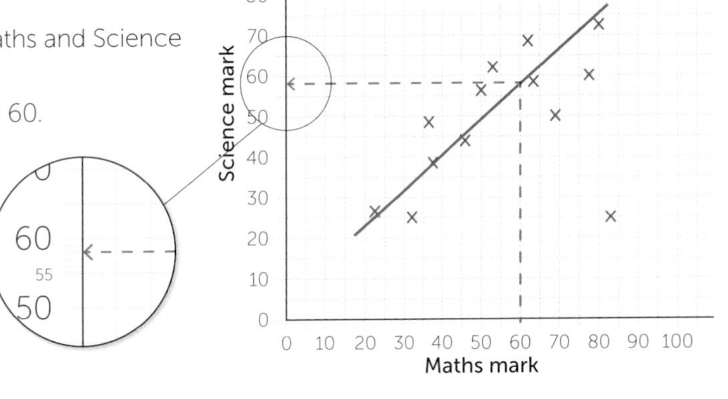


## Practice questions

1. Describe the type of correlation that the following types of data would have:
   a) Height of a person compared to their shoe size.
   b) The distance a child lives from school compared to their English exam mark.
   c) The monthly earnings of a person compared to the value of their car.
   d) The sales of winter coats compared to the temperature.
   e) The distance people live from London compared to the number of visits they make to London per year.

2. The following data was collected regarding earnings and the number of meals cooked per month:

| Earnings (£1000s per year) | 33 | 27 | 80 | 68 | 36 | 25 | 53 | 45 | 70 |
|---|---|---|---|---|---|---|---|---|---|
| Meals cooked (per month) | 8 | 9 | 2 | 3 | 8 | 10 | 5 | 5 | 4 |

   a) Complete the sentence: "the more people earn,..."
   b) Comment on the correlation.
   c) Draw a line of best fit.
   d) Use your line of best fit to estimate how many meals a person earning £60,000 per year will cook in a month.
   e) Give a reason for the relationship between earnings and meals cooked.

3. The following data was collected regarding the temperature and visitors (to the nearest 10) to a beachside café:

| Temp (°C) | 13 | 15 | 8 | 4 | 16 | 19 | 11 | 18 | 6 |
|---|---|---|---|---|---|---|---|---|---|
| Customers | 350 | 380 | 120 | 50 | 390 | 470 | 200 | 460 | 70 |

   a) Comment on the correlation.
   b) Draw a line of best fit.
   c) One day there were 300 customers. Use your graph to estimate the temperature on that day.
   d) Tom uses the line of best fit, and estimates that when the temperature reaches 22°, the number of visitors will be 560. Comment on Tom's estimate.

4. The scatter graph has been drawn using data comparing age and savings in the bank.

   a) Describe the correlation.

   b) Is the correlation strong or weak? Explain your answer.

   c) Another person says that they have £50 000 in savings. Estimate the age of this person.

   d) A 20-year-old wants to use the scatter graph to estimate how much in savings they should have. Why is this graph not appropriate for them to use?

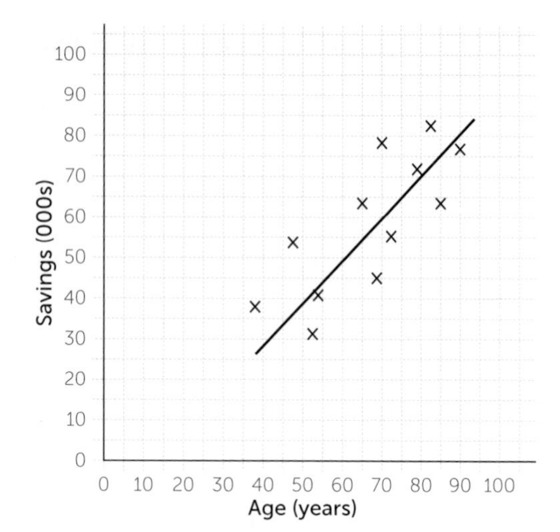

5. Rob collects data on sunglasses sales and ice cream sales.

   His scatter diagram shows a strong, positive correlation.

   Rob says that the sale of sunglasses increases the sale of ice creams.

   Comment on Rob's conclusion.

SECTION 22

# PROBABILITY DIAGRAMS

P5

## 22.1 SAMPLE SIZE

### Objectives

**Compare relative frequencies from samples of different sizes, including by graphing sample size against relative frequency**

The relative frequency of an event is found from experimental data.

The experiment is repeated a number of times and the outcomes are recorded.

This data is sample data.

Relative frequency = $\frac{\text{frequency of outcome}}{\text{total frequency}}$

Relative frequency gives an estimate of the probability of an outcome occurring.

The larger the sample, the better the estimate of the probability.

**e.g.** A fair six-sided dice is rolled 10 times and 3 fives are recorded.

The relative frequency of getting a five from this data = $\frac{3}{10}$ or 0.3

The dice is rolled a further 10 times and this time only 1 five is scored.

The relative frequency of getting a five from all this data = $\frac{4}{20}$ or 0.2

The theoretical probability of getting a five on a fair six-sided dice = $\frac{1}{6}$ or $0.1\dot{6}$

The estimate of the probability based on relative frequency has moved closer to the theoretical probability as the number of trials, or the sample size, has increased.

**Understand that increasing sample size generally leads to better estimates of probability and population characteristics**

Here is a graph showing the relative frequency of throwing a five plotted against the number of trials (sample size).

The graph shows how the relative frequency changes as the sample size increases. It gets gradually closer to the theoretical value. (The graph will be different every time the experiment is carried out.)

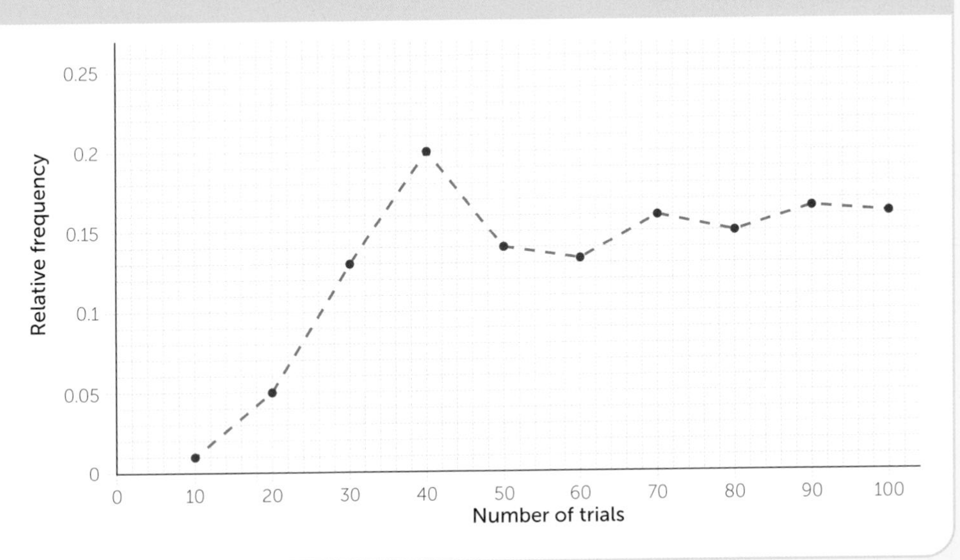

## Practice questions

1. A six-sided dice is rolled 120 times. It lands on 4 on 25 of the rolls.
   a) Using the experimental data, estimate the probability of landing on 4 on the next roll of this dice.
   b) The dice is rolled for another 120 rolls and this time it lands on 4 a further 15 times.
   Estimate the probability of the dice landing on 4 on the next roll, based on all the experimental data.
   c) Comment on the two estimates made.

2. An experiment is conducted using a bag of 100 coloured counters. 10 counters are drawn at random with each counter being replaced before the next one is drawn. In total, 4 blue, 3 green, 2 yellow and 1 red counters are drawn.
   a) Estimate the probability of getting each colour of counter on the next draw.
   b) Another sample is taken at random giving; 5 blue, 2 green, 2 yellow and 1 red.
   Give new estimates of the probability of getting each colour of counter based on all the data collected.

3. Three students carry out a survey; asking others if they went abroad last year.
   Here are the results:

| | Student 1 | Student 2 | Student 3 | Total |
|---|---|---|---|---|
| **Yes** | 4 | 29 | 47 | |
| **No** | 16 | 21 | 53 | |
| **Number asked** | 20 | 50 | 100 | |

   a) Complete the last column of the table.
   b) Work out the relative frequency for those who answered "Yes" in each student's survey.
   Give your answers as decimals to 2 decimal places.
   c) Work out the relative frequency of "Yes" answers for all the people surveyed.
   d) Which student has the most reliable results? Give a reason for your answer.
   e) Draw a graph plotting relative frequency against sample size for those who answered "Yes".
   Plot one point for the data from each survey and one for the total number of people surveyed.
   f) Use your graph to estimate the probability of someone saying 'Yes' in the survey.

4. A survey was taken at a café on Day 1 to estimate how many drinks of each type were sold. Out of 100 drinks sold were; 35 tea, 22 coffee, 33 soft drinks and 10 other.
   a) Work out the relative frequency for each type of drink in the sample for Day 1.
   On three other days, more surveys were taken. Here are the results.

| | Tea | Coffee | Soft drink | Other | Total |
|---|---|---|---|---|---|
| **Day 2** | 18 | 10 | 19 | 3 | 50 |
| **Day 3** | 45 | 22 | 38 | 15 | 120 |
| **Day 4** | 28 | 17 | 27 | 8 | 80 |

   b) Work out the relative frequency for sales of tea on each of Days 2, 3 and 4.
   c) Combine the results for sales of tea across all 4 days and work out the relative frequency for sales of tea using all the data.
   d) Draw a graph plotting relative frequency against sample size for sales of tea.
   Plot one point for each day's data and one for the overall data.
   e) Use your graph to estimate the probability of someone buying tea at the café.

# 22.2 TREE DIAGRAMS

## Objectives

**Use tree diagrams to calculate the probability of two independent events**

**Tree diagrams** are used in probability to show all the possible outcomes of one or more events.

The **probability** of an outcome is written on the branch.

The **outcome** is written at the end of the branch.

**e.g.** This is a tree diagram showing the event of rolling a fair six-sided dice.
The branches show the probability of getting a '4' and 'not a 4'.

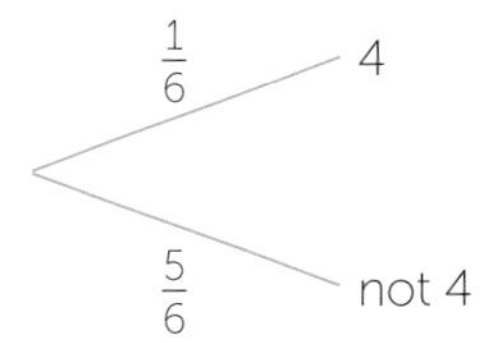


### Combined events

Tree diagrams can be used to show the outcomes of **combined events**.

**e.g.** This tree diagram shows the combined events of throwing a fair coin, two times.
Each route through the tree diagram leads to one of the four possible outcomes: (H, H) (H, T), (T, H) and (T, T)

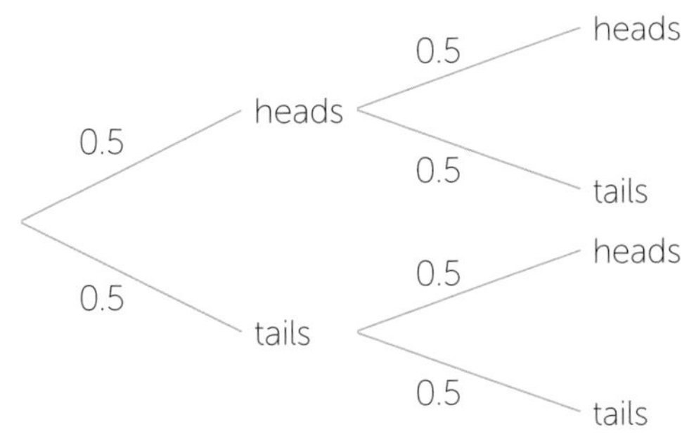


Since these outcomes are equally likely, the probability of each of these outcomes is **0.25**

### Calculating probabilities

The probability of an outcome can be found by **multiplying** the probabilities along the branches, leading to that outcome.

**e.g.** In the example shown above, the probability of getting (head and head) is given by **0.5 × 0.5 = 0.25** (as we have already seen).
To find the overall probability of more than one outcome, the probabilities of each outcome are added together.

**e.g.** The probability of getting exactly one head = P(H, T) + P(T, H)
= (0.5 × 0.5) + (0.5 × 0.5)
= 0.25 + 0.25
= **0.5**

## Practice questions

1. The probability of rain today is 0.3; what is the probability of no rain today?

2. A biased coin is thrown. The probability of getting a head is 0.6.
   Draw a tree diagram to show the outcomes of the event of throwing this coin.

3. Rachael travels to work each day either by bus or by walking.

   The probability of her going to work by bus is 0.3.

   a) Copy and complete this tree diagram to show the combined events of her going to work over two days.

   b) Use your tree diagram to work out the probability that she walks both days.

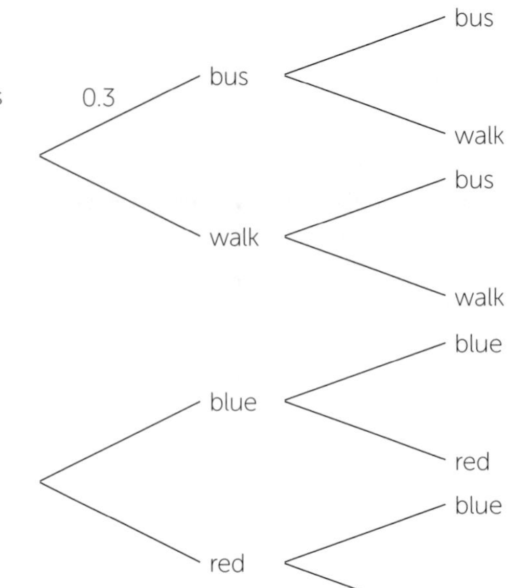


4. A bag contains 6 blue and 4 red counters.

   A counter is chosen at random, its colour noted and then replaced.

   This is then repeated a second time.

   a) Copy and complete this tree diagram to show the possible outcomes.

   b) What is the probability that both counters chosen are red?

   c) What is the probability that the first counter is blue and the second red?

   d) What is the probability that one counter of each colour is chosen?

5. Here is a tree diagram showing the possible outcomes of two combined events.

   a) Suggest what the two combined events might be.

   b) Use the tree diagram to work out the probability of getting a head on a fair coin and a red on a four-sided spinner with one red side and three non-red sides.

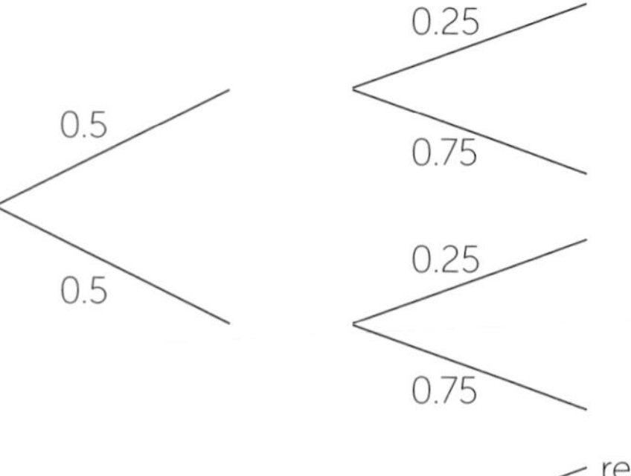

6. Katie has a five-sided spinner which has 1 red side and 4 green sides.

   She spins the spinner twice and records the colour each time.

   She draws a tree diagram to show the outcomes of the combined events.

   Katie has made a mistake.

   a) Explain her mistake and draw out the correct tree diagram.

   b) Use your tree diagram to work out the probability of the spinner landing on one of each colour.

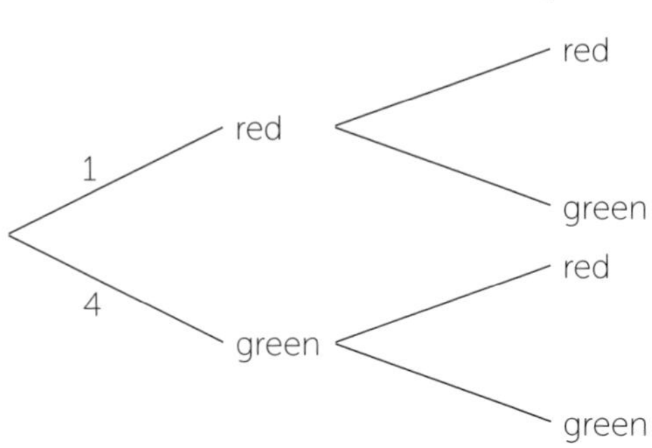

## 22.3 VENN DIAGRAMS

### Objectives

**Use a Venn diagram to represent real-life situations and 'abstract' sets of numbers**

**Use union and intersection notation**

A Venn diagram is a pictorial way of representing information.

It has sets shown as circles, which can overlap.

Each set is defined with a certain property.

If a number belongs to two sets, it is written in the overlap or **intersection** of the sets.

If a number does not belong to either of the sets, it is written outside the circles.

ℰ is the universal set symbol. The universal set is the set of all elements under consideration

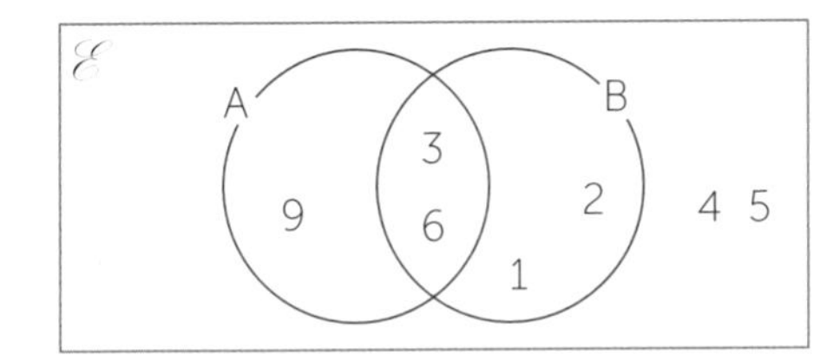


**e.g.** Set A is the set of multiples of 3

Set B is the set of factors of 6

3 and 6 belongs to both sets so are shown in the intersection

4 and 5 do not belong to either set so are shown on the outside.

**Elements** (numbers) which belong in a set are shown using these brackets { }

ℰ = {1, 2, 3, 4, 5, 6, 7, 8, 9}

A = {3, 6, 9}

B = {1, 2, 3, 6}

The **intersection** of two sets is shown using the symbol ∩

A ∩ B = {3, 6}

The **union** of two sets, is the combination of all elements in both sets. It is shown using the symbol ∪

A ∪ B = {1, 2, 3, 6, 9}

**Work out probabilities from a Venn diagram**

The probability of an outcome occurring can be worked out using a Venn diagram.

**e.g.** Here is a Venn diagram showing the number of students in a class taking French or Spanish.

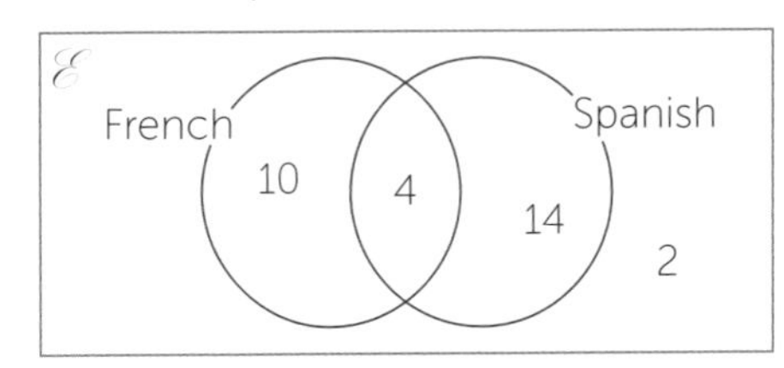


ℰ = {All students in a class}

French = {students taking French}

Spanish = {students taking Spanish}

There are 4 + 14 = **18 students** taking Spanish.

Two students take neither French nor Spanish.

There are 10 + 4 + 14 + 2 = **30 students** altogether.

The probability of choosing a student at random from the class who takes Spanish is: **P(Spanish) =** $\frac{18}{30}$

### Practice questions

1. ℰ = {1, 2, 3, 4, 5, 6, 7, 9}

   Set A is the set of odd numbers.

   Set B is the set of factors of 12.

   a) Copy and complete the Venn diagram.

   b) Write down any elements which are in the intersection of Set A and Set B.

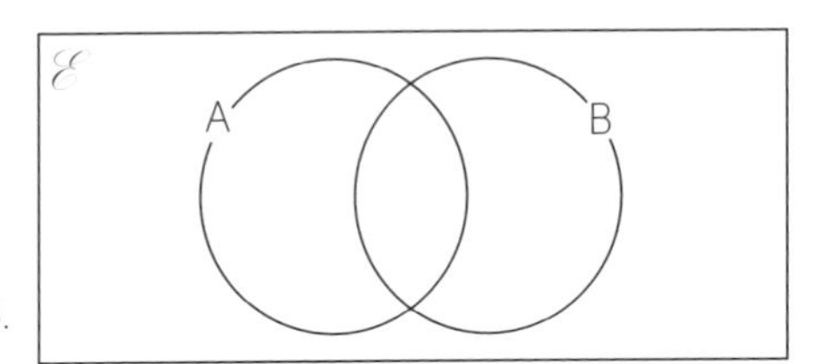

2. Here is a Venn diagram showing the number of students in a class who like pizza or pasta.

   $\mathscr{E}$ = {students in a class}

   a) How many students like pizza and not pasta?
   b) How many students like both?
   c) How many students do not like either?
   d) How many students are in the class altogether?

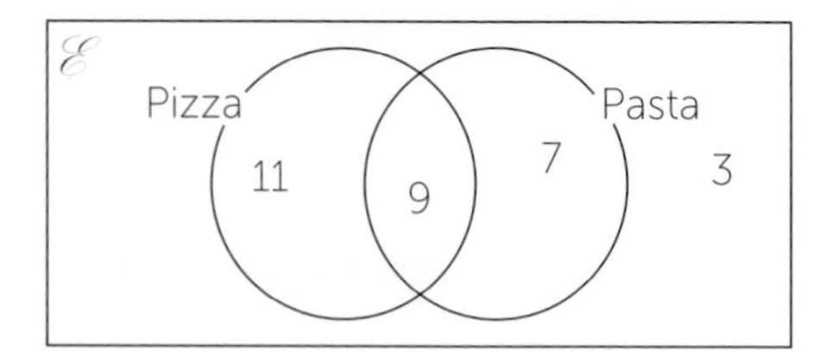


3. Here is a Venn diagram showing the type of events students in a class did on sports day.

   $\mathscr{E}$ = {students in a class} T = {number of students who did track event}

   F = {number of students who did field event}

   a) How many students are in the class?
   b) How many students did a field event?
   c) What is the probability that a student chosen at random did a field event?
   d) How many students are in Set T ∩ F?
   e) Describe what these students did on sports day.

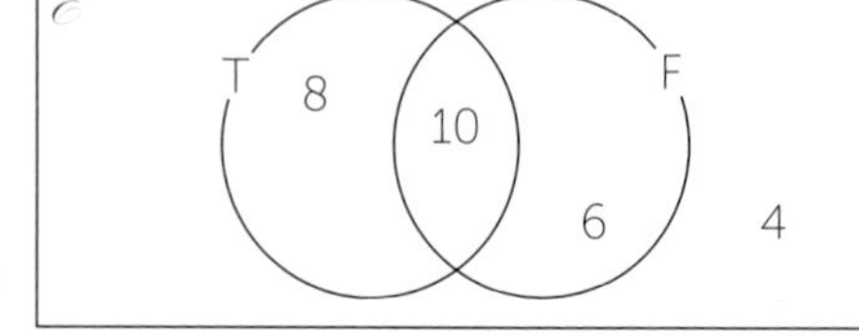


4. Some people were asked about sport. Set P is the set of people who play sport. Set W is the set of people who watch sport.

   $\mathscr{E}$ = {people asked about sport} P = {number of people who play sport}

   W = {number of people who watch sport}

   44 people were asked altogether. 22 said they watched sport.

   a) Copy and complete the Venn diagram.
   b) How many people are in Set P ∪ W?
   c) Describe what this group of people do.

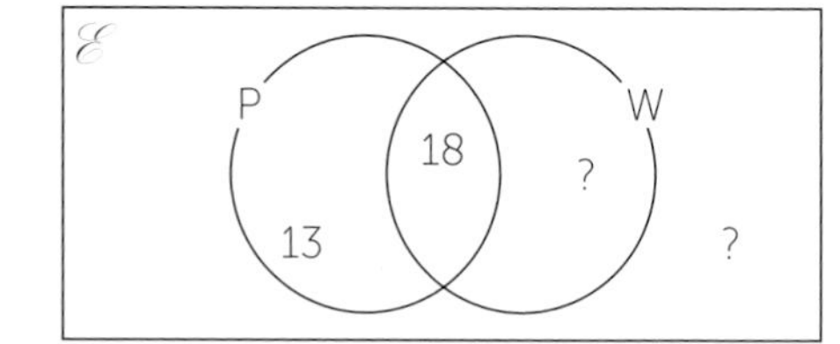


5. 80 planes landed at an airport one afternoon. 30 planes arrived from the USA and 12 planes were late. 7 of the late planes were from the USA.

   $\mathscr{E}$ = {planes landed at airport} USA = {planes arrived from the USA}

   Late = {planes that were late}

   a) Copy and complete the Venn diagram for the data.

   A plane is picked at random.

   b) What is the probability that the plane was not from the USA?
   c) What is the probability that the plane was from the USA and was on time?

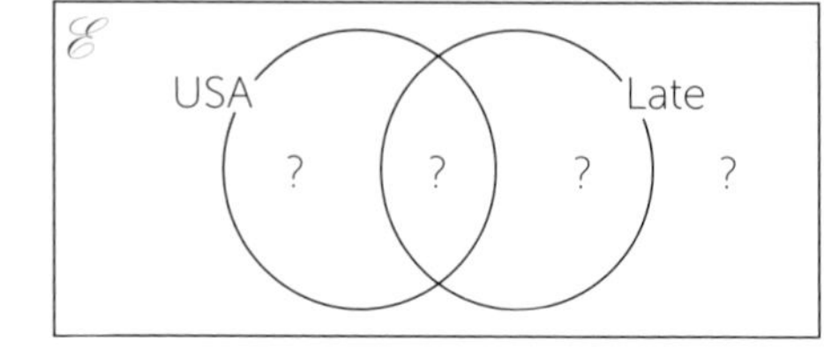


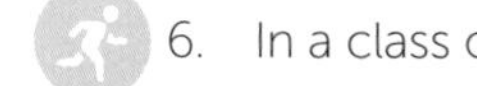

6. In a class of 28 students, 9 students are in the choir (C), 7 are in the school band (B) and 3 are in both.

   $\mathscr{E}$ = {class of students} C = {students in Choir} B = {students in Band}

   a) Draw a Venn diagram to represent this information.

   A student is chosen from the class at random.

   b) Describe the students who are in Set (C ∪ B).
   c) Find P(C ∪ B).

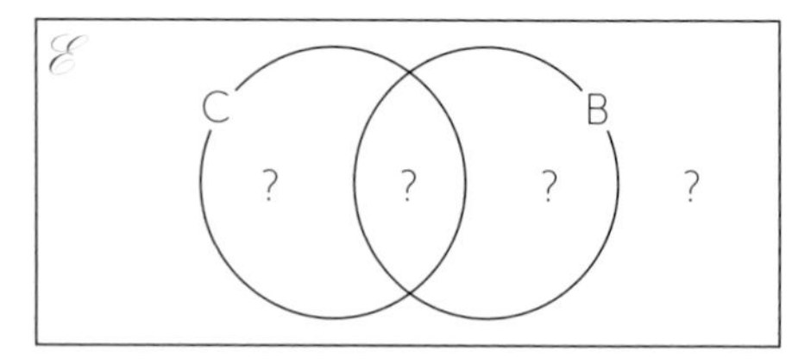

7. 32 students were asked about their siblings. There were 19 with a brother, 20 with a sister and 8 with a sister but not a brother.

   $\mathscr{E}$ = {32 students} Brother = {students with a brother}

   Sister = {students with a sister}

   a) Archie draws a Venn diagram to show this information.

      He has made some mistakes. Find his mistakes and correct them.

   b) Work out the probability that a student chosen at random is an only child.

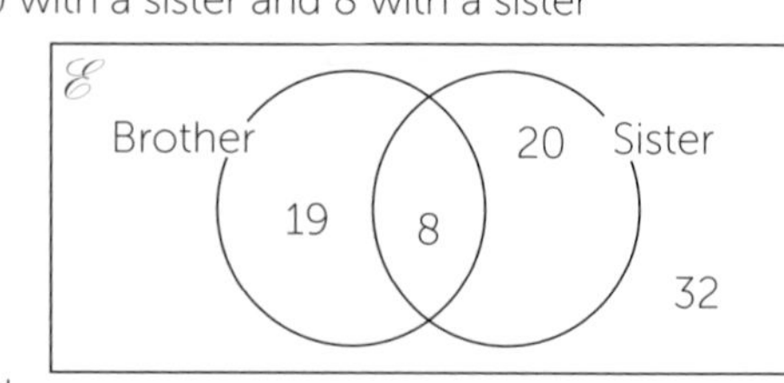

## 22.4 DEPENDENT EVENTS

### Objectives

**Use tree diagrams to calculate the probability of two dependent events**

**Dependent events** are events where the outcome of one event affects the probability of the outcomes of the other event.

**e.g.** The event of rain today could affect the probability of rain tomorrow.

If so, these events; rain today and rain tomorrow, are dependent events.

With dependent events, the probabilities on the second set of branches on the tree diagram will be different to the probabilities on the first set of branches.

**e.g.** A box of chocolates contains 8 milk chocolates and 6 plain chocolates.

Two chocolates are taken at random from the box without being replaced.

Here is a tree diagram showing the possible outcomes.

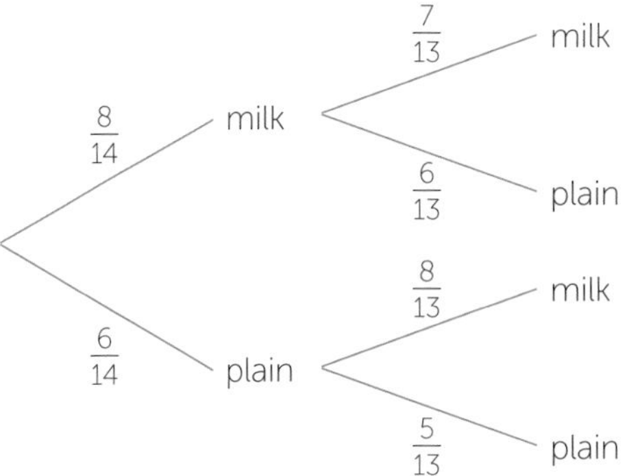


Notice how the second set of branches have different probabilities on them to the first set.

Notice also how the two second sets of branches are different from each other.

P (both milk chocolates) is P(milk, milk) = $\frac{8}{14} \times \frac{7}{13} = \frac{4}{13}$

P(one milk, one plain) is P(milk, plain) + P(plain, milk) = $\frac{8}{14} \times \frac{6}{13} + \frac{6}{14} \times \frac{8}{13} = \frac{48}{91}$

### Practice questions

1. A bag contains 3 white and 2 black balls. A black ball is picked and not replaced.
   What is the probability that the next ball chosen will be white?

2. A class has 14 boys and 16 girls.
   Two students are chosen at random. The first child chosen is a boy.
   What is the probability that the second child chosen is also a boy?

3. A fruit bowl contains 5 apples and 3 oranges.
   Sarah eats one piece of fruit at random and then chooses a second piece of fruit.
   Copy and complete the tree diagram to show the possible outcomes and the probabilities.

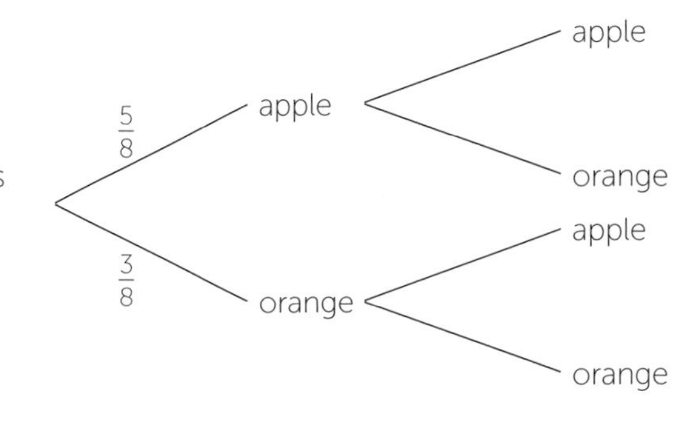


4. In a box there are 10 pens; 2 red and 8 black.
   A pen is taken at random and then a second pen is taken at random.
   a) Draw a tree diagram to show the possible outcomes and the probabilities.
   b) Work out the probability that both pens chosen are black.

5. Jasmine has 6 blue and 6 black socks in a drawer.
   She randomly picks out two socks from her drawer.
   She says that since she has the same number of each colour, the probability of the socks being the same colour is $\frac{1}{2}$.
   Is she correct? Show your working.

6. There are 52 boys and 38 girls in a year group.
   One morning 20 students were late.
   Four-fifths of the late students were boys.
   Use this information to complete the blank frequency tree.

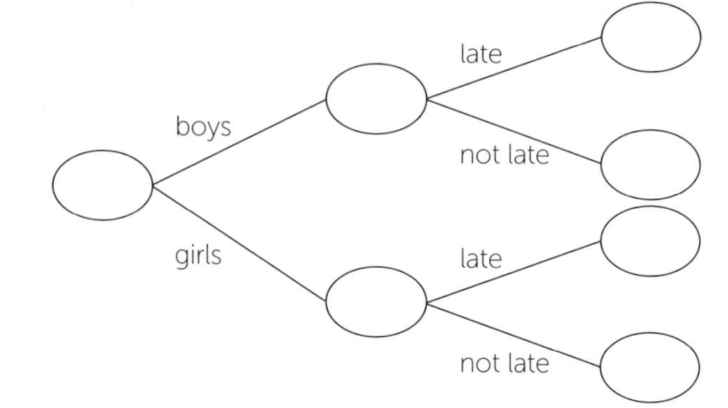

7. The probability that it will rain tomorrow is 0.2.
   If it rains, the probability that Marco will walk to school is 0.4.
   If it does not rain, the probability that Marco will walk to school is 0.85.
   a) Copy and complete the tree diagram.
   b) What is the probability that Marco will walk to school tomorrow?

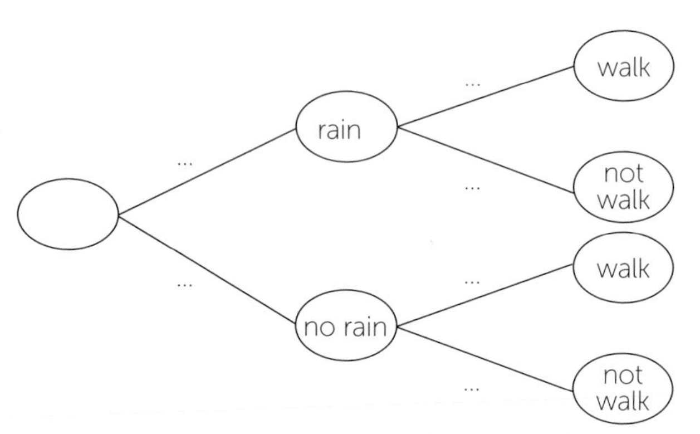

8. Two numbered counters are chosen from a bag, without replacement.
   The counters in the bag have the following numbers on them:

   1, 1, 1, 1, 1, 2, 2, 2, 2, 2, 2, 2, 2, 2, 2

   a) Draw a tree diagram to show the possible outcomes and the probabilities.
   b) What is the probability that the two counters will have the same number on them?
   c) What is the probability that the total of the numbers on the two counters will be 3?

# 22.5 USING TABLES AND DIAGRAMS

## Objectives

**Use two-way tables and Venn diagrams to calculate the probability of two combined events**

A two-way table is used to present and sort data which is classified in two different ways.

**e.g.** In a class of 30 students there are 14 boys.
9 of the students in the class play an instrument;
5 of them are boys.
First put the information given into the table.
Then work out the missing values, using addition and subtraction.

| | Boys | Girls | Total |
|---|---|---|---|
| **Play** | 5 | | 9 |
| **Not play** | | | |
| **Total** | 14 | | 30 |

Probabilities can be found from the data.

**e.g.** The probability of choosing a student at random who does not play an instrument is $\frac{21}{30}$
The probability of choosing a girl at random who plays an instrument is $\frac{4}{16}$

| | Boys | Girls | Total |
|---|---|---|---|
| **Play** | 5 | 4 | 9 |
| **Not play** | 9 | 12 | 21 |
| **Total** | 14 | 16 | 30 |

**Venn diagram**

Information from a two-way table can be presented in a Venn diagram too.

**e.g.** Look at the sets for the Venn diagram and work out how the information has been transferred from the two-way table.

| | Tennis | Not tennis | Total |
|---|---|---|---|
| **Hockey** | 5 | 4 | 9 |
| **Not hockey** | 9 | 12 | 21 |
| **Total** | 14 | 16 | 30 |

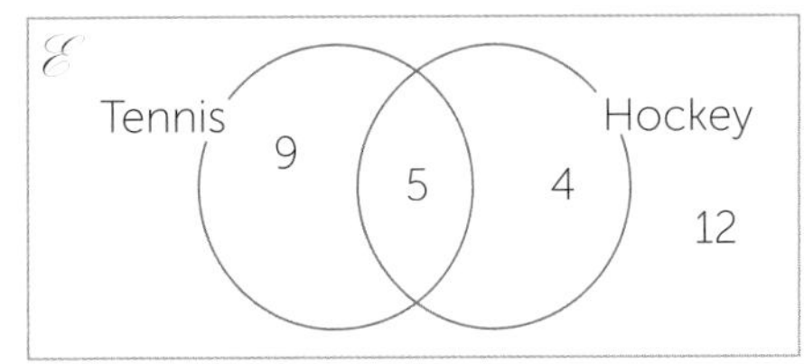


The Venn diagram does not show the totals, but they can be worked out.
Information from a Venn diagram can usually be transferred to a two-way table too.

## Practice questions

1. 96 students each study one of three sciences. The two-way table shows some information about these students and their subjects.
   a) Copy and complete the two-way table.
   b) Find the probability that a student picked at random studies Physics.
   c) Find the probability that a student picked at random is male.

| | Biology | Chemistry | Physics | Total |
|---|---|---|---|---|
| **Female** | 17 | | | 46 |
| **Male** | | 23 | | |
| **Total** | 37 | 41 | | |

2. A group of adults are asked if they have ever posted a video on YouTube.
   Some of the results of the survey are shown in this two-way table.
   a) Copy and complete the two-way table.
   b) What fraction of people surveyed have posted on YouTube?
   c) What is the probability that a person chosen at random is a man?
   d) Someone chosen at random has never posted on YouTube. What is the probability this person is a woman?

| | Yes | No | Total |
|---|---|---|---|
| **Men** | | 143 | |
| **Women** | 124 | | 352 |
| **Total** | 273 | | |

3. Some students were asked if they prefer regular fizzy drinks or sugar-free drinks. Here are some of the results.

   a) Copy and complete the two-way table.

   b) What is the probability that a student chosen at random is a boy?

   c) What is the probability that a student chosen at random prefers sugar-free drinks?

   d) A girl is chosen at random; what is the probability they prefer regular fizzy drinks?

| | **Regular** | **Sugar-free** | **Total** |
|---|---|---|---|
| **Girls** | 21 | 46 | |
| **Boys** | | | |
| **Total** | 64 | | 137 |

4. A class of 25 pupils are asked about their different mobile devices.

   There are 6 pupils with a mobile and a laptop, 12 with a mobile but no laptop and 4 with neither a mobile or a laptop.

   a) Copy and complete the two-way table.

   b) Copy and complete the Venn diagram.

| | **mobile** | **no mobile** | **total** |
|---|---|---|---|
| **laptop** | | | |
| **no laptop** | | | |
| **total** | | | |

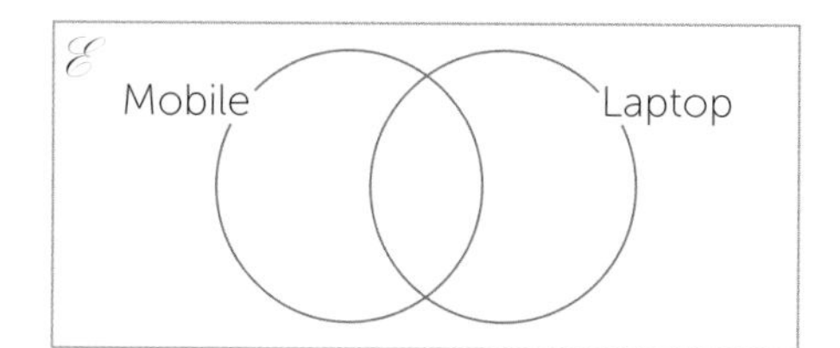


   c) What is the probability that a pupil chosen at random owns just a laptop?

   d) A pupil chosen at random owns a laptop. What is the probability that they also own a mobile phone?

5. In a class of 28 students, 8 are in the football team, 9 are in the chess club and 3 are in both.

   a) Represent this information on a Venn diagram.

   b) How many students are in neither the football team or the chess club?

   c) What is the probability that a student chosen at random is in the chess club?

   d) A student in the football team is chosen at random. What is the probability that they are also in the chess club?

6. A vet surveys some of her clients about their pets.

   The results are shown in the Venn diagram.

| | **Cat** | **No cat** | **Total** |
|---|---|---|---|
| **Dog** | | | |
| **No dog** | | | |
| **Total** | | | |

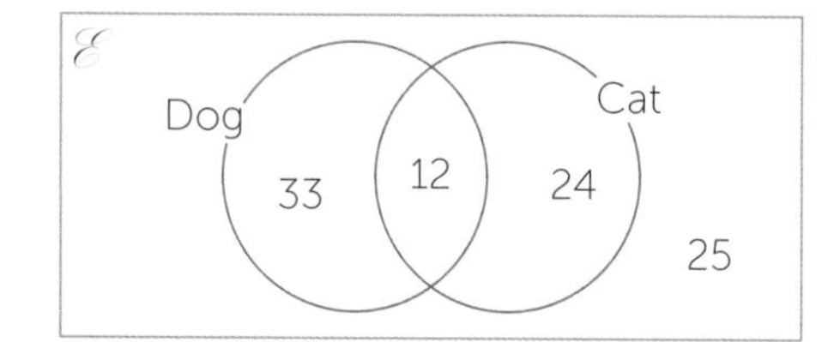


   a) How many clients did she survey?

   b) Copy and complete the two-way table.

   c) Work out P(C), P(C ∩ D) and P(C ∪ D).

## 22.6 SOLVING PROBABILITY PROBLEMS

### Objectives

**Solve problems involving probability that link with other areas such as statistics and algebra**

Probability can occur in many situations and can involve different mathematical topics.

**e.g.** 9 people on a committee choose 3 of them to be president, treasurer and secretary.

Each person is equally likely to fill any of the posts.

What is the probability that Amy is president, Ben is treasurer and Carl is secretary?

There are 9 ways to choose the president, then 8 ways to choose the treasurer (1 person is already allocated) and 7 ways to choose the secretary (since 2 people are already allocated).

This means there are $9 \times 8 \times 7 = 504$ different ways to choose the three posts.

Amy, Ben and Carl filling the posts in this order, is only **one** of these ways.

The probability that this happens is $\frac{1}{504}$

### Practice questions

1. Bolts of a certain length are made in a factory.

   It is found that 2% of bolts are made too short and 5% of the bolts are made too long. A bolt is selected at random. What is the probability that it will be:

   a) too short? b) too long? c) the correct size?

2. A new design for a machine is produced and it is discovered that two possible defects occur with the machine.

   Defect A occurs in 10% of machines while Defect B occurs in 15% of machines.

   4% of machines have **both** defects.

   a) Draw a Venn diagram to represent this information.

   A machine is chosen at random. What is the probability that has:

   b) Only Defect A?

   c) Only Defect B?

   d) No defects?

3. There are four times as many sticks of coloured chalk in a pot than sticks of white chalk.

   Max thinks that the probability of picking a stick of white chalk at random is $\frac{1}{4}$.

   Is he correct? Give a reason for your answer.

4. In a sample of people, 20% are left-handed and 30% are blonde.

   Being left-handed and blonde are independent of each other.

   a) Draw a tree diagram to show the outcomes of these combined events.

   An individual is selected at random from the sample. Work out:

   b) the probability of choosing someone who is left-handed and blonde.

   c) the probability of choosing someone who is either left-handed or blonde, or both.

5. Three fair six-sided dice are rolled.

   Jess needs to work out the probability of getting three 5s from the dice.

   Here is her working: $P(5) = \frac{1}{6}$   $P(5, 5, 5) = \frac{1}{6} + \frac{1}{6} + \frac{1}{6} = \frac{3}{6}$

   Is Jess correct? Give a reason for your answer.

6. Two pots contain some coloured balls. In Pot 1, 5% of the balls are white and in Pot 2, 10% of the balls are white.
   A pot is chosen at random and a ball drawn randomly from that pot.
   a) Complete the tree diagram below.

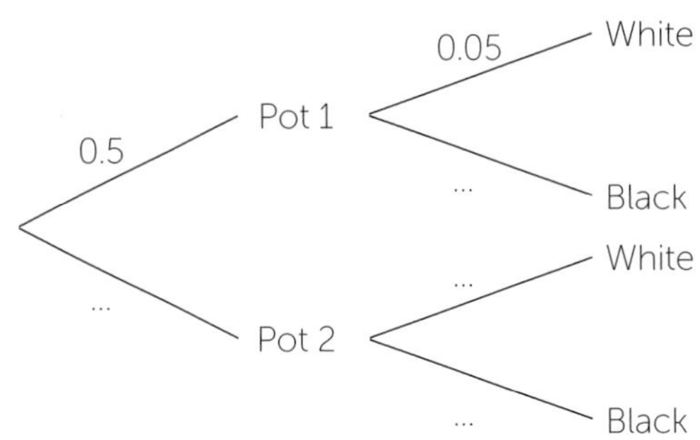


   b) Work out the probability of getting a white ball.

7. The Royal Mail estimates that 34% of letters are posted first-class and 66% are posted second-class.
   It claims that 92% of first-class letters and 96% of second-class letters arrive on time.
   a) Complete the tree diagram below.

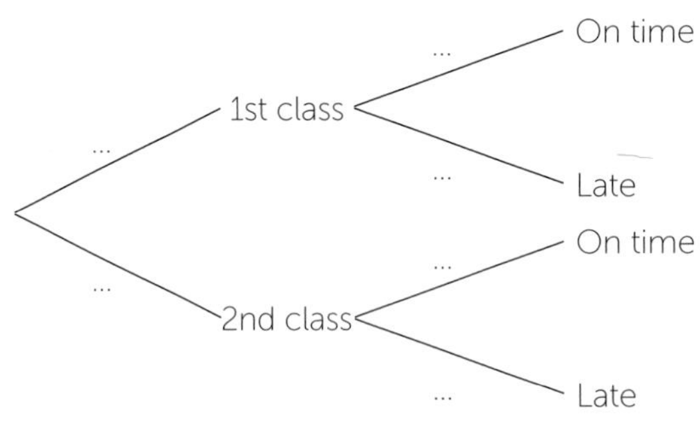


   b) Lily says that she has less than a 95% chance of a letter arriving on time. Do you agree with her?
   Give a reason for your answer.

8. a) Write down all the different possible combinations of boys and girls that could be born into a family of 3 children.

   e.g. BGG means first child boy, second child girl, third child girl.

   BGG .......... .......... .......... .......... .......... .......... ..........

   b) What is the probability of a family of three children having 3 girls?

   c) What is the probability of a family of three children having 2 boys and 1 girl, in any order?

9. There are 35 cars in a car park.
   12 cars are red, 5 cars have 2 doors and 6 cars are red with 4 doors.
   17 cars have 5 doors but there are no red cars with just 2 doors.
   a) Put this information into a two-way table.
   b) A car is chosen at random. Work out the probability that the car has 4 doors.
   c) A red car is chosen at random. Work out the probability that the car has 5 doors.

| | **Red** | **Not red** | **Total** |
|---|---|---|---|
| **2 door** | | | |
| **4 door** | | | |
| **5 door** | | | |
| **Total** | | | |

# SECTION 23
# MENSURATION

## 23.1 VECTORS

G24 G25

A **vector** is a quantity that is determined by a length and a direction.

Vectors are used to describe a displacement from one point to another.

All vectors have both a **magnitude** (size) and **direction**.

**Be able to represent information graphically, given column vectors**

### Objectives

**Calculate using column vectors**

Vectors can be represented

- by single bold letters
- using displacements in the $x$ and $y$ directions written as a column $\begin{pmatrix} x \\ y \end{pmatrix}$
- as a movement from one point to another, for example $\overrightarrow{AB}$
- as an arrow showing the translation from one point to another.

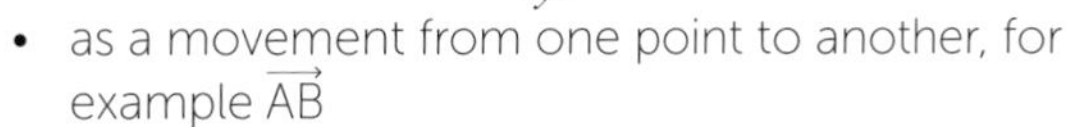

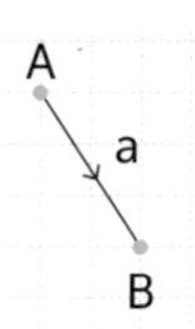


$\overrightarrow{AB} = \begin{pmatrix} 2 \\ -3 \end{pmatrix}$

**Represent graphically, the sum and difference of two vectors**

Vectors can be added (or subtracted) by adding the respective $x$ and $y$ elements.

e.g. Point A has coordinates (2, 3) and point B has coordinates (5, 7).

The vector **a** is the displacement from A to B, denoted $\overrightarrow{AB}$.

a) Write **a** as a column vector.

$a = \begin{pmatrix} 3 \\ 4 \end{pmatrix}$

The point C is such that the displacement $\overrightarrow{BC}$, $\mathbf{b} = \begin{pmatrix} -1 \\ 4 \end{pmatrix}$.

b) What are the coordinates of C?

C = (4, 11)

c) What is the column vector $\mathbf{c} = \mathbf{a} + \mathbf{b}$?

$c = \begin{pmatrix} 3 \\ 4 \end{pmatrix} + \begin{pmatrix} -1 \\ 4 \end{pmatrix} = \begin{pmatrix} 2 \\ 8 \end{pmatrix}$

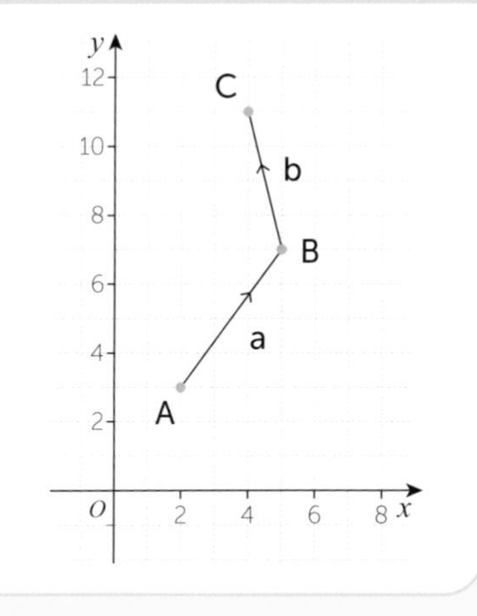


### Practice questions

1. On squared paper, draw and label the following vectors:
   a) $\mathbf{p} = \begin{pmatrix} 2 \\ 4 \end{pmatrix}$ b) $\mathbf{q} = \begin{pmatrix} -3 \\ 2 \end{pmatrix}$ c) $\mathbf{r} = \begin{pmatrix} 6 \\ -1 \end{pmatrix}$ d) $\mathbf{s} = \begin{pmatrix} -2 \\ 0 \end{pmatrix}$ e) $\mathbf{t} = \begin{pmatrix} -1 \\ -5 \end{pmatrix}$

2. A knight on a chessboard always moves 2 squares horizontally and 1 square vertically, or 1 square horizontally and 2 squares vertically.
   Write down as column vectors all of the eight possible moves a knight can make.

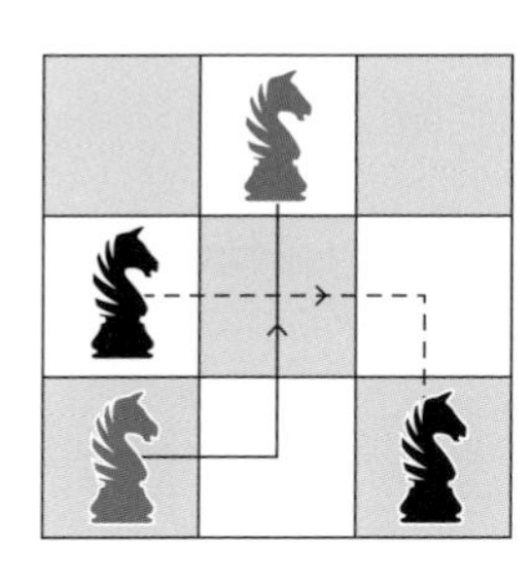

3. The point M is (–2, 4), N is (0, 5) and P is (3, –1).
   a) Write the column vectors for $\overrightarrow{MN}$, $\overrightarrow{NP}$, and $\overrightarrow{PM}$.
   b) Write the column vectors for $\overrightarrow{NM}$, $\overrightarrow{PN}$, and $\overrightarrow{MP}$.
   c) What do you notice about the two sets of vectors?

4. HJKL is a square of side 3 units, where H has coordinates (3, 1) and K has coordinates (6, 4).
   a) Draw the square HJKL (with HJ horizontal) on squared paper and label the vertices.
   b) Write down the column vectors for $\overrightarrow{HJ}$ and $\overrightarrow{JK}$.
   c) Write down the column vectors for $\overrightarrow{HL}$ and $\overrightarrow{LK}$.
   d) What do you notice about the vectors $\overrightarrow{HJ}$ and $\overrightarrow{LK}$?
   e) What do you notice about the vectors $\overrightarrow{HL}$ and $\overrightarrow{JK}$?

5. Points X, Y, Z are the vertices of a triangle with X = (1, 1), $\overrightarrow{XY} = \begin{pmatrix} 5 \\ -1 \end{pmatrix}$, $\overrightarrow{YZ} = \begin{pmatrix} -2 \\ 4 \end{pmatrix}$.
   a) Draw triangle XYZ on squared paper.
   b) Write $\overrightarrow{XZ}$ as a column vector.
   c) The vector $\overrightarrow{YW}$ is the same as $\overrightarrow{XZ}$. What are the coordinates of W?
   d) What shape is XYWZ?
   e) What can you say about vectors $\overrightarrow{XY}$ and $\overrightarrow{ZW}$?

6. A parallelogram has vertices A, B, C, D. Point A has coordinates (2, 0).
   $\overrightarrow{AB} = \begin{pmatrix} 3 \\ -2 \end{pmatrix}$ and $\overrightarrow{AD} = \begin{pmatrix} 1 \\ 3 \end{pmatrix}$.
   a) Draw the parallelogram on squared paper.
   b) What is the column vector for $\overrightarrow{BC}$?
   c) What are the coordinates of point C?

7. $\overrightarrow{AB} = \begin{pmatrix} 5 \\ 2 \end{pmatrix}$ and $\overrightarrow{BC} = \begin{pmatrix} 3 \\ 4 \end{pmatrix}$. What is the column vector for $\overrightarrow{AC}$?

8. $\mathbf{h} = \begin{pmatrix} 2 \\ 4 \end{pmatrix}$, $\mathbf{j} = \begin{pmatrix} 3 \\ 0 \end{pmatrix}$ and $\mathbf{k} = \begin{pmatrix} -4 \\ -1 \end{pmatrix}$.
   a) Work out $\mathbf{h} + \mathbf{j}$ and $\mathbf{j} + \mathbf{h}$. Explain what you notice.
   b) Work out $\mathbf{h} + (\mathbf{j} + \mathbf{k})$ and $(\mathbf{h} + \mathbf{j}) + \mathbf{k}$. Explain what you notice.
   c) Work out $\mathbf{h} - \mathbf{j}$ and $\mathbf{j} - \mathbf{h}$. Explain what you notice.

9. On squared paper mark the points A (3, 2) and B (5, 1).
   a) What is the column vector $\mathbf{a} = \overrightarrow{AB}$?
   b) If $\mathbf{b} = \overrightarrow{BC} = \begin{pmatrix} -2 \\ 3 \end{pmatrix}$, find the coordinates of C.
   c) Calculate $\overrightarrow{AC}$. What is this in terms of **a** and **b**?

10. Let A be the point (1, 2) and B the point (4, 3). $\overrightarrow{AB} = \mathbf{a}$ and $\overrightarrow{BC} = \begin{pmatrix} 1 \\ 2 \end{pmatrix} = \mathbf{b}$.
   a) What are the coordinates of C?
   b) Calculate $\mathbf{a} + \mathbf{b}$. What does this represent graphically?
   c) Write down the column vector for $-\mathbf{b} = \overrightarrow{CB}$.
   d) If $-\mathbf{b} = \overrightarrow{BD}$, what are the coordinates of D?
   e) If $\overrightarrow{AD} = \overrightarrow{AB} + \overrightarrow{BD}$, write $\overrightarrow{AD}$ as a column vector.

11. $\overrightarrow{MN} = \mathbf{m} = \begin{pmatrix} 1 \\ 4 \end{pmatrix}$, $\overrightarrow{NP} = \mathbf{n} = \begin{pmatrix} -2 \\ 1 \end{pmatrix}$, $\overrightarrow{PM} = \mathbf{p} = \begin{pmatrix} 1 \\ -5 \end{pmatrix}$. Three points M, N and P are to be plotted.
   a) If M is the point (3, 1), draw the points M, N and P on squared paper.
   b) If $\overrightarrow{MQ} = \mathbf{m} - \mathbf{n}$, what are the coordinates of Q?
   c) If $\mathbf{p} - \mathbf{m} = \overrightarrow{PR}$, what are the coordinates of R?

# 23.2 PARALLEL VECTORS

## Objectives

### Calculate using column vectors

A column vector is a vector whose components are listed vertically in a single column.

Given vector $\mathbf{p} = \begin{pmatrix} 2 \\ 5 \end{pmatrix}$, then $\mathbf{p} + \mathbf{p} = \begin{pmatrix} 2 \\ 5 \end{pmatrix} + \begin{pmatrix} 2 \\ 5 \end{pmatrix} = \begin{pmatrix} 4 \\ 10 \end{pmatrix} = 2\mathbf{p}$.

Graphically this represents a vector twice as long as **p** and in the same direction.

A negative multiple will be a vector in the opposite direction.

**e.g.** The vector $\overrightarrow{AB} = \mathbf{s} = \begin{pmatrix} 3 \\ -1 \end{pmatrix}$.

Draw the vectors $\overrightarrow{BC} = 3\mathbf{s}$ and $\overrightarrow{AD} = -2\mathbf{s}$.

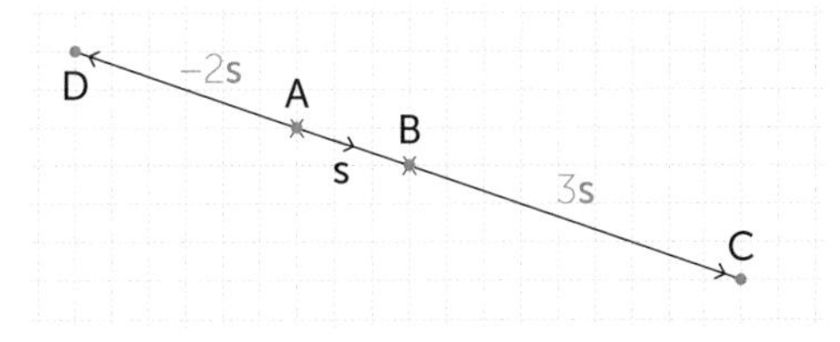


### Represent graphically a scalar multiple of a vector

### Identify two column vectors which are parallel

Vectors have both magnitude and direction.

Vectors are parallel if the $x$ and $y$ components of the vectors are in the same ratio.

**e.g.** All the vectors in the diagram below are parallel.

What is the ratio between the two components of each of these vectors?

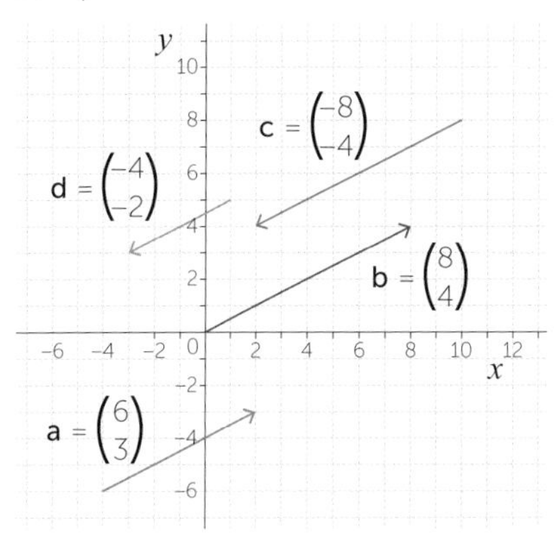


## Practice questions

1. The vector $\mathbf{a} = \begin{pmatrix} 3 \\ 5 \end{pmatrix}$. Calculate each of the following:

   a) $4\mathbf{a}$ b) $0.5\mathbf{a}$ c) $-2\mathbf{a}$ d) $-0.1\mathbf{a}$

2. Here is the vector $\mathbf{b} = \begin{pmatrix} 4 \\ -1 \end{pmatrix}$.

   Draw each of the vectors:

   a) $2\mathbf{b}$ b) $-\mathbf{b}$ c) $\frac{1}{2}\mathbf{b}$ d) $-3\mathbf{b}$

   b

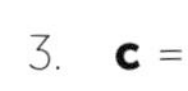

3. $\mathbf{c} = \begin{pmatrix} -2 \\ 3 \end{pmatrix}$ and $\mathbf{d} = \begin{pmatrix} 5 \\ 2 \end{pmatrix}$. Write as column vectors:

   a) $3\mathbf{c}$ b) $-4\mathbf{c}$ c) $6\mathbf{d}$ d) $0.5\mathbf{c}$ e) $-\mathbf{d}$ f) $2.5\mathbf{d}$

4. $\overrightarrow{AB}$ joins A (1, 1) to B (2, 5). $\overrightarrow{CD}$ joins C (3, −2) to D (5, 6)

   a) Write down $\overrightarrow{AB}$ and $\overrightarrow{CD}$ as column vectors.

   b) If $\overrightarrow{CD} = \mathbf{p}$, then write $\overrightarrow{AB}$ as a multiple of **p**.

   c) What does this tell you about $\overrightarrow{AB}$ and $\overrightarrow{CD}$?

5. The point H has coordinates (2, 1), the point J is at (4, 0) and K is at (−4, 7). Find the coordinates of L such that $\overrightarrow{KL} = 5\overrightarrow{HJ}$.

6. Here are six vectors:

   $\mathbf{a} = 4\mathbf{p} + 2\mathbf{q} - 2\mathbf{p} + \mathbf{q}$ $\mathbf{b} = 6\mathbf{q} - 2\mathbf{p} - 2\mathbf{q} + \mathbf{p}$ $\mathbf{c} = 7\mathbf{p} + 4\mathbf{q} - 2\mathbf{p} - 5\mathbf{q}$

   $\mathbf{d} = 4\mathbf{p} + 8\mathbf{q} - 7\mathbf{p} + 4\mathbf{q}$ $\mathbf{e} = 5\mathbf{p} + 8\mathbf{q} - \mathbf{p} - 2\mathbf{q}$ $\mathbf{f} = 5\mathbf{p} - 6\mathbf{q} + 5\mathbf{p} + 4\mathbf{q}$

   a) Simplify each vector.

   b) Pair up the vectors which are parallel.

   c) Write one of each pair as a multiple of the other.

# 23.3 ARCS AND SECTORS

## Objectives

**Calculate arc lengths, angles and areas of sectors of circles**

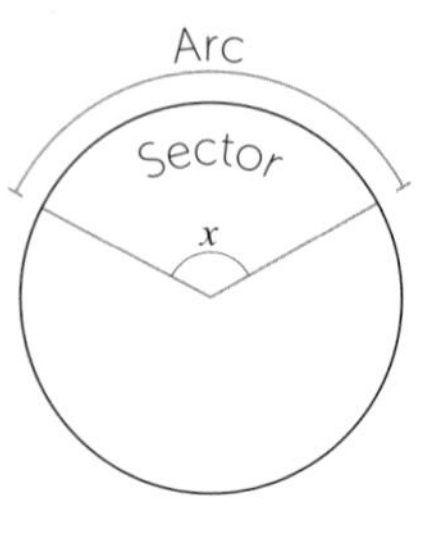


An **arc** is a length, being part of the circumference of a circle.

A **sector** is the area formed from an arc and two radii.

The perimeter of a sector is the arc length plus the two radii.

Arc length $= \frac{x}{360°} \times 2\pi r$ Area of sector $= \frac{x}{360°} \times \pi r^2$

**e.g.** Calculate the perimeter and area of a sector of a circle of angle 60° and radius 4 cm.

Perimeter $= \frac{60}{360} \times (2 \times \pi \times 4) + (4 + 4)$ Area $= \frac{60}{360} \times \pi \times 4^2$

$= \left(8 + \frac{4\pi}{3}\right)$ cm $= \frac{8\pi}{3}$ cm$^2$ (Answers left in terms of $\pi$)

## Practice questions

**For practice questions 1 – 8, leave your answers in terms of $\pi$.**

1. Work out the lengths of the following arcs.
   a) Angle 90°, radius 10 cm b) Angle 180°, radius 5 cm c) Angle 30°, radius 9 m
2. For each arc in question 1, work out the perimeters of the sectors they form.
3. Work out the lengths of the following arcs.
   a) Angle 40°, radius 8 cm b) Angle 90°, radius 14 m c) Angle 15°, radius 24 cm
   d) Angle 108°, radius 6.4 m e) Angle 48°, radius 520 mm f) Angle 144°, radius 38.6 cm.
4. The shape in the diagram has lengths PQ = PR = 12 cm and angle ∠QPR = 90°
   QR is the arc of a circle, centre P. Calculate the area of the shape.

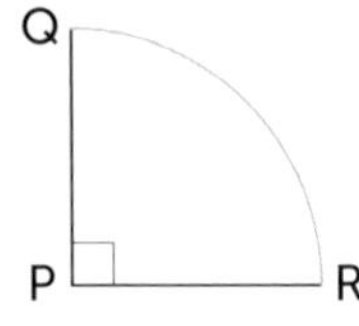


5. A security light scans an angle of 150° to a distance of 15 m.
   What is the area of the sector covered by the security light?
6. A patio is to be covered with decking. The patio is in the shape of a sector bordered by two fences of 10 m each.
   a) If the area of decking required is $20\pi$ m$^2$, what is the angle between the fences?
   b) A hedge is being laid along the curved edge of the patio. What length hedge is required?
7. A sector of angle 30° has radius $x$ cm. If the perimeter of the sector is $\left(\frac{5\pi}{6} + 10\right)$ cm, what is the radius and the arc length?

8. AB is the arc of a circle, radius $p$ cm. ∠AOB = 120°
   a) What is the area of the sector AOB in terms of $\pi$ and $p$?
   b) What is the arc length AB in terms of $\pi$ and $p$?

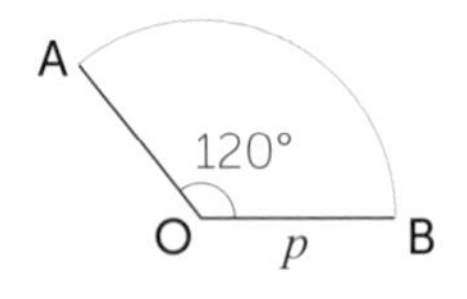

9. A sector of a circle of radius 12 m has an area of $112\pi$ m$^2$.
   a) What is the angle the sector makes at the centre of the circle? b) What is the arc length of the sector?

10. The rear windscreen wiper on a car rotates on an arm 48 cm long through an angle of 115°.
    The rubber wiper blade is 35 cm long.
    a) Calculate the area covered by the whole arm.
    b) Calculate the area not cleared by the blade.
    c) Hence work out the area of windscreen that is cleared. Give each answer to the nearest cm$^2$.

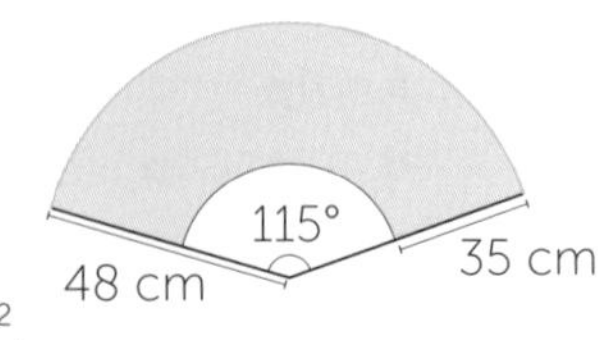

# 23.4 COMPOSITE CIRCLE SHAPES

## Objectives

**Calculate perimeters and areas of composite shapes made from circles and parts of circles**

Knowing the formulae for the circumference and the area of a circle allows the calculation of perimeters and areas of shapes made from circles and parts of circles.

Recall that: Circumference = $2\pi r$ and Area = $\pi r^2$, where $r$ is the radius.

The shape shown is made up of 3 semi-circles on a straight line with given diameters.

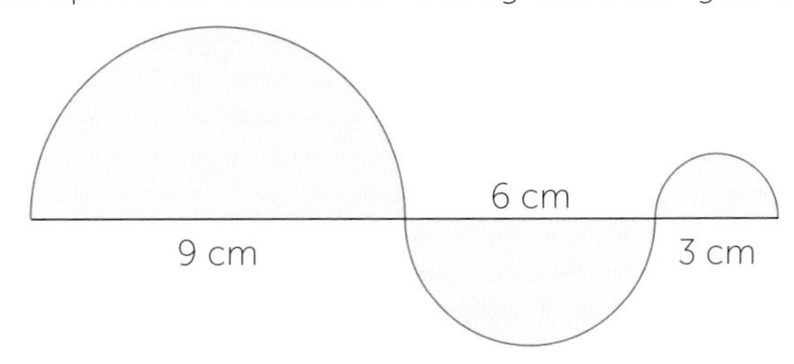


a) What is the perimeter of the shape?

Perimeter = $\frac{9}{2}\pi + \frac{6}{2}\pi + \frac{3}{2}\pi + 9 + 6 + 3 = (9\pi + 18)$ cm

b) What is the area of the shape?

Area = $\frac{1}{2}(\pi \times 4.5^2 + \pi \times 3^2 + \pi \times 1.5^2) = \frac{1}{2}\left(\frac{81}{4}\pi + 9\pi + \frac{9}{4}\pi\right) = \frac{63}{4}\pi \text{ cm}^2$

## Practice questions

1. Calculate the area of the grey shaded area in this shape.
   Leave your answer in terms of $\pi$.

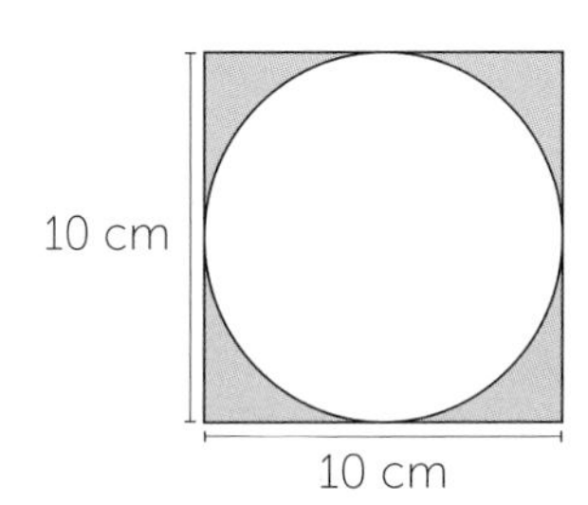


2. Calculate the perimeter and area of this shape formed from 3 semi-circles.

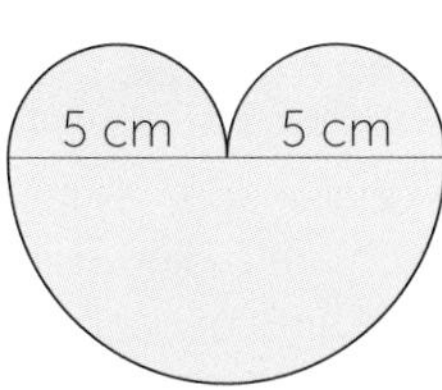


3. This shape is made from 2 circles of radii 7 m and 9 m.
   What is the area of the outside ring?

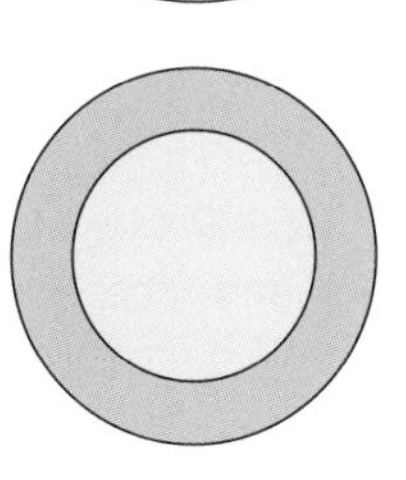

4. This shape is made from 2 semi-circles of radii 8 cm and 4 cm.
   Calculate its perimeter.

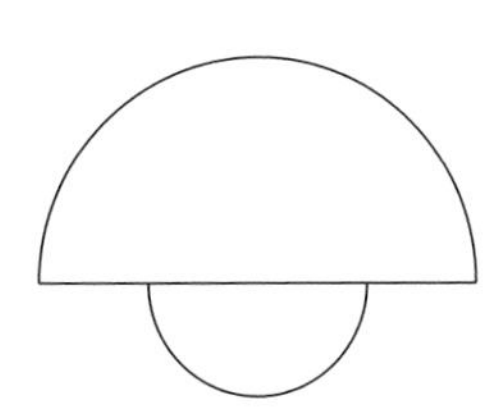

5. a) Work out the area of a semi-circle of radius 16 m.
   b) Work out the area of 2 semi-circles of radius 8 m.
   c) Work out the area of 4 semi-circles of radius 4 m.
   d) Using the answers to parts (a) to (c) above, deduce the area of 8 semi-circles of radius 2 m.

6. Calculate the perimeter and area of this shape made from 2 equal quadrants.

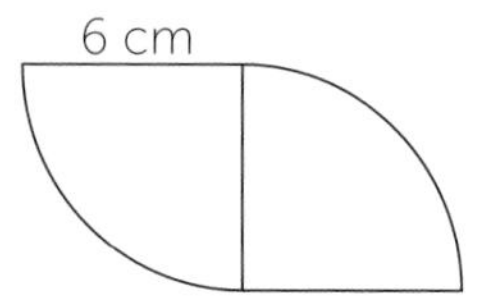

7. A table top is made from a rectangle of wood 80 cm x 50 cm and 2 semi-circles each of diameter 50 cm.
   a) What is the perimeter of the table?
   b) What is the area of the table?

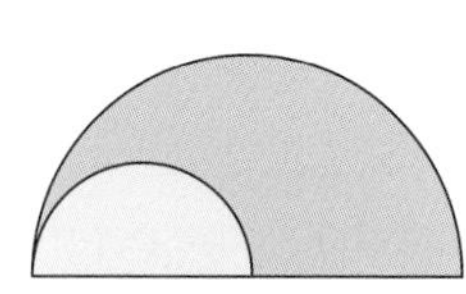

8. This shape is formed from 2 semi-circles of diameters 18 m and 9 m.
   What is the perimeter and area of the darker shaded shape?

9. This shape is made by placing two circles of radius 25 cm onto a circle of diameter 100 cm.
   a) What is the area of the darker shaded portion?
   b) How many times greater is the darker area than one of the small circles?

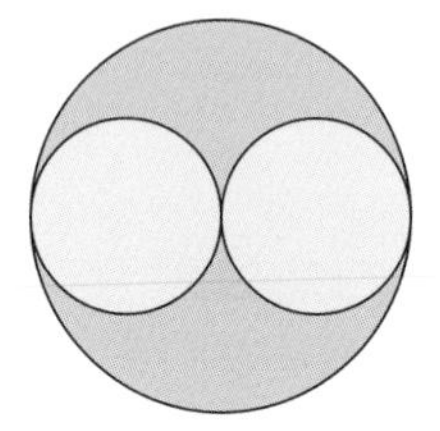

## 23.5 SURFACE AREA

G17 | A5

### Objectives

**Find the surface area of a prism; using rectangles and triangles**

A 3-dimensional shape has faces and surfaces which are 2-dimensional. By calculating the area of each of these, the surface area of the 3D shape can be found.

Triangular prisms have faces that are rectangles and triangles, so use:
'length x width' and '$\frac{1}{2}$ base x height'

**Find the surface area of a cylinder**

Cylinders have circular ends which are calculated using $\pi r^2$, while the curved surface has an area equal to:
circle circumference x cylinder length.

**Find the surface areas of spheres, pyramids, cones and composite solids**

Surface area of a sphere = $4\pi r^2$.

For a cone, the base is a circle, while the curved surface is given by $\pi r l$, where $l$ is the slant length of the cone.

The surface area of a pyramid is the area of the base plus the area of a number of triangles.

**Worked examples**

1. What is the surface area of a cube of side length 5 cm?
   The cube has 6 square faces, each 5 cm × 5 cm, total surface area = 6 × 25 = 150 cm$^2$.
2. What is the surface area of the cone, radius 3 cm as shown?
   Surface area of the cone
   = circle base + curved surface
   = $\pi \times 3^2 + \pi \times 3 \times 5 = 24\pi$ cm$^2$.

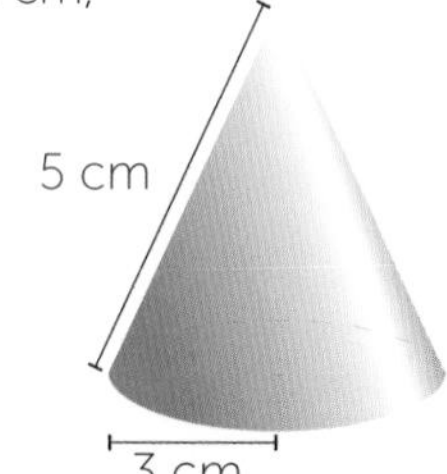

## Practice questions

1. A wooden door wedge has a square end of 3 cm, a rectangular base of length 4 cm, a sloping rectangular face of length 5 cm and two triangular sides.

   What is the surface area of the door wedge?

2. A chocolate product is sold in a box in the shape of a triangular prism. The triangular ends each have an area of 15.588 cm$^2$ and an edge length of 6 cm. The box is 30 cm long.

   What is the total surface area of the box?

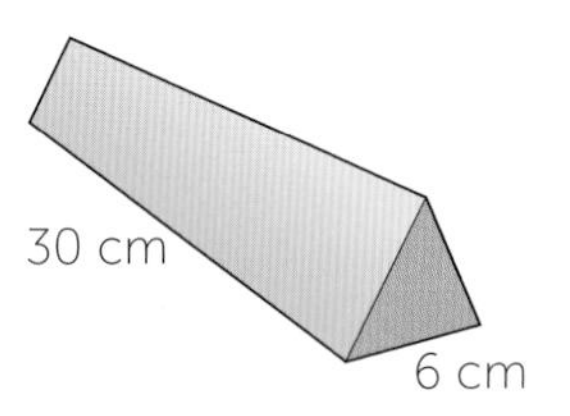


3. Find the total surface area of each of these cylinders. (Leave your answers in terms of $\pi$.)

   a) Height 4 cm, radius 6 cm
   b) Height 10 cm, radius 3 cm
   c) Height 15 m, radius 2 m
   d) Height 45 cm, radius 7.5 cm

4. Which of these solid cylinders has the greater surface area, one of height 8 cm and radius 5 cm, or one of height 5 cm and radius 8 cm?

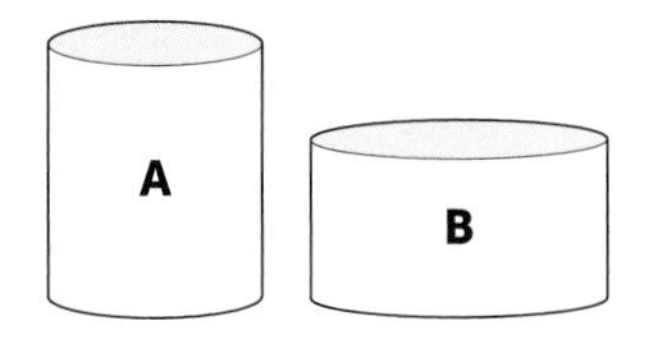


5. What is the surface area of a football measuring 30 cm in diameter?

6. A pyramid has a square base of side 180 m, and 4 triangular sides each of slant height 140 m. What is the total surface area of the sloping sides?

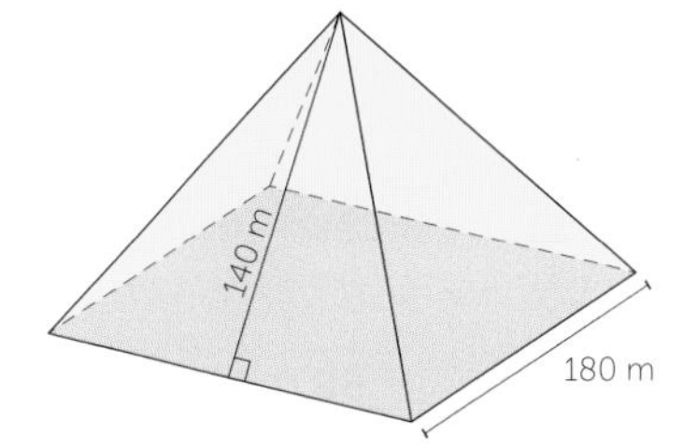

7. A sector of a circle of radius 18 cm and arc length $10\pi$ cm is folded and joined to make a cone. The base of the cone has a circumference of $10\pi$.

   a) What is the radius of the base of the cone?
   b) What is the curved surface area of the cone?
   c) What is the total surface area of the cone?

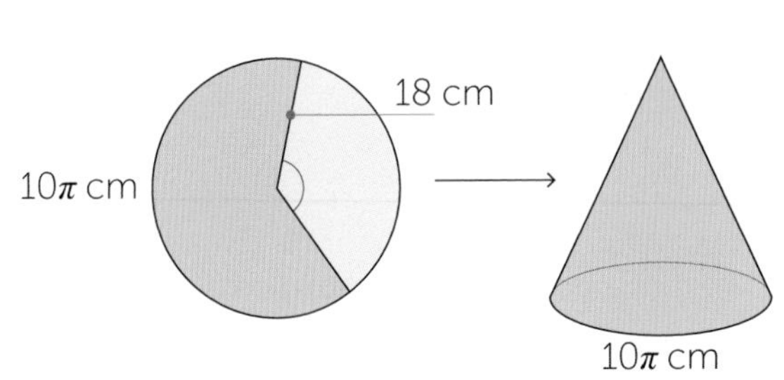

8. A plastic component for a design of a rotating movable joint is made from a cone of radius 16 cm joined by its base to the flat face of a hemisphere of the same radius. The curved surface area is given by $\pi rl$, where $l$ is the slant length of the cone.

   The slant length of the cone is 24 cm.

   a) What is the curved surface area of the cone?
   b) What is the curved surface area of the hemisphere?
   c) What is the total surface area of the component?

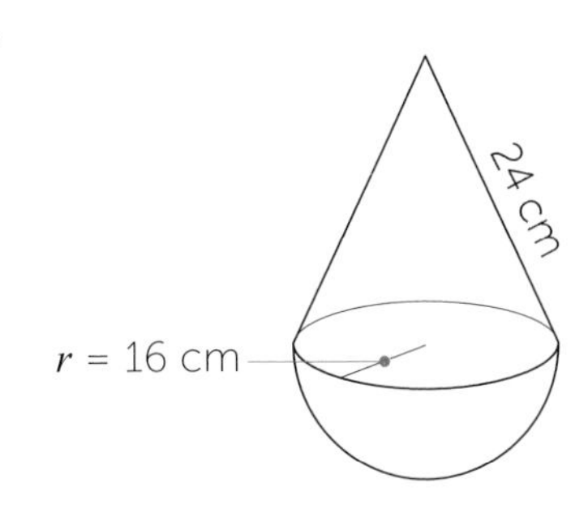

# 23.6 VOLUME OF COMPOUND SOLIDS

## Objectives

**Find the volumes of cylinders, spheres, cones, pyramids and composite solids**

Volume of a cylinder, radius $r$, height $h$

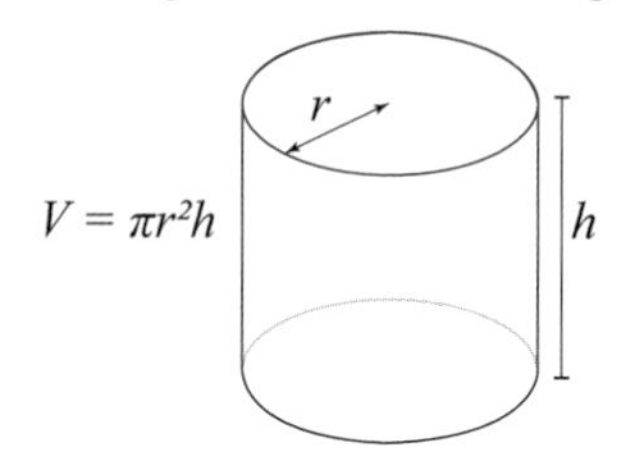


Volume of sphere, radius $r$

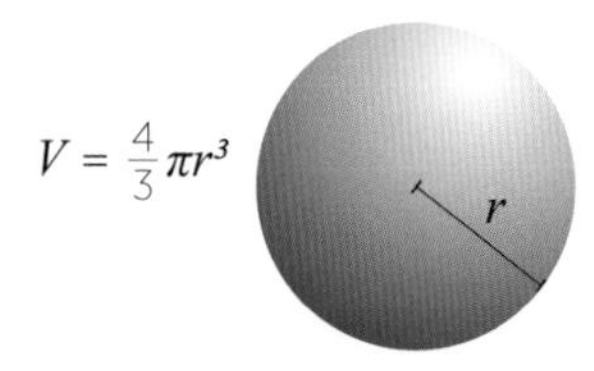


Volume of a cone, base radius $r$, perpendicular height $h$

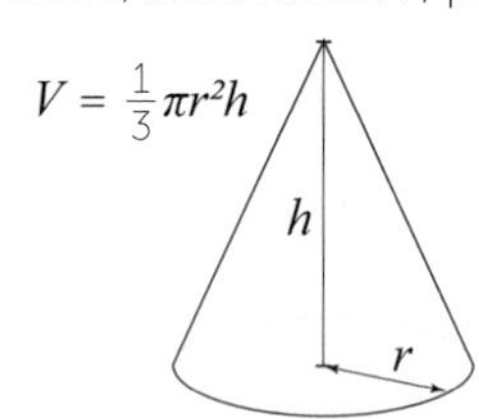


Volume of a pyramid, perpendicular height $h$:

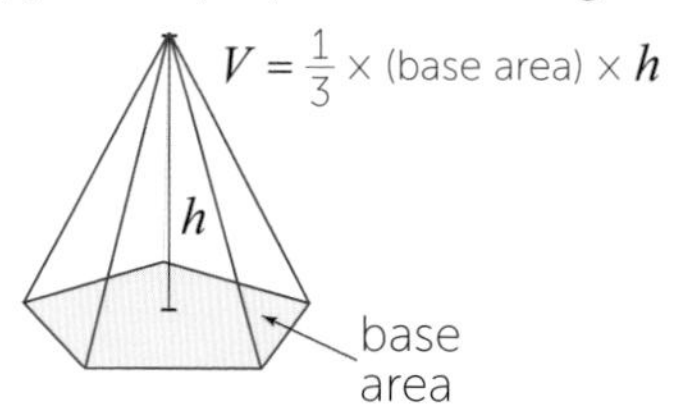


The volume of a composite solid is the sum of the volumes of the individual parts.

**e.g.** A rubber part for a child's toy is made from a cylinder diameter 4 cm with a cone, height 2 cm on one end and a hemisphere at the other.

Calculate the volume of rubber needed to make this part.

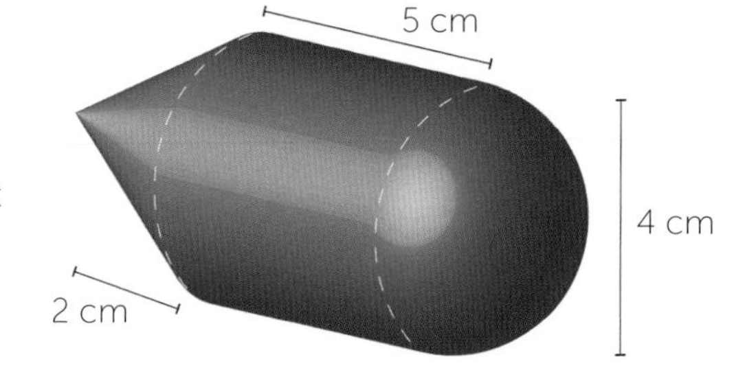


Vol. of cone $= \frac{1}{3} \times \pi \times 2^2 \times 2$
$= 8.37... \text{ cm}^3$

Vol. of cylinder $= \pi \times 2^2 \times 5$
$= 62.83... \text{ cm}^3$

Vol. of hemisphere $= (\frac{4}{3} \times \pi \times 2^3) \div 2$
$= 16.75... \text{ cm}^3$

Total Volume $= 8.37... + 63.83... + 16.75...$
$= \mathbf{88.0 \text{ cm}^3 \text{ (1 d.p.)}}$

## Practice questions

1. Find the volume of these cylinders:
   a) radius 4.5 m and height 3.6 m
   b) radius $3p$ cm and height $5q$ cm

2. Sarah has a full glass of lemonade. She drinks some of it.
   She says she still has three-quarters of her lemonade left.
   Is she correct? Give a reason for your answer.

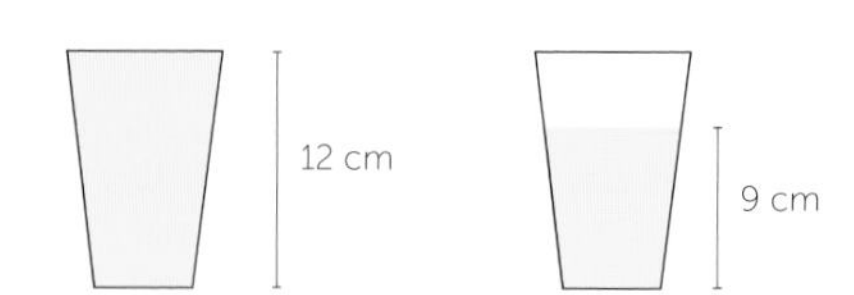


3. A cylindrical marble block is 12.6 cm high and has a diameter of 5.8 cm.
   a) What is the volume of the stone column?
   b) If marble costs £96 per 1000 cm$^3$, how much would the cylindrical block cost?

4. A can of drink is in the shape of cylinder. It has radius 5 cm and height 18 cm.
   Find the volume of the can.

5. An estimate for the radius of the Earth is 6370 km.

   Find an estimate for the volume of the Earth.

6. One of the pyramids in Egypt has a square base of side 230 m and a vertical height of 139 m.

   Find the volume of the pyramid.

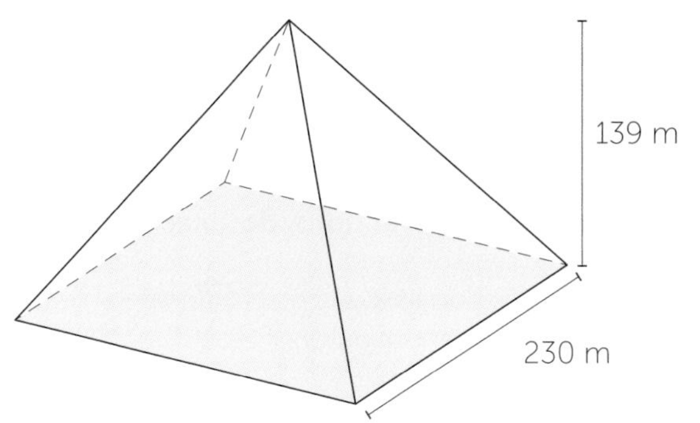


7. A wooden cone has a diameter of 30 cm and a vertical height of 36 cm.

   a) What is the volume of the cone?

   A horizontal cut is made through the cone 12 cm from the top to form a small cone and a frustum.

   b) What is the volume of the small cone which is cut off?

   c) What is the volume of the frustum?

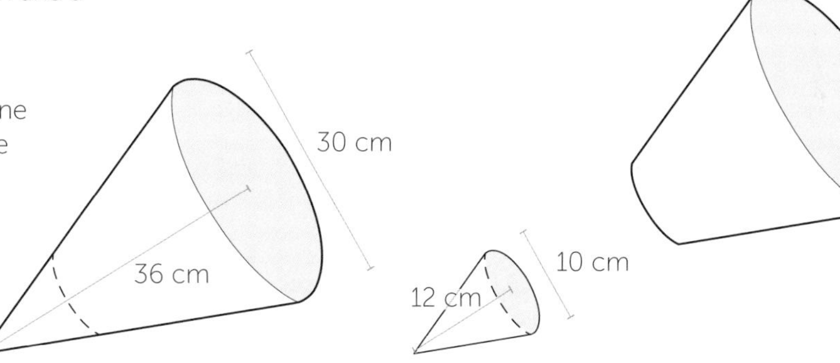

8. A decorative block is made by fixing 4 hemispheres, each of radius 15 cm, onto the top face of a cuboid measuring 60 cm by 60 cm and 10 cm high.

   The plan view and side elevation of the block are shown.

   a) What is the volume of the decorative block?

   b) What is the surface area of the block?

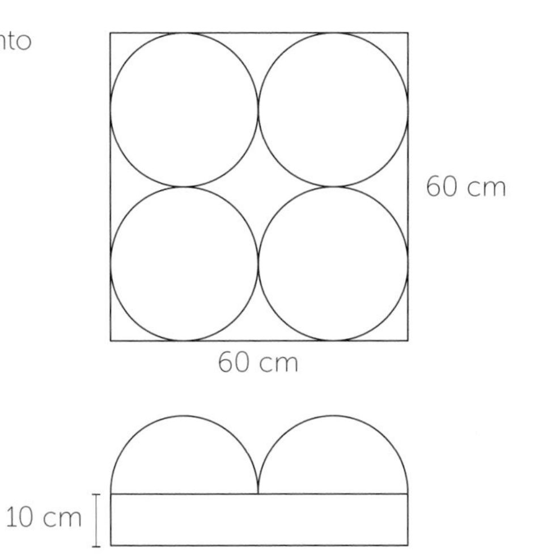

SECTION 24

# APPLICATIONS OF RATIO

## 24.1 TERMINATING AND RECURRING DECIMALS

R9 N10 N12

### Objectives

**Convert between fractions, decimals and percentages**

Convert a fraction to a decimal by finding an equivalent fraction with a denominator of 10, 100 or 1000.

**e.g.** $\frac{7}{20} = \frac{35}{100} = 0.35$

Convert a fraction to a decimal by dividing the numerator by the denominator.

**e.g.** $\frac{3}{8} = 3 \div 8 = 0.375$ or $8\overline{)3.{}^{3}0{}^{6}0{}^{4}0}$ = 0.375

Convert a decimal to a percentage by multiplying by 100.

**e.g.** 0.35 = 0.35 × 100 = 35% 1.3 = 1.3 × 100 = 130%

**Recognise terminating decimals and recurring decimals**

A terminating decimal has a finite number of decimal places.

**e.g.** 0.678 is a terminating decimal

A recurring decimal has a repeating pattern in its decimal places which does not end.

**e.g.** 0.678 787 878... is a recurring decimal.

Recurring decimals can be written using a dot to indicate the recurring digits.

**e.g.** 0.676 767... = $0.\dot{6}\dot{7}$ 0.188 8 ... = $0.1\dot{8}$

**Convert fractions such as 3/7, 1/3 and 2/3 into recurring decimals with and without a calculator**

To convert a fraction into a decimal, divide the numerator by the denominator.

When a recurring decimal results, spot the pattern and stop the division.

**e.g.** $\frac{5}{9} = 5 \div 9$

$\frac{5}{9}$ **= 0.555 5... or** $0.\dot{5}$ $9\overline{)5.{}^{5}0{}^{5}0{}^{5}0}$ = 0.555

### Practice questions

1. Convert these decimals to fractions. Give each answer in its simplest form.
   a) 0.57 b) 0.15 c) 0.7 d) 0.125 e) 0.239
   f) 0.058 g) 0.017 h) 0.109 i) 1.38 j) 3.8

2. Convert these percentages to decimals.
   a) 43% b) 80% c) 70% d) 23% e) 6%
   f) 34.5% g) 5.3% h) 20.5% i) 123% j) 180%

3. Convert these percentages to fractions. Give each answer in its simplest form.
   a) 30% b) 25% c) 45% d) 58% e) 75%
   f) 65% g) 72% h) 12.5% i) 37.5% j) 120%

4. Copy and complete

a) $\frac{7}{50} = \frac{(\ldots\ldots\ldots)}{100} = $ ____%
b) $\frac{9}{25} = \frac{(\ldots\ldots\ldots)}{100} = $ ____ %
c) $\frac{7}{20} = \frac{(\ldots\ldots\ldots)}{100} = $ _____%
d) $\frac{13}{40} = \frac{(\ldots\ldots\ldots)}{100} = $ ____%

5. Match the equivalent decimals and fractions.

**0.8 36% 0.375 5% 12.5% 0.12 0.225 0.96**

$\frac{9}{25}$ $\frac{9}{40}$ $\frac{1}{20}$ $\frac{3}{25}$ $\frac{3}{8}$ $\frac{4}{5}$ $\frac{12}{25}$ $\frac{1}{8}$ $\frac{24}{25}$

6. Use a calculator to change the following fractions to decimals and percentages.
Give your answer correct to 2 decimal places.

a) $\frac{3}{7}$ b) $\frac{19}{30}$ c) $\frac{13}{28}$ d) $\frac{14}{45}$ e) $\frac{21}{32}$

7. Convert these fractions to decimals, without using a calculator.
One of them is the odd one out. Identity which one and give a reason why.

a) $\frac{7}{12}$ b) $\frac{5}{8}$ c) $\frac{1}{9}$ d) $\frac{1}{6}$

8. Write in a symbol <, >, = to make correct number statements.

a) $\frac{13}{16}$ ____ 0.76
b) 0.65 ____ $\frac{13}{21}$
c) 5.25 ____ 5.25%
d) 18.75% ____ $\frac{3}{16}$
e) $\frac{3}{40}$ ____ 7.5%
f) 58.5% ____ $\frac{7}{12}$

9. The theatre is full. $\frac{3}{8}$ of the audience are over 18. 38% of the audience are 14 to 17 year olds.
Are there more 14 to 17 year olds or more over 18 year olds?
Show your working.

10. A 1 litre jar is exactly half way between $\frac{5}{8}$ and $\frac{6}{8}$ full.
Find a decimal number that shows the amount of liquid (in litres) that is in the jar.

11. Investigate the recurring decimals that are equivalent to $\frac{1}{11}$, $\frac{2}{11}$, $\frac{3}{11}$, $\frac{4}{11}$, $\frac{5}{11}$

a) Describe the pattern you see.
b) Now predict the value of $\frac{7}{11}$.

Use a calculator to check your answers.

# 24.2 DENSITY AND PRESSURE

## Objectives

### Change freely between related units and compound units

**Units of measurements**

| Length | Mass | Capacity |
|---|---|---|
| 1 km = 1000 m | 1 tonne = 1000 kg | 1 litre = 1000 ml |
| 1 m = 100 cm | 1 kg = 1000 g | 1 litre = 100 cl |
| 1 cm = 10 mm | | |

e.g. 5 km = 5 × 1000 m = 5000 m    3890 g = 3890 ÷ 1000 kg = 3.89 kg

**Compound measures**

A **compound measure** is one which measures two different types of units at the same time.

e.g. **Density** is a compound unit as it measures mass per unit volume.

**Pressure** is a compound unit as it measures force per unit area.

### Use compound units such as density and pressure

**Density**

Density is a measure of how compact a substance is.

Density = $\frac{mass}{volume}$

So mass = density × volume and volume = $\frac{mass}{density}$

e.g. The mass of a substance is 165 g and its volume is 15 $cm^3$

density = $\frac{mass}{volume}$    density = $\frac{165}{15}$ = **11 g/$cm^3$**

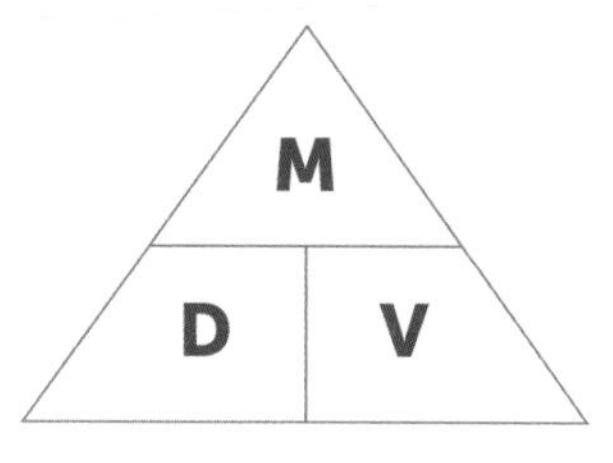


**Pressure**

Pressure is the force applied per unit area of surface.

Pressure = $\frac{force}{area}$

So force = pressure × area    and    area = $\frac{force}{pressure}$

e.g. The force applied to a button is 10 N. The button has an area = 5 $cm^2$

Pressure applied to the button = $\frac{10}{5}$ = **2 N/$cm^2$**

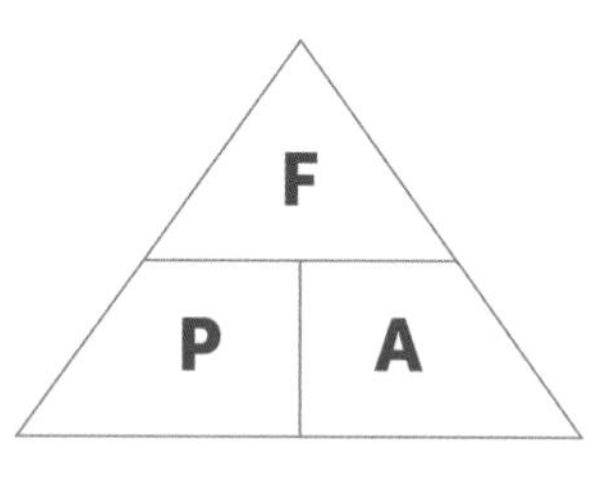


## Practice questions

1. Copy and complete these statements.
   a) 3.8 km = ____ m
   b) 4.2 tonne = ____ kg
   c) 3400 ml = ____ cl
   d) 6700 g = ____ kg
   e) 1800 mm = ____ cm
   f) 0.85 litres = ____ ml

2. True or false? Correct the statements that are false.
   a) 790 m = 7.9 km
   b) 3.5 tonne = 3500 kg
   c) 800 mm = 8 cm
   d) 3900 cm = 39 m
   e) 6700 ml = 6.7 litres
   f) 83 000 g = 83 kg

3. Copy and complete the missing units.
   a) 5 litres = 5000 ___
   b) 8900 ___ = 8.9 km
   c) 2500 g = 2.5 ___
   d) 4300 ___ = 430 cm
   e) 600 ___ = 6 litres
   f) 9.7 ___ = 9700 kg

4. Find the density for each substance.

| Substance | Mass (g) | Volume ($cm^3$) | Density ($g/cm^3$) |
|---|---|---|---|
| A | 7600 | 200 | |
| B | 6510 | 150 | |
| C | 9800 | 140 | |

5. A tin of 5.6 litres of paint has a density of 1.875 kg/litre.
   a) Find the mass of the tin.
   b) Find the number of litres of paint that would have a mass of 15 kg.

6. Copy and complete the table.

| Force (N) | Area ($m^2$) | Pressure ($N/m^2$) |
|---|---|---|
| 39 | 13 | |
| 624 | | 13 |
| | 16 | 4 |
| 1 449 | 80.5 | |

7. A force of 185.6 N is applied to a square of side length 4 cm.
   a) Work out the pressure on the square.
   b) The same force is applied to a rectangle of measuring 8 cm by 2 cm.
   Fern thinks that the pressure will not change.
   Is she correct? Show your working.

8. A cube with side length of 8 cm has a mass of 819.2 g.
   Find the density of the cube in $g/cm^3$.

9. Angela is wearing heeled shoes. Each heel has an area of 1 $cm^2$.
   Angela's heels are applying a force of 550 N to the ground.
   a) Work out the pressure onto the ground.
   b) Will the pressure on the ground be higher or lower if she wears trainers instead?

10. The density of 10 $cm^3$ of gold is 19.32 $g/cm^3$.
    Hannah thinks that the density of 1 $cm^3$ of gold is $19.32 \div 10 = 1.932$ $g/cm^3$.
    Is Hannah correct? Explain why.

11. 30 $cm^3$ of substance A and 70 $cm^3$ of substance B are mixed together to make a new substance C.
    The density of substance A = 18 $g/cm^3$ and density of substance B is 16 $g/cm^3$.
    Ben finds the mass of substance C like this:
    
    Volume of substance C = $30 + 70 = 100$ $cm^3$
    
    Density of substance C = $\frac{16 + 18}{2} = 17$ $g/cm^3$
    
    Mass of substance C = volume × density = $100 \times 17 = 1700$ g
    
    Ben is not correct. Explain why and find the mass of substance C.

## 24.3 COMPARING LENGTHS, AREAS AND VOLUMES

### Objectives

**Write lengths, areas and volumes of two shapes as ratios in their simplest form**

The ratio of lengths is the length of one shape written as a ratio of the equivalent length of the other shape.

**e.g.** Here are two rectangles. One is an enlargement of the other.

The ratio of lengths is **6 : 12** or **1 : 2** in its simplest form.

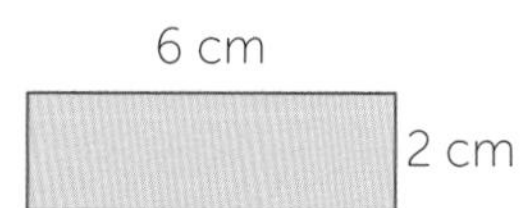


**Convert between metric units of area, volume and capacity**

**Ratio of areas**

The ratio of areas is the area of one shape written as a ratio of the area of the other shape.

**e.g.** In the rectangles above the area of one rectangle is $6 \times 2 = 12$ cm$^2$ and the area of the other rectangle is $4 \times 12 = 48$ cm$^2$

The ratio of areas of the rectangles is **12 : 48** or **1 : 4** in its simplest form.

**Ratio of volumes**

The ratio of volumes is the volume of one shape written as a ratio of the volume of the other shape.

**Units of area, volume and capacity**

The conversion between units of area and units of volume are not the same as the conversion between units of length.

**e.g.**

| Area | Volume | Capacity |
|---|---|---|
| 1 cm$^2$ = 100 mm$^2$ | 1 cm$^3$ = 1000 mm$^3$ | 1 litre = 1000 ml |
| 1 m$^2$ = 10 000 cm$^2$ | 1 m$^3$ = 1 000 000 cm$^3$ | 1 ml = 1 cm$^3$ |

### Practice questions

1. Nadine draws two straight lines whose lengths are in the ratio 4 : 3
   One of the lines is 12 cm.
   How long is the other line? Choose **two** possible answers.
   A) 11 cm  B) 9 cm  C) 16 cm  D) 8 cm

2. Shape B is an enlargement of shape A.
   Find the length of the missing sides.

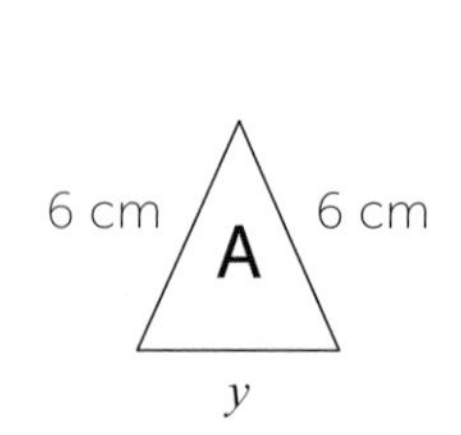


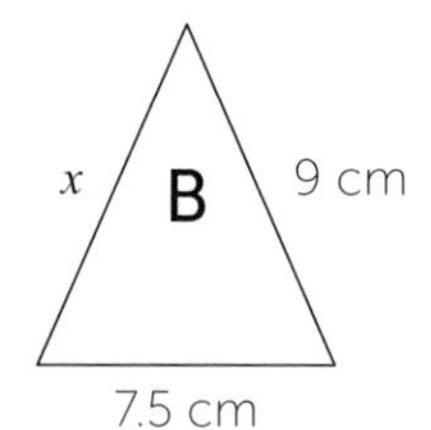


3. Copy and complete each of these statements.
   a) 5 m$^3$ = ____ cm$^3$  b) 8000 cm$^2$ = ____mm$^2$  c) 25 000 cm$^2$ = ____ m$^2$
   d) 3500 ml = ____ litres  e) 23 000 mm$^2$ = ____ cm$^2$  f) 520 cm$^3$ = ____ ml

4. Sharon makes two cubes using plasticine. The volume of the first cube is 5 times greater than the volume of the second cube.

   Write down the ratio of volumes of the cubes.

5. Match the measurements that are equal to each other.

   50 000 cm$^2$    5000 mm$^2$    500 000 cm$^3$    5000 mm$^3$    50 000 cm$^3$

   5 cm$^2$    50 cm$^2$    5 m$^2$    0.5 m$^3$    5 cm$^3$

6. Two cubes are have side lengths 2 cm and 5 cm.

   a) Write down the ratio of lengths of the cubes.

   b) Write down the ratio of volumes of the cubes.

7. A rectangular tank measuring 8 cm by 12 cm by 6 cm contains 0.288 litres of water.

   Delphi thinks that the tank is half full.

   Is she correct?

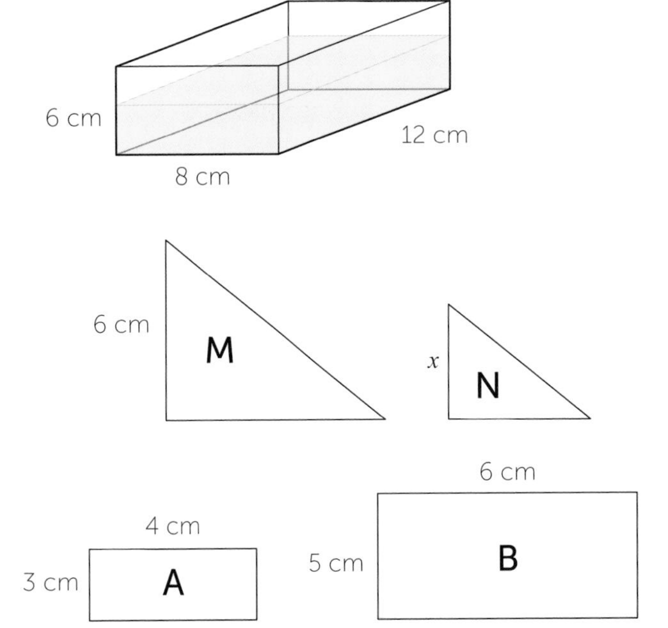


8. Triangle N is enlarged to give triangle M.

   The ratio between the lengths is 1 : 3

   Work out the missing length.

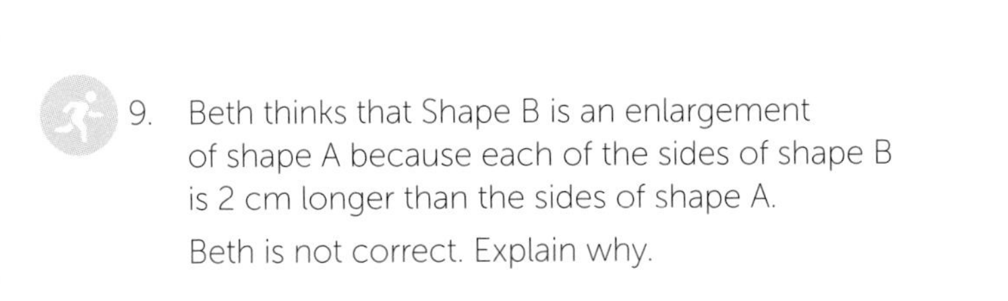

9. Beth thinks that Shape B is an enlargement of shape A because each of the sides of shape B is 2 cm longer than the sides of shape A.

   Beth is not correct. Explain why.

10. A litre of paint will cover 22 m$^2$ of wall. Helen is painting 2 walls in her room.

    Her walls are rectangular measuring 300 cm by 350 cm and 2.8 m by 350 cm.

    Helen has 2 tins of paint. Does she have enough for 2 coats of paint?

11. Every second, 200 ml of water flows into a tank.

    The tank is in the shape of a cube with sides of length 40 cm.

    Harry says that it takes 5 minutes and 20 seconds to fill up the tank.

    Is Harry correct? Show your working.

# 24.4 INTERPRETING GRADIENTS

## Objectives

**Draw speed time graphs**

**Interpret gradient as the rate of change in speed time graphs**

**e.g.** A robot starts its journey at rest. Its speed increases constantly up to 10 m/s in the first 5 seconds. It maintains this speed for the next 2 seconds and then accelerates for the next 5 seconds reaching a speed of 15 m/s.

Here is a speed time graph of the journey.

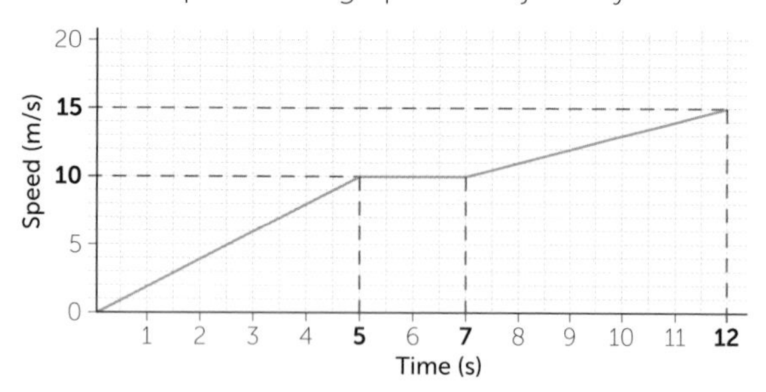


**Interpret graphs of containers filling and emptying including gradients of the graphs**

The gradient of a line plotted against time shows how fast a quantity is changing.

A straight line means a constant rate of change.

A horizontal line means there is no change over that time.

**e.g.** Liquid is poured into these two containers at a constant rate.

Graph A shows the depth of water changing over time for container 1

Graph B shows the depth of water changing over time for container 2

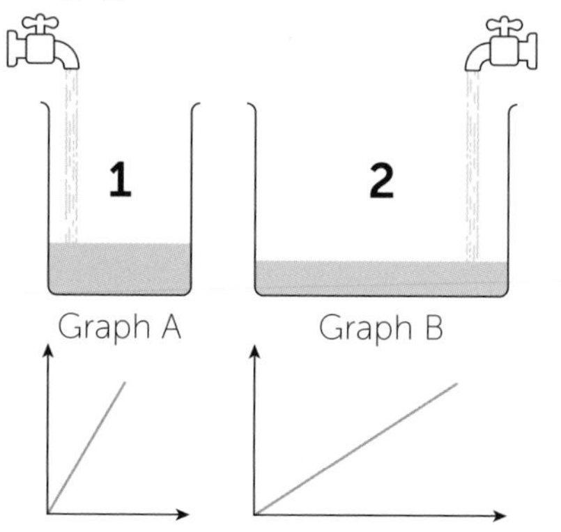


## Practice questions

1. The graphs show the journeys of three students walking to school. Match each description to a graph.
   a) A walks at a constant speed.
   b) B starts slowly, stops to meet a friend and then they walk faster together.
   c) C starts off fast, then walks even faster towards the finish.

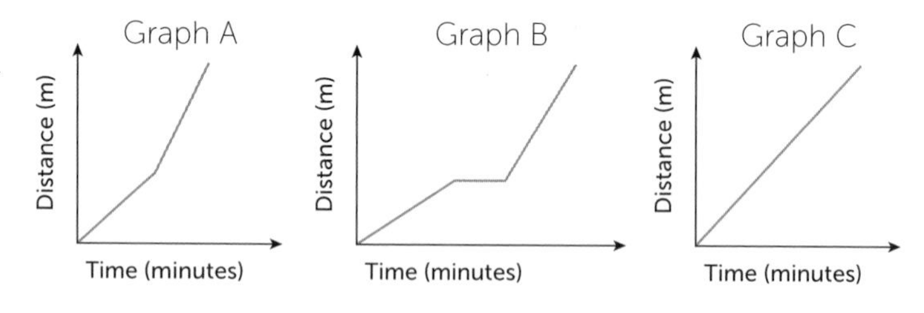


2. Alma, Bella and Cameron took part in a 10 000 km race.
   This distance time graph shows each of their races.
   a) How long did each person take to complete the race?
   b) Work out the speed of each runner in km/minutes.

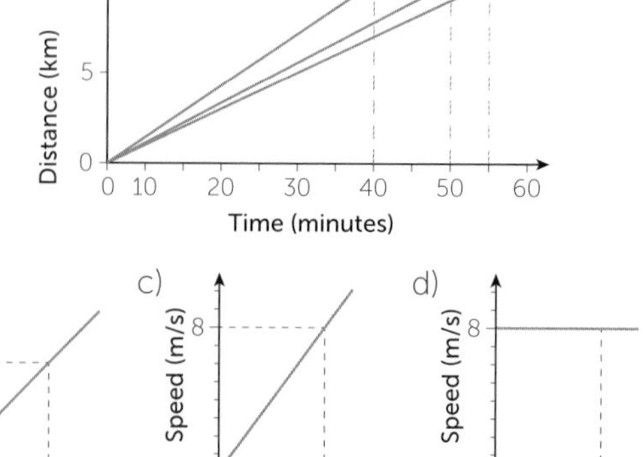


3. Work out the acceleration shown in each graph.
   Place the accelerations in ascending order.

a)

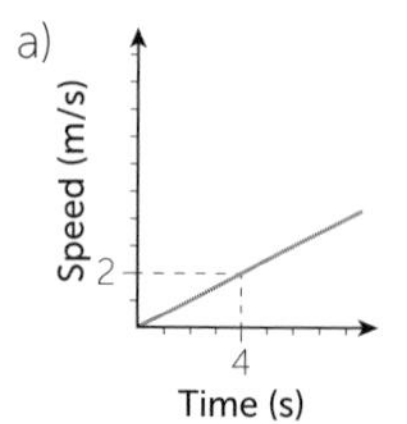


b)

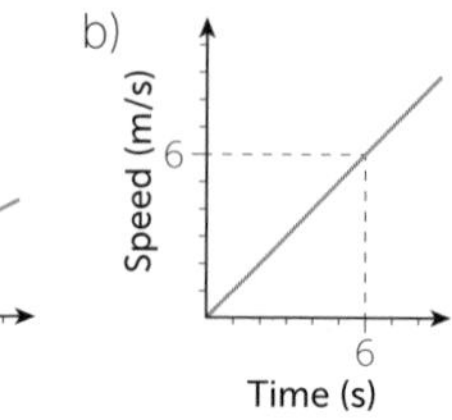


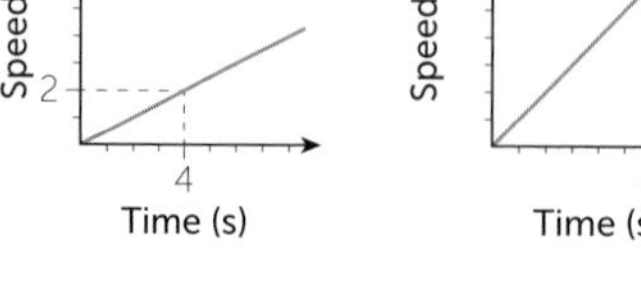

c) Speed (m/s) 8, Time (s) 6

d) Speed (m/s) 8, Time (s) 6

4. A car travelling at a speed of 30 m/s slows down steadily over 15 seconds and stops.

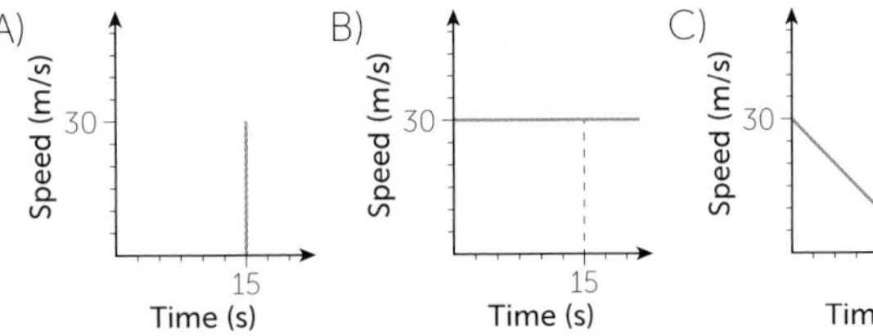
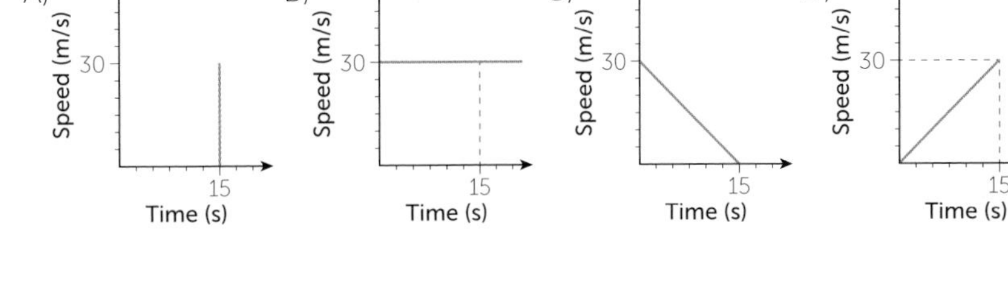


   a) Which of these graphs shows the journey of the car?
   b) Which of the following values is the acceleration of the car in the last 15 seconds?

   $2\ \text{m/s}^2$ $-2\ \text{m/s}^2$ $15\ \text{m/s}^2$

5. The graph shows the average monthly high temperature in Moscow last year.
   a) What was the highest average monthly temperature? In which month did it occur?
   b) In how many months was the average high temperature below 0°C?
   c) Jake says that the average monthly high temperature rose in April at the same rate as it fell in September.
      Is he correct? Give a reason for your answer.

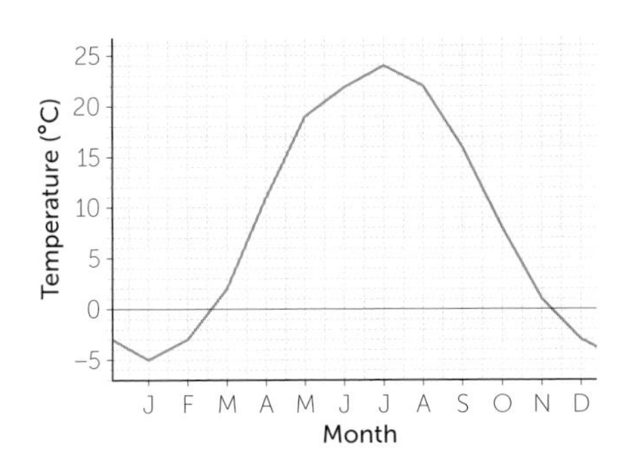


6. The graphs show the population growth of two colonies of ants over time.
   Which of the ant colonies grew faster?
   Explain your reasoning.

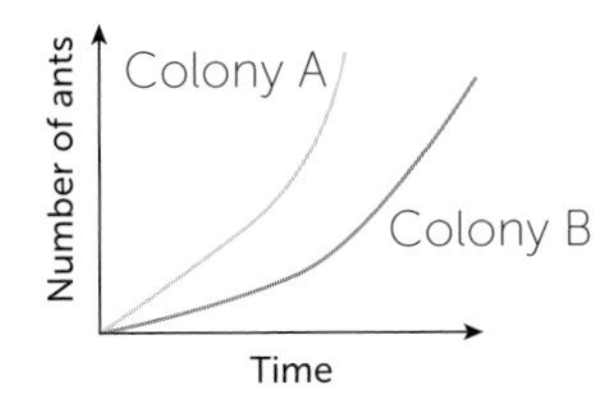


7. The graph shows the amount of fuel in a tank and the number of hours a car travels.
   a) How many litres of fuel are used each hour?
   b) The fuel tank is full. After how many hours of driving is it half full?

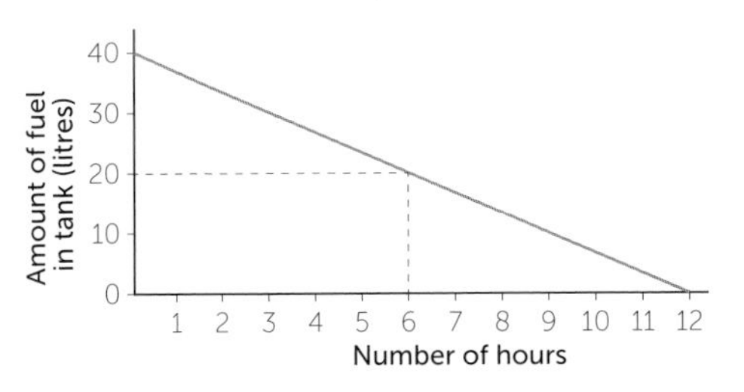


8. Which of the graphs (i), (ii) and (iii) show the depth of water against time when the water flows at a constant rate into the vase A?

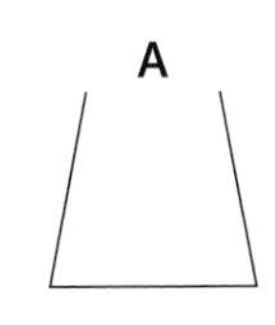

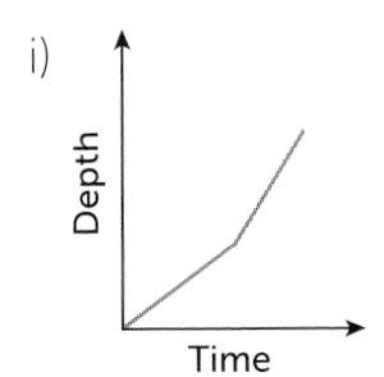

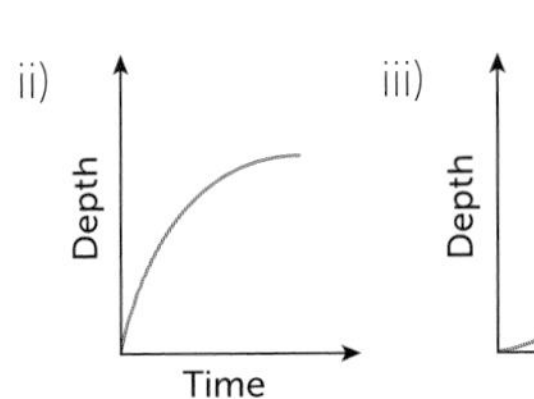

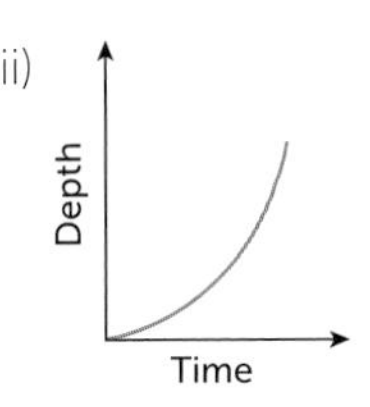

9. Look at these graphs.
   Describe a similarity between what the two graphs show.
   Describe a difference between what the two graphs show.

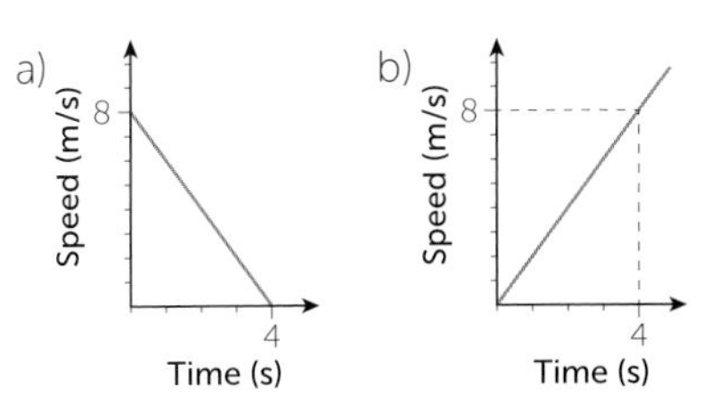

10. Here is a distance time graph showing the journeys of two people walking into town using the same route.
    Some of the statements below are false. Explain why.
    a) Selim travelled at a faster pace than Lira.
    b) Lira took the same time as Selim to get to town.
    c) Lira caught up with Selim after 10 minutes.

# 24.5 DIRECT AND INVERSE PROPORTION

## Objectives

### Understand direct proportion

Two variables are in direct proportion when they increase or decrease in the same ratio.

**e.g.** The cost of a bag of apples (C) is directly proportional to the weight of the bag (W).

A graph of variables in direct proportion is a straight line, passing through the origin.

The gradient of the graph is the ratio between the variables.

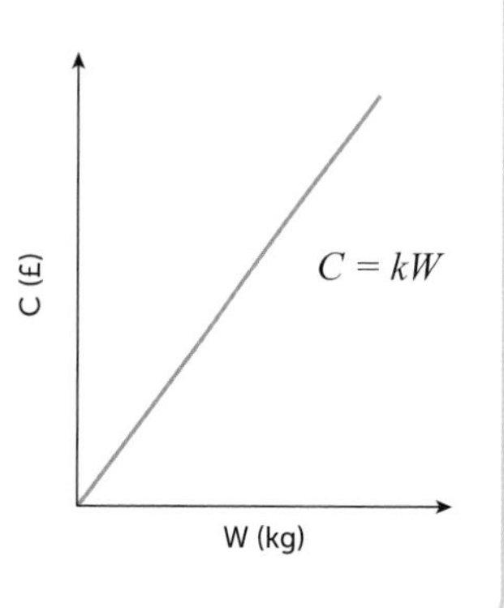


### Understand inverse proportion

Two variables are in inverse proportion when one increases at the same rate as the other decreases.

**e.g.** The number of people sharing a prize ($x$) is inversely proportional to the share of the prize they each get ($y$).

A graph of variables in inverse proportion is a reciprocal curve.

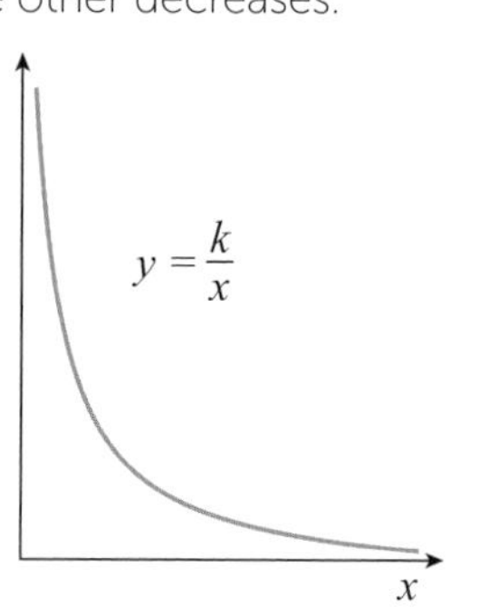


### Understand that if $x$ is inversely proportional to $y$, $x$ is proportional to $1 / y$

### Interpret equations that describe direct and inverse proportion

## Practice questions

1. If 5 pens have a mass of 135 g, find the mass of:
   a) 1 pen b) 9 pens c) 11 pens

2. The cost of 6 bottles of water is £1.50. Find the cost of:
   a) 1 bottle b) 9 bottles c) 12 bottles.

3. Joe uses these ingredients to make 25 coconut biscuits.

| 500 g flour |
|---|
| 200 g coconut butter |
| 150 g sugar |

   a) Work out how much flour he needs to make 40 biscuits.
   b) Work out how much coconut butter he needs to make 60 biscuits.
   c) Work out how much sugar he needs to make 50 biscuits.

4. Which of A, B, C and D is the odd one out? Explain your answer.
   A The amount of time studying for an exam, the number of errors in the exam.
   B The number of guests at the party, the cost of having the party.
   C The number of builders building a wall, the time it takes to build the wall.
   D The amount of time spent training for a marathon, the amount of time to complete the marathon.

5. The conversion graph shows the relationship between miles and kilometres.

   Use the graph to answer the questions.

   a) Are miles and kilometres in direct proportion? Give a reason for your answer.

   b) Convert 5 miles into kilometres.

   c) Convert 6 km into miles.

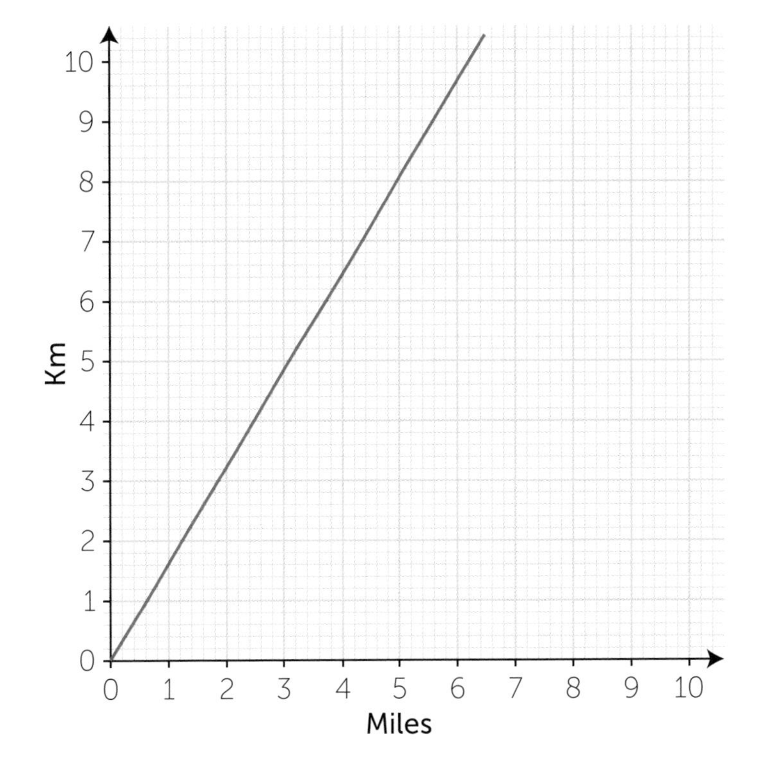


6. Four graphs are shown below:

A

B

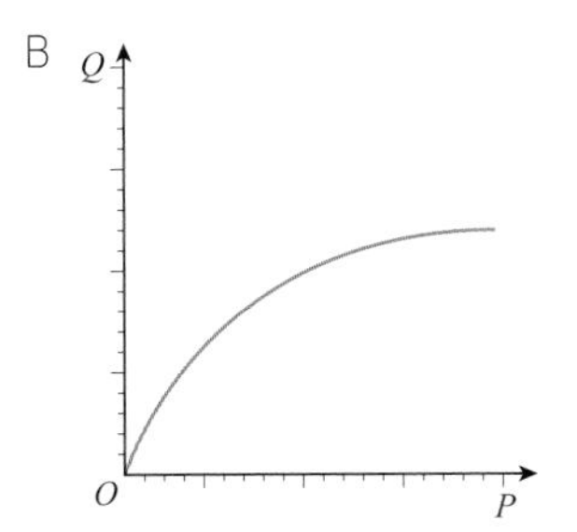


C

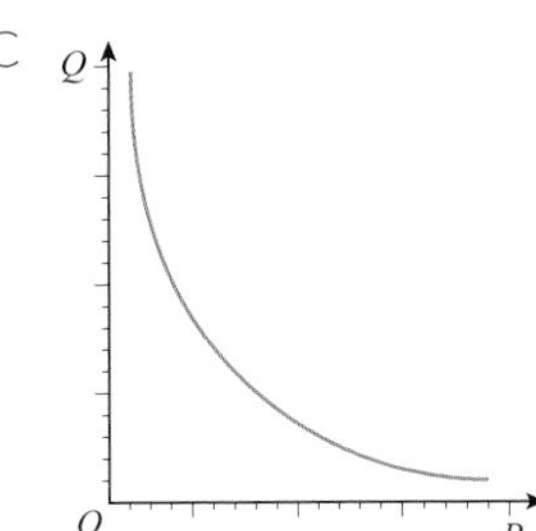


D

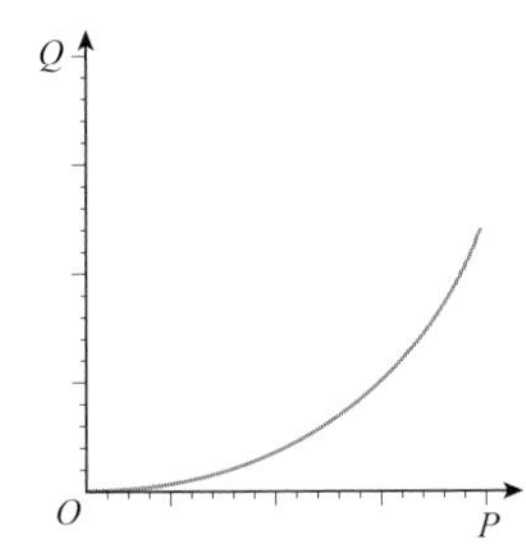


Which of these graphs shows that P is inversely proportional to Q?

7. It will take 7 days for 3 dressmakers to make a wedding dress.

   a) How many days will it take 1 dressmaker to make the dress by herself?

   b) How many days will it take 6 dressmakers to make 6 similar wedding dresses?

   c) What assumptions did you make in your calculations?

8. It will take George and his 4 brothers two days to paint an entire house.

   a) How long will it take George to paint the house by himself?

   b) How many people are needed for the job if they want to finish it in one day?

   c) What assumptions are you making in your calculations?

9. $D$ is proportional to $T$ so that $D = kT$. When $D = 10$, $T = 4$.

   a) What is $D$ when $T = 6$? b) What is $T$ when $D = 35$?

10. $P$ is in direct proportion to $Q$ so that $P = kQ$. Are the following statements always, sometimes or never true? Give examples to justify your answer.

   a) When $P$ decreases by 1 unit, $Q$ decreases by 1 unit.

   b) When $P$ doubles in value, $Q$ doubles in value too.

## 24.6 COMPOUND INTEREST

### Objectives

**Use compound interest**

Money earns **simple interest** when the interest is taken out of the account each year.

**e.g.** Invest £1000 for 4 years. Simple interest rate = 3%

Interest earned in 1 year = 3% of £1000 = £30

Interest earned in 4 years = £30 × 4 = **£120**

Money earns **compound interest** when the interest is left in the account and it also earns future interest.

**e.g.** Invest £1000 for 3 years at a compound interest rate of 2%

At the end of the first year,
account value = £1000 × 1.02 = £1020

At the end of the 2nd year,
account value = £1020 × 1.02 = £1040.40

At the end of the 3rd year, account value = £1040.40 × 1.02 = **£1061.21 (to 2 d.p.)**

**Set up, solve and interpret solutions to compound interest problems**

**e.g.** Toby has £4000 to invest in a savings account. He wants to save for 2 years.

He is choosing between two accounts: 4% simple interest or 3.8% compound interest.

| Simple interest | Compound interest |
|---|---|
| Interest after 1 year = 4000 × 0.04 = £160 | Amount after 1 year = £4000 × 1.038 = £4152 |
| Interest after 2 years = 2 × 160 = £320 | Amount after 2 years = £4152 × 1.038 = **£4309.78** |
| Account value after 2 years = **£4320** | |

The simple interest account is better for the 2 years that Toby requires.

### Practice questions

1. Work out the amount in each account after the given number of years.

| | Original amount | Compound interest (%) | Number of years |
|---|---|---|---|
| a) | £1000 | 5% | 3 |
| b) | £2000 | 4% | 4 |
| c) | £10 000 | 3.5% | 5 |

2. Kit invests £2000 in a bank account that pays 2% per year compound interest.
   How much is in the account after three years?
   A) £2120  B) £122.42  C) £2080  D) £2122.42

3. Keshena invests £3750 in a bank account that pays 3.2% per year compound interest.
   How much interest does she earn after:
   a) 1 year  b) 3 years

4. £3000 is invested for 7 years at 2% compound interest.
   Which of these answers shows the total interest earned over 7 years?
   a) £446.06  b) £3446.06  c) £3410  d) £410.10

5. A company invests money into an account. The company uses the formula £50 000 × 1.027 × 1.027 × 1.027 to find the amount of money that will be in the account after the investment matures.
   a) How much money was invested?
   b) What is the interest rate?
   c) How long was the investment for?

6. Connor works out the compound interest earned by £800 invested for 5 years at a rate of 5%. Here are his workings:

Interest in 1 year $= 5\% \times £800 = £40$ Interest for 5 years $= £40 \times 5 = £200$

a) Why are Connor's working incorrect? b) What should the answer be?

7. £6500 was invested in a compound interest account for 2 years at 2% per year.
Which of these calculations should NOT be used to find the money in the account after 2 years?
A) $6500 \times 0.02 \times 2$ B) $6500 \times 1.02 \times 1.02$ C) $6500 \times 1.2 \times 1.2$

8. Poppy deposits £3000 in a bank account. Compound interest is paid at 4% per year.
She leaves the money in the account until it exceeds £4000. What is the shortest period of time she could have left the money in the account?
A) 6 years B) 7 years C) 8 years D) 9 years

9. Cameron invests some money in a bank account. Compound interest is paid at 4% per annum. After two years there is £1622.40 in the account. How much did Cameron invest? Show your workings clearly.

10. Sam has £1900 to invest. He finds the interest rate for three accounts.

| Account A | Account B | Account C |
|---|---|---|
| 0.2% per month compound interest | 4.5% compound interest | 4.7% simple interest |

Sam wants to invest the money for 2 years. Which account should he choose?

11. Lois has £70 000 to invest. She finds the interest rates for two accounts.

**Account A**

3.6% per year compound interest

**Account B**

| | |
|---|---|
| Year 1 | 4.0% interest |
| Year 2 | 3.7% interest |
| Year 3 | 3.5% interest |
| Year 4 | 3.3% interest |

a) Lois wants to invest the money for 4 years. Which account will have the most money after 4 years?
b) Would the answer change if she invested the money for 2 years only?

## 24.7 GROWTH AND DECAY PROBLEMS

### Objectives

**Make calculations involving repeated percentage change**

More than one multiplier can be used to work out the effect of repeated percentage change.

**e.g.** House prices increase by 2% in May and then fall by 1% in June.

What affect does this have on a house priced at £250 000 in April?

Use the multiplier 1.02 to find the increase of 2%

Use the multiplier 0.99 to find the decrease of 1%

Value of house in June =
£250 000 × 1.02 × 0.99 = **£252 450**

Use these formula to find the percentage increase and percentage decrease.

$$\% \text{ increase} = \frac{\textit{actual increase}}{\textit{original number}} \times 100\%$$

$$\% \text{ decrease} = \frac{\textit{actual decrease}}{\textit{original number}} \times 100\%$$

**e.g.** The population of an island was 10 200 people last year and is 13 108 this year. Work out the percentage increase.

Actual increase = 13 108 – 10 200 = 2908

Percentage increase = $\frac{2908}{10\,200}$ × 100% = **28.51% (2 d.p.)**

**Set up, solve and interpret the answers in growth and decay problems**

**Depreciation** is when the value or quantity of an item decreases.

**Growth** is when the value or quantity of an item increases.

**e.g.** A car is bought for £14 000

Each year the value of the car depreciates by 20%.

Work out its value after 3 years.

Use the multiplier 0.8 to find a decrease of 20%

Use the multiplier three times for the 3 years

Value of car after 3 years =
£14 000 × 0.8 × 0.8 × 08 = **£7168**

### Practice questions

1. Write down the multiplier to find each of these percentage increases or decreases.
   a) 3% increase  b) 30% increase  c) 7% decrease  d) 30% decrease

2. True or false? Give the correct multipliers for the statements which are false.
   a) To find a 5% increase followed by 3% increase use multipliers: 1.05 × 1.03
   b) To find 4% increase followed by 10% decrease use multipliers: 1.04 × 0.9
   c) To find 12% decrease followed by 9% increase use multipliers: 0.12 × 1.09
   d) To find 7% increase followed by 8% increase use multipliers 1.07 × 1.8

3. A house is worth £137 500. Its value increases by 4% each year.
   Work out to the nearest pound, the value of the house after:
   a) 1 year  b) 3 years  c) 10 years

4. The population of a town is 65 000. The population increases at a rate of 13% per year.
   a) Work out the population of the town after 4 years.
   b) How many more people live in the town after 4 years compared to now?

5. The value of a machine when new is £8000. The value of the machine depreciates by 10% each year.
   Which of these shows its value after 3 years?
   a) £8242.41  b) £5832  c)£5600  d) £7760

6. An iPad battery lasted for 20 hours. The improved version of the iPad battery lasted for an extra 50 minutes.

   Work out the percentage increase in the life of the iPad battery.

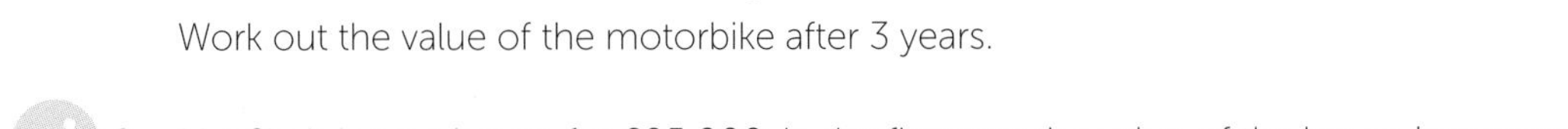

7. A motorbike is worth £8500. Each year the value of the motorbike depreciates by 28%.

   Work out the value of the motorbike after 3 years.

8. Mrs Shah buys a house for £95 000. In the first year the value of the house increases by 15%.

   In the second year the value of the house decreases by 4% of its value.

   Mrs Shah says the value of the house has increased by 11% since she bought it. Mrs Shah is not correct. Explain why and give the correct percentage increase.

9. A delivery company bought a new van. Each year the value of the van depreciates by 20%.

   The value of the van can be multiplied by a single number to find its value at the end of 4 years. Which of these is the single number as a decimal?

   A) 0.4096 B) 0.016 C) 0.8 D) 0.2

10. The number of people running in the Summer Marathon has increased from 8500 last year to 10 000 this year. The organisers said that this is 15% increase.

    Explain why the organisers are not correct.

11. Jonathan has an app which tells him the number of hours he spends on the Internet each day and compares the values from one day to another.

    On Sunday the app said that there was a 30% increase in the amount of time he used on the Internet compared to Saturday. On Monday the app said there was a 30% decrease.

    Does Jonathan use the same, more or less time on the Internet on Monday compared to Saturday? Give a reason for your answer.

SECTION 25

# FURTHER EQUATIONS

## 25.1 LINEAR EQUATIONS

A17 A21

### Objectives

**Revise solving linear equations including those involving brackets and variables on both sides**

Expand brackets and balance both sides of the equation.

Collect variables on one side and numbers on the other.

| | |
|---|---|
| **e.g.** Solve | $3(x + 4) = 4(2x - 3)$ |
| Expand the brackets | $3x + 12 = 8x - 12$ |
| Subtract $3x$ from both sides | $12 = 5x - 12$ |
| Add 12 to both sides | $24 = 5x$ |
| Divide both sides by 5 | $x =$ **4.8** |

**Derive and solve an equation from a given situation and interpret the solution**

Read the question carefully; think "what is the problem asking me to do?"

**e.g.** The area of this rectangle is 96 cm$^2$; find $x$

(5$x$ – 2) cm

6 cm

| | |
|---|---|
| First form an equation: | width × length = area |
| | $6(5x - 2) = 96$ |
| Then solve the equation | |
| Expand the brackets | $30x - 12 = 96$ |
| Add 12 to both sides | $30x = 108$ |
| Divide both side by 30 | $x =$ **3.6** |

### Practice questions

1. A number has 5 added to it and is then multiplied by 3 to give the answer 27.
   Trevor writes this using algebra as $3x + 5 = 27$
   Explain what Trevor has done incorrectly and rewrite the equation for him.

2. Solve the following equations.
   a) $18 - 3y = 6$ b) $4(a + 1) = 22$ c) $6x + 5 = 4x - 2$
   d) $2(n - 3) = 9 - n$ e) $\frac{2p + 5}{3} = 7$

3. Look at this straight line.
   Find the value of $y$.

   $(y + 46)$ $(3y - 18)$

4. Iris is solving the equation $25 - 4q = 8$. Here are her workings.

   Add 25 to both sides $4q = 33$

   Divide both sides by 4 $q = 8.25$

   Explain what Iris has done wrong and find the correct answer.

5. The area of this rectangle is 36cm$^2$
   Find the value of $p$.

$(5p - 1)$ cm

4 cm

6. Kelly and Ian want to solve the equation $23 - 3x = 2(x + 4)$. Here are their solutions.

| **Kelly' solution:** | **Ian's solution:** |
|---|---|
| Expand the bracket: | Expand the bracket: |
| $23 - 3x = 2x + 4$ | $23 - 3x = 2x + 8$ |
| Subtract 23 from both sides: | Subtract 23 from both sides: |
| $-3x = 2x - 19$ | $3x = 2x - 15$ |
| Subtract 2x from both sides: | Subtract 2x from both sides: |
| $-5x = -19$ | $x = -15$ |
| Divide both sides by −5: | |
| $x = 3.8$ | |

Comment of both these solutions and correct them where appropriate.

7. Lines AB and CD are parallel.
   Find the value of $n$.

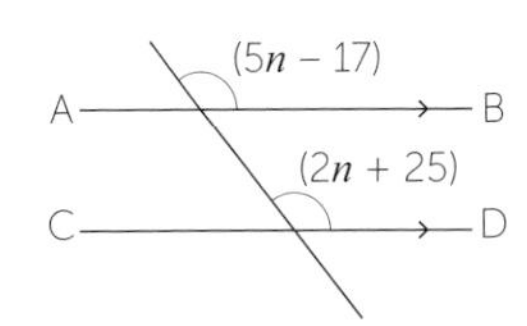


8. Triangle ABC is isosceles.
   Find the sizes of all the angles in the triangle.

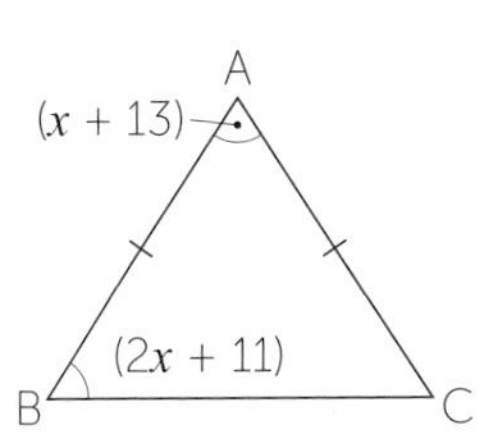


9. Briony is 8 years older than Angie. Celia is three times as old as Angie.
   Celia's age is equal to the sum of Angie and Briony's ages.
   How old is Angie?

10. The area of this triangle is double the area of the rectangle.

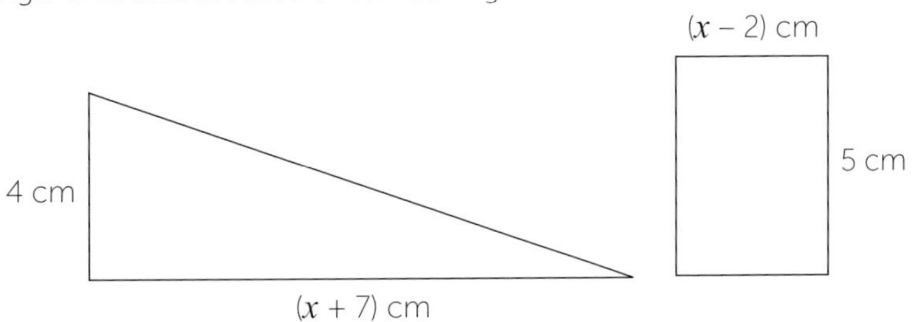


Form an equation and solve it to find the value of $x$.

## 25.2 SIMULTANEOUS EQUATIONS 1

### Objectives

**Solve simultaneous equations by elimination**

**Simultaneous equations** are two equations which contain more than one variable.

They are called simultaneous equations because both equations are solved at the same time.

Simultaneous equations can be solved by **eliminating** one of the variables to find the other.

Once one variable is found, it is used to find the other.

**e.g.** Solve $3x + 2y = 21$

$5x - 2y = 19$

Add the two equations to eliminate the $y$ terms:

$8x = 40$ so $x = 5$

Use the value of $x$ to find the value of $y$ by substituting into one of the original equations.

$3 \times 5 + 2y = 21$

$15 + 2y = 21$

$2y = 6$ $y = 3$

**Scale equations up**

Sometimes one or both equations need to be scaled up so that a variable can be eliminated.

**e.g.** Solve $5a + 2b = 3$

$4a + 3b = 8$

Make the '$b$' terms equal in each equation.

Scale the first equation up × 3 and the second equation up × 2:

$15a + 6b = 9$

$8a + 6b = 16$

Eliminate the '$b$' terms by subtracting the two equations:

$7a = -7$ $a = -1$

Substitute $a = -1$ into: $5a + 2b = 3$

$-5 + 2b = 3$ $b = 4$

**Form and solve a simultaneous equation to represent a situation**

Read the question carefully and choose a letter to use for each variable in the question.

### Practice questions

1. John wants to solve these simultaneous equations.

   $2x + y = 8$

   $3x + y = 11$

   Here is his answer: *Subtract the two equations* $x = 3$

   Comment on his answer and correct it where necessary.

2. Solve these simultaneous equations.

   a) $x + y = 8$, $x - y = 2$

   b) $2a + 3b = 7$, $a + 3b = 8$

   c) $4p - 3q = 11$, $4p - q = 9$

   d) $3m + n = 3$, $3m - 3n = -3$

3. Look at these two rectangles.

   Rectangle A has a perimeter of 22 cm.

   Rectangle B has a perimeter of 32 cm.

   a) Form an equation for the perimeter of Rectangle A.

   b) Form an equation for the perimeter of Rectangle B.

   c) Solve the equations simultaneously to find the values of $x$ and $y$.

4. 3 coffees and 2 teas cost £9.40. 3 coffees and 1 tea cost £8.00.

   Work out the total cost of 1 coffee and 1 tea.

5. Solve the following simultaneous equations:

   a) $3x + 2y = 16$
      $x - y = 2$

   b) $2m + 3n = 8$
      $m + 2n = 6$

   c) $3p - 3q = 9$
      $4p - q = 6$

   d) $5j - k = 38$
      $3j + 3k = 12$

6. Look at these rods.
   Calculate the length of rod $a$ and of rod $b$.

   | $a$ | $a$ | $a$ | $b$ | $b$ |
   |---|---|---|---|---|

   45 cm

   | $a$ | $a$ | $b$ | $b$ | $b$ | $b$ |
   |---|---|---|---|---|---|

   46 cm

7. Two families go to a theme park.
   Two adults and two children cost £104.
   One adult and three children cost £80.
   How much does it cost for one adult and one child to go the theme park?

8. Solve the simultaneous equations.

   a) $3a + 2b = 11$
      $2a + 3b = 9$

   b) $5x + 2y = 23$
      $2x - 3y = 13$

   c) $7t - 3u = 13$
      $4t - 2u = 7$

   d) $5n - 2m = 3$
      $4m - 3n = -13$

9. Look at these shapes.
   Shape A has a perimeter of 61 cm.
   Shape B has a perimeter of 53 cm.
   Find the values of $x$ and $y$.

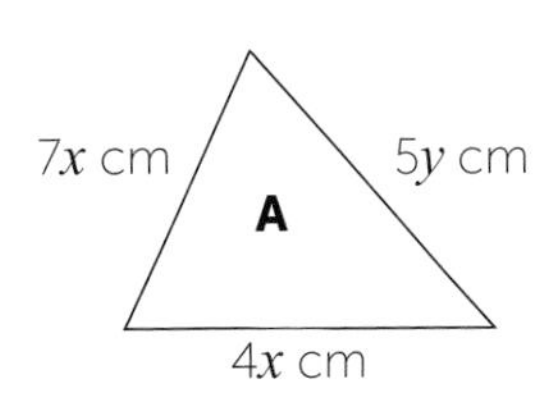


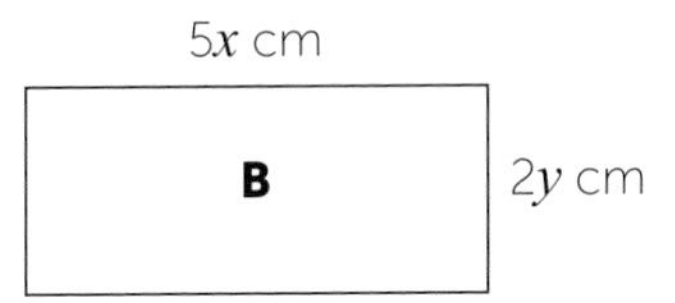

10. Solve these simultaneous equations:
    $2y = 3x + 9$
    $5x + 3y = 4$

## 25.3 SIMULTANEOUS EQUATIONS 2

**A19** **A21**

### Objectives

**Solve simultaneous equations by substitution**

Simultaneous equations can be solved by substitution as well as elimination.

Substitution is when one variable is swapped in for the other variable.

The equation can then be solved.

**e.g.** Solve $2x + 3y = 14$
$y = 4x$

Substitute $y$ for $4x$ in the first equation: $2x + 3(4x) = 14$

$2x + 12x = 14$ so $x = 1$

Substitute $x = 1$ into $y = 4x$ $\quad y = 4 \times 1 \quad y = 4$

**Form and solve simultaneous equations from given situations**

## Practice questions

1. Solve by substitution the simultaneous equations below:

   a) $x = y$
   $x + 3y = 12$

   b) $p = 2q$
   $p - q = 7$

   c) $a = 3b$
   $a + b = 8$

   d) $n = 5m$
   $m + n = 6$

2. Dylan is solving these equations by substitution.

   $y = 2x$ and $3x + 2y = 14$

   He starts with: Substitute: $3x + 2x = 14$

   Explain what he has done wrong. Correct his workings and solve the equations.

3. Solve by substitution the simultaneous equations below:

   a) $a = b + 2$
   $a + 5b = 26$

   b) $p = q - 3$
   $4p - q = 9$

   c) $x = 5 - y$
   $2x + 5y = 4$

   d) $u = t + 3$
   $3t - 7u = -13$

4. Look at these rods.

   How long is an individual rod $a$ and an individual rod $b$?

| $a$ | | |
|---|---|---|
| $b$ | $b$ | 5 cm |

| $a$ | $a$ | $b$ | $b$ | $b$ |
|---|---|---|---|---|
| 38 cm | | | | |

5. There are two types of prize available at a lucky dip: red and blue.

   Christine weighed the prizes.

   How much does each prize weigh?

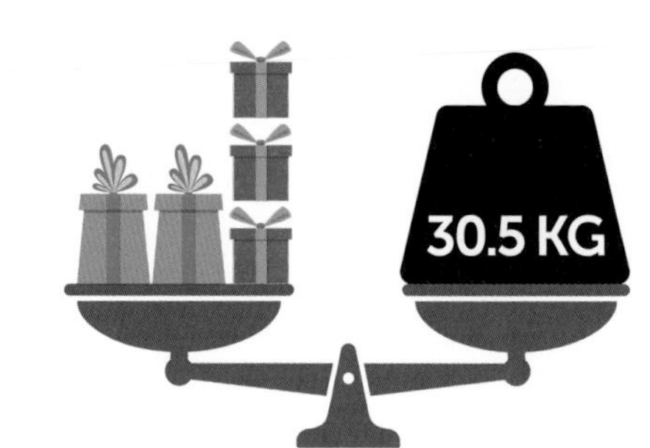

6. Denis thinks that $y = 5x + 7$, Lois thinks that $y = 2x - 5$

   Both Denis and Lois are correct.

   a) Find the values of $x$ and $y$.

   b) Denis and Lois draw the graphs of $y = 5x + 7$ and $y = 2x - 5$
   Give the coordinates of the point at which their graphs intersect.

# 25.4 SOLVING QUADRATIC EQUATIONS

## Objectives

### Solve quadratic equations by factorising

Recall that to **factorise** a quadratic means to put it into brackets.

Usually there are **two** brackets needed when factorising a quadratic expression.

e.g. Factorise $x^2 + 5x + 6$

Look for factors of 6, which add or subtract to make 5

Factor pairs of 6 are (1, 6), (−1, −6), (2, 3), (−2, −3)

(2, 3) combine to make 5

$x^2 + 5x + 6 = (x + 3)(x + 2)$

Rearrange the equation (if necessary) to equal zero.

Factorise the quadratic and then use the fact that when two things multiply to give zero, **one or other of them must be equal to zero**.

| | |
|---|---|
| e.g. Solve | $x^2 + 5x + 6 = 0$ |
| Factorise the quadratic | $(x + 3)(x + 2) = 0$ |
| Put each bracket = 0 | $x + 3 = 0$ **or** $x + 2 = 0$ |

So $x = -3$ or $x = -2$

### Link the solutions of a quadratic equation to the roots of a quadratic function

The solutions to a quadratic equation show where the quadratic graph cuts the x-axis.

These values are called the roots of the quadratic graph.

e.g. Here is the graph of $y = x^2 + 5x + 6$

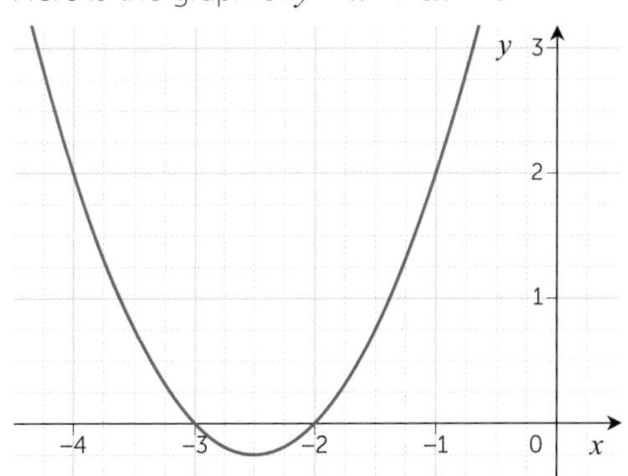


When $y = 0$ $\quad x^2 + 5x + 6 = 0$

$x = -3$ or $x = -2$

Notice how these are the points where the graph cuts the $x$-axis.

## Practice questions

1. Solve these quadratic equations by factorising.
   a) $y^2 + 4y + 3 = 0$ b) $a^2 - 5a + 4 = 0$ c) $n^2 + 6n + 8 = 0$
   d) $d^2 + 10d + 24 = 0$ e) $x^2 - 7x + 12 = 0$

2. Harriet is trying to solve the quadratic equation $x^2 - 7x + 10 = 0$ by factorising.
   Her solution is: $(x - 2)(x + 5) = 0$ So $x = 2$ and $-5$
   Harriet is wrong. Explain why and correct her solution.

3. Here is the graph of $y = x^2 - 8x + 15$
   a) Where does the graph meet the $x$-axis?
   b) Use the graph to solve $x^2 - 8x + 15 = 0$

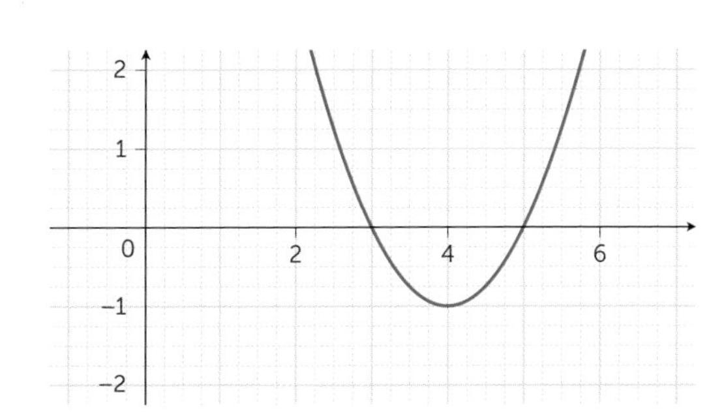


4. Solve these quadratic equations by factorising.
   a) $q^2 + 3q - 10 = 0$ b) $t^2 + 5t = 0$ c) $h^2 - h - 6 = 0$ d) $y^2 + 9y + 20 = 0$

5. Tyson is solving the quadratic equation $n^2 - 5n - 14 = 0$
   Tyson's solution is $(n + 7)(n - 2) = 0$ so $n = 7$ and $-2$
   Is his solution correct? Give a reason for your answer.

6. This rectangle has an area of 36 cm$^2$
   a) Show that $x^2 + x - 42 = 0$
   b) Solve $x^2 + x - 42 = 0$
   c) Explain why only one of the solutions is sensible.

$(x + 3)$ cm

$(x - 2)$ cm

7. Here is a quadratic equation: $y^2 - 2y - 8 = 7$
   Rearrange this quadratic equation so that one side of the equation equals zero.
   Solve the quadratic equation $y^2 - 2y - 8 = 7$
   Show each stage of your solution.

8. Find the coordinates where the graph $y = x^2 - x - 30$ intersects with the $x$-axis and the $y$-axis.
   Show your working.

## 25.5 WORKING WITH FORMULAE

A5

### Objectives

#### Recall and use common formulae

There are some formulae you are expected to know and use.

**e.g.** Area of a trapezium $= \frac{1}{2}(a + b)h$

Circumference of a circle $= \pi d$ or $2\pi r$

Area of a circle $= \pi r^2$

Volume of a prism = area of the cross-section × length

Equation of a line: $y = mx + c$

Pythagoras' theorem: $a^2 + b^2 = c^2$

Trigonometric formulae:

$\sin\theta = \frac{\text{opp}}{\text{hyp}}$ $\cos\theta = \frac{\text{adj}}{\text{hyp}}$ $\tan\theta = \frac{\text{opp}}{\text{adj}}$

speed $= \frac{\text{distance}}{\text{time}}$

density $= \frac{\text{mass}}{\text{volume}}$

#### Change the subject of a formula where the subject appears on both sides

The 'subject' of a formula is the variable (letter) that everything else is equal to.

**e.g.** In the formula $y = mx + c$, the letter $y$ is the subject.

When rearranging the letters to change the subject, use inverse operations.

**e.g.** Make $x$ the subject of $y = mx + c$

Subtract $c$ from both sides $y - c = mx$

Divide both sides by $m$, $\frac{y - c}{m} = x$

i.e. $x = \frac{y - c}{m}$

#### Change the subject of a formula where the subject had an index or root

The inverse of a square is square root; the inverse of a square root is square.

**e.g.** Make $r$ the subject of $A = \pi r^2$

Divide both sides by $\pi$: $\frac{A}{\pi} = r^2$

Take square root of both sides: $\sqrt{\frac{A}{\pi}} = r$

Swap the sides of the equation: $r = \sqrt{\frac{A}{\pi}}$

## Practice questions

1. A delivery company uses this formula to calculate the cost of a delivery:

   $c = 5 + 0.1d$ where $c$ is the total cost in pounds (£) and $d$ is the distance the delivery is in kilometres

   Work out:

   a) The total cost of a delivery of distance 60 kilometres.

   b) The total cost of a delivery of distance 125 kilometres.

   c) The total cost of a delivery of distance 233 kilometres.

   d) The distance of a delivery that costs £9.

   e) The distance of a delivery that costs £15.20.

2. Work out the area of this trapezium.

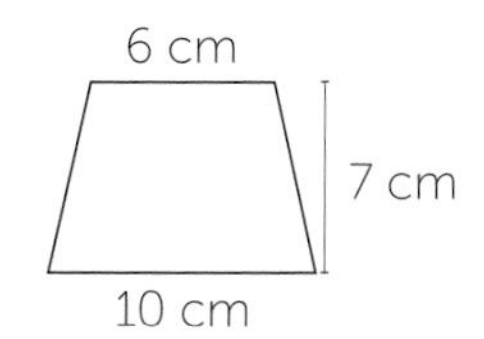


3 A catering company uses this formula to work out how many cakes to bake.

   $T = \frac{n}{4} + 5$ where T is the total number of cakes required

   $n$ is the number of people at an event

   Work out:

   a) The total number of cakes required for an event with 20 people attending.

   b) The total number of cakes required for an event with 56 people attending.

   c) The number of people 13 cakes will feed.

   d) Gladys says that the number of cakes needed for 40 people is double the number needed for 20 people. She is not correct; explain why.

4. Jasmine has rearranged the formula $t = 4n - 3$ to make $n$ the subject.

   She has written:

   Add 3 to both sides: $t + 3 = 4n$

   Divide both sides by 4: $t + \frac{3}{4} = n$

   Explain what Jasmine has done incorrectly. Rearrange the equation correctly.

5. Rearrange each of these formula to make the letter in brackets the subject.

   a) $d = 7t$ [$t$]  b) $m = n + 9$ [$n$]  c) $c = 6h + 10$ [$h$]

   d) $h = 10 - g$ [$g$]  e) $y = \frac{1}{2}x - 5$ [$x$]

6. A decorating company uses this formula to work out the cost of painting a wall.

   $c = xy + 12h$

   Where: $c$ is the cost of decorating the wall in pounds (£).
   $x$ and $y$ are the length and width in metres of the wall that needs painting.
   $h$ is the time it takes to paint the wall (hours).

   Rearrange the formula to make $h$ the subject.

7. Make $x$ the subject of each of these formulae.

   a) $x + y = 5x - 2$  b) $7x + y = 3y - 2x$  c) $x^2 = y$

   d) $4x^2 + y = 7$  e) $3x^2 + 2y = 5y + 1$

8. The formula to find the area of the trapezium is $A = \frac{1}{2}(a + b)h$
Find the formula for calculating the side $b$ given the length of side $a$, the height and the area of the trapezium.

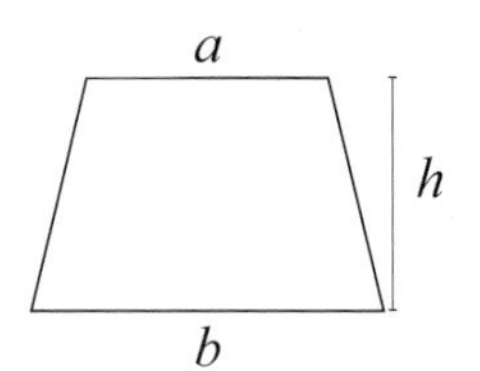

9. The area of a semi-circle is calculated using the formula $A = \frac{1}{2}\pi r^2$
Given the area, write a formula to work out the radius of the semi-circle.

10. Make $x$ the subject of the formula: $2x^2 + 6\pi + 7y = 10\pi + x^2$

## 25.6 LIMITS OF ACCURACY

N15 N16

### Objectives

#### Use inequality notation to write an error interval

When measures are rounded, the original amount falls between two intervals; these are called **error intervals**.

e.g. A number has been rounded to 400, the nearest hundred.

The smallest the number could be is 350

The largest the number could be is $449.\dot{9}$ or anything up to, but not including 450

This range of values can be written as an inequality:
$350 \leq$ actual measurement $< 450$

e.g. A number has been rounded to 40.3 to 1 decimal place.

The smallest the number could be is 40.25

The largest the number could be is anything up to, but not including, 40.35

The error interval for 40.3 (1 d.p.) is
$40.25 \leq$ actual measurement $< 40.35$

#### Apply and interpret degrees of accuracy

When giving a rounded answer, it should be given to a **sensible degree of accuracy**.

This means the answer should be as accurate as is meaningful.

e.g. A bag of sweets weighing 500 grams is shared equally between 13 children.

Each child will get $500 \div 13 = 38.461\,538\,46...$ g

It is not sensible to give the answer so accurately.

A reasonable degree of accuracy in this context is 38 g (to the nearest whole gram) or even 40 g (to the nearest 5 g).

#### Truncated values

Sometimes when giving an approximate answer, decimal numbers are **truncated**.

To truncate a decimal number to a given degree of accuracy, simply **delete** any more accurate decimal values.

Digits are **not** rounded up when truncating.

e.g. The number 38.46153846.... truncated to 1 decimal place is 38.4

Note how the 4 tenths is **not rounded** up to 5 tenths

## Practice questions

1. Round 18 452.7 to the nearest:

   a) one  b) ten  c) hundred  d) thousand  e) ten thousand

2. Gerry, Paul and Natalie want to round 1473.5928 to 2 significant figures.

   Gerry says the answer is 15  Paul says the answer is 1473.59  Natalie says the answer is 1400

   All three of them are wrong.

   Explain what they have done incorrectly and find the correct answer.

3. What is the smallest number that will round to 140 to the nearest ten?

4. A number rounds to 400 when rounded to the nearest one, ten and hundred.

   What could the number have been?

5. A journey is timed as lasting 43 minutes to the nearest minute.

   What is the longest the journey could have been? Give your answer in minutes and seconds, to the nearest second.

6. Find the error intervals of these approximations. Write your answers as inequalities.

   a) 80 seconds (rounded to the nearest 10 seconds)
   b) 12 300 metres (rounded to the nearest 100 metres)
   c) 34 centilitres (rounded to the nearest centilitre)
   d) 12.3 centimetres (rounded to 1 decimal place)

7. George needs to buy a shed for his bicycle.

   His bicycle is measured as being 150 centimetres long, rounded to the nearest 10 centimetres. The shed is 2 metres long, rounded to the nearest metre.

   Will the bicycle definitely fit into the shed? Show your working.

8. A class of 29 students has collected 600 £1 coins for charity.

   They are going to share them equally between the charities chosen by each student in the form.

   a) Work out 600 ÷ 29; write all the figures on your calculator display.
   b) Round your answer to part a) to 1 decimal place.
   c) Truncate your answer to part a) to the nearest unit.
   d) Which of your answers to parts b) and c) is the most sensible degree of accuracy to give? Give a reason for your answer.

9. Clare wants to share her quiz winnings of £90 between the 7 members of the team.

   She decides to round the value to the nearest pound.

   Explain the problem with her doing this and suggest a more appropriate degree of accuracy, explaining why you have done so.

10. Here is a rectangle.

    Each length has been rounded to the nearest centimetre.

    Using inequalities, write the error interval for the area of the rectangle.

    7 cm

    12 cm

# ANSWERS

## Section 1

### 1.1 Integers and place value

1. a) 6 000 b) thirty thousand, two hundred and nine c) 2 001 036
2. a) –10 **–9** –8 –7 **–6** –5 **–4** –3 **–2** –1 **0** **1** **2** **3** 4 5 6 7 **8** 9 10

   b) 8, 3, 2, 1, 0, –2, –4, –6, –9
   c) It will have no effect on the order since they are all twice as big
3. a) 97 431 b) 49 731 c) 31 479
4. a) 4443, 30 400, 44 033, 44 104, 400 300
   b) 888, 8882, 28 888, 208 088, 280 880
5. –4, –2, 0, 1, 2
6. 480, 360, 240, 120
7. 772 or 722
8. 523 026
9. a) 269 350 and 225 350
   b) 225 350, 236 350, 247 350, 258 350, 269 350, 280 350
10. 345 and 321, 456 and 432, 567 and 543, 678 and 654, 789 and 765.

### 1.2 Negative integers

1. a) –7 b) 8 c) –12 d) –12 e) –10 f) –3
   g) 2 h) –11
2. a) 6 b) –11 c) –17 d) –10 e) –9 f) 3
   g) 16 h) –23
3. a) –61 b) –272 c) –2110 d) –13 169
4. a) –3 + 7 b) 7 – –11 c) –11 – –3 d) –3 – 7 – –11
5. Subtracting 24 means to move down the number line from –10, so the answer will be smaller than – 10. The answer is – 34.
6. The temperature fell by 10°C
7. a) –168 b) –26 c) 1366 d) –1339
8. a) 81°C b) 145°C c) Mercury at 619°C
9.

| | | |
|---|---|---|
| 1 | –14 | –8 |
| –16 | –7 | 2 |
| –6 | 0 | –15 |

10. a)
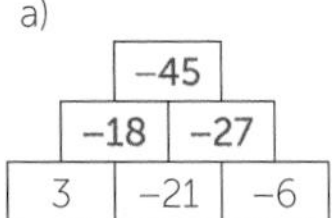

b)
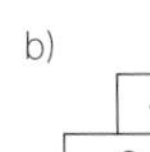
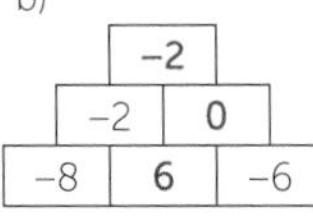

c)
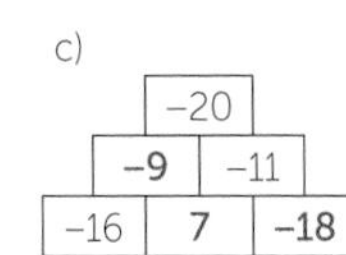


### 1.3 Calculating with negative integers

1. a) –18 b) –8 c) 54 d) –12 e) –12 f) –3
   g) 48 h) 20
2. a) False –54 b) True c) False –4 d) False 28
3. a) –2242 b) 10 710 c) –37 d) 118
4. –18°C
5. a) –48 b) –12 c) –12 d) 48 e) 12
6. a) –132 b) –6 c) –25 d) 0 e) –100
   –132, –100, –25, –6, 0
7. a) –6 × –7 = 42 b) 9 × –7 = –63

### 1.4 Multiplication

1. a) 34 b) 0 c) 129 d) –8
2. a) 7 b) 6 c) 12 d) 12
3. a) –25 and 10 b) –25 and –5 c) –60 and –5
4. a) 646 b) 350 c) 1854 d) 2728
5. £425
6. 64 × 12 = 768 p = £7.68
7. 18 × 12 = 216 $m^2$
8. a)

| × | 40 | 7 |
|---|---|---|
| 20 | 800 | 140 |
| 5 | 200 | 35 |
| total | 1000 | 175 |

   b) 1175
9. a) She can subtract 50 from 1000 b) 50 × 19 = 950
10. He has not put a zero in the units column for the second row (so he has multiplied 62 by 3 rather than by 30).
    The correct answer is 62 × 37 = 2294

### 1.5 Division

1. a) 45 lots of 19 is 855, so 855 divided by 19 equals 45
   b) The product of 36 and 42 is 1512, so 1512 divided by 42 equals 36
   c) 58 multiplied by 24 equals 1392, so 1392 divided by 24 equals 58
2. a) 35 b) 43 c) 37 d) 21
3. a) 25 b) 38 c) 30 d) 52
4. a) 21 r 5 b) 14 r 6 c) 21 r 17 d) 28 r 3
5. The remainder must be less than the divisor
6. 33 packets
7. 18 people
8. a) 21 pieces b) 45 cm
9. 12 coaches
10. 1258
11. 544 divided by 16 = 34 remainder 0
12. 221 (17 × 13)

### 1.6 Priority of operations

1. a) 15 b) 8 c) 7 d) 8
2. a) 42 b) 3 c) –28 d) 98
3. a) 5 b) 29 c) 5 d) 5
   part b) is the odd one out.
4. Tamin is wrong; 6 + 5 × 8 = 6 + 40 = 46.
   He has done the addition first and then the multiplication.
5. a) (20 – 10) ÷ 2 = 5 b) 2 × (7 + 2) = 18
   c) 24 ÷ (8 – $2^2$) = 6 d) 10 × (12 – 8) + 2 = 42
6. a) 4 + 6 × 3 < (4 + 6) × 3 b) 5 × 6 ÷ 2 = 5 × (6 ÷ 2)
   c) 10 – 3 × 2 < (10 – 3) × 2 d) 9 + $1^2$ < $(9 + 1)^2$
7. c) 20 – (3 × 2 + 5)
8. The calculation only multiplies 13 by 2; she needed to add 13 and 18 first before multiplying by 2.
   The correct calculation is (18 + 13) × 2 ( = 62)
9. a) 45 b) 4 c) 4 d) 5
10. a) (3 + 0) × (10 – 7) = 9 b) 3 × 1 ÷ (10 – 7) = 1
    c) 3 ÷ (0 + 10 – 7) = 1 d) 3 – (0 × 10 – 7) = 10

## Section 2

### 2.1 Square numbers

1. a) 25 b) 49 c) 121 d) 169
2. a) 6 b) 9 c) 12 d) 10
3. a) 16 b) 2 c) 13 and –13 d) 64
4. 8
5. a) $14^2 > 144$ b) $\sqrt{121} < 15^2$
   c) $3^2 = \sqrt{81}$ d) $-6 <$ the negative square root of 25
6. $(-2)^2 = 4$ $\sqrt{100} = -10$ $(-3)^2 = 9$ $0^2 = 0$
7. Answers to a) and c) are not integers
8. a) 10 b) 16 c) 9, 81
9. a) the square root of 16 is 4 cm not 8 cm
   b) the length of the square = 4 cm and the perimeter is 4 × 4 = 16 cm
10. a) sometimes true
    e.g. $1^2 = 1$ (square is not greater) but $2^2 = 4$ (square is greater)
    b) sometimes true e.g. $2^2 = 4$ (is even) but $3^2 = 9$ (is odd)
    c) always true e.g. $\sqrt{0} = 0$, $\sqrt{25} = 5$ or –5 (so a square root can be positive, negative or zero)
    d) never true e.g. anything squared is always positive

### 2. 2 Index notation

1. a) $3^3$ b) $4^5$ c) $2^4$ d) $5 \times 10^2$
2. a) 1 b) 125 c) 1000 d) 64
3. a) 2 b) 3 c) 1 d) 4
4. a) 200 is the odd one out.
   All numbers are square numbers apart from 200
   b) 9 is the odd one out. All numbers are cube numbers apart from 9
5. a) $5^2$ (25) is larger than 4 b) $10^3$ (1000) is larger than 27
   c) $4^5$ (1024) is larger than 625 d) $10^1$ (10) is larger than 3
6. $2^5 = 32$ $\sqrt[3]{1000} = 10$ $\sqrt{196} = 14$ $\sqrt{121} = 11$ $\sqrt[3]{1} = 1$
7. a) 169 b) 64 c) 3 d) 3
8. a) False $1^{10} < 10^1$ b) false $2 \times 2 \times 2 = 2^3$
   c) false $\sqrt[3]{1} = 1$ only d) true
9. a) e.g. $\sqrt[3]{1} = \sqrt{1}$ b) e.g. $64 = 8^2 = 4^3$
10. Square number = 196 (= $16^2$) and
    Cube number = 27 (= $3^3$). 196 – 27 = 169 (= $13^2$)

### 2.3 Laws of indices

1. a) $4^5$ b) $10^6$ c) $5^6$ d) $11^8$
2. a) $3^3$ b) $2^5$ d) $11^5$ d) $7^0$
3. a) $2^6$ b) $2^{12}$ c) $3^{15}$ d) $3^{30}$
4. part a) $(10^2)^4 = 10^8$ and part c) $5^3 \times 2^3 = 1000$; do not equal $10^6$
5. a) $2^5 \times 5^2$ b) $3^5 \times 5^4$ c) $5^4 \times 2^7$ d) $10^5$
6. Part d) is the odd one out – as it is the only one which is correct.
7. a) No, she is not correct. The answer is $2^5$
   b) No, he is not correct. The answer is $8^8$
8. a) * = 5 b) * = 5 c) * = 4
9. a) Mistakes: $(2 + 1)^3 = 3^3 = 27$ d) Mistakes: $5^6 \div 5^6 = 5^0 = 1$
10. a) $10^0 = 1^5$ b) $9^3 = 3^6$ c) $2^4 = 4^2$

### 2.4 Prime numbers

1. 37 and 79 are prime
2. 57 | 24, 36 | 90, 20 | 65
3. a) 49 (odd number) b) 51 (not divisible by 5)
   c) 28 (not divisible by 6) d) 3 (not divisible by 9)
4.

| | Divisible by 2 | Divisible by 3 |
|---|---|---|
| Prime Numbers | 2 | 3 |
| Not prime numbers | 10, 32 | 9, 45, 63 |

5. a) False b) true c) true d) true
6. a) False. 2 is prime number and is even.
   b) False. 2 is the smallest prime
   c) False. 97 is the largest prime under 100. 99 is not prime.
   d) True
7. a) integer answer b) not integer answer; not divisible by 9
   c) integer answer d) not integer answer; not divisible by 3
8. 90
9. a) sometimes true b) sometimes true
   c) never true d) never true
10. 2, 5, 11

### 2.5 Factors

1. a) $2^6$ b) $2^3 \times 3^2$ c) $2 \times 3^4$ d) $2^2 \times 3 \times 5^2$
2. a) 28 b) 36 c) 40 d) 180
3. a) 96; 8, 12; 4, 2, 3, 4; 2, 2, 2, 2 b) 90; 6, 15; 2, 3, 3, 5
4. 4 and 1 are not prime factors. 24 = 2 × 2 × 2 × 3
5. a) 40 = 2 × 2 × 2 × 5 b) 32 = 2 × 2 × 2 × 2 × 2
   c) 52 = 2 × 2 × 13 d) 105 = 3 × 5 × 7
6. Isabella is correct. Jake has not used index form and Ben has not multiplied.
7. $48 = 2 \times 2 \times 2 \times 2 \times 3 = 2^4 \times 3$
8. a) 33 = 3 × 11 and b) 1 is not a prime factor
9. a) true since 2 × 3 = 6
   b) true since $2^2 = 4$
   c) false; it is a multiple not a factor
   d) true since $2^3 \times 5 = 40$ and so is a factor of itself.
10. a) e.g. $32 = 2^5$ b) e.g. $500 = 2^2 \times 5^3$
    c) e.g. $48 = 3 \times 2^4$ and $700 = 2^2 \times 5^2 \times 7$

### 2.6 Multiples and LCM

1. a) 4, 8, 12, 16, 20 b) 10, 20, 30, 40, 50
   c) 15, 30, 45, 60, 75 d) 30, 60, 90, 120, 150
2. Multiples of 5: 5, 10, 20 | 30 | Multiples of 6: 6, 18
3. 24
4. a) 18 = 2 × 3 × 3 24 = 2 × 2 × 2 × 3
   b) 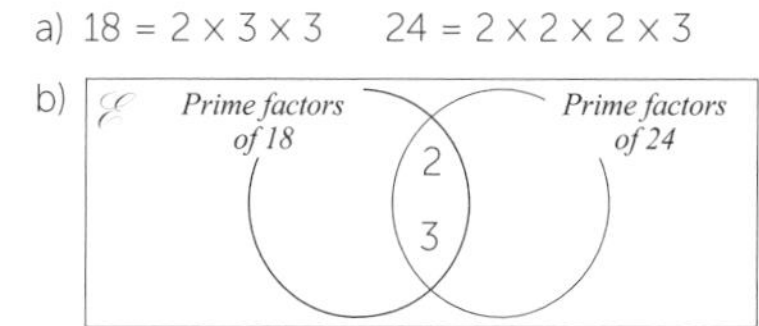


   c) LCM (18, 24) = 72
5. a) LCM (6, 8) = 24 b) LCM (9, 12) = 36
   c) LCM (10, 25) = 50 d) LCM (13, 7) = 91
6. LCM (2, 3, 5) = 30
7. e.g. 9 or 18 or 45
8. e.g. 7 or 14 or 42
9. 13 times including midnight and noon.
10. 5, 7, 13
11. Each 30 seconds, Sandi has done 5 and Joel has done 4. So 5 × 60 ÷ 30 = 10 times.
12. Mr Green bakes 15 batches of brownies.
    Miss Silver bakes 7 batches of cookies. They will make 210 of each.

## 2.7 Factors and HCF

1. a) 2 b) 10 c) 6 d) 12
2. a) HCF = 10, LCM = 150 b) HCF = 22, LCM = 132
   c) HCF = 12, LCM = 420 d) HCF = 50, LCM = 300
3. a) HCF = 2, LCM = 70 b) HCF = 6, LCM = 180
   c) HCF = 10, LCM = 350 d) HCF = 42, LCM = 252
4. a) 54 = 2 × 3 × 3 × 3 45 = 5 × 3 × 3 b) HCF = 3 × 3 = 9
   c) LCM = 5 × 3 × 3 × 3 × 2 = 270
5. a) HCF = 3, LCM = 45 b) HCF = 4, LCM = 96
   c) HCF = 10, LCM = 300 d) HCF = 15, LCM = 225
6. HCF = 5, LCM = 150
7. 30, 50, 70
8. 25 cm
9. LCM of 18 and 30 is 90. Buy 5 packs of textbooks and 3 packs of workbooks.
10. a) 12 b) Yes you could have e.g. 4 and 30 or 20 and 6
11. True – the HCF is the product of the overlapping (repeated) prime factors and the LCM is the product of the unique factors. Multiplying these together must give all the factors of both numbers and hence their product.

# Section 3

## 3.1 Algebraic notations

1. a) $g$ and $h$ b) 5 and 3 c) $5g$ and $3h$
2. a) $5p$ b) $\frac{y}{2}$ c) $3a$ d) $n^2$
3. Francis is incorrect; it should be $h^5$
4. a) $a + a$ b) $p - p + p$ and $q - q$ c) $5m - 2m$ and $3n + 3n$
   d) $v + v + 3v$, $2w - w$ and $-x + 4x$ e) $y - 5y$ and $+ 3 + 7$
   f) $3g - 5g$
5. $6x$
6. $x + 10$
7. $2x + 6$
8. a) $4n$ b) $3a$ c) $2x + 2y$ or $2(x + y)$
9. You don't know the value of $x$ or $y$; e.g. if $x = 5$ and $y = 1$ then the term $4x$ would be larger.
10. a) $n^2$ b) $3y$ c) $\frac{1}{2}mn$ or $\frac{mn}{2}$

## 3.2 Expressions as functions

1. a) $n + 4$ b) $n - 3$ c) $5n$ d) $7 - n$
   e) $\frac{n}{5}$ f) $\frac{20}{n}$ g) $10 + n$ h) $\frac{n}{6}$
2. $a = 20$ $b = 9$ $5x$ $c = 3$ $d = 16$ $y - 7$
3. a) 10 b) 9 c) 11 d) 4 e) 9
4. No, it should be $15 - n$
5. a) $2n + 3$ b) $\frac{n}{2} + 3$ c) $3n - 1$ d) $2n - 5$
6. a) a = 14, b = 6, c = −7 b) $3n + 2$
7. a) 13 b) 10 c) 3 d) 8 e) 48
8. No. 12 − 3 × −2 = 12 + 6 = 18
9. $5x - 3$ (The function machine multiplies by 5 and then subtracts 3.)
10. He has not used "order of operations" correctly; should be 5 × 16 + 1 = 81

## 3.3 Simplifying algebraic expressions

1. a) $4a$ b) $3y$ c) $8n$ d) $4p$
2. $14a$
3. $q^2$ and $q$ are not like terms; it should be $2q^2 + 4q$
4. a) $n^3$ b) $x^8$ c) $y^4$ d) $p^8$ e) $a^4$
5. $9n^2$
6. a) $a + 7b$ b) $4p - 3q$ c) $n^2$ d) $x^3 + 3x^2 + 2x$
7. a) $4c + 4d$ b) $5a + 8b$ c) $21x - 9y$ d) $11p - 3q$
8. a) no b) yes c) yes d) no e) yes
9. No, she is wrong. Yolanda has multiplied the powers; it should be $28a^5b^9$
10. a) $24a^2$ b) $49n^2$ c) $10ab^2$

## 3.4 Simple equations

1. a) formula b) equation c) formula d) expression
2. a) $x = 5$ b) $y = 16$ c) $n = 7$ d) $a = 45$ e) $t = 4$
3. $x = 7$ cm
4. Clem is 13 years older than Petra.
5. $y + 8 = 13$ so $y = 5$
6. a) $a = 3$ b) $y = 9$ c) $t = 3.5$ d) $q = 4$
7. $x = 6$
8. It should be $-3x = 15$ so $x = -5$
9. $3x - 7 = 11$ so $x = 6$
10. George is 11 years old.

## 3.5 Simple formulae

1. a) $A = p - 2$ b) $B = 3p$ c) $S = p + 5$ d) $z = \frac{p}{4}$
2. a) C = 19 b) C = 35 c) C = 3 d) C = −5 e) C = 17
3. Kelly has forgotten the negative; should be C = 10 − 3 × −2 = 16
4. a) C = 11 b) C = 13.5 c) h = 38
5. a) $x = T - 5$ b) $x = \frac{T}{4}$ c) $x = 8 - T$
   d) $x = 3T$ e) $x = \frac{24}{T}$
6. a) P = 16 cm b) P = 40 cm c) $l$ = 11 m d) $w$ = 5.5 m
7. $C = 1.5a + 1.8b$
8. $x = \frac{y + 2}{5}$
9. a) 59°F b) 25°C
10. a) £25 b) 8 km c) C = 5 + 0.4n or C = 500 + 40n
    d) due to the £5 standing charge

## 3.6 Brackets and common factors

1. a) $2x + 2$ b) $3y - 6$ c) $5b + 15$ d) $35 + 7n$ e) $40 - 10m$
2. No; it should be $x^2 + 6x$
3. $8(y - 3)$ and $8y - 24$
4. a) $2(x + 2)$ b) $3(a - 4)$ c) $5(g - 6)$ d) $3(2p + 3)$ e) $4(2t - 5)$
5. $x + 2$
6. a) $5x + 8$ b) $9y + 11$ c) $5n + 7$ d) $2p - 24$
7. He added the number outside the bracket to each coefficient instead of multiplying. Should be $12x + 15y$
8. a) $12x - 20$ b) $y^2 + 7y$ c) $pq - 2p$
   d) $n^2 - 3mn$ e) $3a^2 + 5ab - 2ac$
9. a) $a(b - 5)$ b) $y(2x - 7)$ c) $5q(p + 3)$
   d) $3n(2 - 7m)$ e) $4(2xy + 4 - 5xz)$
10. Area = $5(x + 5) - 4(x - 3) = x + 37$

## 3.7 Powers and roots

1. a) $x$, $4x$ and $3x^2$, $5x^2$   b) $7y^2$, $y^2$ and $-3y^3$, $-2y^3$
2. Kim is correct – it cannot be simplified.
3. $17n^2$
4. a) $5c^7$   b) $8w^6$   c) $2p^4$   d) $6d^3$
5. a) no = 2   b) yes   c) no = 1   d) yes
6. Expression = $-p^3 + 5p^2 - 5p$
7. Difference between the areas = $n(n + 5) - n(n - 2) = 7n$
8. a) $5x^2 + 6x$   b) $6p^2q^2 - 18pq^3$   c) $6m^3 + 33m^2 - 20m$

# Section 4

## 4.1 Decimal place value

1. a) 0.13 < 0.2   b) 0.07 > 0.065

0.13 — 0 0.1 0.2   0.065 — 0.05 0.06 0.07

c) 0.1000 = 0.1   d) 1.329 < 1.4

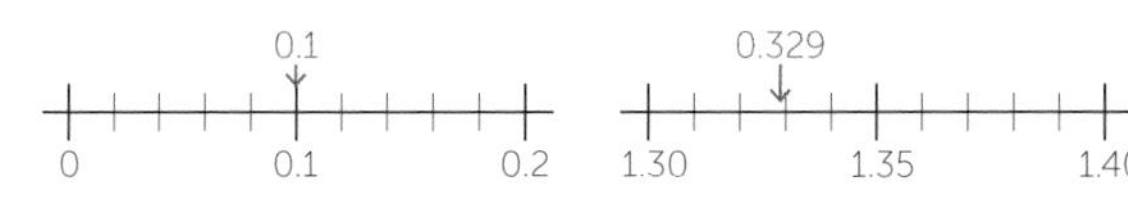


2. a) 2.709   b) −1.4   c) −2.35
3. a) 4.09, 4.12, 4.2, 4.31, 4.35, 4.8   b) 12.34, 12.5, 13.61, 13.7, 14.85
   c) −0.43, −0.27, −0.11, −0.06, 0.35
   d) −1.29, −1.281, −1.243, −1.203, 1.2
4. a) 0.41, 0.2405, 0.24, 0.214, 0.2104   b) 8.8501, 8.7, 8.52, 8.367, 8.1457
   c) −0.8, −0.63, −0.27, −0.25, −0.16
   d) −0.76, −0.74, −0.72, −0.713, −0.701
5. a) 0.6 < _0.63___ < 0.7   b) 0.87 < _0.872___ < 0.88
   c) 0.31 < ___0.32_ < 0.329   d) 0.4 < __0.43___ < 0.479
6. Rosie Dixon – 1st;  Keshena Mustafa – 2nd;  Josephine Rosberg – 3rd
7. 2.413
8. a) False; True; True; True
   b) Tirana Athens Budapest Milan Sofia Moscow Oslo
9. There are many possible answers, e.g. 1.234, 1.235, 1.236...
10. a) −6.7; −6.6   b) −6.07; −6.528; −6.6; −6.7; −6.84; −6.9

## 4.2 Calculating with decimal numbers

1. a) 11.16   b) 20.139   c) 6.91   d) 6.83
2. a) 34.7   b) 89.1   c) 0.725   d) 0.004 31
   e) 4895   f) 6.78   g) 0.348   h) 0.078 3
3. a) 0.28 + 0.22 = 0.5   b) 0.743 − 0.493 = 0.25
   c) 2.859 + 0.6 = 3.459   d) 3.073 − 0.37 = 2.703
4. a) 34.865   b) 9.529   c) 5.564   d) 8.969
5. a) 407   b) 324   c) 792   d) 18
6. a) 0.58 × 10 = 5.8   b) 9.24 ÷ 100 = 0.092 4
   c) 0.336 7 × 10 = 3.367   d) 7.25 ÷ 1000 = 0.007 25
   e) 0.039 × 100 = 3.9   f) 43.1 ÷ 1000 = 0.043 1
7. a) 6.265 + 3.735 = 10   b) 11 − 0.38 = 10.62
   c) 6.265 − 3.735 = 2.53   d) 3.735 + 0.38 = 4.115
8. a) 0.306   b) 0.030 6   c) 3.06   d) 1.7
9. a) 16 × 160 = 2560   b) 30 × 160 = 4800
   c) 15 × 161 = 2415   d) 15 × 8 = 120
10. a) 8g   b) 0.064 3 minutes
11. She is incorrect. 39 × 25 = 1000 − 25 = 975
12. £8.01

## 4.3 Rounding and decimal places

1. a) 200 (nearest 100)   b) 2360 (nearest 10)
   c) 1600 (nearest 100)   d) 16 000 (nearest 1000)
   e) 34 010 (nearest 10)   f) 20 000 (nearest 1000)
2. a) 108 (nearest unit)   b) 8.5 (1 d.p.)
   c) 80 (nearest unit)   d) 4.4 (1 d.p.)
3. a) 34.7 m   b) 100.3 m   c) 78.1 m   d) 9.9 m
4. a) 0.67   b) 0.45   c) 0.89   d) 0.59
5. a) 0.71   b) 0.63   c) 2.17   d) 1.22
6. 2.646
7. No. She is not correct. The answer is 0.23
8. a) Aina weighs the most.
   b) Both cubs weigh 1.75 kg. (2 decimal places)
   c) 1.752 kg and 1.746 kg. (3 decimal places)
9. 10.45, 10.46, 10.47, 10.48, 10.49
10. a) e.g. 0. 745 or 0.754 rounds to 0.75 to 2 decimal places.
    b) e.g. 0. 63 rounds to 0.6 to 1 decimal place.

## 4.4 Significant figures

1. a) 3000   b) 20 000   c) 0.03   d) 0.004
2. a) 13 000 (2 s.f.)   b) 102 000 (3 s.f.)   c) 4 700 000 (2 s.f.)
   d) 2.68 (3 s.f.)   e) 0.035 8 (3 s.f.)   f) 0.070 (2 s.f.)
3. a) 12.45 ≤ 12.5 < 12.55   b) 10.35 ≤ 10.4 < 10.45
   c) 2.995 ≤ 3.00 < 3.005   d) 0.425 ≤ 0.043 < 0.435
4. a) 0.182   b) 0.433   c) 4.83
5. a) 100 000 000   b) 140 000 000   c) 144 000 000
6. The answer should be 8.0
7. a) 2000  b) 3
8. a) 947.75 ≤ a < 947.85   b) 877.5 ≤ v < 878.5
9. 50% of 20 000 x 60 = 50% of 1 200 000 = £600 000
10. a)

| Planet | Earth | Mercury | Venus | Mars |
|---|---|---|---|---|
| Radius (km) Rounded value (1 s.f.) | 6000 | 2000 | 6000 | 3000 |

| Planet | Neptune | Jupiter | Uranus |
|---|---|---|---|
| Radius (km) Rounded value (1 s.f.) | 20 000 | 70 000 | 30 000 |

b) The radius of **Neptune** is approximately 10 times bigger than the radius of **Mercury**.

11. The error interval for the bookshelf b, is 74.5 ≤ b < 75.5cm
    No. The bookshelf could be 753 or 754mm.

## 4.5 Multiplying decimal numbers

1. a) 95.04   b) 0.950 4   c) 0.095 04   d) 0.095 04
2. a) 96   b) 0.96   c) 0.009 6   d) 2.16
3. a) 2.64   b) 0.863 9   c) 18.17   d) 0.906 28
4.

| × | 20 | 2 | 0.2 |
|---|---|---|---|
| 9 | 180 | 18 | 1.8 |
| 0.9 | 18 | 1.8 | 0.18 |
| 0.09 | 1.8 | 0.18 | 0.018 |

5. a) 0.7 × 0.6 = 0.42   b) 1.4 × 0.3 = 0.42   c) 0.02 × 2.1 = 0.042
   d) 0.12 × 3.5 =0.42   e) 0.06 × 6.7 = 0.402
6. a) True   b) False   c) True
7. a) 14   b) 20   c) 15   d) 12
   e) 15   f) 24   g) 20   h) 10
8. 15.75 × 7 = 110.25 hours in a week
9. 120 × 1.06 = 127.20 Euros
10. a) Area of A = 4 × 3 = 12cm²   Area of B = 6 × 2 = 12cm²
    b) Area of A = 4.32 × 3.05 = 13.176 cm²
    Area of B = 5.95 × 1.98 = 11.781 cm²

## 4.6 Dividing decimal numbers

1. a) 8 b) 4000 c) 0.05 d) 0.03
2. a) $\frac{239}{14}$ and c) $\frac{2.39}{0.14}$
3. a) 1.2 b) 27 c) 46 d) 162
4. a) 5.7 ÷ 1.5 < 5.7 b) 1.54 ÷ 2.8 = $\frac{7.7}{4}$
   c) 6.24 ÷ 0.6 > 6.24 d) 0.204 ÷ 0.4 < 5
5. b) 2.28 / 0.24 | 9.5 c) 0.078 / 0.12 | 0.65 d) 13.975 / 4.3 | 3.25
6. Shop A £1.80 ÷ 5 = £0.36 / pen or 36p / pen
   Shop B £1.05 ÷ 3 = £0.35 / pen or 35p / pen
7. 4 × 0.5 = 2
8. a) 35 ÷ 1.21 < 35 b) 13.8 ÷ 0.89 > 13.8
   c) 38.75 ÷ 1.52 > 20 d) 27.89 ÷ 2.93 > 9
9. A bag of 2.58 kg of sand costs £1.25. Velma is correct.
   £1.25 ÷ 2.50 = £0.5/kg. 2.58 > 2.50 therefore the price will be less than 50p per kg.
10. Sometimes. When dividing by a number less than 1, the answer becomes bigger. 12 ÷ 2 = 6. 12 ÷ 0.2 = 60

# Section 5

## 5.1 Estimate answers

1. a) 20 b) 7 c) 100 d) 0.03 e) 0.007
2. a) 14.5 b) 3.0 c) 0.2 d) 13.8 e) 0.0
3. a) 9 b) (10 × 4) ÷ 6 ≈ 7
   c) (10 × 10) divided by 20 = 5 d) 10
4. a) 10 b) 84 c) 10
5. 88 cm (or estimate 10 × 8 = 80 cm)
6. 12.6 m (or estimate 14 m)
7. 35.65 seconds ≤ actual time < 35.75 seconds
8. 4.8 km ≤ actual length < 4.9 km
9. a) i) 29.47359043 ii) 29.47 b) i) 1.690325444 ii) 1.69
   c) i) 0.6239428306 ii) 0.62 d) i) 12.92093288 ii) 12.92
10. 28.3 m/s

## 5.2 Scale diagrams

1. b) AC = 11.7 cm c) 11.7 m
2.

| Length on diagram (mm) | Actual length (m) |
|---|---|
| 24 | **7** |
| **60** | 17.5 |
| **96** | 28 |
| 84 | **24.5** |

3. a) 19.2 m b) 21.25 cm
4. 120 m
5. Jasmine not correct because a scale of 5 cm to 15 km is the same as 1 cm to 3 km (not 2.5 km).
6. 2 m
7. 1 cm to 0.4 km or 1 cm to 400 m
8. 1 cm to 2.5 m
9. a) 20 cm b) 80 cm c) 25 m
10. 2.24 m (accept 2.15 to 2.35 m)

## 5.3 Bearings

1. a) 66° b) 128° c) 300°
2. a) b) c) d)



3. 135°
4. 000° or 360°
5. b) 7.1 km
6. a) 252° b) 244°
7. 111°
8. 087°

## 5.4 Plans and elevations

1. Front, side and plan elevations are all the same – a square of side 5 cm.
2. Plan elevation – rectangle 8 cm by 5 cm
   Front elevation – rectangle 8 cm by 4 cm
   Side elevation – rectangle 5 cm by 4 cm
3. Cube drawn with 3 cm length sides.

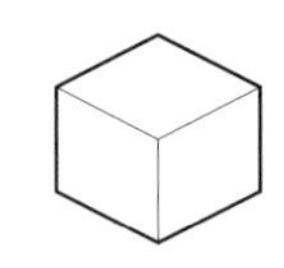

4. a) b) Plan Front Side

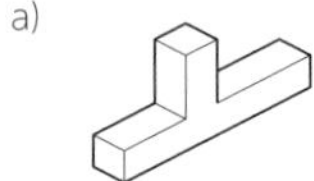

5. a) Plan Front Side b)
6. a) b)
7.
8. a) b) c)
9. a) Plan Front Side b) Plan Front Side
   c) Plan Front Side
10. a) b) c)

## 5.5 Perimeter

1. 78 m
2. 180 cm
3. 17 cm
4. 14 m
5. 42 m
6. Becky is correct – the square and triangle have the same perimeter (20 cm) but the rectangle is 22 cm.

7. Draw 3 rectangles with $l + w = 9$ cm (e.g. $2 \times 7$, $3 \times 6$, $4 \times 5$)
8. 60 cm
9. 48 cm
10. 192 m/min (3.2 m/s)
11. Length = width = 14 cm

### 5.6 Area of simple shapes

1. 221 $m^2$
2. 19 cm
3. 8.5 cm
4. 20 cm
5. 216 $cm^2$
6. 390 $cm^2$
7. 20 m
8. No, the parallelogram is 99 $cm^2$ , the triangle is 100 $cm^2$.
9. $6 \times 4$ (smallest perimeter), $8 \times 3$, $12 \times 2$, $24 \times 1$
10. Large square = 9 sq units, small square = 1 sq unit, triangle = 1 sq unit, rectangle = 2 sq units, trapezium = 3 sq units, parallelogram = 2 sq units.

### 5.7 Volume of simple shapes

1. 22 500 $cm^3$
2. No, the volume is 3 $m^3$ (5 m $\times$ 3 m $\times$ 0.2 m)
3. 0.216 $m^3$
4. 4 cm
5. $(10 \times 7 \times 19) + (8 \times 3 \times 19) = 1786$ $cm^3$
6. 120 crates
7. 240 $cm^3$
8. 9.6 $m^3$
9. 60 $cm^2$
10. a) 210 $m^2$ b) 210 000 $m^3$
    c) 252 000 000 kg or 252 000 tonnes of soil

## Section 6

### 6.1 Equivalent fractions

1. a) $\frac{3}{4}$ b) $\frac{5}{8}$ c) $\frac{1}{3}$ d) $\frac{2}{5}$
2. a) $\frac{2}{11}$ b) $\frac{3}{5}$ c) $\frac{1}{2}$ d) $\frac{1}{2}$
3. a) $\frac{9}{3} = 3$, others reduce to $\frac{1}{3}$ b) $\frac{9}{12}$ reduces to $\frac{3}{4}$, others reduce to $\frac{2}{3}$
   c) $\frac{40}{100}$ reduces to $\frac{2}{5}$, others to $\frac{4}{5}$ d) $\frac{60}{140}$ reduces to $\frac{3}{7}$, others to $\frac{6}{7}$
4. Valma should divide numerator and denominator by the same number, not subtract. $\frac{32}{36} = \frac{16}{18} = \frac{8}{9}$
5. a) 15 b) 30 c) 20 d) 48
6. $\frac{1}{\mathbf{4}} = \frac{3}{\mathbf{12}} = \frac{75}{300} = \frac{\mathbf{8}}{32}$
7. $\frac{9}{16}$ of the shape is shaded.
8. e.g. $\frac{1}{\mathbf{2}} = \frac{3}{6}$ $\frac{2}{\mathbf{3}} = \frac{8}{\mathbf{12}} = \frac{10}{\mathbf{15}} = \frac{6}{\mathbf{9}}$ $\frac{1}{\mathbf{2}} = \frac{\mathbf{4}}{8} = \frac{\mathbf{11}}{22}$ $\frac{5}{\mathbf{7}} = \frac{\mathbf{10}}{14}$
9. Answer – French and Spanish are both equal to $\frac{5}{6}$
10. a) $\frac{1}{6}$ b) $\frac{1}{10}$
11. a) Sometimes. Only divide.
    b) Never unless it is the same fraction.
    c) Sometimes. $\frac{1}{6}$ is a fraction in its simplest form but 6 is not a prime number.
12. a) $\frac{45}{18 + \mathbf{36}} = \frac{5}{6}$ b) $\frac{30 - \mathbf{3}}{30 + \mathbf{33}} = \frac{3}{7}$
    c) $\frac{12}{21 - \mathbf{5}} = \frac{\mathbf{3}}{4}$ d) $\frac{\mathbf{10} + 12}{50 - \mathbf{24}} = \frac{11}{13}$ or $\frac{\mathbf{21} + 12}{50 - \mathbf{11}} = \frac{11}{13}$

### 6.2 Proper and improper fractions

1. a) $\frac{11}{4}$ the only improper fraction b) $\frac{6}{7}$ the only proper fraction
   c) $\frac{4}{7}$ the only fraction less than 1 d) $\frac{2}{5}$ the only one less than 1
2. $1\frac{1}{5} = \frac{6}{5}$ $4\frac{1}{4} = \frac{34}{8}$ $1\frac{1}{3} = \frac{4}{3}$ $2\frac{1}{7} = \frac{15}{7}$ $3\frac{3}{8} = \frac{27}{8}$
3. a) $3\frac{2}{3}$ b) $2\frac{4}{5}$ c) $5\frac{3}{4}$
   d) $1\frac{1}{5}$ e) $2\frac{7}{8}$ f) $4\frac{2}{7}$
4. a) $\frac{13}{5} > 1\frac{3}{5}$ b) $\frac{7}{3} < 3\frac{1}{3}$ c) $2\frac{1}{10} > \frac{13}{10}$
   d) $\frac{25}{2} = 12\frac{1}{2}$ e) $\frac{16}{9} < 2\frac{2}{9}$ f) $2\frac{1}{13} > \frac{16}{13}$
5. a) $\frac{5}{8} > \frac{7}{16}$ b) $\frac{4}{9} > \frac{5}{12}$ c) $\frac{9}{24} > \frac{3}{24}$
   d) $\frac{5}{6} > \frac{3}{4}$ e) $\frac{5}{8} < \frac{7}{10}$ f) $\frac{11}{14} < \frac{18}{21}$
6. a) $\frac{5}{12}, \frac{3}{4}, \frac{5}{6}$ b) $\frac{3}{5}, \frac{5}{8}, \frac{7}{10}$ c) $\frac{2}{3}, \frac{7}{9}, \frac{5}{6}$ d) $\frac{1}{6}, \frac{2}{11}, \frac{2}{9}$
7. a) $1\frac{11}{12}$ b) $2\frac{1}{2}$ c) $\frac{27}{13}$
8. Jake: $\frac{7}{9}$
9. When comparing fractions, Gwen needs to change the denominators first so that they are the same, then compare the numerators.
   $\frac{8}{15} = \frac{88}{165}$ and $\frac{6}{11} = \frac{90}{165}$ Gwen is not correct.
10. $1\frac{9}{20}$
11. a) $2\frac{3}{4} < 2\frac{\mathbf{7}}{8}$ b) $4\frac{\mathbf{5}}{6} > 4\frac{5}{8}$ c) $3\frac{\mathbf{3}}{4} = \frac{45}{12}$ d) $\frac{1}{2} < \frac{\mathbf{5}}{\mathbf{8}} < \frac{\mathbf{13}}{\mathbf{20}} < \frac{2}{3}$

### 6.3 Adding and subtracting fractions

1. a) $\frac{4}{5}$ b) $\frac{2}{3}$ c) 1 d) $\frac{4}{10} = \frac{2}{5}$
   e) $\frac{6}{8} = \frac{3}{4}$ f) $\frac{6}{12} = \frac{1}{2}$ g) $\frac{4}{10} = \frac{2}{5}$ h) $\frac{7}{9}$
2. a) $\frac{5}{8}$ b) $\frac{1}{9}$ c) $\frac{11}{20}$ d) $\frac{5}{12}$ e) $\frac{7}{8}$ f) $\frac{1}{12}$ g) $\frac{13}{15}$ h) $\frac{1}{2}$
3. a) and i) b) and iv) c) and iii) d) and v)
4. a) $\frac{2}{3} + \frac{1}{5} = \frac{10}{15} + \frac{3}{15} = \mathbf{\frac{13}{15}}$ b) $1 - \frac{9}{12} = \frac{3}{12} = \mathbf{\frac{1}{4}}$
   d) $\frac{2}{10} - \frac{1}{8} = \frac{1}{2}$ $\mathbf{\frac{8}{40} - \frac{5}{40} = \frac{3}{40}}$
5. a) $\frac{1}{2}$ b) $\frac{5}{16}$ c) $\frac{7}{18}$ d) $\frac{2}{9}$
6. a) $\frac{5}{6} + \frac{7}{9} = 1\frac{11}{18}$ b) $\frac{5}{6} - \frac{7}{9} = \frac{1}{18}$
7. $\frac{14}{15}$ of an hour or 56 minutes.
8. a) $\frac{23}{30}$ b) $\frac{1}{2}$ c) $1\frac{1}{6}$ d) $\frac{23}{36}$
9. $\frac{11}{20}$
10. a) $\frac{2}{3}$ b) $\frac{7}{16}$ c) $\frac{1}{60} + \frac{1}{60} = \frac{1}{30}$ d) $\frac{1}{2} - \frac{1}{6} = \frac{1}{3}$
11. $\frac{1}{4} + \frac{1}{3} + \frac{1}{2} > 1$
12. $\frac{2}{3} + \frac{1}{6} = \frac{5}{6}$ $\frac{3}{2} - \frac{2}{3} = \frac{5}{6}$

## 6.4 Mixed numbers

1. a) $\frac{25}{18} = 1\frac{7}{18}$ b) $1\frac{9}{20}$ c) $2\frac{1}{3}$ d) $3\frac{3}{10}$
   e) $1\frac{1}{3}$ f) $\frac{11}{12}$ g) $1\frac{3}{4}$ h) $1\frac{2}{9}$
2. a) $6\frac{3}{8}$ b) $2\frac{1}{3}$ c) $4\frac{7}{12}$ d) $2\frac{1}{10}$
   e) $\frac{11}{15}$ f) $1\frac{1}{20}$ g) $\frac{13}{18}$ h) $2\frac{12}{35}$
3. a) $2\frac{10}{15}$ and $7\frac{1}{3}$ b) $2\frac{1}{10}$ and $1\frac{3}{5}$ c) $1\frac{1}{2}$ and $1\frac{1}{7}$ d) $7\frac{1}{3}$ and $1\frac{1}{7}$
4. a) is the odd one out. All the others are $\frac{52}{15}$
5. a) $2\frac{1}{2} + 1\frac{4}{5} = 5\frac{1}{2} - 1\frac{1}{5}$ b) $\frac{2}{3} + 3\frac{1}{11} = 5 - 1\frac{8}{33}$
   c) $2\frac{1}{9} - 1\frac{5}{6} < 1\frac{1}{6} + \frac{2}{3}$ d) $1\frac{1}{8} + \frac{5}{12} < 1\frac{5}{8} + \frac{1}{12}$
6. a) b)

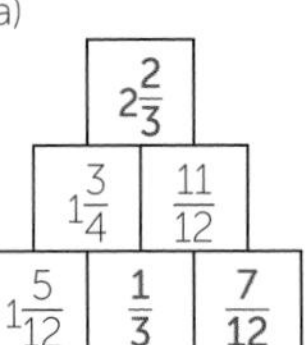


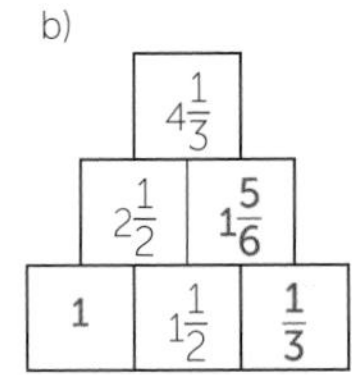


7. $\frac{9}{10}$ kg of soil mixture
8. a) $4\frac{2}{5}$ b) $2\frac{13}{20}$ c) $2\frac{13}{16}$ d) $\frac{29}{30}$
   a) has the greatest answer.
9. a) $\frac{11}{12}$ m
10. a) $3\frac{7}{30}$ km b) $\frac{13}{15}$ km

## 6.5 Multiplying fractions

1. a) $\frac{2}{5}$ b) 14 c) $\frac{2}{9}$ d) $\frac{3}{7}$ e) $1\frac{1}{2}$ f) 21
2. a) $\frac{17}{18}$ b) $1\frac{2}{33}$ c) $\frac{7}{8}$ d) $16\frac{1}{2}$ e) $33\frac{1}{3}$ f) $\frac{5}{6}$
3. a) $2\frac{1}{7}$ b) $7\frac{5}{6}$ c) $13\frac{1}{3}$ d) $7\frac{5}{9}$ e) 7 f) 5
4. $\frac{1}{2}$ with 2, $\frac{2}{9}$ with $4\frac{1}{2}$, $\frac{3}{5}$ with $1\frac{2}{3}$
5. a) $\frac{7}{16}$ b) $\frac{9}{20}$ c) 2 d) $\frac{3}{4}$
6. Yes. The perimeter = $13\frac{1}{3}$ m Area = $11\frac{1}{9}$ m$^2$
7. $15\frac{5}{9}$ kg of oranges
8. a) True b) True c) Sometimes
9. When multiplying mixed numbers, change the fractions to improper fraction, then multiply them.
   $1\frac{3}{4} \times 4\frac{1}{3} = \frac{7}{4} \times \frac{13}{3} = \frac{91}{12} = 7\frac{7}{12}$
10. $\frac{7}{9}$ km
11. No. It is $\frac{1}{5}$ of the school garden.
12. $\frac{1}{5}$ of the marbles are pink and small.

## 6.6 Dividing fractions

1. a) $1\frac{3}{5}$ b) $\frac{7}{30}$ c) $\frac{3}{5}$ d) $3\frac{1}{2}$ e) $2\frac{1}{2}$ f) 16 g) $\frac{1}{48}$ h) $\frac{3}{5}$
2. a) $\frac{1}{8}$ b) $1\frac{4}{5}$ c) $\frac{1}{3}$ d) $4\frac{1}{2}$ e) 10 f) $\frac{3}{22}$ g) $\frac{11}{42}$ h) $\frac{5}{58}$
3. a) $\frac{3}{4}$ b) $1\frac{4}{11}$ c) $2\frac{1}{10}$ d) $\frac{37}{130}$ e) $1\frac{7}{10}$ f) $\frac{5}{6}$ g) 6 h) $12\frac{1}{2}$
4. $1\frac{8}{15} \div 5$ with $4\frac{3}{5} \div 15$ both these come to $\frac{23}{75}$
5. a) $\frac{4}{9} \div \frac{1}{2}$ b) $2\frac{7}{10} \div 1\frac{1}{3}$ c) $3\frac{1}{8} \div 1\frac{1}{4}$
6. 9 pieces that are $\frac{1}{2}$m long.
7. She needs to divide $\frac{2}{3}$ by $\frac{3}{4}$ to find the mass of 1m. She has divided the fractions incorrectly. When dividing fractions, she needs to multiply the first fraction with the reciprocal of the second one.
   Correct calculation to find mass of 1 m of pipe is $\frac{2}{3} \times \frac{4}{3} = \frac{8}{9}$ kg
8. $4\frac{1}{2}$ smaller plots.
9. a) $\frac{19}{60}$ kg b) $1\frac{7}{12}$ kg
10. a) $1\frac{1}{20}$ kg of flour b) $\frac{7}{10}$ kg of flour

# Section 7

## 7.1 Working with coordinates

1. a) A (2, 2) B (3, 5) C (6, 4) b) D (5, 1)
2. a) L b) M c) V d) U e) P f) G
3. a) (−6, 4) b) (6, 7) c) (−8, 9) d) (9, 2) e) (−8, −6) f) (4, −2)
4. a) YC (CY) b) MW (WM) c) LX (XL) d) AD (DA)
5. a) (6, 0.5) b) (8, 2) c) (0.5, 4) d) (−3, −7.5)
6. a) (5, 3) b) e.g (6, 4) and (4, 2) , (7, 5) and (3, 1)
7. a) (3, 4) b) (−2, 5) c) (0, 5) d) (−4, 3.5)
8. a) (1, −1) b) Both midpoints are at (0,1)
   c) the midpoints are the same – which means the diagonals of a rhombus bisect each other.
9. $p = -5$ and $q = -1$

## 7.2 Equations of lines

1. a) vertical b) horizontal c) perpendicular d) parallel
2. a) $(3, n)$ b) $(n, 7)$ c) $(n, -2)$ d) $(-1, n)$
3. The line has equation $y = 4$ as it intersects the $y$-axis at (0,4) and is parallel to the x-axis.
4. 
5. A: $x = -6$ B: $x = 5$ C: $y = 1$ D: $y = -4$
6. $x = 3$ and $y = -4$
7. Yes; they all lie on the line $y = -x$
8. a) (−3, 8); all the others lie on the line $x = 3$
   b) (−4, 4); all the others lie on $y = x$
9. a) $y = 3$ b) $y = -x$
10. $x = 6$, $x = -6$, $y = 3$, $y = -1$, $y = x$ and $y = -x$

## 7.3 Plotting graphs 1

1. a) and d) are equations of straight lines
2. A = (3, 7)
3.

| $x$ | 0 | 1 | 2 | 3 | 4 |
|---|---|---|---|---|---|
| $y$ | 2 | 3 | 4 | 5 | 6 |

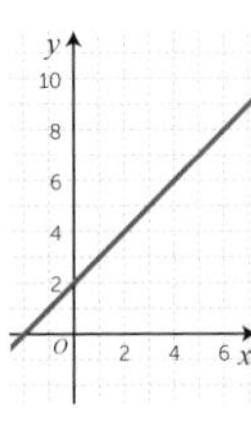

4.

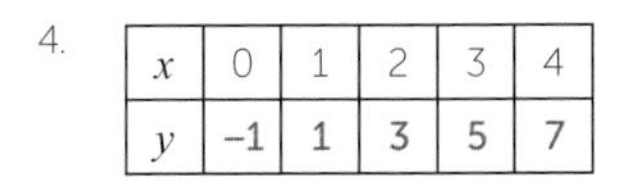

| $x$ | 0 | 1 | 2 | 3 | 4 |
|---|---|---|---|---|---|
| $y$ | −1 | 1 | 3 | 5 | 7 |

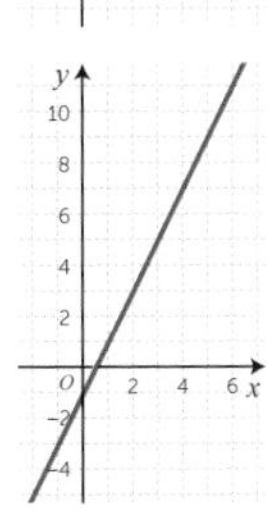

5. Yes, because 2 × 2 + 1 = 5
6.

| $x$ | −2 | −1 | 0 | 1 | 2 |
|---|---|---|---|---|---|
| $y$ | −8 | −5 | −2 | 1 | 4 |

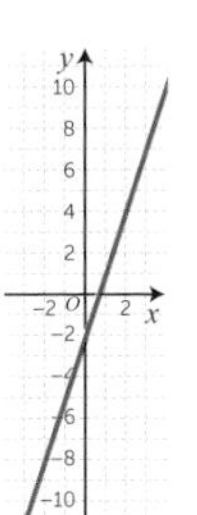

7.

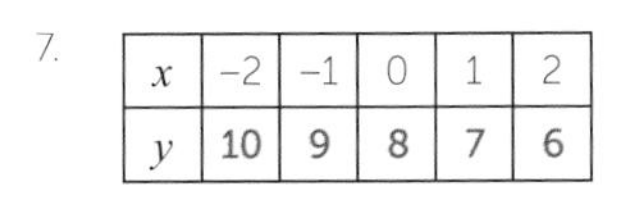

| $x$ | −2 | −1 | 0 | 1 | 2 |
|---|---|---|---|---|---|
| $y$ | 10 | 9 | 8 | 7 | 6 |

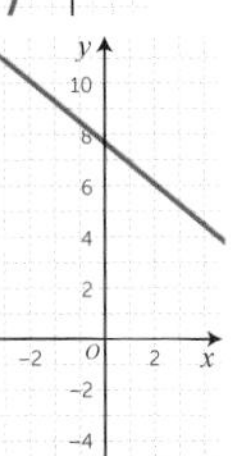

8.

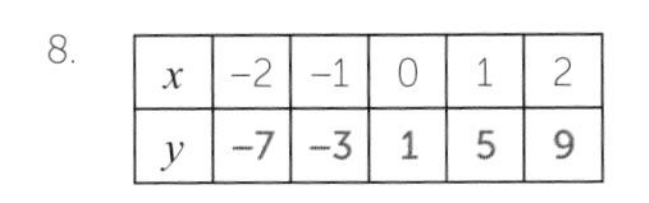

| $x$ | −2 | −1 | 0 | 1 | 2 |
|---|---|---|---|---|---|
| $y$ | −7 | −3 | 1 | 5 | 9 |

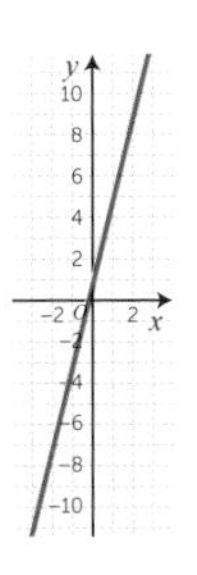

9. They are both correct: 2 × (−2) − 3 = − 7 and 6 × (−2) + 5 = − 7
10. a) $y = x + 1$ b) $y = x - 2$ c) $y = 2x + 3$ d) $y = 6 - x$

## 7.4 Plotting graphs 2

1. a) and d)
2. a) (3, 0) and (0, 3) b) (−2, 0) and (0, −2)
   c) (5, 0) and (0, 10) e) (4, 0) and (0, 3)
3.

| $x$ | 0 | 1 | 2 | 3 | 4 |
|---|---|---|---|---|---|
| $y$ | 4 | 3 | 2 | 1 | 0 |

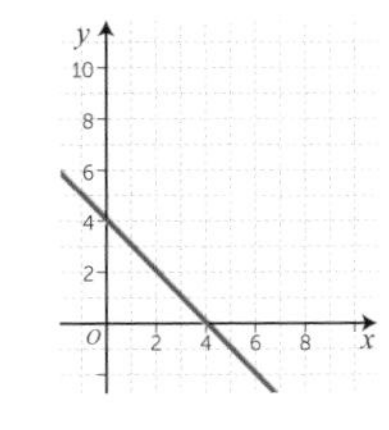

4.

| $x$ | 0 | 1 | 3 | 5 | 7 |
|---|---|---|---|---|---|
| $y$ | 7 | 6 | 4 | 2 | 0 |

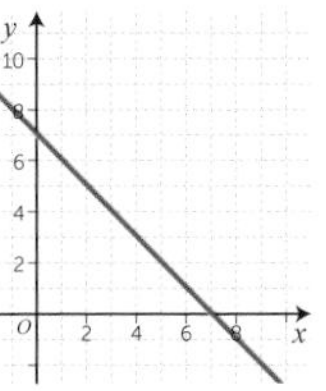

5. e.g. (0, 10), (2, 8) (10, 0)

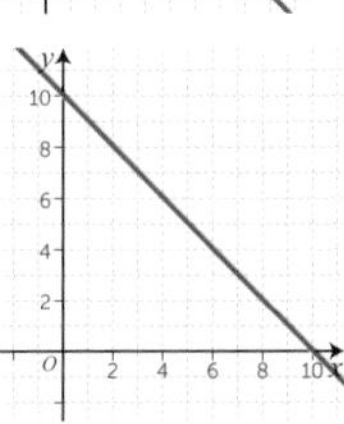

6. a) Intercepts are (5, 0) and (0, 2) b)

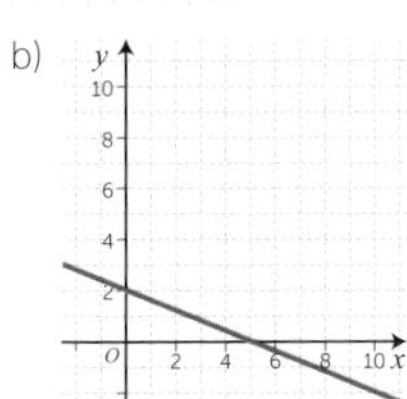

7.

| $x$ | 0 | 1 | 2 | 3 | 4 |
|---|---|---|---|---|---|
| $y$ | 4 | 2 | 0 | −2 | −4 |

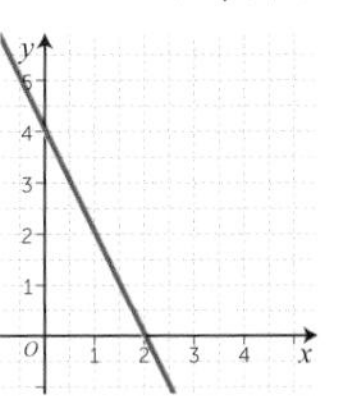

8.

| $x$ | 0 | 3 | 6 | 9 |
|---|---|---|---|---|
| $y$ | 4 | 2 | 0 | −2 |

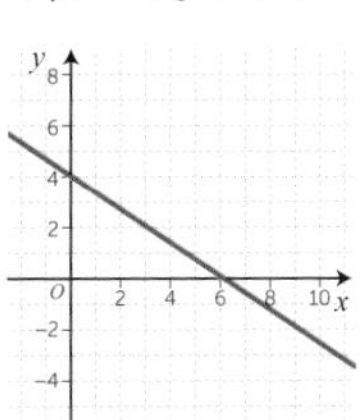

9.

| $x$ | 0 | 2 | 4 | 6 | 8 |
|---|---|---|---|---|---|
| $y$ | −4 | −3 | −2 | −1 | 0 |

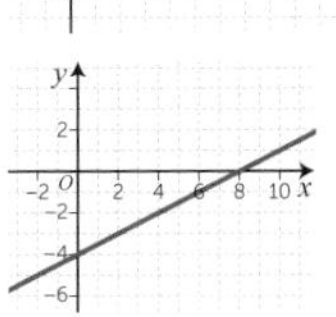

10. a) $x + y = 3$ b) $x + y = 5$ c) $2x + y = 8$

## 7.5 Gradients of straight lines

1. a) 2 b) 1 c) $\frac{9}{2}$ d) $\frac{1}{3}$ e) $\frac{1}{2}$
2. Harriet; since 3.5 > 3
3. a) $-\frac{7}{8}$ b) −3 c) $-\frac{1}{2}$ d) $-\frac{1}{4}$ e) −2
4. Sally; since 4 > 3
5. a) $-\frac{5}{2}$ b) $\frac{3}{2}$ c) $-\frac{3}{2}$ d) $\frac{4}{5}$ e) $-\frac{2}{3}$
6. a) 2 b) −4 c) 1 d) 1 e) $\frac{1}{2}$
7. a) 2 b) $-\frac{1}{4}$
8. The gradient should be calculated like this: $\frac{(-14) - (-2)}{-3 - 1} = \frac{-12}{-4} = 3$
9. Edie's gradient: $\frac{12}{8} = \frac{3}{2}$ Stan's gradient: $\frac{-9}{-6} = \frac{3}{2}$
   They are the same
10. Alex's gradient: $\frac{8}{6} = \frac{4}{3}$ Chloe's gradient: $\frac{-10}{-8} = \frac{5}{4}$
    Chloe's gradient is closer to 1

## 7.6 Equation of a straight line 1

1. a) (0, −2) and 3  b) (0, −9) and 2
   c) (0, 5) and 4  d) (0, 8) and 3
2. No; it should be (0, 0) and (1, 4)
3. 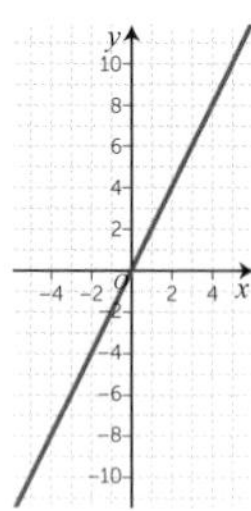
4. 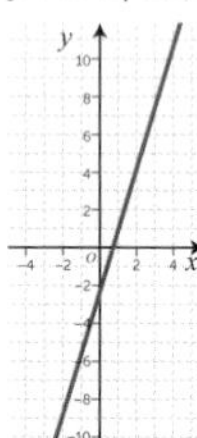
5. $y = x - 3$
6. No; the $y$-intercept should be (0, 3) but the other coordinate is correct
7. $y = 4 - 3x$
8. 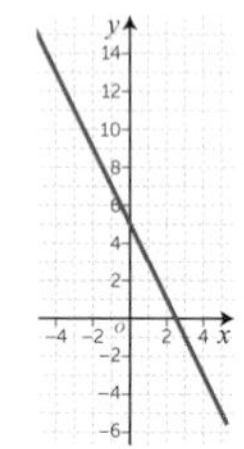
9. 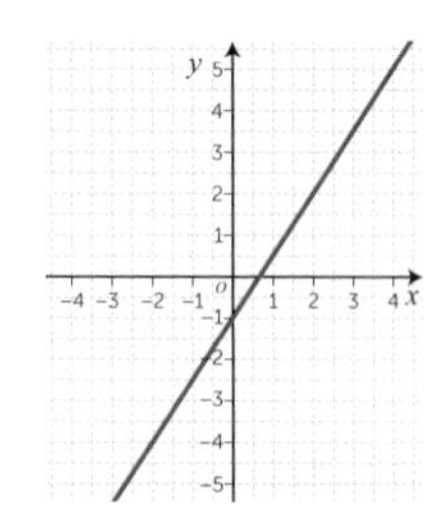
10. A: $y = -0.5x - 1$  B: $y = 3x + 1$  C: $y = 0.5x + 1$

## 7.7 Equation of a straight line 2

1. a) negative  b) positive  c) negative  d) negative  e) positive
2. Jamie is correct. Gradient = $\frac{15-3}{5-1} = \frac{12}{4} = 3$
3. $y = 6x$
4. a) $y = 2x - 1$  b) $y = 5x - 14$  c) $y = -3x + 16$  d) $y = \frac{1}{2}x - 4$
5. a) $y = 3x$  b) $y = -3x$  c) $y = 2x + 2$  d) $y = -3x - 1$
6. The gradient is negative; equation should be $y = -x + 3$
7. $y = x + 1$
8. David's workings should be:  Gradient: $\frac{-2}{4} = -\frac{1}{2}$
9. $y = -\frac{3}{4}x + 3$
10. a) $y = x + 3$  b) $y = -3x + 13$  c) $y = 2x - 4$
    d) $y = \frac{1}{4}x + 1$  e) $y = -2x + 3$

# Section 8

## 8.1 Decimals to fractions

1. a) $\frac{3}{10}$  b) $\frac{3}{100}$  c) $\frac{3}{1000}$
   d) $\frac{9}{100}$  e) $\frac{9}{1000}$  f) $\frac{19}{10\,000}$
2. a) $\frac{3}{5}$  b) $\frac{8}{25}$  c) $\frac{3}{20}$
   d) $\frac{4}{5}$  e) $\frac{9}{20}$  f) $\frac{13}{250}$
3. a) and ii)  b) with i)  c) with iv)  d) with iii)
4. a) $\frac{1}{3}$  b) $2\frac{1}{10}$  c) 0.040  d) 6 ÷ 10 000
5. c) $0.050 = \frac{1}{20}$  f) 0.006 = 6 ÷1000
6. a) $\frac{3}{50}$  b) $\frac{1}{50}$  c) $\frac{3}{25}$  d) $\frac{3}{500}$
   e) $\frac{6}{25}$  f) $\frac{3}{25}$  g) $\frac{3}{50}$
   a and g are equivalent. c and f are equivalent.
7.

| 0.4 | 0.40 | 0.04 | 0.004 | 0.008 | **0.02** |
|---|---|---|---|---|---|
| $\frac{2}{5}$ | $\frac{2}{5}$ | $\frac{1}{25}$ | $\frac{1}{250}$ | $\frac{1}{125}$ | $\frac{1}{50}$ |

8. $\frac{1}{6} = 0.1666...$
   $0.2 > \frac{1}{6}$ Matilda is incorrect. She ran further than she cycled.
9. $\frac{27}{40}$ km
10. Lily spends $\frac{3}{20}$ of an hour doing her homework.
11. Tom has made two mistakes and accidentally found the right answer.
    $0.005 = \frac{5}{1000} = \frac{1}{200}$
12. $0.48 = \frac{12}{25}$

## 8.2 Fractions to decimals

1. a) $\frac{3}{10}$, 0.3  b) $\frac{1}{50}$, 0.02  c) $\frac{4}{25}$, 0.16  d) $\frac{1}{5}$, 0.2
2. a) $\frac{3}{5} = \frac{6}{10} = 0.6$  b) $\frac{11}{50} = \frac{22}{100} = 0.22$
   c) $\frac{7}{25} = \frac{14}{50} = \frac{28}{100} = 0.28$  d) $\frac{27}{30} = \frac{9}{10} = 0.9$
3. a) 0.92  b) 0.875  c) 0.225  d) 0.04  e) 0.9375
4. $0.164 = \frac{41}{250}$  $0.02 = \frac{1}{50}$
5. a) 0.07  0.27  3/8  $\frac{11}{25}$  0.615  $\frac{13}{20}$
   b) 0.46  $\frac{3}{5}$  0.66  $\frac{7}{10}$  $\frac{32}{40}$  $\frac{17}{20}$
6. No. Dan is incorrect. $\frac{7}{20} = 0.35$  $\frac{3}{8} = 0.375$
7. a) Basketball  b) Hockey
8. $\frac{3.5}{5} = \frac{7}{10} = 0.7$
9. $\frac{3}{20}$
10. No. The answer is 0.625.
11. b) $\frac{11}{200}$  d) $\frac{1}{20}$  e) $\frac{7}{40}$
12. $\frac{37}{125} = 0.296$

## 8.3 Percentages 1

1. a) b) c) d)

2. a) $\frac{13}{25} = \frac{52}{100} = 52\%$ b) $\frac{56}{80} = \frac{7}{10} = \frac{70}{100} = 70\%$
   c) $\frac{72}{120} = \frac{12}{20} = \frac{60}{100} = 60\%$ d) $\frac{100}{125} = \frac{4}{5} = \frac{80}{100} = 80\%$

3. a) 65% b) 78% c) 60% d) 55%

4. a) 44.68% of students use ipads b) 41.94% of students are vegans
   c) 85.71% of girls achieved grade 9 d) 84.67% of men like cats

5. a) False. 26 out of 65 = 40% 21 out of 35 = 60%
   b) False. 20 out of 50 = 40% 10 out of 40 = 25%
   c) True. 36 out of 48 = 75 out of 100 = 75%
   d) False. 55 out of 60 = 91.67% 23 out of 24 = 95.83%

6. a) $25\% = \frac{1}{4}$ b) $8\% < \frac{8}{10}$ c) $\frac{3}{20} = 15\%$
   d) $\frac{1}{50} = 2\%$ e) $\frac{2}{3} > 60\%$

7. No. $\frac{1}{25} = \frac{4}{100}$ and $\frac{1}{50} = \frac{2}{100}$

8. Strictly Comedy = $\frac{960}{1200} = 80\%$

   Leeds got talent = $\frac{1130}{1500} = 75.33\%$

   Strictly Comedy is the most popular show.

9. 52.44% of the laptops owned by doctors are less than 3 years old.
   47.56% of laptops owned by doctors are aged 3 years or more.

   39.06% of the laptops owned by accountants are less than 3 years old.
   60.94% of laptops owned by accountants are aged 3 years or more.

10. 71.76% of people that took the test for the first time passed.
    85.6% of people that took the test for the second time, passed.
    Based on these results it is more likely to pass the test the second time.

11. $\frac{5}{8} < 64\% < \frac{7}{8}$

## 8.4 Percentages 2

1. a) $60\% = 0.6 = \frac{3}{5}$ b) $36\% = 0.36 = \frac{9}{25}$
   c) $40\% = 0.4 = \frac{2}{5}$ d) $12.5\% = 0.125 = \frac{1}{8}$

2. a) 0.35 b) 0.21 c) 0.186
   d) 0.04 e) 0.002 f) 0.025

3. a) 6% b) 80% c) 0.9%
   d) 85% e) 120% f) 1250%

4. a) 32 m b) £4 c) 4 kg
   d) 25 ml e) 172.2 cm f) 1.5 g

5.

| Fraction | $\frac{1}{5}$ | $\frac{1}{2}$ | $\frac{1}{4}$ | $\frac{3}{4}$ | $\frac{4}{5}$ | 1 |
|---|---|---|---|---|---|---|
| Decimal | 0.2 | 0.5 | 0.25 | 0.75 | 0.8 | 1 |
| Percentage | 20% | 50% | 25% | 75% | 80% | 100% |

6. a) 0.18%, 0.18, $\frac{18}{20}$ b) 34%, $\frac{3}{4}$, 0.8
   c) $\frac{3}{7}$, 69%, 0.7 d) 38%, $\frac{3}{5}$, 3.8

7. a) $\frac{1}{2}$ of 256 b) 95% of 160 c) 2.5% of 400

8. a) $\frac{2}{25}$, = 0.08 = 8 % b) $\frac{63}{250}$ = 0.252 = 25.2 %
   c) $\frac{3.9}{50} = \frac{39}{500}$ = 0.078 = 7.8 % d) $\frac{1.2}{50} = \frac{3}{125}$ = 0.024 = 2.4 %

9. a) $\frac{172}{175}$ b) 98.29%

10. Cycle = 8.4 km ran = 13 km swim = 2.6 km

11. Beth, Charlie, Ella, Dina, Anthony

12. a) i) Divide 0.01 by 2, multiply 0.01 by 2, multiply 0.01 by 4.
    ii) 0.005, 0.02, 0.04

    b) i) Divide 0.25 by 2 to find $\frac{1}{8}$, then multiply the answer by 3 or 7.
    ii) 12.5%, 37.5%, 87.5%

13.

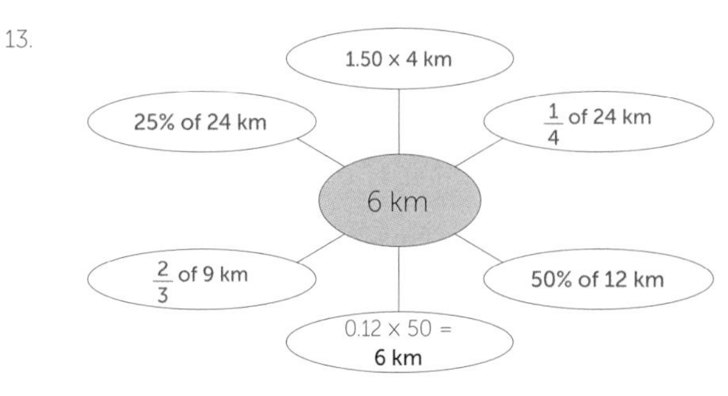


## 8.5 Problems involving percentages

1. 36%
2. 72%
3. 9 hours
4. 35.71 %
5. 76.92%
6. a) The total should be 100%. b) Short stories = 23%
7. 18 out of 20 = 90%
   21 out of 24 = 87.5%
   Evelyn did better in Maths.
8. Saturday
9. 25% of 40% = 10%
10. 36%
11. a) 10% = 31, number = 310 b) 20% = 350. 80% = 1400
12. 20% = travelling. 40% = food and bills
    40% = £200 10% = £50, 100% = £500

## 8.6 Percentage increase

1. a) £144 b) 354 kg c) 6 km d) 187.5 kg
2. £944
3. 20%
4. a) 1.05 b) 1.8 c) 2.5 d) 1.005
5. a) $x = 50$ b) $y = 90$. c) $z = 60$.
6. Will is not correct. 10% of 90 = 9, and 90 + 9 = 99
7. a) e.g. *"Jessie's method is easier to understand and it shows the amount of VAT added, which can be checked on the invoice"* or, *"Tom's method is quicker to calculate"*
   b) £360 × 1.2 = £432
8. 1.1 ÷ 6.42 × 100 = 17.13%
9. £750
10. 1.14m (2 d.p.)
11. Start = 100% Year 2 = 102% Year 3 = 0.98 ×102% = 99.96%

## 8.7 Percentage decrease

1. a) £120 b) 187.5 cm c) 48 kg d) 78 km
2. a) 1.12 b) 0.92 c) 1.15% d) 0.73
   e) 2.35 f) 1.001 g) 0.775 h) 0.9625
3. £17.50.
4. £11 700
5. a) False. 60 decreased by 40% = 60 – (0.4 × 60)
   b) False. 0.35 decreased by 90% = 0.35 × 0.1
   c) False. 110 decreased by 22.5% = 110 × 0.775
6. a) 10% b) 20% c) 9.09% d) 24.55%
7. 230 000 – 104 700 = 125 300 125 300 ÷ 230 000 × 100 = 54.48%
8. a) £1540 b) £1309

9. £106.67 (2 d.p.)

10. Now = £720 Deposit = £240, Ten payments = £580, Total = £820. Paying now is £20 cheaper.

11.

| **Original Price (£)** | 40 | 64 | 80 |
|---|---|---|---|
| **Discount (%)** | 20 | 50 | 60 |

12. The original price of an item is £100.
90% of 100 = 90 90% of £90 = £81.

# Section 9

## 9.1 Measuring probability

1. a) certain b) likely c) unlikely d) impossible

2.

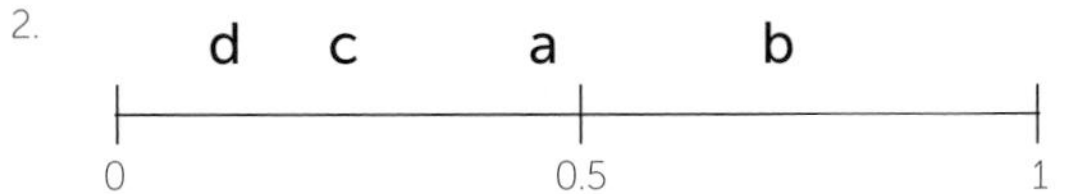


3.

c b d a

0 0.5 1

4. a) 'You will pick a red card' (probability is $\frac{1}{2}$. Probability of throwing a six is $\frac{1}{6}$.)
   b) 'two fair coins will land on heads'

5. Examples of impossible events.

6. Examples of certain events.

7. Equally likely since the probability of the first child born being a girl is 0.5

8. Unlikely since the possibilities are: GG, GB, BG, BB; so actually 0.25

9. **d)** Someone you know will win 'Britain's Got Talent', **b)** You will talk to your Headteacher next week, **c)** The next two cars you see will both be red. **a)** It will rain during the next month.

10. Amy is correct – any of the other numbers could be drawn next.

## 9.2 Listing systematically

1. 1, 2, 3, 4, 5, 6, 7, 8, 9, 10, 11, 12, 13, 14, 15, 16, 17, 18, 19, 20.

2. Lemon/sprinkles, lemon/no sprinkles, blueberry/sprinkles, blueberry/no sprinkles, chocolate/sprinkles, chocolate/no sprinkles

3. AB, AC, AD, AE, BC, BD, BE, CD, CE, DE

4. HHH, HHT, HTH, HTT, THH, THT, TTH, TTT

5. PM, PA, PC, MM, MA, MC, NM, NA, NC

6.

| BB | BG | BY | BR |
|---|---|---|---|
| GB | GG | GY | GR |
| YB | YG | YY | YR |
| RB | RG | RY | RR |

7. a)

| 1,1 | 1,2 | 1,3 | 1,4 | 1,5 | 1,6 |
|---|---|---|---|---|---|
| 2,1 | 2,2 | 2,3 | 2,4 | 2,5 | 2,6 |
| 3,1 | 3,2 | 3,3 | 3,4 | 3,5 | 3,6 |
| 4,1 | 4,2 | 4,3 | 4,4 | 4,5 | 4,6 |
| 5,1 | 5,2 | 5,3 | 5,4 | 5,5 | 5,6 |
| 6,1 | 6,2 | 6,3 | 6,4 | 6,5 | 6,6 |

   b) Possible totals are 2, 3, 4, 5, 6, 7, 8, 9, 10, 11, 12

8.

| 1C | 2C | 3C | 4C | 5C | 6C |
|---|---|---|---|---|---|
| 1D | 2D | 3D | 4D | 5D | 6D |
| 1H | 2H | 3H | 4H | 5H | 6H |
| 1S | 2S | 3S | 4S | 5S | 6S |

9. FFF FFP FFS FPP FSS FPS PSS SPP SSS PPP

10.

| | | **Set A** | | | |
|---|---|---|---|---|---|
| | | 1 | 2 | 3 | 4 |
| **Set B** | 1 | 0 | 1 | 2 | 3 |
| | 3 | 2 | 1 | 0 | 1 |
| | 5 | 4 | 3 | 2 | 1 |
| | 7 | 6 | 5 | 4 | 3 |

## 9.3 Theoretical probability

1. 0.5 or $\frac{1}{2}$

2. 400

3. a) HH, HT, TH, TT b) $\frac{1}{4}$ c) $\frac{2}{4}$ or $\frac{1}{2}$ d) 90 times

4. a) $\frac{5}{8}$ b) 2.25 so 2 times

5. 35

6. a) $\frac{1}{4}$ b) $\frac{2}{4}$ or $\frac{1}{2}$ c) 20 times

7. a) $\frac{2}{6}$ or $\frac{1}{3}$ b) $\frac{3}{6}$ or $\frac{1}{2}$ c) 50

8. a) 2, 3, 4, 5, 6, 7, 8, 9, 10, 11, 12 b) $\frac{1}{36}$
   c) $\frac{3}{36}$ or $\frac{1}{12}$ d) $\frac{6}{36}$ or $\frac{1}{6}$
   e) Dixie is right... they both have a probability of $\frac{2}{36}$ or $\frac{1}{18}$

9. 100

10. No, not a fair dice since that would expect just 100 3's in 600 rolls.

11. a) 400 times b) P(11) = 0.05

## 9.4 Experimental data

1. a) Experiment b) should be around 0.5

2. a) Experiment b) should be around $\frac{1}{6}$
   c) Would expect similar but not the same

3. a) $\frac{10}{60} = \frac{1}{6}$ b) $\frac{30}{60} = \frac{1}{2}$ c) estimate 10 times

4. a)

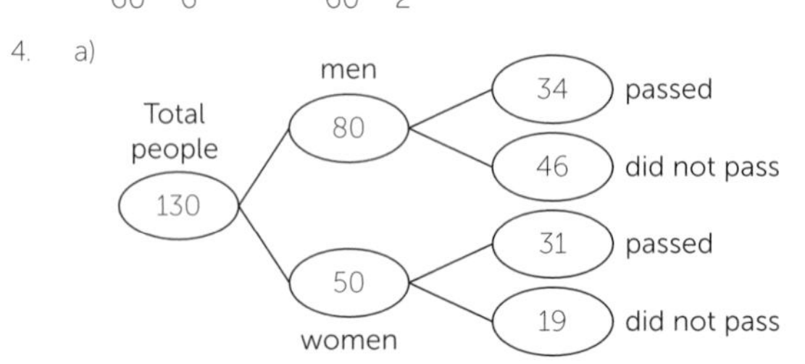


   b) P(man who did not pass) = $\frac{46}{80}$

5. a)

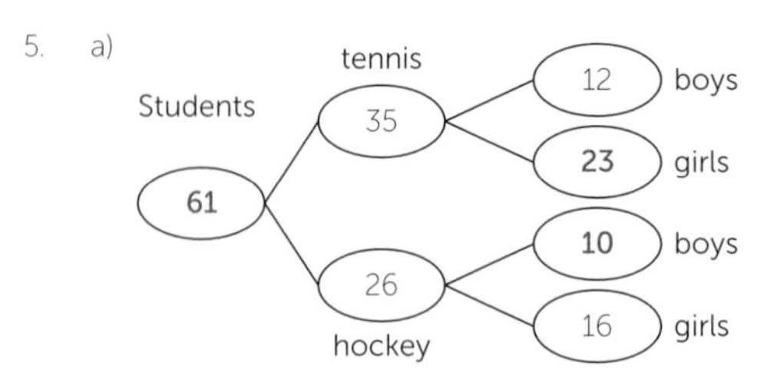


   b) P(tennis) = $\frac{35}{61}$ c) P(boy and hockey) = $\frac{10}{61}$

6. a)

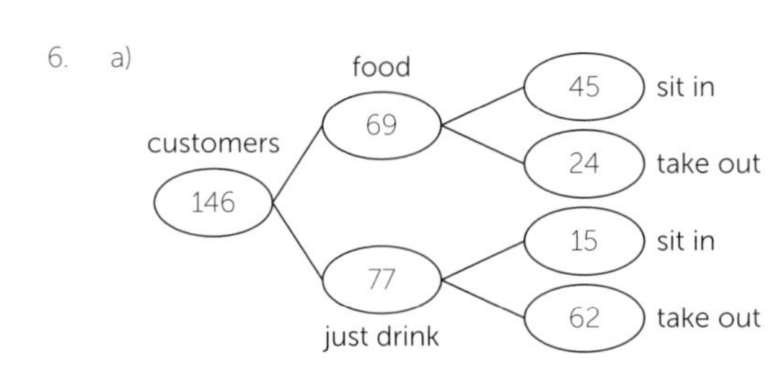


*(Note students may swap over the last two set of branches which is also correct).*

b) P(take out) = $\frac{86}{146} = \frac{43}{73}$ c) $\frac{15}{77}$

## 9.5 Probability experiments

1. 0.485
2. a) 133 b) $\frac{267}{400}$ c) $\frac{133}{400}$
3. a) $\frac{130}{400} = 0.325$ b) Black
4. a) 0.75 b) $(750 + 640) \div 2000 = 0.695$
5. a) 0.15 b) $\frac{1}{6}$ or 0.167
   c) the experimental value is slightly lower. Experimental results are rarely exactly the same as theoretical probability, and the two results are close enough to assume the dice is 'fair'.
6. a) 200 times b) 0.105 c) V
   d) The experimental number for L (10) is very close to what the expected number would be (8).
7. a) Blue 0.25, Yellow 0.35, Green 0.4
   b) 0.19 as it comes from a much larger number of trials.
8. a) 18 b) 0.4

## 9.6 Mutually exclusive events

1. a) not b) Mutually exclusive c) not
   d) mutually exclusive e) not
2. a) $\frac{5}{6}$ b) $\frac{1}{2}$ c) $\frac{48}{52} = \frac{12}{13}$
3. No because there is also the probability of a draw to consider.
4. 0.19
5. a) 0.75 b) 0.45 c) 0.2
6. a) P(tram) = 0.2 b) P(not by car) = 0.8 c) 6
7. a) P(Joy wins) = 0.35 b) P(Kit does not) = 0.85
8. a) P(pip) = 0.3 b) P(Hammy) = 0.35
   c) P(not Nibbles) = 0.65 d) 9 times

## 9.7 Combined events

1. 0.5
2. a) 0.55 b) 0.6
   c) 1; because these are the only choices possible.
3. a) 0.5 b) 0.7 c) 0.2
4. a) 0.65 b) 1; because these are the only choices
5. a) 0.55 b) 0.7
6. $\frac{1}{8}$
7. $\frac{1}{100} = 0.01$
8. $\frac{8}{10} \times \frac{8}{10} = \frac{64}{100} = 0.64$
9. $0.85 \times 0.85 \times 0.85 = 0.614\,125$
10. a) $0.6 \times 0.6 = 0.36$
    b) P(Same colour) = $0.36 + (0.3 \times 0.3) + (0.1 \times 0.1) = 0.46$.
    So, no he is not correct as $0.46 < 0.5$

# Section 10

## 10.1 Ratio notation

1. a) 2 : 1 b) 4 : 3
2. a) 7 : 9 b) 1 : 7 c) 3 : 1 d) 3 : 1
3. a) 3 : 4 b) 2 : 3 c) 4 : 5 d) 2 : 7
4. a) 1 : 2 b) $1 : \frac{1}{3}$ c) $1 : \frac{5}{8}$ d) $1 : \frac{6}{5}$
5. a) 5 : 6 b) $\frac{5}{11}$
6. a) 25 b) $\frac{2}{7}$
7. a) 1: 1 b) $\frac{1}{2}$
8. a) Latin : French = 1 : 2 b) Latin : Italian = 4 : 5
   c) German : Latin = 5 : 2 d) Spanish : French = 6 : 7
   e) Spanish : Latin = $1 : \frac{7}{12}$ f) Latin : Italian = $\frac{4}{5} : 1$
9. a) 16: 24 = 2 : 3 b) 24 : 60 = 2 : 5
   c) 42 : 49 = 6 : 7 d) 84: 56 = 3 : 2
   e) $3 : 7 = 1 : \frac{7}{3}$ f) $5 : 4 = \frac{5}{4} : 1$
10. 11 stamps in total. Grace has 6 stamps and Josh has 5 stamps.

## 10.2 Dividing in a given ratio

1. a) 6 kg : 14 kg b) 80 g : 100 g
   c) 30 cm : 50 cm : 40 cm d) £60: £270 : £120
2. 72 stickers
3. a) 5 sunny days b) 5 more windy days than rainy days
4. 45 teachers
5. 250 g of oats
6. a) 6 : 3 : 1 b) 36 red roses
7. 264 cm
8.

| | Rae | Mo | Lara | Total |
|---|---|---|---|---|
| Small shells | 8 | 4 | 8 | 20 |
| Big shells | 8 | 12 | 20 | 40 |
| Total | 16 | 16 | 28 | 60 |

Rae has 8 small and 8 big shells. Mo has 4 small and 12 big shells. Lara has 8 small and 20 big shells.

9. The width of a rectangle is 10 cm.
   The length of the rectangle = 14 cm. Area = 140 cm$^2$
10. a) 3 is less than half of 7, but 90 is bigger than half of 170.
    b) 31 more blue marbles than pink marbles

## 10.3 Ratio and proportion

1. £14
2. a) 40p b) £3.20 c) £4.40 d) 15 pens
3. a) 24p b) 72p c) £1.44 d) 40 bars
4. 12 year old and 18 year old
5. 300 m tall
6. c) 28 cm
7. £392
8. $(192 \div 16 \times 12) + (128 \div 8 \times 10) = £304$
9. $1710 \div 360 \times 8 = 38$ bags
10.

| School 5 | School 1 | School 3 | School 4 | School 2 |
|---|---|---|---|---|
| 7 : 43 | 3 : 27 | 2 : 23 | 1 : 14 | 1 : 17 |

11. 9 scooters
12. Ratio of A : B : C = 1 : 6 : 6

## 10.4 Solving problems with ratio

1. 5 bananas, 3 apples and 4 pears.
2. 12 : 11 ≠ 3 : 2
3. $\frac{4}{5}$ of 75 = 60 lollies Dan = 10 lollies, Kesh = 20 lollies, Polly = 30 lollies
4. a) 250 ml of water b) 10 rolls
5. a) 350 ml water b) 600 ml cordial
6. $\frac{2}{5} \neq \frac{3}{7}$
7. Keira = £75, Fred = £125. Keira gives Joe £15. Fred gives Joe £75. Keira has £60 left, Fred has £50 left, Joe has £90. The ratio of the money that Keira, Fred and Joe have is 6: 5: 9
8. No. $\frac{5}{7}$ of Max's drink = water. $\frac{2}{7}$ of Lars drink = water.
9. Sandi needs 40 kg of cement, 120 kg of sand and 200 kg of gravel. She doesn't have enough cement.
10.

| | Pink icing | Yellow icing | Total |
|---|---|---|---|
| **Square** | 6 | 18 | 24 |
| **Circle** | 8 | 8 | 16 |
| **Total** | 14 | 26 | 40 |

$\frac{13}{20}$ of the biscuits have yellow icing.

11. Angles in a triangle add up to 180°. The smallest angle should be 30°.
12. There are more boys in the drama club than in the science club.

## 10.5 Ratio as a linear function

1. b) The more builders work the quicker the house will be built.
2. a) and c)
3. Only b) is true.
4. a) and d)
5. The greater the $x$, the smaller the $y$.
6. $x : y = 1 : 4$ with b) $y = 4x$ $y : x = 1 : 4$ with a) $x = 4y$
   $y : x = 1 : 2$ with d) $y = 0.5x$
7. b) 4 : 9 d) 2 : 4.5
8. a) $y$, $O$, $x$, (2,5)
   b) 2.5 Gradient represents ratio of raspberries to strawberries
9. a)

| Number of people | 2 | 4 | 6 | 8 | 10 |
|---|---|---|---|---|---|
| **Cost per meal (£)** | 50 | 100 | 150 | 200 | 250 |

b) 1: 25

## 10.6 Map scales

1. a) 2 cm b) 4 cm c) 0.5 cm d) 1.5 cm
2. a) 10 km b) 24 km c) 32.5 km d) 61 km
3. 31 km
4. 19 cm.
5. a) and ii) b) and i) c) and v) d) and iv)
6. a) 1: 200 000 b) 1: 50 000 c) 1: 30 000 d) 1: 20 000
7. 30 : 600 000 = 1: 20 000
8. a) 700 km b) 495 km c) 315 km
9. 10 cm by 13 cm. The scale could be 1 : 10 000
10. 1 : 10 000. The actual dimensions are 1.2 km and 1.5 km.
    1.2 km = 1200 m = 120 000 cm and 1.5 km = 1500 m = 150 000 cm

# Section 11

## 11.1 Angle properties of lines

1. a) $t$: ∠PQS or ∠SQP $u$: ∠SQR or ∠RQS
   b) $v$: ∠VTW or ∠WTV $w$: ∠VTU or ∠UTV
   c) $x$: ∠ADC or ∠CDA $y$: ∠ABC or ∠CBA
2. a) $x = 64°$ (angle on a line)
   b) $y = 45°$ (angle on a line)
   c) $x = 61°$ (angle on a line)
   d) $v = 58°$ (angle on a line) $t = 58°$ (vertically opposite) $s = 122°$ (vertically opposite)
   e) $q = 117°$ (angle on a line) $p = 63°$ (vertically opposite) $r = 70°$ (angles that meet around a point)
   f) $x = 60°$ (angle on a line) $y = 60°$ (vertically opposite) $z = 120°$ (vertically opposite)
3. a) $a = 73°$ (alternate)
   b) $b = 114°$ (corresponding)
   c) $c = 56°$ (angle on a line) $d = 56°$ (corresponding)
   d) $e = 108°$ (vertically opposite) $f = 108°$ (corresponding)
   e) $g = 46°$ (alternate)
   f) $h = 77°$ (angle on a line)
4. a) $a = 53°$ (angle on a line) $b = 53°$ (corresponding) c = 127° (alternate)
   b) $e = 60°$ (angle on a line) $f = 60°$ (alternate) $d = 33°$ (alternate)
   c) $a = 22 + 15 = 37°$ (alternate angles)

## 11.2 Angle properties of shapes

1. a) $x = 83°$ b) $y = 52°$ c) $z = 58°$
2. a) $p = 102°$, $q = 139°$, $r = 41°$
   b) $s = 105°$, $t = 75°$, $u = 72°$, $v = 75°$
3. a) $d = 32°$, $e = 74°$ b) $f = 52°$, $g = 52°$ (c) $h = 71°$, $i = 71°$
4. 104°
5. 54°
6. Not a regular pentagon as not all the angles are equal. $m = n = 150°$
7. 135°
8. If the given angles are opposite, the others will both be 100°
   If the given angles are adjacent, the others will be 86° and 114° or 46° and 154°
9. Sum of angles = 1440°, so in a regular decagon each angle = 144°
10. 145°
11. 360° divided by 12 = 30°
12. 18

## 11.3 Translations and rotations

1. a) $\begin{pmatrix}4\\3\end{pmatrix}$ b) $\begin{pmatrix}-3\\-5\end{pmatrix}$ c) $\begin{pmatrix}7\\0\end{pmatrix}$
2. a) 6 units in the $x$ direction and −3 units in the $y$ direction.
   b) −8 units in the $x$ direction.
   c) −3 units in the $x$ direction and 4 units in the $y$ direction.
   d) 5 units in the $y$ direction.
3. Translating a shape by $\begin{pmatrix}-1\\7\end{pmatrix}$
4. $\begin{pmatrix}-5\\2\end{pmatrix}$
5. A to B is $\begin{pmatrix}0\\7\end{pmatrix}$, A to C is $\begin{pmatrix}4\\5\end{pmatrix}$, A to D is $\begin{pmatrix}-4\\3\end{pmatrix}$
6.

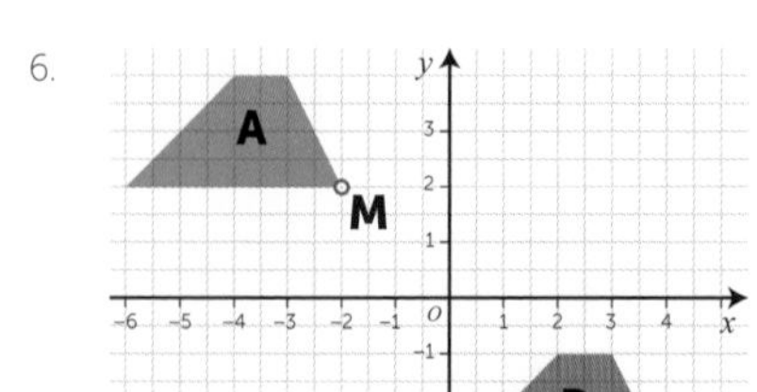


b) $\begin{pmatrix}6\\-5\end{pmatrix}$

7. a) Rotation 90° clockwise, b) Centre of rotation (2, 1)
   c) Rotation 90° anticlockwise d) Centre of rotation (5, 4)

8. a) b) (1, 3), (3, 3), (3, 6) and (1, 6)

9. a) A to B: rotation 180°, centre of rotation (1, 2)
   b) B to C: rotation 90° anticlockwise, centre of rotation (–1, –1)
   c) C to D: rotation 180°, centre of rotation (1, 1)
   d) D to A: rotation 90 degrees clockwise, centre of rotation (4, 4)

## 11.4 Transformations and reflections

1. 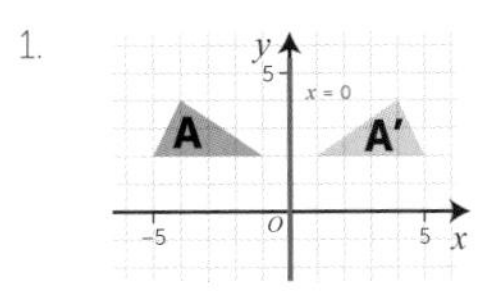


2. 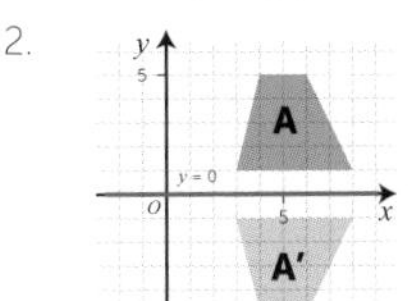


3. 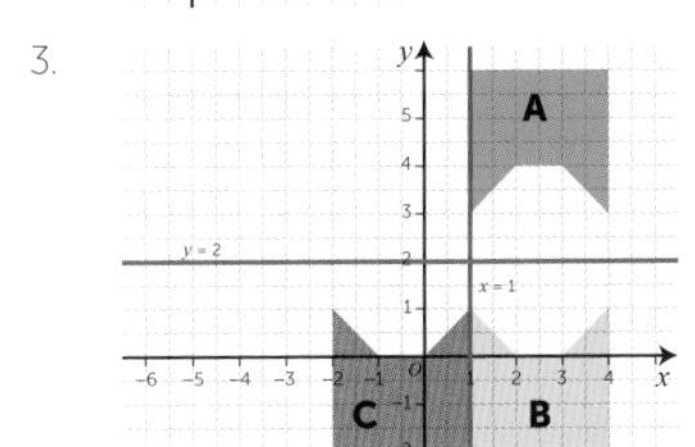

A rotation of 180°, centre of rotation (1, 2)

4. 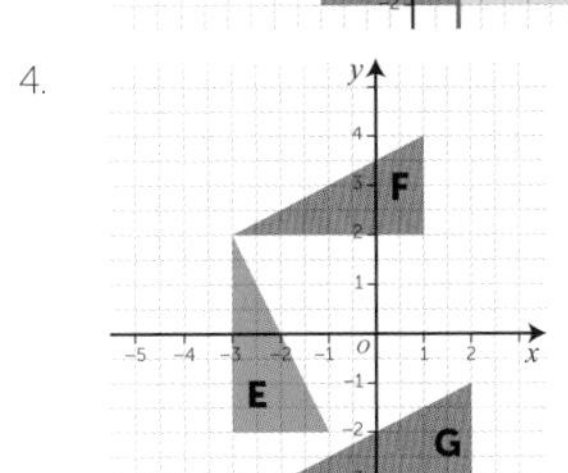

A rotation of 90° clockwise, centre of rotation (0, 0)

5. 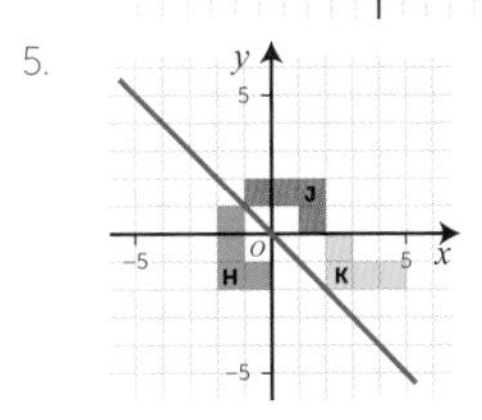

J has coordinates (-1 1), (-1, 2), (2, 2), (2, 0), (1, 0), (1, 1). K has coordinates (2, 0), (3, 0), (3, –1), (5, –1), (5, –2), (2, –2).
Reflection in $x$ = 0, then rotation 90°, centre of rotation (2, –2).

6. 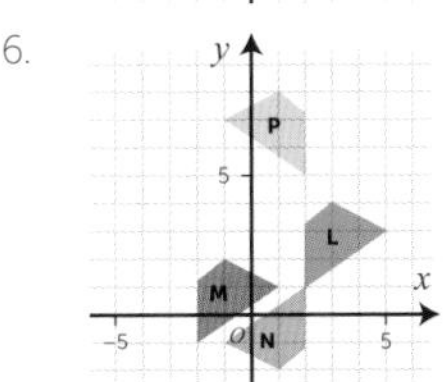

P has coordinates (2, 5), (2, 7), (1, 8), (–1, 7).

## 11.5 Enlargements and similarity

1. Length 96 cm, width 60 cm.
2. 36 cm by 54 cm.
3. Scale factor 3.
4. a) D from C has scale factor 2. b) E from C has scale factor 4.
   c) F from C has scale factor 8. d) F from D has scale factor 4.
5. Length 210 cm, height 80 cm, smaller angles 47°.
6. G to J would have scale factor 8.

7. 2 m, 4 m, 4 m, 6 m

8. a) Not similar since 24/11 ≠ 16/8 b) Similar because 15/5 = 6/2

## 11.6 Further enlargements

1. a) $\frac{1}{4}$ b) $\frac{1}{4}$ c) $\frac{1}{2}$

2. a) 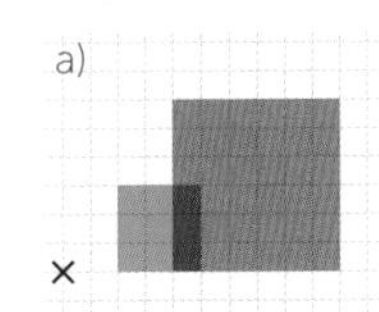 b) 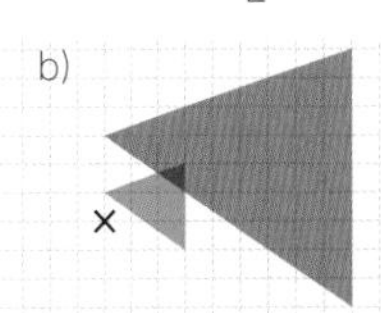 c) 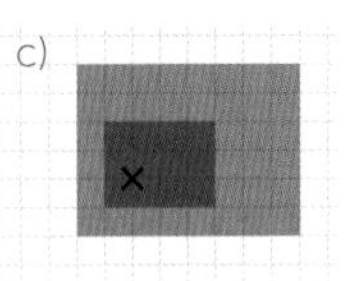

3. 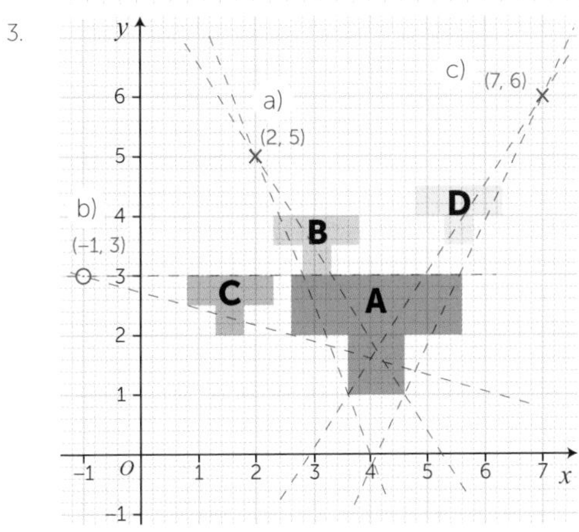


4. 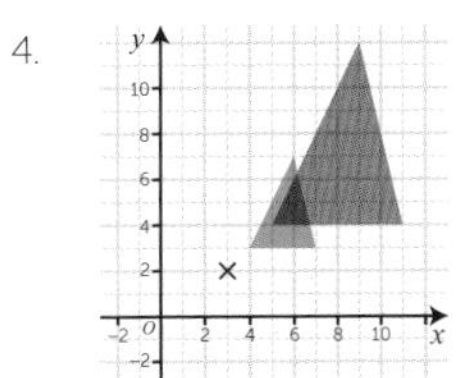
Coordinates of the vertices of the enlargement are: (5, 4), (11, 4) and (9,12)

5. 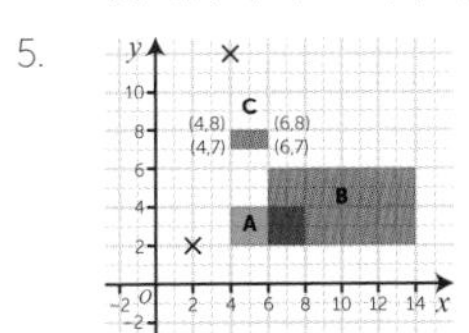

C has coordinates (4, 8), (6, 8) (4, 7) and (6, 7)

6. 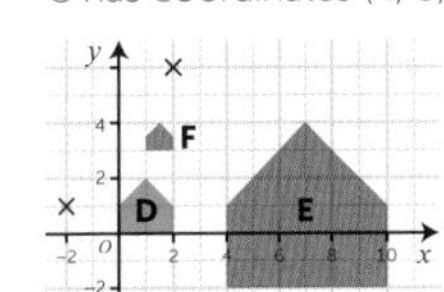

F has coordinates (1, 3), (2, 3), (2, 3.5), (1.5, 4), (1, 3.5)

# Section 12

## 12.1 Introducing sequences

1. a) 5 b) 15 c) 4 d) –3
2. a) 17, 20
   b) The terms are 2 more than the 3 times table, but 30 is not
3. a) + 4 b) + 2 c) – 3 d) + 5
4. a) 19, 23 b) 23, 25 c) 18, 15 d) 7, 12
5. 7, 12, 17, 22, 27, ...

6. Position to term rule is n + 6
(Term 1 = 1 + 6 = 7, Term 2 = 2 + 6 = 8, Term 3 = 3 + 6 = 9, Term 4 = 4 + 6 = 10)
Term-to-term rule is Add 1
Sequence 7 8 9 10
7. No because the term-to-term rule is 'add 5' but the sequence is not the 5 times table
8. 409; term-to-term rule is 8 so we need to add 16 to 393
9. Chloe is correct as long as the input is a whole number; multiplying by 2 always produces an even number then you add an odd number which produces an odd.
10. Both methods are correct but Bryan's will take a long time!

## 12.2 Linear sequences and $n$th term

1. a) 3, 5, 7, 9, 11 b) 2, 5, 8, 11, 14 c) 8, 13, 18, 23, 28
d) 11, 15, 19, 23, 27 e) −4, −2, 0, 2, 4
2. It should be $4n + 1$
3. a) $2n - 1$ b) $3n + 1$ c) $4n + 4$ d) $9n$
4.

| **Term** | 1 | 2 | 3 | 10 | 20 | 100 |
|---|---|---|---|---|---|---|
| **Sequence** | 5 | 12 | 19 | 68 | 138 | 698 |

5. 153 is in the 3 times table but the sequence is one more than the 3 times table
6. a) 0, −5 b) −5 c) −50 d) $-5n + 25$
7. a) $-15n + 115$ b) $-3n + 53$ c) $-4n + 49$
8. £34
9. 14 months
10. 332 terms

## 12.3 Special sequences

1. a) Pattern 4 b) 11 c) no; it will have 13 matchsticks in it
2. a) 26, 37 b) 15, 21 c) 50, 72
3. a) one less than $n^3$ c) double $n^3$ d) half $n^3$
4. a) Pattern 4 b) 31 c) 16
5. a) $n^2 + 2$ b) $n^2 - 3$ $5n^2$
6. a) 15 b) 78
7. a) 14 b) 32 c) Yes, because the sequence is the three times table add 2 and 48 is in the three times table
8. a) 18 (14 white and 4 black) b) 96 (30 black and 66 white)
c) pattern number 54
9. There are always four more white than black tiles
10. a) 18 b) 102 c) pattern 14

## 12.4 Geometric sequences

1. a) 20, 24 b) 64, 128 c) 8, 13
d) 6.25, 3.125 e) 81, 243
2. a) + 4 b) × 2 c) add 2 previous terms
d) ÷ 2 e) × 3
3. a) geometric b) arithmetic c) neither (Fibonacci)
d) geometric e) neither
4. The term-to-term rule is the same, but it is 'multiply by 2'
5. a) 14, 19 b) 7, 11, 19 c) 17, 11 d) −1, 3, 11
6. a) 4, 16 b) 30, 15 c) 6, 54, 486 d) 8, 32, 128
7. 324
8. 62
9. 2.5
10. Kate: 3072 Dan: 87

## 12.5 Quadratic sequences

1. A, C, D
2. Multiplied by 2 instead of squaring
3. a) 5, 8, 13, 20, 29 b) 0, 3, 8, 15, 24
c) 2, 8, 18, 32, 50 d) 0, 2, 6, 12, 20
4. A, B and D are quadratic
5. 2, 5, 10, 17; the differences are 3, 5, 7 which go up in 2, so it is a quadratic sequence. Or, sequence is $n^2 + 1$
6. a) 3, 8, 15, 24, 35 b) 3, 9, 19, 33, 51 c) 0, 7, 18, 33, 52
7. She has multiplied by 3 and then squared which is the wrong order. The 1st term should be 5.
8. b, c; both have a difference pattern which goes up by a constant
9. Yes; the differences go up in 4s
10. Yes; the differences go up in 1s

## 12.6 Inequalities

1. a) > b) < c) < d) =
2. 3.4 is not an integer; the answer is 3
3. d) $x > -3$
4.

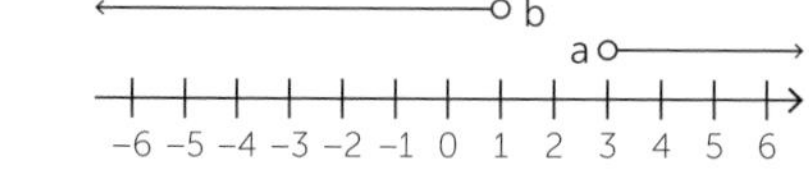


5. a) 4 b) −4 c) −2
6. a) 1 b) −4 c) −3
7. a) −2, −1, 0, 1, 2, 3 b) $-2 \le x < 4$
8. He has missed 0 and should not have included 4
9. a) 3, 4 b) 0, 1, 2, 3 c) 0, 1,2,3 d) −5, −4, −3, −2
10. a) 1, 2, 3 b) 2, 3, 4 c) 2, 3, 4 d) −1, 0, 1, 2, 3

## 12.7 Solving linear inequalities

1. a) $x > 2$ b) $x \le 8$ c) $x \ge 5$
d) $x < 50$ e) $x < 1$
2. She should have subtracted 4 from each side to get $y < -5$
She also gave her answer incorrectly; it should be −6
3. 5
4. a) $x \le 5$ b) $x > 6.5$ c) $x > 3.5$ d) $x < 33$
5. The inequality sign is the wrong way around and not negative; solution should be: $6 \ge 2x$ giving $x \le 3$
6. a) $5 < x \le 10$ b) $-9 < y < -4$ c) $1 \le n < 4$ d) $-2 \le n \le 2$
7. a) $n \le 2$ b) $n > -4$ c) −3, −2, −1, 0, 1, 2
8. $2.7x \le 75$ $x \le 27.777...$ so the max integer value of $x$ is 27
9.

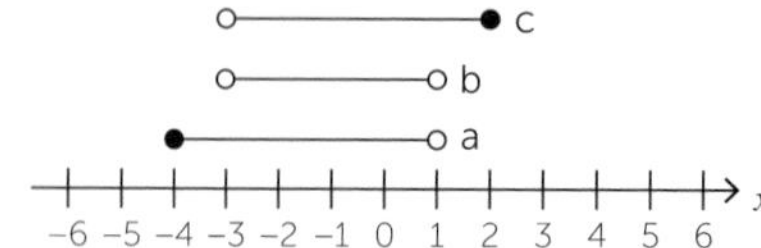


10. $30x + 10 \le 500$ so $x \le 16.333...$ Hence the van can carry 16 boxes.

# Section 13

## 13.1 Conversion of currency

1. 945 francs
2. 4125 rand
3. a) £1 = 1.50 US Dollars b) £80

4. A
5. D
6. B
7.

| rupees | 0 | 800 | 1600 | 2000 | 3200 | 4000 |
|---|---|---|---|---|---|---|
| yen | 0 | 1300 | 2600 | 3250 | 5200 | 6500 |

8. a) £5 = €6

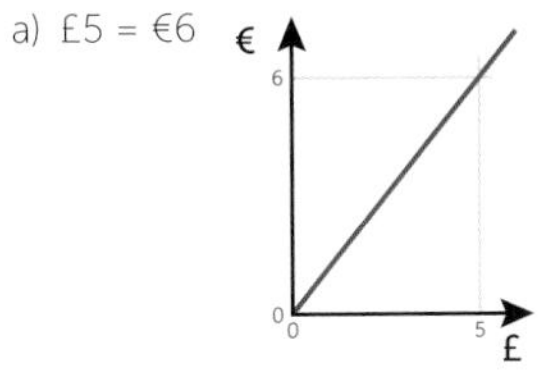


b) £150 = €180 £100 = €120
c) 1.2
d) £1 = €1.2

9. Bank B. For the same amount of £ he will get more Euros.
10. a) £195.62 b) £4

## 13.2 Best buy problems

1. 6 bottles of orange juice for £4.80
2. Deal 3, Deal 2, Deal 1.
3. £7.50 for 6 kg of apples
4. a) Offer A – 1 pencil for 15p, Offer B – 1 pencil for 18p
   b) Offer A
5. a) They are both correct. They both get the correct answer.
   b) £3.91 for 1 kg of grapes.
6. For up to 60 minutes Phones abroad is better value as you get more minutes for the same amount of money.
7. £4.16
8. a) False. Beaumont Road car park is cheaper if you park the car for more than 5 hours.
   b) False. They are the same price.
   c) False. It is a fixed price.
   d) False. City car park is cheaper if you park the car for less than 5 hours.
   e) False
9. 10 boxes for £3.70

## 13.3 Standard units

1. a) 1 m = 100 cm b) 2 km = 2000 m
   c) 3000 ml = 3 litre d) 6400 mm = 6.4 m
   e) 10.1 kg = 10 100 g f) 10 cl = 0.1 litre
2. a) 1 t = 1000 kg b) 1 m = 1000 mm c) 100 cl = 1 litre
3. a) mm b) cm c) kg
4. 1 km = 1000 m 1 l = 1000 ml 1 m = 1000 mm
   1 m = 100 cm 1 g = 1000 mg 1 t = 1000 kg
5. a) 3890 mg, 259 g, 0.6 kg, 0.07t
   b) 67 mm, 12.6 cm, 890 mm, 0.89 km
   c) 0.7 l, 71 cl, 790 ml, 7.09 l
   d) 0.01 cl, 1 ml, 10 ml, 0.1 l
6. a) True. b) False. 1 kg = $\frac{1}{1000}$ of 1 tonne
   c) True. d) False. 1 cm = $\frac{1}{100}$ of 1 m
7. a) 2.4 m + 50 cm < 300 cm
   b) 2250 m – 1.5 km = 750 m
   c) 489 g + 568 g < 1 kg 580 g
   d) 2 litre – 348 ml < 1 litre + 3480 ml
8. b) 270 l
9. £7.50
10. 25.5 cups
11. 1 cm = 10 mm Both cubes have the same volume.
12. a) 3.628 litres b) 4.7 m

## 13.4 Units of time

1. a) 120 seconds = 2 minutes b) 3 days = 72 hours
   c) 0.5 hours = 30 minutes d) 43 years = 4.3 decades
2. a) 342 seconds b) 216 minutes c) 108 hours d) 380 years
3. a) 150 minutes = 2 hours 30 minutes
   b) 400 seconds = 6 minutes 40 seconds
   c) 40 hours = 1 day 16 hours
   d) 50 days = 7 weeks 1 day
4. a) 3 hours 20 minutes
5. a) 1000 days > 15 months
   b) 10 days 10 hours = 250 hours
   c) 2800 milliseconds < 28 seconds (2800 ms = 2.8 seconds)
   d) 10 millennia = 10 000 years
6. a) To convert 100 hours to days, divide by 24.
   b) To convert 100 days to weeks, divide by 7.
   c) To convert 100 centuries to years, multiply by 100.
   d) To convert 100 millenniums to years, multiply by 1000.
7. a) £130 b) $7\frac{1}{3}$ hours or 7 hours 20 minutes.
8. 23 minutes 23 seconds
9. Bolt takes 9.58 ÷ 60 = 0.16 minutes, not hours, to run 100 m.
10. 3 hours 47 minutes
11. 1 decade = 10 years
    $\frac{1}{24}$ of a day = 1 hour or, $\frac{1}{60}$ of a day = 0.4 hours = 24 minutes

## 13.5 Metric and imperial conversions

1.

| Length | Mass | Volume |
|---|---|---|
| Feet | Ounces | Pints |
| Inches | Pounds | Gallons |
| Miles | Stones | |
| | | |

2. a) pint b) ounce c) stone
3. a) 1 stone > 1 kilogram b) 1 inch >1 cm c) 1 metre > 1 foot
   d) 1 litre > 1 pint e) 1 ounce >1 g f) 1 mile > 1 kilometre
4. a) 12.5 miles b) 320 km c) 1.6 km
5.

| 1 litre | 5 litres | 10 litres | 25 litres | 50 litres |
|---|---|---|---|---|
| 1.76 pints | 8.8 pints | 17.6 pints | 44 pints | 88 pints |

6. a)

feet
70
60
50
40
30
20
10
0
10
20
metres

b) 15.2 metres c) 19.7 feet.

7. 4.2 × 4.55 = 19.11 litres 19.11– 6.5 = **12.61 litres**
8. a) 1.60 m (2 d.p.) b) 63.09 inches (2 d.p.)
9. £2.72 + £1.02 = £3.74
10. Mia's car drives 72 km for 4.55 litres. 100 km = 6.32 litres. Yes Mia does have enough petrol in her car.

## 13.6 Direct proportion problems

1. £2475
2. a) True. b) False c) True.
3. d) 32p
4. **A** The ratio of the 2 sides of all the other triangles is 1.5.
5. 160 g butter 200 g sugar 4 eggs 176 g flour
6. No. It doesn't start from (0,0).
7. Costs €124.80 in UK so cheaper to buy in UK.
8. 2019 school trip: 40 students: 4 teachers = 10 students: 1 teacher
   2020 school trip: 300 students: 15 teachers = 20 students: 1 teacher
9. 15 pancakes
10. a) €504.36 b) £442.42
11. 1 litre of milk = 1.76 pints.
    1 pint of milk in Copenhagen costs 5.681 krone = 67.6p
    One pint of milk in UK is 64p, so cheaper than Copenhagen.

# Section 14

## 14.1 Categorical and discrete data

1.

| Colour | Tally | Frequency |
|---|---|---|
| Blue | 𝍸 I | 6 |
| Red | 𝍸 | 5 |
| Green | III | 3 |
| Pink | II | 2 |
| Orange | IIII | 4 |

2.

| Team | Tally | Frequency |
|---|---|---|
| **MU** (Manchester United) | 𝍸 III | 8 |
| **MC** (Manchester City) | 𝍸 | 5 |
| **C** (Chelsea) | 𝍸 | 5 |
| **A** (Arsenal) | II | 2 |
| **L** (Liverpool) | II | 2 |
| **O** (Other) | II | 2 |

Manchester United are the most popular team.

3. a) 2 b) 2 c) 24
4. a) 7 b) 4 c) 27
5.

£0

£5

Key:

represents 3 students

£10

£15

£20

£25 +

6. Jennifer could give options to use a tally.
   Jennifer doesn't need the name of the respondents
7. a) 12 b) 5 c) 46
8. Kenny should add an 'Other' option for those who don't like those options. Kenny should do 𝍸 to make his tally easier to count
9. a) £3 b) £4 c) £42
10. Gina needs a key. Gina's envelopes ought to all be the same size

## 14.2 Data, charts and graphs

1.

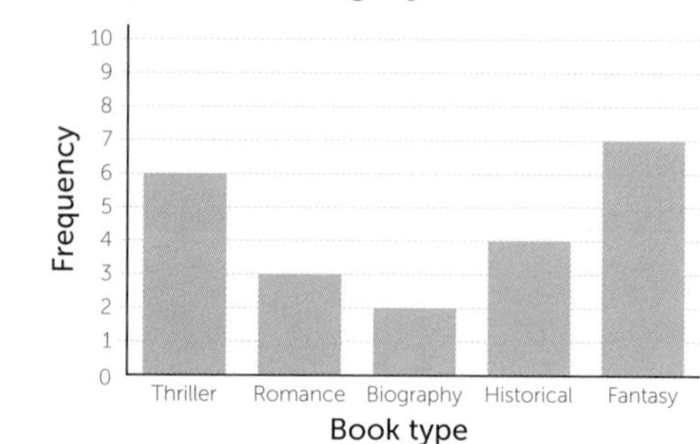


2. a) 24
   b)

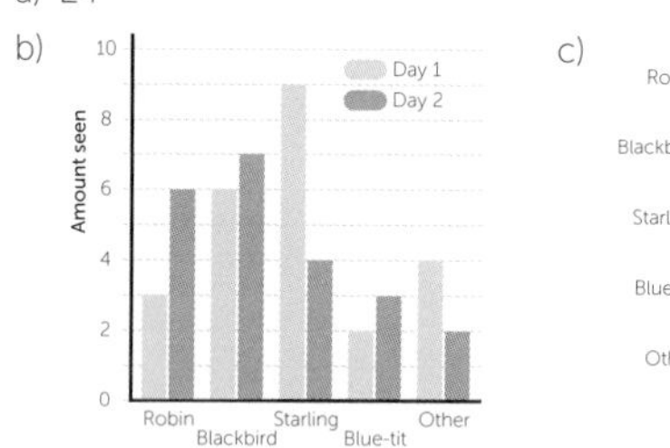


c)

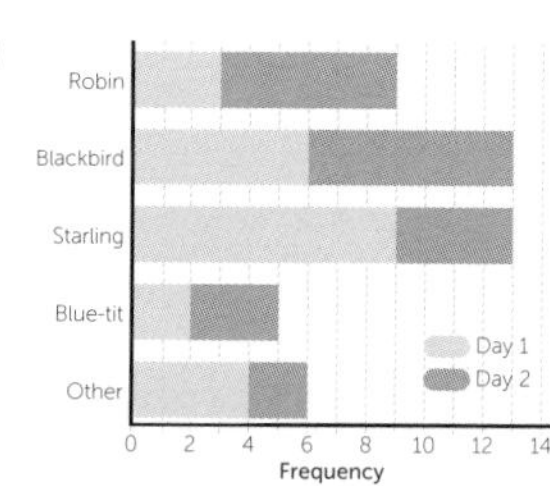


3. 9°
4. a) 12 students,
   b) $\frac{80}{360} \times 72$ students = 16 so 4 extra students come by bike.
5. a) Cheese b) 6 c) Cheese d) 16
6. 72
7. 65 people
8. 25 people
9. a) 18 students b) 14 students c) 3 students
   d) 10F: 31 students; 10G: 33 students; 10G has more students
   e)

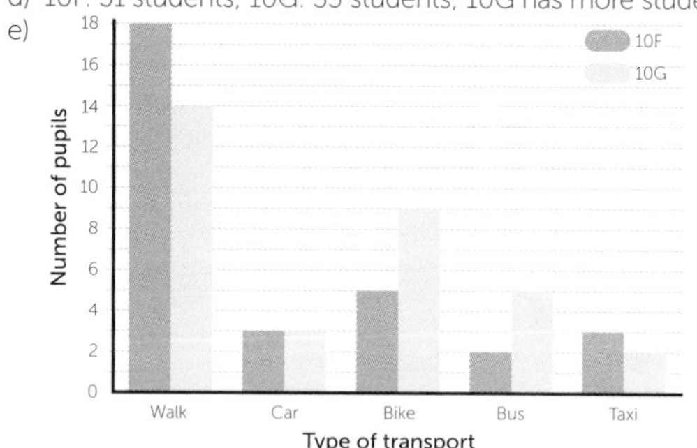


10.

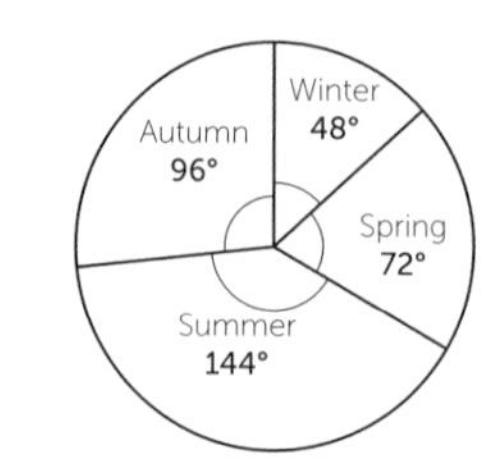


## 14.3 Continuous data

1. a) for example, 1 - 5, 6 - 10, 11 - 15 and 16 - 20
   b) for example, 1 - 25, 26 - 50, 51 - 75, 76 - 100
   c) for example, 321 - 340, 341 - 360, 361 - 380, 381 - 400
2. a) for example, $20 \leq x < 40$, $40 \leq x < 60$, $60 \leq x < 80$, $80 \leq x < 100$
   b) for example, $0 \leq x < 15$, $15 \leq x < 30$, $30 \leq x < 45$, $45 \leq x < 60$
   c) $300 \leq x < 305$, $305 \leq x < 310$, $310 \leq x < 315$, $315 \leq x < 320$
3. The groups have different widths, some 20 and some 10.
   Values of 20, 30 40 and 50 are ambiguous - they can fit into more than one group.
   Better groups are:
   $0 \leq x < 20$, $20 \leq x < 30$, $30 \leq x < 40$, $40 \leq x < 50$, $50 \leq x < 60$ and $60 \leq x < 70$

4.

| Height ($h$ centimetres) | Tally | Frequency |
|---|---|---|
| $140 \leq h < 150$ | ~~||||~~ \| | 6 |
| $150 \leq h < 160$ | \|\|\| | 3 |
| $160 \leq h < 170$ | ~~||||~~ | 5 |
| $170 \leq h < 180$ | \|\|\|\| | 4 |
| $180 \leq h < 190$ | \|\| | 2 |

5.

| | Canteen | Packed Lunch | Total |
|---|---|---|---|
| **10A** | 19 | 13 | 32 |
| **10B** | 20 | 10 | 30 |
| **Total** | 39 | 23 | 62 |

a) $\frac{2}{3}$ b) 37%

6.

| | Sketching | Painting | Totals |
|---|---|---|---|
| **School A** | 18 | 17 | 35 |
| **School B** | 23 | 17 | 40 |
| **Total** | 41 | 34 | 75 |

51.4% attended the sketching workshop

7.

| | Canoeing | Windsurfing | Surfing | Total |
|---|---|---|---|---|
| **Girls** | 12 | 16 | 12 | 40 |
| **Boys** | 14 | 11 | 15 | 40 |
| **Total** | 26 | 27 | 27 | 80 |

59.3% windsurfing are girls

## 14.4 Mean, mode, median and range

1.

| | Mean | Median | Mode | Range |
|---|---|---|---|---|
| **a.** | 5.9 | 6 | 3 | 6 |
| **b.** | 20.8 | 21 | 21 | 6 |
| **c.** | 36.9 | 37 | 39 | 9 |

2. She needs to press "equals" before dividing or put the addition in brackets. Answer should be 12.4
3. a) 6 b) 16 c) 8.5
4. Mode; the data is not numeric, so you cannot use the other averages
5. Mean: 4.9 Median: 5 Mode: 6 Range: 4
6. Grant needs to get at least 78%
7. Ian should divide by 20, not 5; answer should be 1.5
   Stephanie's answer should be 1 because that has the largest frequency.
   David's answer should be 1 because the halfway point of the frequency is in that group.
   Susan's answer should be 4 because you subtract the smallest number of meals out from the largest
8. Clare's mean (52.8) is larger than Vanessa's (46) but Vanessa's median (48) is larger than Clare's (13).
9. 7 goals
10. $a = 12$, $b = 13$ and $c = 20$

## 14.5 Estimating averages

1. a) 15 b) 6 c) 15 d) 22.5 e) 22.5
2. a) $30 < t \leq 40$ b) $20 < t \leq 30$ c) 5, 15, 25, 35, 45
   d) 24.5 e) 50
3. a) $30 < t \leq 60$ b) $20 < t \leq 30$ c) 1680 minutes d) 56 minutes
4. a) The answer should be within the range £0 to £50.
   $\frac{5 \times 6 + 15 \times 7 + 25 \times 4 + 35 \times 1 + 45 \times 2}{20} = 18$
   b) The frequency goes over halfway in the $10 \leq m < 20$ class.
   c) Matthew could have found the difference between the largest value in the highest group and the largest value in the smallest group.
5. 52.8 minutes
6. £1075

## 14.6 Statistical diagrams and tables

1. a) 2 b) 4 c) 0 d) 4
2.

| | |
|---|---|
| 1 | 7 9 |
| 2 | 0 5 6 7 8 8 8 9 |
| 3 | 1 2 2 6 |
| 4 | 1 |

Key: 1 | 7 = represents 17 years old

3. a) 1 year old b) 51 years old c) 50 years
   d) 21 years old e) 21 years old
4. a) 4 b) 4 c) 3 d) 58 e) 2.9
5. a) 37 b) 51 c) 22 d) 33 e) 31
6. a) 1 b) 1.04
7. a) 12.9 seconds b) 16.1 seconds c) 3.2 seconds
   d) 14.7 seconds e) 14.1 seconds
8. a) 1.77 brothers/sisters b) 1.39 brothers/sisters
   c) 1.59 brothers/sisters

## 14.7 Comparing populations

1. 11B have a higher median but are less consistent than 11A as their range is larger.
2. 9T's estimated mean spend was approx. £11.50.
   9T's range: £21. 9T spent less, on average and their range is smaller.
3. a) The modal class for 10R is $10 \leq x < 20$ and for 10S is $20 \leq x < 30$
   b) 10R: 21 minutes 10S: 26.6 minutes c) 10S
4. a) 4 for both b) 4 for both
   c) 10F: 4.1 and 10G: 3.74 d) 10F: 4 and 10G: 3
   e) The median and mode are the same for both classes but the mean for 10F is higher. The range for 10F is also larger .
5. a) Road A: 30mph; Road B: 23mph
   b) Road A: 28mph; Road B: 32mph
   c) Road A has a higher median and more cars over 30mph

# Section 15

## 15.1 Quadrilaterals

1. a) Square, rhombus b) Square, rectangle, rhombus, parallelogram
2. Perimeter 34 cm, area 66 cm$^2$
3. 28 cm
4. 4.5 cm
5. 75°, 105°, 105°
6. 612 cm$^2$
7. 49 cm$^2$
8. 98°
9. 48 cm
10. 105°
11. 56
12. a) 24 cm$^2$ b) 8 cm c) 12 cm

## 15.2 Tessellations

1. Yes, because each interior angle is 120° and 3 × 120° = 360°
2. No, because each interior angle is 108° which is not a factor of 360°
   Quadrilaterals and triangles will tessellate, and some non-regular polygons.
3. a) $a = 80°$ b) $b = 133°$
4. Yes. An octagon has interior angle 135° and a square has 90°, so they will tessellate by joining 2 octagons and a square at each point since 135° + 135° + 90° = 360°
5. Ray is correct. Any quadrilateral will tessellate since it can be split into 2 triangles and all triangles tessellate.

6. a) (−5, 5) b) (2, 1) c) (2, -3) d) (−3, −2)
7. a) (−5, 3) or (−1, 3) b) (0, 1) c) (9, 5) d) (−2, −1) e) (−6, −2)

## 15.3 Solving geometrical problems

1. a) (2, −2) b) Possible coordinates (10, 0) or (0, 6)
2. a) $g = h = i = 60°$ b) $j = k = 76°$
3. $m = 70°$, $n = 74°$
4. 95° (35° + 60°)
5. $p = 158°$, $q = 120°$ Isosceles triangle, equilateral triangle, angles on a straight line
6. a) $d = 37°$ (alternate angles), $e = 66°$ (corresponding angles)
   b) $i = 41°$ (alternate angles), $k = 76°$ (corresponding angles), $j = 76°$ (alternate angles), $h = 63°$ (angles in a triangle = 180°)

## 15.4 Circumference

1. a) 44.0 cm b) 157.1 cm c) 102.4 m d) 1477.8 mm
2. a) 37.7 cm b) 142.3 m c) 259.5 mm d) 451.4 cm
3. 40 212 km
4. 21.0 cm
5. a) 6.3 cm b) 8.2 cm c) 5.3 cm d) 7.5 cm
6. 1055.6 cm
7. 25.2 m
8. a) 1979.2 mm b) 4750 m
9. 94 m
10. 53.6 cm
11. a) 219.9 mm b) 172.8 mm c) 52.2 mm
12. $(3\pi + 6)$ cm

## 15.5 Circle areas

1. a) 88.2 cm² b) 1779.5 mm² c) 94 024.7 cm² d) 4026.4 m²
2. a) $1024\pi$ cm² b) $64\pi$ m² c) $361\pi$ mm²
   d) $103\,041\pi$ cm²
3. a) 3.9 cm b) 7.2 m c) 9 mm d) 27 m
4. a) 10.2 m b) 20.4 mm c) 26 cm d) 64 m
5. a) 3.1 cm² b) 5.3 cm² c) 2.3 cm² d) 4.5 cm²
6. 1809.6 cm²
7. 46.6 m²
8. $(5 + \frac{25}{2}\pi)$ cm²
9. $(64 - 16\pi)$ cm² = 13.7 cm²
10. $(\frac{225}{4}\pi - 108)$ cm²

## 15.6 3D shapes

1. Square-based pyramid, for example in Egypt.
2. a) Triangular prism has 5 faces, 9 edges and 6 vertices.
   b) Hexagonal prism has 8 faces, 18 edges and 12 vertices.
3. Two circular faces. One curved surface.
4. Pentagonal pyramid, it has 6 faces, 10 edges and 6 vertices.
5. a) 864 cm³ b) 648 cm²
6. Cuboid has 6 faces, 12 edges and 8 vertices.
   Tetrahedron has 4 faces, 6 edges and 4 vertices.
   Octagonal prism has 10 faces, 24 edges and 16 vertices.
7. They all have curved surfaces. Cylinders have 2 faces, cones have one face and a sphere has none.
   A cone has one edge and one vertex. A cylinder has 2 edges.
   Cylinders and spheres have no vertices.
8. a) Cone, cylinder, sphere
   b) cuboid, or pentagonal-based pyramid
   c) e.g. triangular or pentagonal prism

## 15.7 Planes of symmetry and nets

1. a) 9 b) 4 c) 13 d) infinite number
2. a) square, rectangle b) equilateral triangle and rectangle
   c) regular hexagon and rectangle d) rectangle and circle
3. Circle
4. a) e.g.

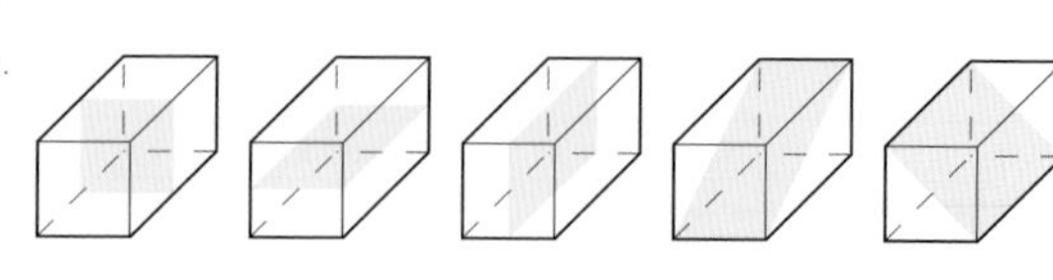

b) e.g.

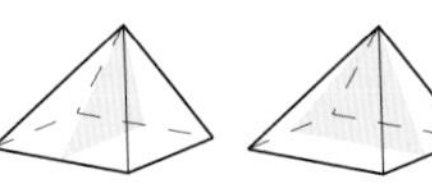
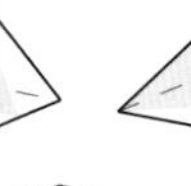
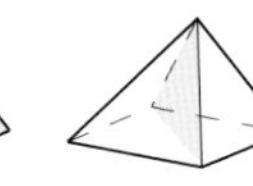

c) e.g.

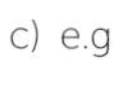
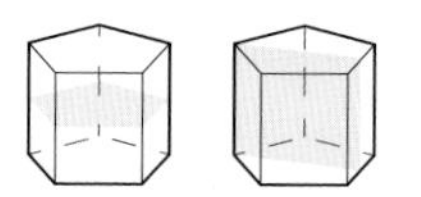

d) e.g.

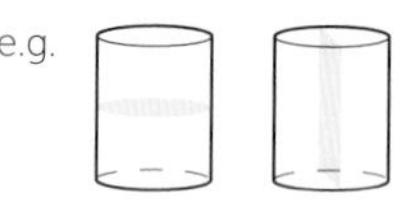

5.

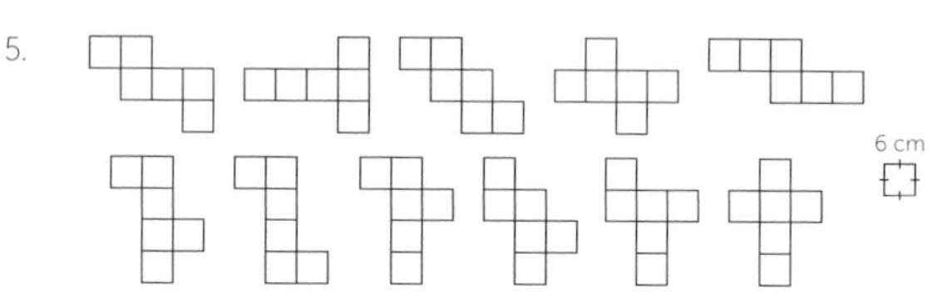


6. a)

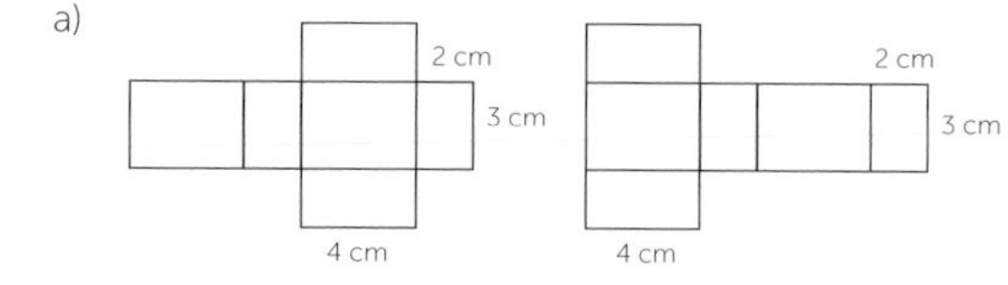


b)

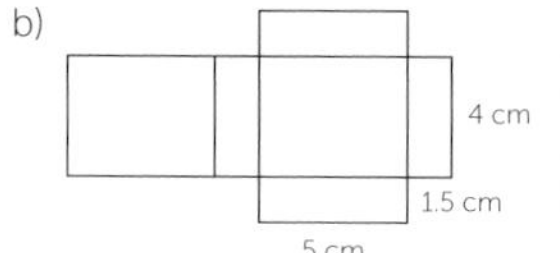

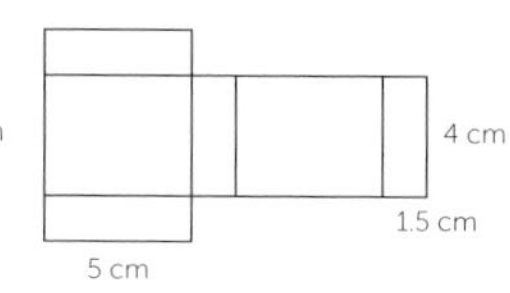


7.

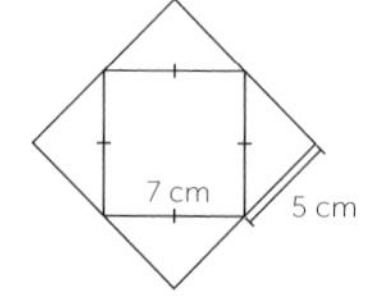


8.

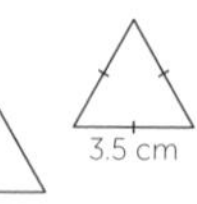


9. a)

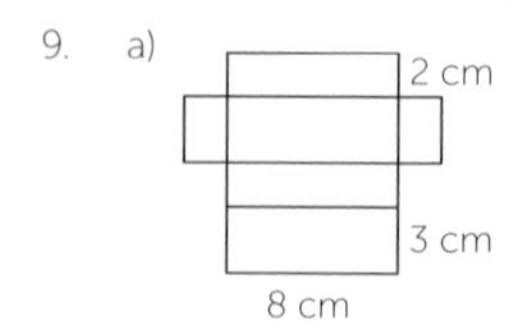

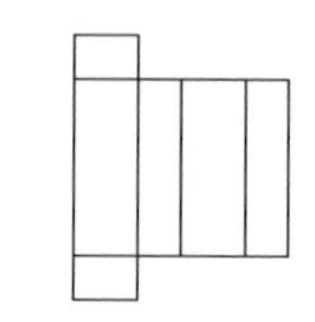

b)

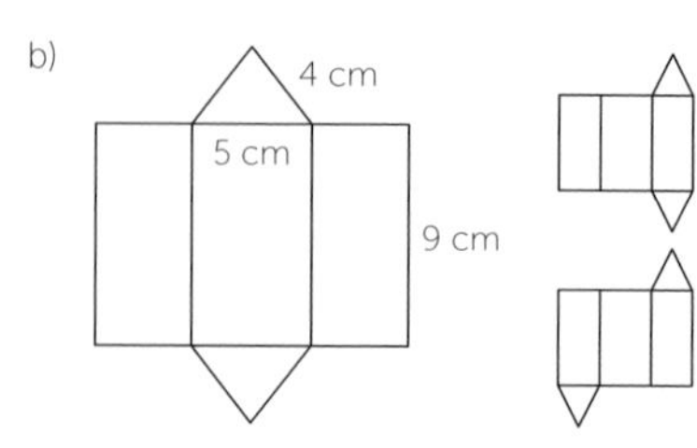

c)

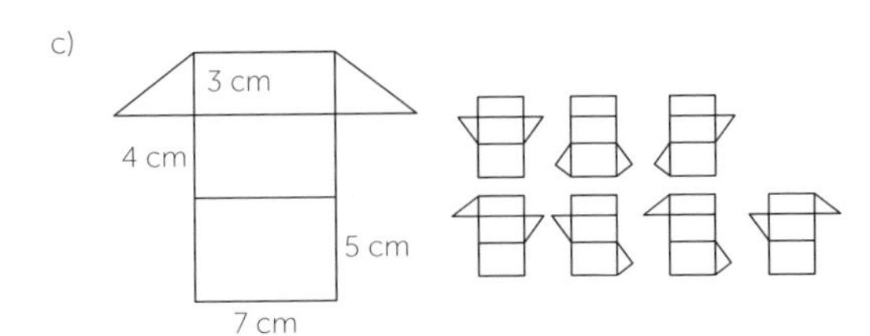


d)

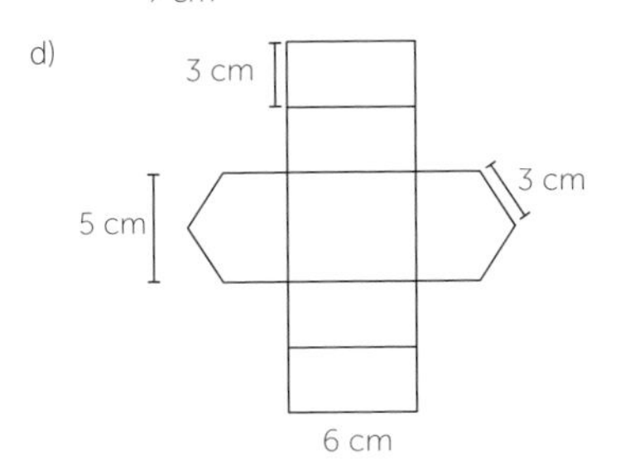


Some other possible nets

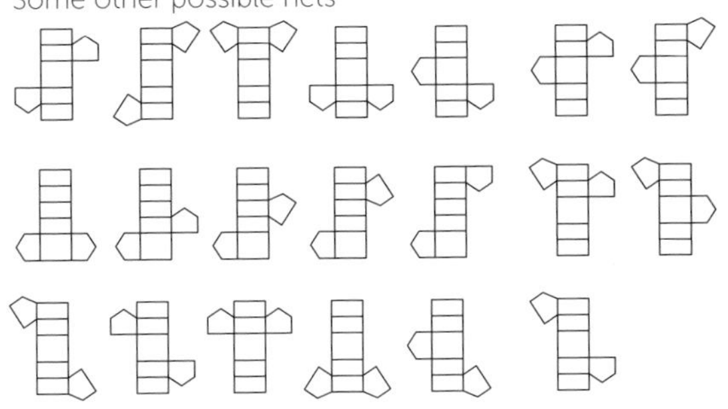

10. a)

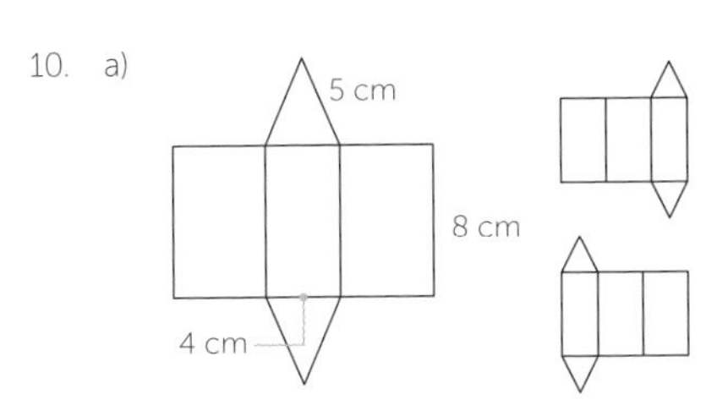


b) 4.6 cm

c) Rectangle 4.6 cm by 8 cm and triangle with 2 sides of 5 cm and base 4 cm.

# Section 16

## 16.1 Dealing with money

1. £22.40
2. £39 330
3. £366
4. 10%
5.

| Item | Packet of cards | Book | Collectable football cards |
|---|---|---|---|
| Buying price | £1 | £6 | £3 |
| Selling price | 50p | £4 | £3.50 |
| Percentage profit or loss | 50% | 33.33% | 16.67% |

6. No. The price of the ribbon is £6.40.
7. Ali has the most money in the account. £19.98 more.
8. a) £24.36 b) £604.36
9. a) £29.25 b) 4.5%
10. The uniform costs £20. It is reduced by £60. This is 75% off.
11. 12 x £16.50 = £198

    $\frac{198 - 185}{185} \times 100 = 7\%$ profit
12.

| | December | January |
|---|---|---|
| Expenses | £100 | 100 |
| Sales | £132 | 92 |
| | 32% profit | 8% loss |

13. Total costs = £32.50
    Sales = 60 x 90p + 40 x 63p = £79.20
    £79.20 - £32.50 = £46.70 profit. £46.70 ÷ 32.50 x 100 = 143.7%

## 16.2 Financial mathematics

1. Martin earns £153 interest in 1 year
2. D £290
3. a) £40.70 b) 2.2%
4. B 520 x 1.2 and D 520 + 520 x 0.2
5. Shop = £96 Online = £93.99 Online is cheaper
6. a) £32.30 b) £882.30
7. a) 20% of (28 000 – 12 500) = £3100 b £2075
8. £1000
9. Tax should be calculated as:
   20% tax rate: £50 000 - £12 500 = £37 500 20% of £37 500 = £7500
   40% tax rate: £62 500 - £50 000 = £12 500 40% of £12 500 = £5000
   Tax paid is £7500 + £5000 = £12 500
10. Interest per year - 3.65% x 1500 = £54.75
    £54.75 x 5 = £273.75
    3 months = $\frac{1}{4}$ of a year Interest after 3 months = £13.69
    £273.75 + £13.69 = £287.44
    He takes out £1787.44

## 16.3 Speed

1. 66 km/h
2. a) 6 m/minute b) 0.1 m/s
3. 364 km
4. a) 900 m/minute b) 3000 m/hour c) 4.8 km/h
   d) 0.2 km/minute e) 166.67 m/minute
5. D is the odd one out. It is the only one that is 180 km/h or 3000 m/minute.
   All the other values are 18 km/h or 300 m/minute.
6. Lucy finds the distance travelled in 60 minutes. Anthon uses the formula speed = distance ÷ time.
7. Time taken = $2 \div 4.2 = \frac{10}{21}$ hours $= \frac{10}{21} \times 60$ minutes ≈ 29 minutes
   Jane will arrive just before 10 am.
8. Bob, Amal, Charlie, Dianne
9. Jake is not correct.
   Total distance = 10 + 20 = 30 km
   Total time = $\frac{10}{30} + \frac{20}{50} = \frac{11}{15}$ hours
   Speed = $30 \div \frac{11}{15} = 40.91$ km/h (2 d.p.)
10. Time taken for 5km run:
    5000 ÷ 3.5 = 1428.57 seconds = 23.81 minutes
    Total time = 40 + 23.81 = 63.81 minutes
    Farah left the library around 9.46 am

## 16.4 Compound measures

1. a)

| Box | Price | Price per egg |
|---|---|---|
| 6 eggs | £1.50 | 25 p |
| 10 eggs | £2.30 | 23 p |
| 12 eggs | £2.64 | 22 p |

b) 12 eggs box is the best value.

2. False. Annie is the fastest. Connor is the slowest.

3.

| Country | Carbon dioxide emissions (millions of tonnes) | Population (millions) | Emission / person |
|---|---|---|---|
| A | 1230 | 110 | **11.18** |
| B | 680 | 79 | **8.61** |
| C | 639 | 70 | **9.13** |

Amal is not correct. Country B has the lowest emission.

4. Tia's method:
Small bag = 55 ÷ 50 = 1.1 p/g
Big bag = 110 ÷ 180 = 0.61 p/g
Big bag is the cheaper per gram.

Michelle's method:
Small bag = 50 ÷ 0.55 = 90.9 g per £1
Big Bag = 180 ÷ 1.1 = 163.64 g per £1
The big bag is better value.

5. a)

| City | Population | Area (km²) | Number of people/km² |
|---|---|---|---|
| Tokyo | 14.1 million | 2191 | 6435 |
| Berlin | 3.56 | 891.8 | 3992 |
| Rome | 4.2 | 1285 | 3268 |

b) Rome

6. a) 13 km/litre  b) 30 × 13 = 390km. Yes he does have.

7. Café = 117 ÷ 18 = £6.50/hour
Supermarket = 148.50 ÷ 22 = £6.75/hour
Paul should keep the supermarket job.

8. a) Lora's tree = 31.25 cm/year Michael's tree = 33.33 cm/year
b) Lora's tree = 25 ÷ 31.25 = $\frac{4}{5}$ of a year.
Michael's tree = 25 ÷ 33.33 = $\frac{3}{4}$ of a year.
c) Michael's tree is the fastest growing.

9. Rate of pay = 175 ÷ 25 = £7/hour
Rate of pay in the weekend = £10.50/hour
Total pay = 26 × 7 + 8 × 10.50 = £266

10. Lucy's car
230 miles = 368 km
6.7 gallons = 30.15 litres
368 ÷ 30.15 = 12.21 km/litre

Bob's car
420 ÷ 25 = 16.8 km/litre

Bob's car is more economical.

## 16.5 Standard form

1. a) 10 000  b) 10  c) 1000  d) 10 000 000 000
2. a) $10^{-3}$ = 0.001  b) $10^{-1}$ = 0.1
c) $10^{-2}$ = 0.01  d) $10^{-4}$ = 0.0001
3. a) 500 = $5 \times 10^2$  b) 9000 = $9 \times 10^3$  c) 7500 = $7.5 \times 10^3$
d) 43 000 = $4.3 \times 10^4$  e) 870 = $8.7 \times 10^2$  f) 653 = $6.53 \times 10^2$
4. a) 0.08 = $8 \times 10^{-2}$  b) 0.000 9 = $9 \times 10^{-4}$
c) 0.004 3 = $4.3 \times 10^{-3}$  d) 0.018 4 = $1.84 \times 10^{-2}$
e) 0.007 53 = $7.53 \times 10^{-3}$  f) 0.000 030 2 = $3.02 \times 10^{-5}$
5. a) 300 000  b) 29 000  c) 807  d) 6 873 000
e) 0.04  f) 0.006 1  g) 0.000 523  h) 0.000 011 05
6. a) $10 \times 10^2$ is not in standard form
b) $11 \times 10^{-2}$ is not in standard form
c) $8 \times 10^{0.5}$ is not in standard form
d) $0.7 \times 10^{-2}$ is not in standard form
7. a) 0.000 7 = $7 \times 10^{-4}$  b) 360 < $36 \times 10^2$
c) 52 000 = $5.2 \times 10^4$  d) $4 \times 10^0 > 3.8 \times 10^0$
8. a) $7.1 \times 10^2$  $7.3 \times 10^2$  $7.5 \times 10^4$  $7 \times 10^7$
b) $2.2 \times 10^{-3}$  $2.6 \times 10^{-3}$  $2.8 \times 10^{-2}$  $2.05 \times 10^{-1}$
c) $8.8 \times 10^{-2}$  $8.7 \times 10^{-1}$  $8.9 \times 10^1$  $8.79 \times 10^2$
d) $5 \times 10^{-2}$  $5.4 \times 10^{-2}$  $5 \times 10^0$  $5.2 \times 10^3$
9. No. 108 000 000 = $1.08 \times 10^8$
$1.08 \times 10^8 > 5.79 \times 10^7$

10. a) bacteria  b) atom
11. a) $8 \times 10^{-1}$ = 0.8, not −8
b) 3200 = $3.2 \times 10^3$, not $32 \times 10^3$
c) $5.3 \times 10^{-2}$ = 0.053, not 0.53

## 16.6 Calculating with standard form

1. a) $8 \times 10^9$  b) $4.5 \times 10^9$  c) $1.2 \times 10^{11}$  d) $1 \times 10^{10}$
2. a) $4 \times 10^7$  b) $2 \times 10^5$  c) $1 \times 10^2$  d) $2.5 \times 10^1$
3. a) $9.96 \times 10^8$  b) $2.9988 \times 10^9$  c) $7 \times 10^4$  d) $2.171875 \times 10^{12}$
4. a) $(3 \times 10^7)^2 = 9 \times 10^{14}$  b) $(6 \times 10^5)^2 = 3.6 \times 10^{11}$
c) $(4 \times 10^6)^2 = 1.6 \times 10^{13}$  d) $(5 \times 10^3)^2 = 2.5 \times 10^7$
5. a) $9 \times 10^7$  b) $1.5 \times 10^8$  c) $1 \times 10^{10}$
d) $4.3 \times 10^4$  e) $3.6 \times 10^3$
6. a) $9 \times 10^2$  b) $9 \times 10^3$  c) $9 \times 10^3$  d) $9 \times 10^3$
7. $2.8 \times 10^{-6} \times 10^6$ = 2.8cm = 28 mm
8. $1.26025 \times 10^{-5}$ mm²
9. Lily added the powers instead of subtracting them.
$(6 \times 10^7) \div (2 \times 10^2) = 3 \times 10^5$
10. Time = $\frac{40\,000 \times 50}{1.2 \times 10^3} = 1.667 \times 10^3$ hours
11. Speed of sound = $\frac{3 \times 10^9}{8.8 \times 10^5} = 3.41 \times 10^3$ m/s (2 d.p)

# Section 17

## 17.1 Real life graphs

1. a) £40  b) £75  c) 13 hours  d) £10
2. a) 0800  b) 1 hr  c) 1 hr
d) 35 km  e) 15 km/h  f) 10 km/h
3. Graph B – the width of the cup is uniform, so the graph is a straight line.
4. a) £50  b) 5 hrs  c) £20
d) The call-out charge means that this won't be the case; a job lasting 8 hours costs £60 and a job lasting 16 hrs costs £100.
5. No. The depth increases more slowly as the bowl widens.
6. For journeys up to 25 km Quick Couriers is cheaper; after that she should choose Speedy Deliveries
7. a) 30 km  b) 1 hr  c) 30 km/h
d) 1 hr  e) 20 km/h
8. They have got the call-out fee and hourly rate the wrong way round.
9. For example: Jennifer turned one hose on for a while then turned on the second hose.
Jennifer then turned the hoses off, got in for a bit then got out.
She then emptied the water at a constant rate.
10. a) 45 km/h  b) 2 hours
c)

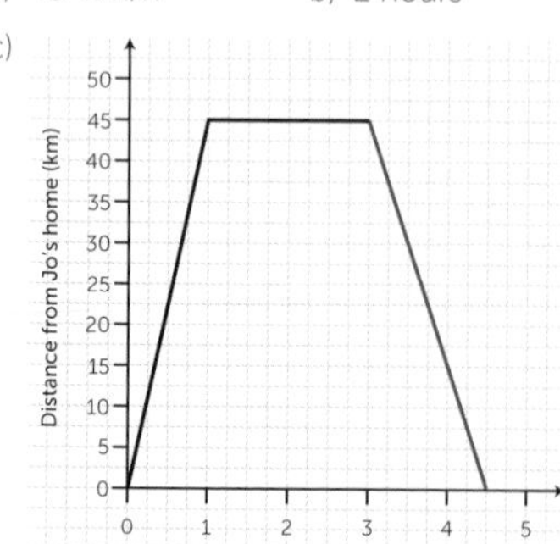

## 17.2 Quadratic functions

1. A and D

2. a)

| $x$ | −3 | −2 | −1 | 0 | 1 | 2 | 3 |
|---|---|---|---|---|---|---|---|
| $y$ | 12 | 7 | 4 | 3 | 4 | 7 | 12 |

b)

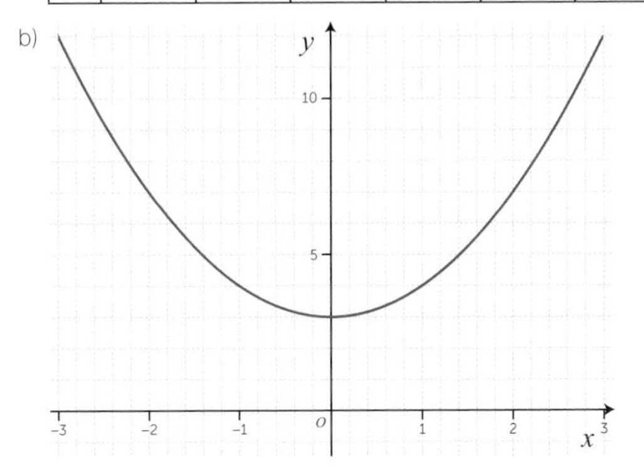

3. a) $y = x^2$ : Graph 1 b) $y = -x^2$ : Graph 4
   c) $y = x^2 + 2$ : Graph 6 d) $y = 2 - x^2$ : Graph 2
   e) $y = x^2 - 3$ : Graph 5 f) $y = -x^2 - 3$ : Graph 3

4. Declan needs to put the negatives in brackets as $(-2)^2 = 4$ not $-4$ and $-4 \times -2 = 8$ not $-8$

5. a)

| $x$ | −3 | −2 | −1 | 0 | 1 | 2 | 3 |
|---|---|---|---|---|---|---|---|
| $y$ | −4 | 1 | 4 | 5 | 4 | 1 | −4 |

b)

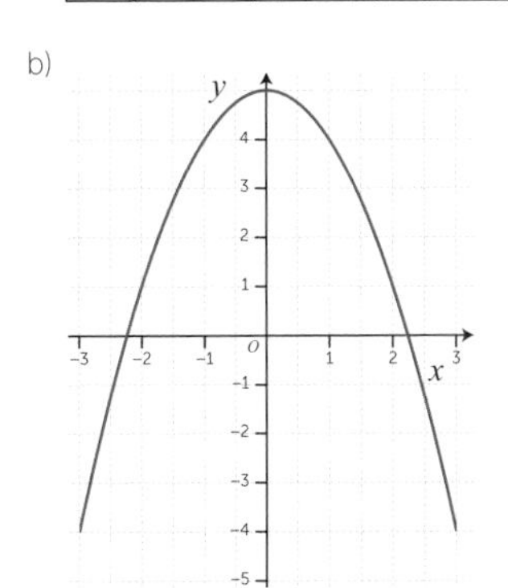

6. a)

| $x$ | −5 | −4 | −3 | −2 | −1 | 0 | 1 |
|---|---|---|---|---|---|---|---|
| $y$ | 15 | 8 | 3 | 0 | −1 | 0 | 3 |

b)

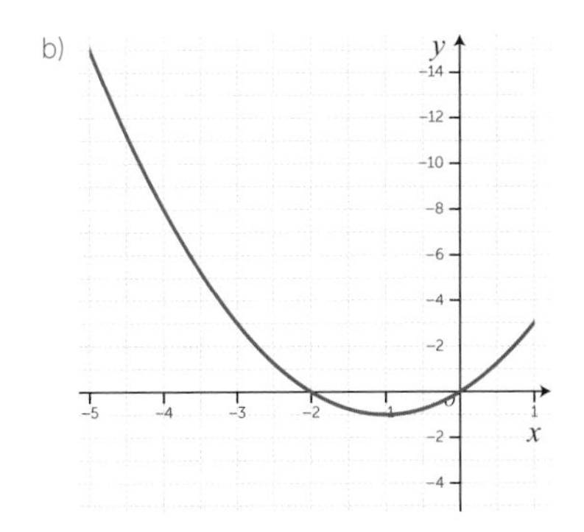

7. a)

| $x$ | −2 | −1 | 0 | 1 | 2 | 3 | 4 |
|---|---|---|---|---|---|---|---|
| $y$ | 10 | 4 | 0 | −2 | −2 | 0 | 4 |

b)

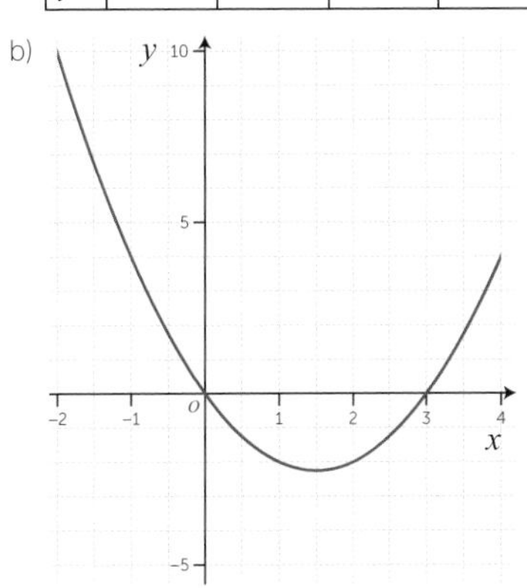

8. a)

| $x$ | −3 | −2 | −1 | 0 | 1 | 2 | 3 |
|---|---|---|---|---|---|---|---|
| $y$ | 4 | 0 | −2 | −2 | 0 | 4 | 10 |

b)

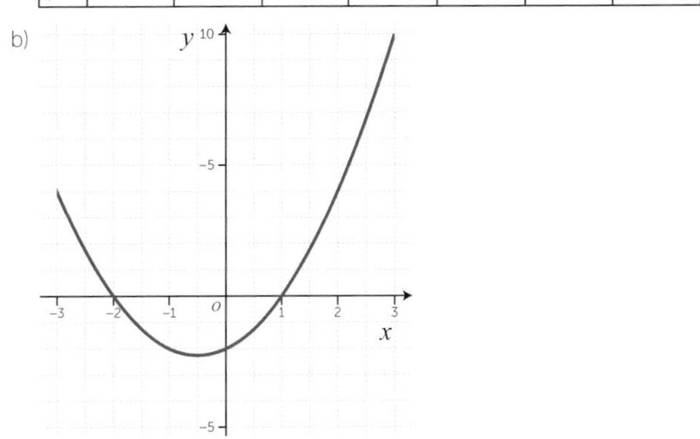

## 17.3 Quadratic graphs

1. a)

| $x$ | −3 | −2 | −1 | 0 | 1 | 2 | 3 |
|---|---|---|---|---|---|---|---|
| $y$ | 5 | 0 | −3 | −4 | −3 | 0 | 5 |

b)

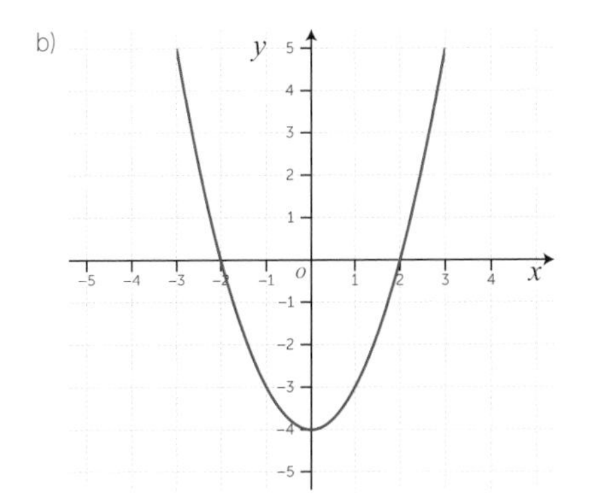

2. a) $x = -2, x = 2$ b) $x = 0$ c) (0,−4)

3. a)

| $x$ | −3 | −2 | −1 | 0 | 1 | 2 | 3 |
|---|---|---|---|---|---|---|---|
| $y$ | 8 | 3 | 0 | −1 | 0 | 3 | 8 |

b)

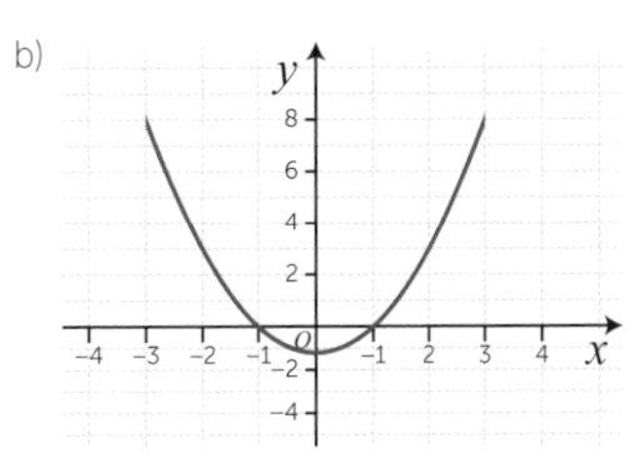

4. a) $x = -1, x = 1$ b) $x = 0$ c) $x = -2, x = 2$

5. a)

| $x$ | −2 | −1 | 0 | 1 | 2 | 3 | 4 |
|---|---|---|---|---|---|---|---|
| $y$ | 8 | 3 | 0 | -1 | 0 | 3 | 8 |

b) 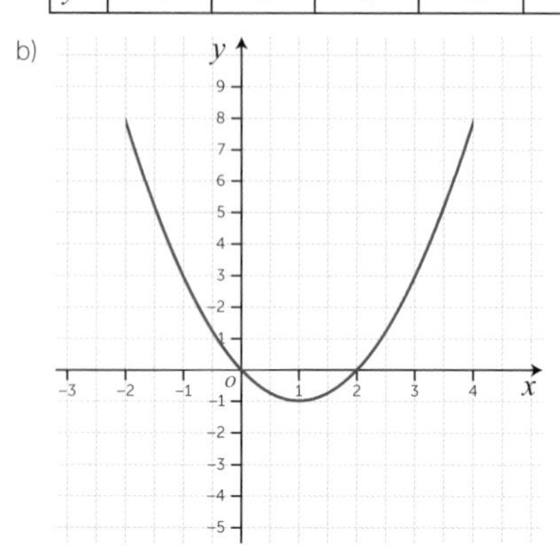

c) $x = 0, x = 2$ d) (1,-1) e) $x = 1$

6. a)

| $x$ | −2 | −1 | 0 | 1 | 2 | 3 | 4 | 5 |
|---|---|---|---|---|---|---|---|---|
| $y$ | 6 | 0 | −4 | −6 | −6 | −4 | 0 | 6 |

b) 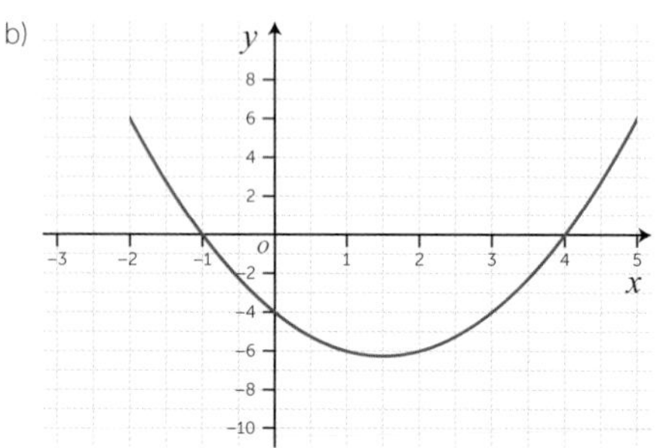

c) It is the '$+c$' of the equation i.e. −4 d) $x = -1, x = 4$ e) $x = \frac{3}{2}$

7. 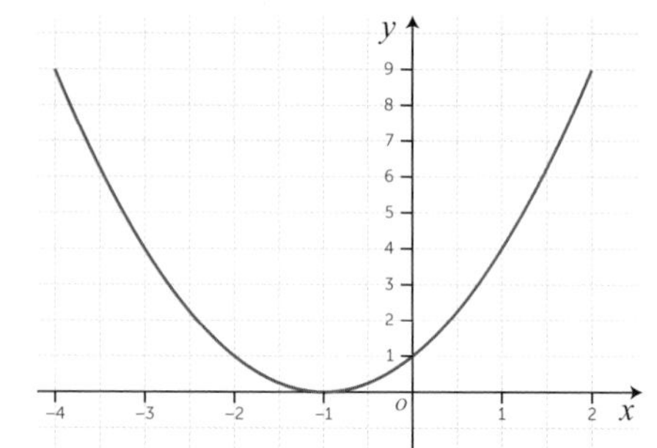

a) $x = -1$ b) $x = -3, x = 1$ c) $x = -2.4, x = 0.4$

8. 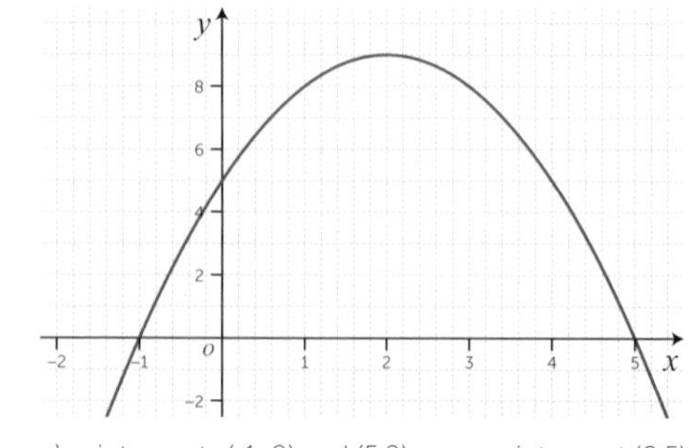

a) $x$ intercepts (-1, 0) and (5,0) $y$ intercept (0,5)

b) $x = -1, x = 5$ c) $x = -0.4, x = 4.4$

## 17.4 Reciprocal and cubic graphs

1. a) linear b) quadratic c) reciprocal d) cubic

2. Graph A: reciprocal Graph B: quadratic
Graph C: linear Graph D: cubic

3. a)

| $x$ | 0.1 | 0.25 | 0.5 | 1 | 2 | 3 | 4 |
|---|---|---|---|---|---|---|---|
| $y$ | 10 | 4 | 2 | 1 | $\frac{1}{2}$ | $\frac{1}{3}$ | $\frac{1}{4}$ |

b) 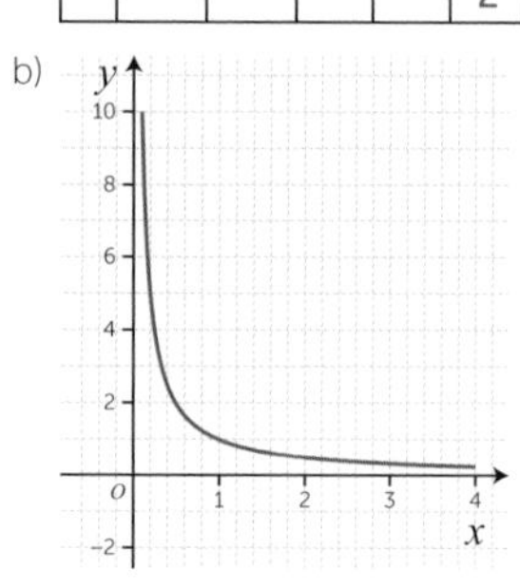

4. a)

| $x$ | −3 | −2 | −1 | 0 | 1 | 2 | 3 |
|---|---|---|---|---|---|---|---|
| $y$ | −27 | −8 | −1 | 0 | 1 | 8 | 27 |

b) 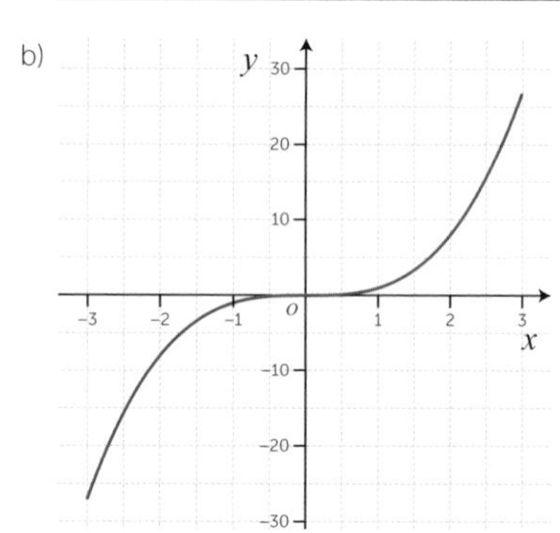

5 a)

| $x$ | −3 | −2 | −1 | 0 | 1 | 2 | 3 |
|---|---|---|---|---|---|---|---|
| $y$ | −24 | −5 | 2 | 3 | 4 | 11 | 30 |

b) 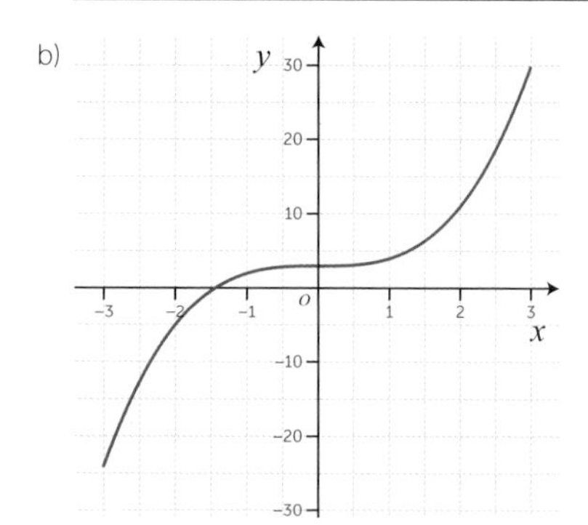

6 a) iii) b) iv) c) ii) d) i)

7. a)

| Speed (mph) | 10 | 20 | 30 | 40 | 50 | 60 |
|---|---|---|---|---|---|---|
| Time (hours) | 3 | 1.5 | 1 | 0.75 | 0.6 | 0.5 |

b) It will be a curve, as it is a reciprocal graph.

c) 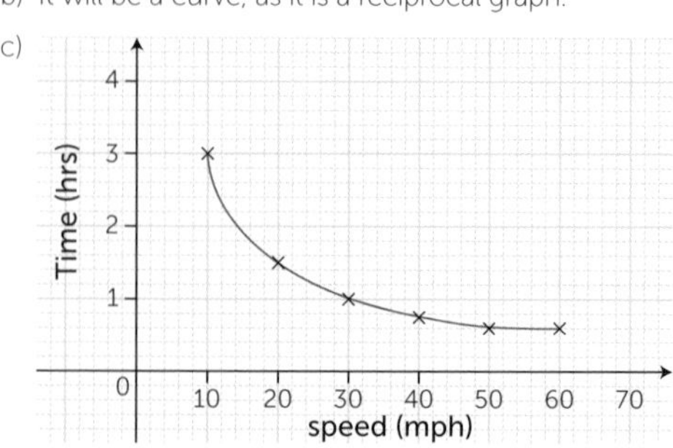

## 17.5 Solving simultaneous equations graphically

1. $x = 2, y = 2$

2. a)

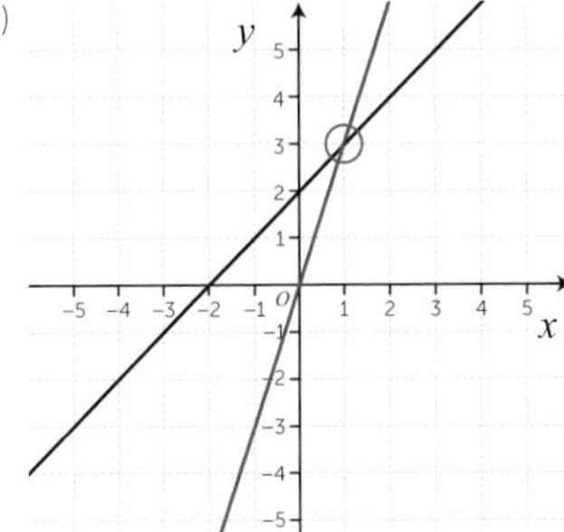

   b) $x = 1, y = 3$

3. a)

| $x$ | 0 | 1 | 2 | 3 |
|---|---|---|---|---|
| $y$ | −1 | 2 | 5 | 8 |

   b)

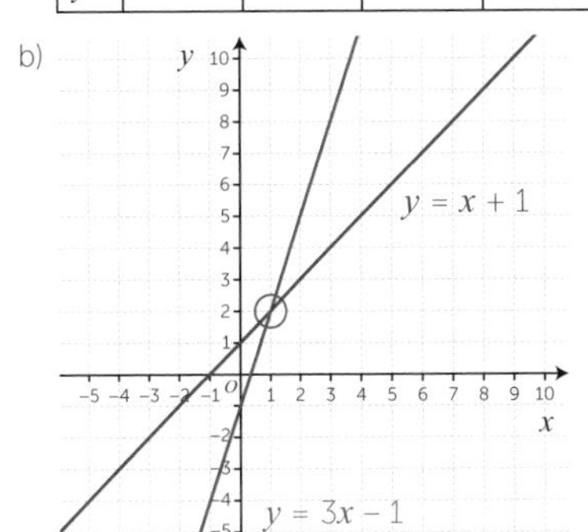


   c) See green line above    d) $x = 1, y = 2$

4.

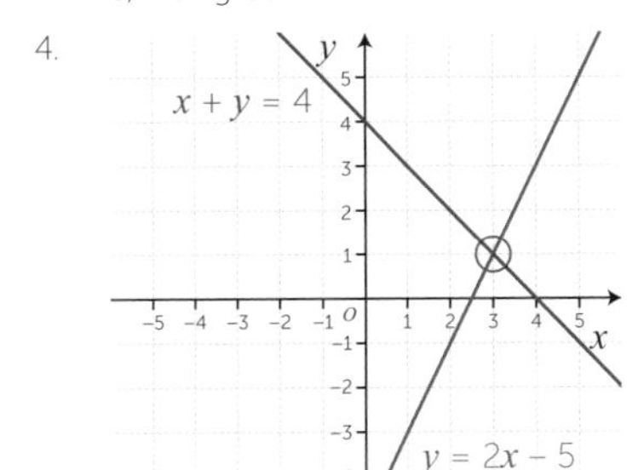


   $x = 3, y = 1$

5. a) and b)

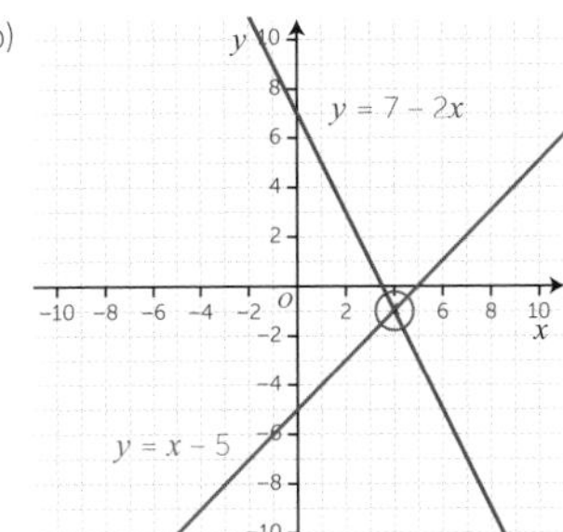


   c) $x = 4, y = -1$

6. a)

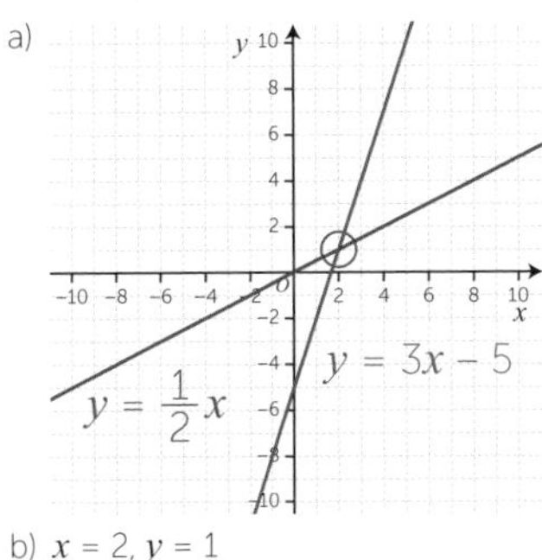


   b) $x = 2, y = 1$

7.

| $x$ | −3 | −2 | −1 | 0 | 1 | 2 | 3 |
|---|---|---|---|---|---|---|---|
| $y$ | 6 | 1 | −2 | −3 | −2 | 1 | 6 |

   a) and b)

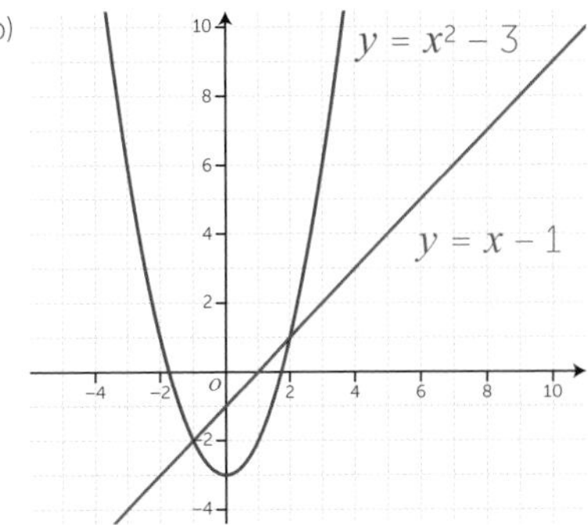


   c) $x = -1, y = -2$ and $x = 2, y = 1$

# Section 18

## 18.1 Constructions

1. a) Draw arc, centre L, radius more than half of LM.
   b) Draw arc, centre M, same radius to cross at E and F.
   c) Join EF which cuts LM at its midpoint.

2. a) Draw arc, centre Q to cross QR at K and QP at L.
   b) Draw arc, centre K, and draw arc, centre L, same radius to cross at M.
   c) Join QM, the angle bisector of PQR.

3. a) Draw 2 arcs, centre X, with the same radius, to cross VW at P and Q.
   b) Draw an arc, centre P, radius more than PX. Draw another arc, same radius, centre Q. The two arcs cross at D.
   c) Draw a line from X through D. This is the required perpendicular.

4. As for question 2.

5. a) Draw a circle of radius 4 cm. Mark a point X on the circle.
   b) Draw two arcs, each centre X, radius 4 cm to cut the circle at Y and Z.
   c) From centre Y, draw an arc radius 4 cm to cut the circle at W.
   d) From centre W, draw an arc radius 4 cm to cut the circle at V.
   e) From centre V, draw an arc radius 4 cm to cut the circle at T.
   f) Join points around the circle to get a hexagon.

6. a) Draw an arc, centre H, to cross JK at P and Q.
   b) Draw an arc, centre P, radius more than half PQ.
   c) Draw an arc, centre Q, same radius as before, the arcs cross at L.
   d) Draw the line through H and L.

7. For side AB:
   a) Draw arcs, centre A, radius more than half AB.
   b) Draw arcs, centre B, same radius as before.
   c) The arcs cross at points D and E. Join DE to get the perpendicular bisector of AB.
   Repeat for sides BC and AC.

8. a) Draw an arc, centre F, to cross DE at P and Q.
   b) Draw an arc, centre P, radius more than half PQ.
   c) Draw an arc, centre Q, same radius as before, the arcs cross at L.
   d) Draw the line through F and L.
   Repeat for sides DF and EF.

9. a) Draw an arc, centre J, radius equal to JK.
   b) Draw an arc, centre K, radius equal to JK. The arcs cross at L.
   c) Join J to L, then ∠LJK = 60°.
   To bisect LJK, see question 2.

10. See question 7.

## 18.2 Loci

1. a) Circle, centre A, radius 3 cm
   b) Circle, centre A, radius 5 cm
   c) Circle, centre A, radius 6.3 cm
   d) Interior of a circle, centre A, radius 4 cm, but not the circle itself.

2. a) Perpendicular bisector of CD
   b) 2 points on the perpendicular bisector, 5 cm from C and from D.
   c) 2 straight line sections parallel to CD at a distance of 6 cm, and two semi-circles centred on C and D, radius 6 cm.

3. a) Circle radius 6 m, centred on the stake.
   b) Circle radius 6 m with a segment cut off 3m from the centre at its closest point.
4. a) Line of symmetry that bisects EF
   b) Top half of the square
   c) Along the diagonal from H to F
   d) Quadrant centre G, radius 8 cm
   e) Remaining part of the square after removing a quadrant, centre H, radius 5 cm.
   f) Remaining part of square after drawing a circle centre G, radius 8. (GE = 11.3 cm)
5. a) A horizontal line parallel to the ground, the height of the wheel centre.
   b) A series of arches from the ground to the height of the wheel.
6. Suppose the top left corner is point A, and the dog is tethered to that corner. Then the area the dog can reach is 3 quadrants of a circle, centre A, radius 4 m, plus a quadrant of radius 1 m, centre the bottom left corner of the shed.

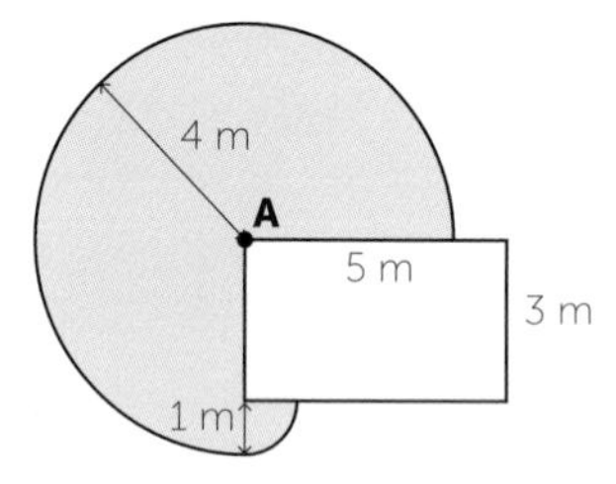


7. The mast can be in the area bordered by two arcs, one centred on Ashbridge, radius 6 km, and one centred on Bedlow, radius 7 km.

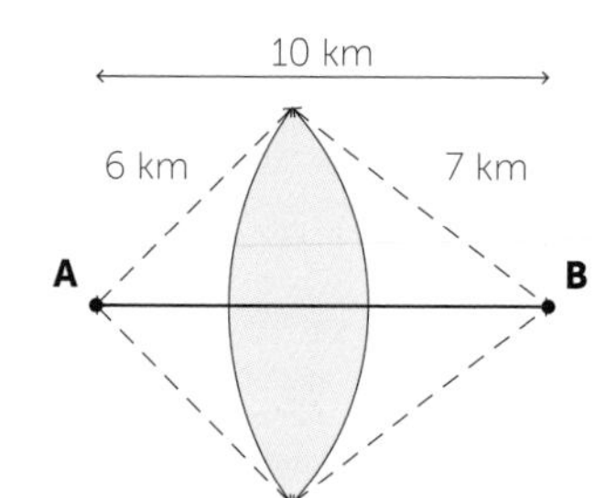


8. 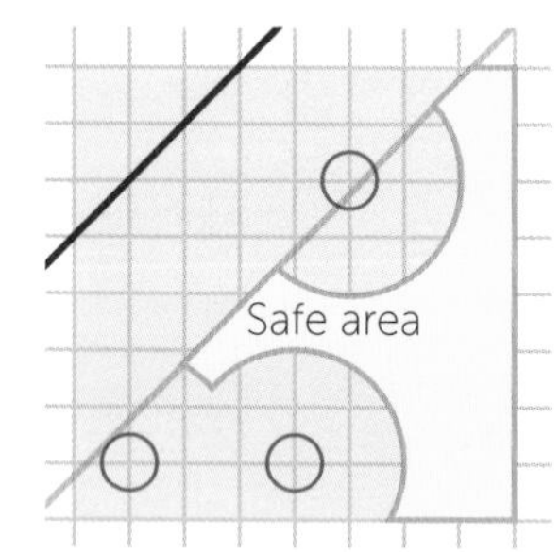


9. 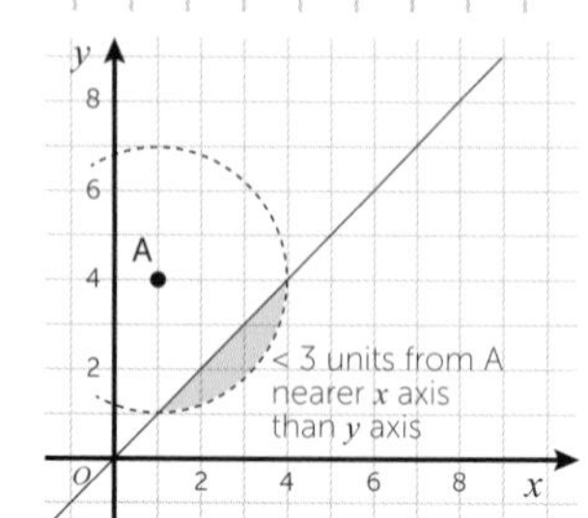


## 18.3 Congruence

1. a) 55°
   b) 70°
2. ∠MLJ = 50°
   ∠LMJ = 100°
3. a) RHS
   b) The two triangles are similar so ∠SPR = ∠QPR.
   Therefore RP bisects ∠SPQ.
4. a) SAS b) SSS
   c) RHS d) ASA
5. a) ASA
   b) Since triangles are congruent, ∠ABD = ∠CBD hence DB bisects ∠ABC.
6. (i) True (ii) True (iii) False
7. a) Yes, they are congruent. ∠ACB = 35 (isosceles triangle), ∠BAC = 110 (Angles in a triangle) so the two triangles are congruent by SAS.
   b) XZ = 8 cm
8. HM = KM and JM = LM (corresponding sides)
   ∠JMH = ∠LMK (opposite angles)
   By SAS, the triangles are congruent.

## 18.4 Similarity

1. a) 30° (corresponds to ∠ABC b) 3 cm (Scale factor 2)
   c) 8 cm
2. Not similar since $14 = 7 \times 2$ but $7 \neq 3 \times 2$
3. $y$ = 4 cm
4. $m$ = 4.5 cm, $n$ = 8 cm
5. $a$ = 110°, $b$ = $e$ = 57°, $c$ = 115° and $d$ = 78°; $x$ = 8 cm, $y$ = 9 cm
6. $p$ = 7 cm, $q$ = 10.5 cm, $r$ = 24.5 cm; The triangles are isosceles.
7. a) ∠ADC = 125°
   b) Quadrilateral LMJK is not similar to the other three. (The angle opposite the right angle should be 70°.)
8. a) SV = 2 cm b) TU = 9 cm
9. a) PM = 3 cm b) NM = 16.5 cm
10. a) B and F are congruent b) E is similar to A

## 18.5 Pythagoras' theorem

1. a) a = 5 cm b) b = 17 cm c) c = 15 cm
2. a) d = 6 m b) e = 15 cm c) f = 12 mm
3. a) No b) Yes c) No
4. 9.99 m
5. 5.86 m
6. 8.49 m
7. 9.90 cm
8. 10 units
9. A 7.81 units, B 10.63 units
10. a) AB = 5 units, BC = 12 units, AC = 13 units.
    b) $AB^2 + BC^2 = AC^2$ so ABC is right-angled.

## 18.6 Geometric proofs

1. a) $a$ = 71°, $b$ = 71° b) $c$ = 74°, $d$ = 32° c) $e$ = 69°, $f$ = 69°
2. a) $a$ = 28°, $b$ = 76°, $c$ = 76° b) $d$ = 73°, $e$ = 107°
   c) $f$ = 74°, $g$ = 32°
3. Interior 60°, exterior 120°
4. Triangle ABF is equilateral, so angle a = 90° + 60° = 150°
   Angle b = 60° + 90° + 45° = 195°
   Angle c = 180° - 45° = 135°
5. $d$ = 17° Isosceles triangle has two equal angles. Angles in a triangle add up to 180 degrees.
6. a) $e$ = 135° b) $f$ = 116° c) $g$ = 130° d) $h$ = 118°
7. a) $k$ = 80° b) $j$ = 75°
   Exterior angle of triangle = sum of opposite interior angles

## Section 19

### 19.1 Knowledge check

1 a) $2x + 14$ b) $6y - 15$ c) $n^2 + 2n$ d) $24 - 18p$

2 a) $4(x + 3)$ b) $5(3a - 2)$ c) $p(p - 8)$ d) $2(3d - 2e + 5f)$

3 a) $x = 3$ b) $y = 15$ c) $n = 8$ d) $g = 1$

4. Kelly has not expanded the bracket correctly; it should be $14x^2 - 35x$

5. Ben is 23 years old and Matthew is 16 years old.

6. $x = 5.5$

7. No, it should be $y(y - 7)$

8. $4n + 5 = 57$ Frieda has 13 sweets

9. $x = 4$ The area is 30 cm$^2$

10. $x = 1.5$

### 19.2 Harder linear equations

1 a) $y = 7$ b) $n = -8$ c) $x = 3$ d) $a = 2.5$ e) $g = -2$

2. $x = 8$

3. It should be:
Subtract 18 from both sides: $-3n = 2n - 15$
Subtract 2n from both sides: $-5n = -15$
Divide both sides by $-5$: $n = 3$

4. a) $x = 5$ b) $x = 8.5$ c) $x = -4$ d) $x = -3$ e) $x = -7$

5. It should be: $5(x + 2) = 3(2x + 1)$
$5x + 10 = 6x + 3$
$x = 7$

6. $x = 8$; oldest person is Dima (29)

7. $n = 6$; range is $22 - 6 = 16$

8. a) $x = 7$ b) perimeter = 78 cm

9. $x = 5$

10. $3x + 7 = 7x - 5$ so $x = 3$
So area of square is 256 cm$^2$, area of triangle is 128 cm$^2$
So $128 = 4(5y - 2)$ and so $y = 6.8$

### 19.3 Product of two binomials

1. a) $y^2 + 5y + 4$ b) $n^2 + 9n + 14$ c) $x^2 - 8x + 12$
d) $p^2 - 4p + 3$ e) $t^2 + 7t + 10$

2. $x^2 + 5x + 6$

3 a) $w^2 - w - 2$ b) $x^2 - 3x - 10$ c) $d^2 - 10d + 21$
d) $y^2 + 3y - 18$ e) $t^2 - 7t - 18$

4. He has forgotten some terms. The answer is $x^2 + 8x + 16$

5. Product of their numbers = $(n - 4)(n + 3) = n^2 - n - 12$

6. Area of square = $(n - 5)^2 = n^2 - 10n + 25$

7 a) $2x^2 + 5x + 2$ b) $3y^2 - 17y + 10$ c) $4n^2 + 20n + 25$
d) $3d^2 + 11d - 4$ e) $6v^2 + 13v - 5$

8 a) Rectangle area = $(5x - 7)(2x + 3) = 10x^2 + x - 21$
b) Square area = $(3x- 5)^2 = 9x^2 - 30x + 25$
c) Triangle area = $\frac{1}{2}(2x - 5)(2x + 3) = 2x^2 - 2x - 7.5$

9. $n - 32$

10. $5x^2 + 12x - 3$

### 19.4 Factorising quadratics

1. a) 1 and 5, −1 and −5
b) 1 and 8, −1 and −8, 2 and 4, −2 and −4
c) 1 and 20, −1 and −20, 2 and 10, −2 and − 10, 4 and 5, −4 and − 5
d) 1 and 12,−1 and − 12, 2 and 6, −2 and − 6, 3 and 4, −3 and −4
e) 1 and 25, −1 and − 25, 5 and 5, −5 and − 5

2. a) $(a + 2)(a + 4)$ b) $(k - 1)(k - 4)$ c) $(n - 4)(n - 5)$
d) $(y + 3)(y + 4)$ e) $(x - 2)(x - 8)$

3. $n^2 + 5n + 4 = (n + 1)(n + 4)$ so an expression for the width is $(n + 1)$

4. a)−1 and 3, 1 and −3
b)−1 and 10, 1 and −10, −2 and 5, 2 and −5
c)−1 and 4, 1 and −4, −2 and 2
d)−1 and 18, 1 and −18, −2 and 9, 2 and −9, −3 and 6, 3 and −6
e)−1 and 24, 1 and −24, −2 and 12, 2 and −12, −3 and 8, 3 and −8, −4 and 6, 4 and −6

5. a) $(t - 1)(t + 5)$ b) $(q - 1)(q + 6)$ c) $(a + 2)(a - 3)$
d) $(x + 5)(x - 3)$ e) $(y - 10)(y + 2)$

6. a) $(x + 8)(x - 8)$ b) $(y + 9)(y - 9)$ c) $(r + 11)(r - 11)$

7. They multiply to make $p^2 - 5p + 6$; the correct answer is $(p + 1)(p - 6)$

8. Georgia has found pairs that multiply to the wrong amount.
Henry has found a pair that multiplies to make +5 not −5
Answer should be : $(n - 5)(n + 1)$

9. $(y + 4)$ and $(2y - 4)$ or $(2y + 8)$ and $(y - 2)$

10. Length = $2x + 1$ and width = $x - 3$

### 19.5 Identities

1. a) $\neq$ b) $\equiv$ c) $\neq$

2. a) identity b) not c) identity
d) not e) identity

3. 5 and 12

4. a) equation b) equation c) identity
d) identity e) identity

5. It should be: $4x - 12 - 2x - 10 = 2x - 22 = 2(x - 11)$
So the identity is $4(x - 3) - 2(x + 5) \equiv 2(x - 11)$

6. Must show $12x - 30 - 5x + 15$

7. $-11x + 24$

8 a) identity b) not c) identity d) identity

9 a) $a = 1, b = 2, c = -15$ b) $a = 2, b = -2, c = 8$
c) $a = 6, c = 3$ d) $b = 2, c = 6$

10. Must show $\frac{5(2x + 3)}{20} - \frac{4(x - 2)}{20}$ or $\frac{10x + 15}{20} - \frac{4x - 8}{20}$

### 19.6 Proving identities

1. a) even b) even c) odd d) even e) odd

2. $2x$ is even so $2x - 1$ is odd

3. Simplifies to $3y + 3 = 3(y + 1)$ which is a multiple of 3

4. a) 5 b) 5 c) 4

5. $2n + 2n + 2 = 4n + 2 = 2(2n + 1)$ This is a multiple of 2, so even

6. He has added three consecutive multiples of 3, instead of three consecutive numbers.
e.g. It should be: $n + (n + 1) + (n + 2) = 3n + 3 = 3(n + 1)$

7. e.g. $6^2 + 3 = 39$ which is not prime.

8. $a^2 + a + 5a - a^2 \equiv 6a \equiv 2 \times 3a$ and hence a multiple of 2 so even

9. She hasn't added even numbers
It should be: $2n + (2n + 2) + (2n + 4) \equiv 6n + 6 = 6(n + 1)$ so a multiple of 6

10. $(2n + 1)^2 \equiv 4n^2 + 4n + 1 = 2(2n^2 + 2n) + 1$
This is an odd number as it is 1 more than a multiple of 2 (even number).

### 19.7 Solving algebraic problems

1. $a = 3$

2. Ariana is 6, Beatrice is 11 and Chloe is 33

3. $x = 39°$

4. Col is 9 and Dea is 27

5. tea = $x$ coffee = $2x$
2 teas and 3 coffees: $2x + 3(2x) = 7.60$
$8x = 7.60$
$x = 95$p
Tea costs 95p and coffee costs 2 × 95 = £1.90

6. Rob has $n$, Steve has $n + 5$ and Tom has $n + 4$
Sum of sweets = $3n + 9 = 3(n + 3)$
Multiple of 3 divides exactly by 3, so yes.

7. $y = 70°$

8. $n = 8$ so numbers are 8, 12, 15 and 25.
Range is 25 − 8 = 17 and median is the midpoint of 12 and 15 = 13.5

9. Area = $(x + 2)(x - 7) = x^2 + 2x - 7x - 14$
Area is given as 14 so $x^2 + 2x - 7x - 14 = 14$
$x^2 - 5x - 28 = 0$

10. $x^2 - 9$
If $x = 100$ then $103 \times 97 = 100^2 - 9 = 9991$

# Section 20

## 20.1 Ratios in similar shapes

1. B is true
2. A and C are true. In regular shapes, all the angles are the same.
3. a) $s = 38°, t = 32°$ b) $j = 3.5$ m, $k = 6.75$ m
c) Scale factor = 3
4. a) $m = 53°, n = 37°, p = 53°$ b) $d = 8$ m
5. $\frac{x}{2} = 0.5 \Rightarrow x = 1$
6. a) 4 cm b) 128 cm$^2$
7. No, she is not correct. All the sides in the same ratio, (SSS) or all the angles the same, (AAA) or two pairs of sides in the same ratio and the angles between them equal (SAS).
8. a) YZ = 32 mm b) $p : q = 5 : 3$
9. a) 4.8 cm b) 6 cm

## 20.2 Trigonometric ratios

1. a) $a$ is the opposite side for angle $\theta$
$b$ is the hypotenuse
$c$ is the adjacent side for angle $\theta$
b) $d$ is the hypotenuse
$e$ is the adjacent side for angle $\alpha$
$f$ is the opposite side for angle $\alpha$
c) $g$ is the adjacent side for angle $\beta$
$h$ is the opposite side for angle $\beta$
$i$ is the hypotenuse
2. $\sin\theta = \frac{b}{a}$ $\cos\theta = \frac{c}{a}$ $\tan\theta = \frac{b}{c}$
$\sin\alpha = \frac{d}{f}$ $\cos\alpha = \frac{e}{f}$ $\tan\alpha = \frac{d}{e}$
$\sin\beta = \frac{i}{h}$ $\cos\beta = \frac{j}{h}$ $\tan\beta = \frac{i}{j}$
3. a) $\sin 32° = 0.530$ b) $\cos 78° = 0.208$ c) $\tan 14° = 0.249$
4. a) sin b) cos c) tan
5. a) TRUE b) TRUE c) FALSE d) FALSE
6. a) 11.92 cm b) 17.43 cm c) 14.10 cm
7. a) 2.56 cm b) 8.01 cm c) 8.72 cm
8. C

## 20.3 Using trigonometric ratios

1. 5.4 m
2. a) $a = 47.0°$ b) $b = 48.6°$ c) $c = 66.2°$
3. a) $a = 26.9°$ b) $b = 42.4°$ c) $c = 49.7°$
4. a) $t = 6.5$ cm b) $u = 16.8$ mm c) $v = 2.1$ m
d) $w = 3.6$ m e) $x = 0.1$ mm f) $y = 18.2$ mm
5. a) $g = 134.2$ m b) $h = 12.2$ m c) $j = 10.4$ m
6. a) i) 11.1 m ii) 10.0 m
b) 6.7 m c) 16.7 m

## 20.4 Trigonometry in context

1. 83.14 m
2. 2195 m
3. 16.1 m
4. 86.4 m
5. a) 68.12 m b) 28.91 m
6. 5.13°
7. 59°
8. a) 615 m b) 541 m
9. 537 m
10. a) 192 m b) 68.7°

## 20.5 Special angles

1. a) 5 m b) $5\sqrt{3}$ m
2. a) 9 m b) 9 m
3. a) $\sin 45° = \frac{1}{\sqrt{2}}$ b) $\cos 60° = \frac{1}{2}$ c) $\tan 30° = \frac{1}{\sqrt{3}}$
d) $\cos 0° = 1$ e) $\tan 45° = 1$ f) $\sin 60° = \frac{\sqrt{3}}{2}$
4. a) $\cos^{-1}\frac{1}{\sqrt{2}} = 45°$ b) $\sin^{-1}\frac{1}{\sqrt{2}} = 45°$ c) $\tan^{-1}\sqrt{3} = 60°$
5. $m = 7$ cm, $n = 15\sqrt{3}$ cm
6. a) $w = 11$ cm, $x = 11\sqrt{2}$ cm b) $y = 3\sqrt{2}$ cm, $z = 3\sqrt{2}$ cm

## 20.6 Trigonometry problems

1. 6.73 m
2. 3.34 km
3. a) 21.81 km b) 32.33 km
4. 4.81 m
5. 72.22 m
6. 4.02 m
7. 10.3°
8. 2.36 m
9. a) 3.99° b) 5.23 m c) 143 years

# Section 21

## 21.1 Knowledge check

1. a) Continuous b) Discrete c) Continuous
d) Continuous e) Discrete
2. No labels on the axes, no gaps between bars
The median cannot be found because the data values cannot be placed in order of size. If the animals had been placed in the chart in a different order (for example alphabetically) the middle value would be different.
3. The mean age will increase to 29.
4. a) 5 b) 4 c) 3.2
5. James needs to calculate the relative frequency for each meal by dividing by 180, the total number of meals. Each relative frequency should then be multiplied by 360 to find the angle in the pie chart. He needs to double each angle.
6. Jonty needs to score 69%
7. a)

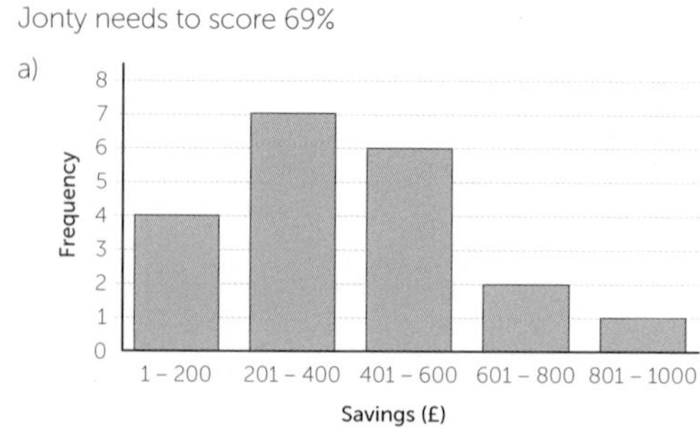


b) Estimation of the mean: £391 (nearest £)
c) The data is not continuous because there are gaps in values between whole number of pence. In pounds, values have at most 2 decimal places.

8. 77 runs

9. a) Road A: 31; Road B: 22 b) Road A: 28; Road B: 27
   c) Road B as the cars are going slower

10. a) 30
    b) Library A: 45.
       Library B: 44.
    c) Minimum possible range = 80 - 19 = 61.

## 21.2 Sampling

1. a) Sample b) Whole population c) Whole population
   d) Sample e) Sample

2. Friends have similar tastes so they are likely to name the same or similar actors; Gretel's friends do not represent the whole population.

3. 40 boys, 60 girls

4. Jimmy could pick one from ten from the first hundred people who arrive at the canteen, or one from each table once everyone has arrived.

5. Those at an entertainment store are likely to buy music quite regularly; this will not give a representative sample of the whole population.

6. Year 7: 11 Year 8: 8 Year 9: 10 Year 10: 9 Year 11: 12

## 21.3 Time series

1.

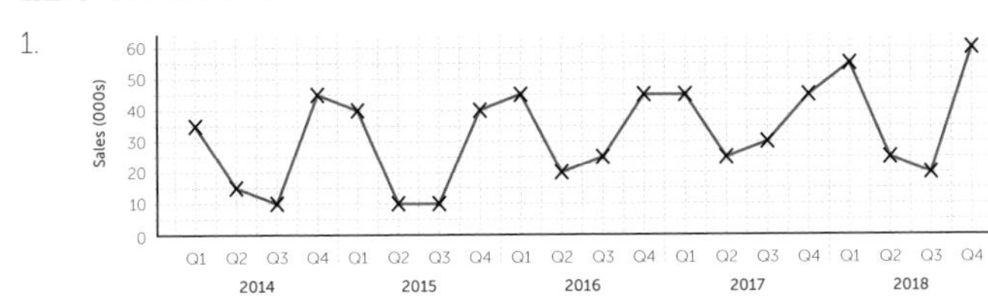


Sales are generally on the rise , with quarters 4 and 1 showing the best sales figures.

2. a) 120 000 b) 930 000
   c) Summer holidays d) December, January, February
   e) They would open for March but then have to close for April

3. a) 4th b) Winter (coldest)
   c) 2nd and 3rd d) Summer (warmest)
   e) Sales have dropped from their initial numbers

4. a) 4.8 million
   b) Sales are increasing faster and faster. Increase was 0.2 million in 2014, and 1.5 million in 2019.
   c) Justin is probably correct. The trend line shows that between 6.6 and 8 million cars is a reasonable estimate.

5. a) The seventh week (between 6 weeks and 7 weeks old). The gradient of the graph as steepest during this period.
   b) 3 weeks old. The graph dipped in the 4th week and then rose again.

## 21.4 Scatter diagrams

1.

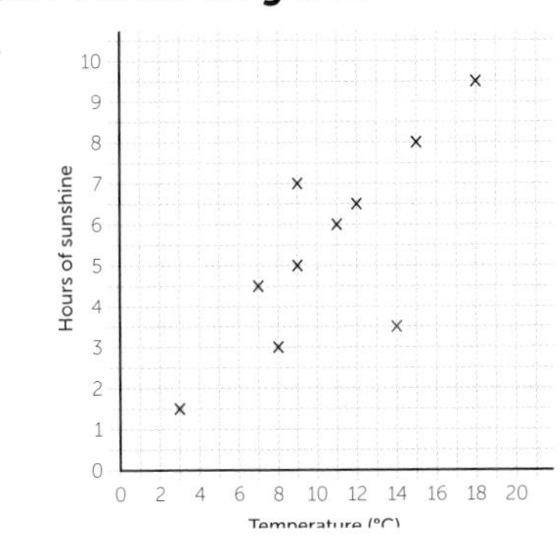


a) The more hours of sunshine, the higher the temperature
b) It is an outlier

2. a)

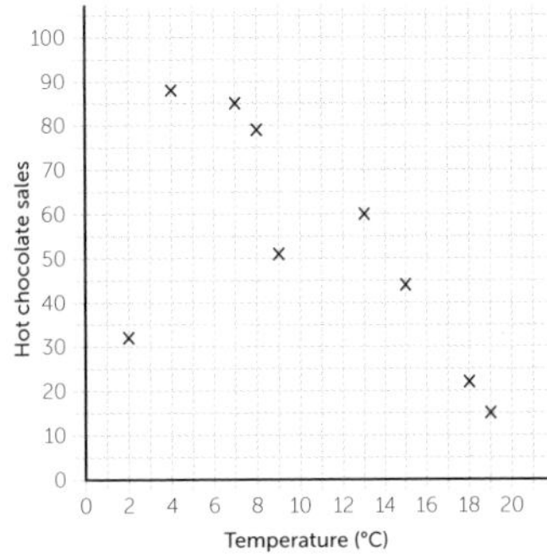


b) The higher the temperature, the fewer hot chocolates the café sells
c) (2,32) is an outlier

3. a)

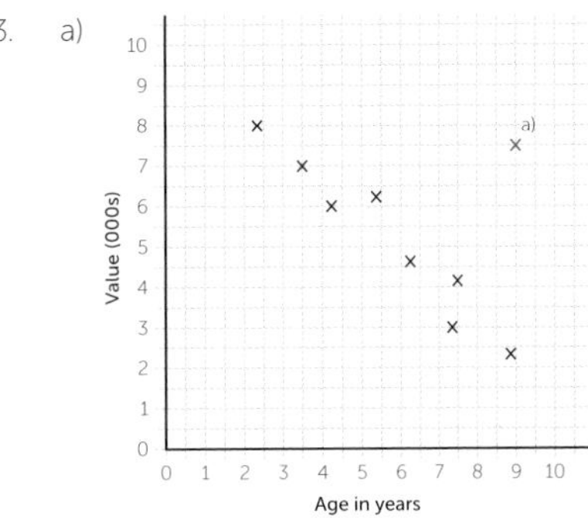


b) Its value is higher than would be predicted by the line of best fit. It is an outlier
c) e.g.
   Its new price may have been much more than that of the other cars
   Its value has started to appreciate, given its age
   It's a more popular make, therefore people will pay more for it
   It is in very good condition for its age.

4. Hattie's line does not need to go through the origin.
   Whilst Hattie's line of best fit has half the points either side it should follow the line of the points.
   The line of best fit should stay within the range of the points.

## 21.5 Correlation

1. a) Positive correlation b) No correlation
   c) Positive correlation d) Negative correlation
   e) Negative correlation

2.

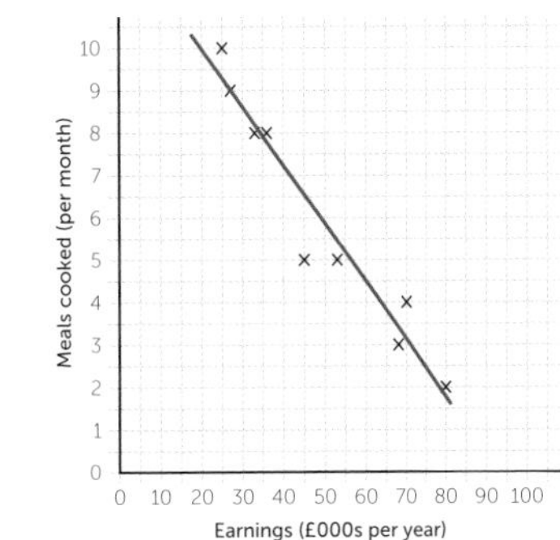


a) The more people earn, the fewer meals they cook per month
b) (Strong) negative correlation
c) See diagram d) Between 4 and 5
e) People who earn more can afford to go out to eat.

3. a) (Strong) positive correlation

b)

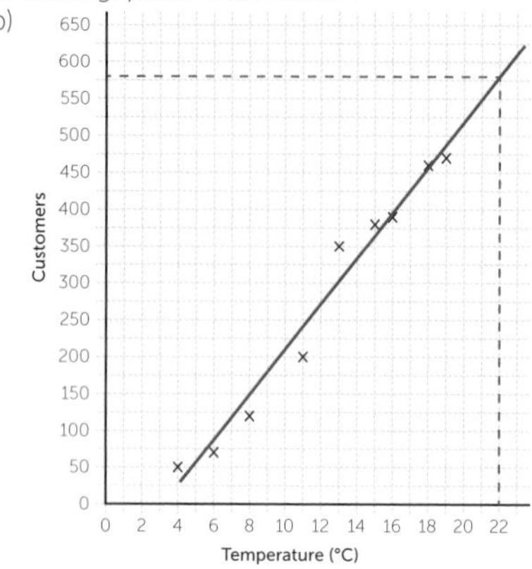


c) 13°C

d) The reading is correct, but it is outside the range of the data therefore unreliable and should not be used.

4. a) Positive correlation

b) Weak as the points aren't very close to the line of best fit.

c) 62 years old

d) They are outside the range of ages on the scatter graph.

5. There is a third variable (outside temperature) that causes both sunglasses sales and ice cream sales to increase.

# Section 22

## 22.1 Sample size

1. a) $\frac{25}{120} = \frac{5}{24}$ b) $\frac{40}{240} = \frac{1}{6}$

c) the second one should be more reliable as it is based on more trials.

2. a) blue = 0.4, green = 0.3, yellow = 0.2 and red = 0.1

b) blue = 0.45, green 0.25, yellow = 0.2 and red = 0.1

3. a)

| | Student 1 | Student 2 | Student 3 | Total |
|---|---|---|---|---|
| **Yes** | 4 | 29 | 47 | 80 |
| **No** | 16 | 21 | 53 | 90 |
| **Number asked** | 20 | 50 | 100 | 170 |

b) Student 1 = $\frac{4}{20}$ = 0.2, student 2 = $\frac{29}{50}$ = 0.58,

student 3 = $\frac{47}{100}$ = 0.47

c) Relative frequency for total surveyed = $\frac{80}{170}$ = 0.47

d) Student 3 as they did the most trials.

e)

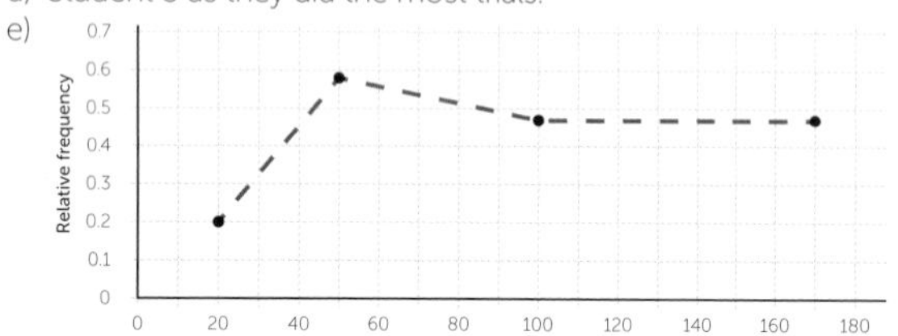


f) 0.47

4. a) Tea 0.35, Coffee 0.22, Soft drinks 0.33, Other 0.1

b) day 2 = 0.36, day 3 = 0.375, day 4 = 0.35

c) total tea = 35 + 18 + 45 + 28 = 126

total sales = 100 + 50 + 120 + 80 = 350

relative frequency for tea across all days = 0.36

d)

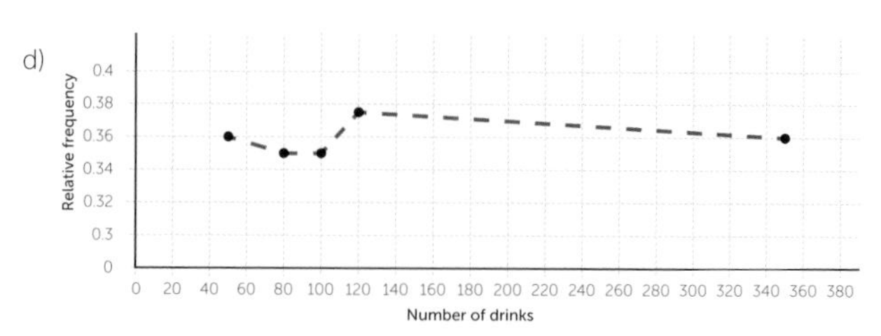


e) estimated probability ≈ 0.36 (using all available data).

## 22.2 Tree diagrams

1. 0.7

2.

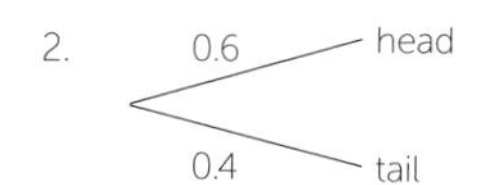


3. a)

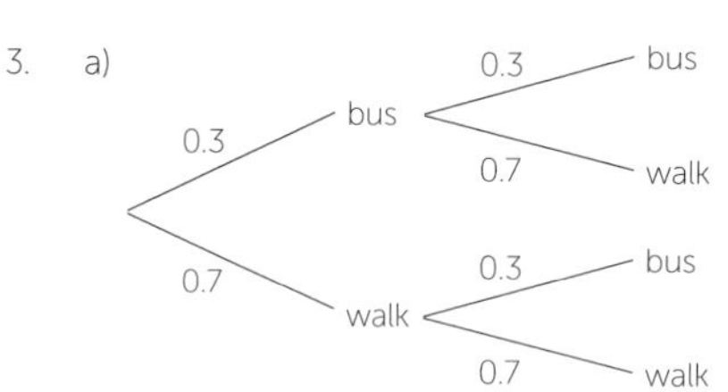


b) 0.7 x 0.7 = 0.49

4. a)

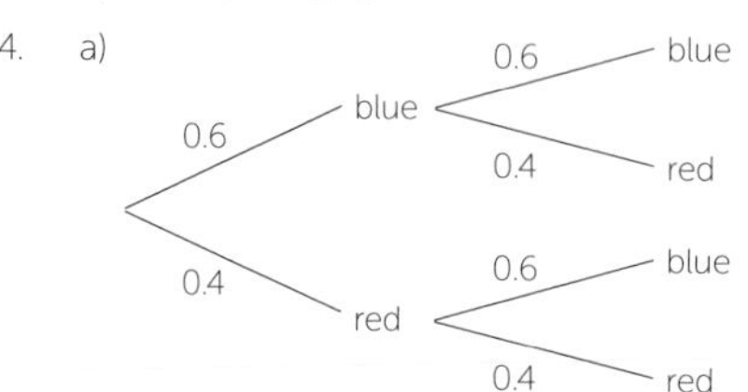


b) 0.4 x 0.4 = 0.16 c) 0.6 x 0.4 = 0.24 d) 0.24 + 0.24 = 0.48

5. a) For e.g. a fair coin and a four-sided spinner, with 1 red side and 3 non-red sides.

b) 0.5 x 0.25 = 0.125

6. a) She has put the number of outcomes on the branches rather than the probabilities. It should look like this:

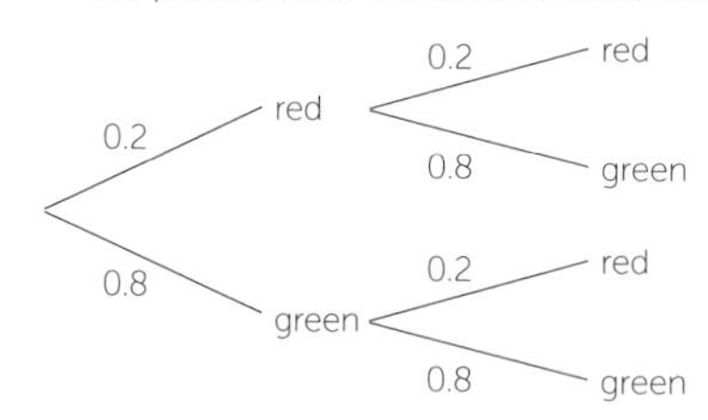


b) 0.2 x 0.8 x 2 = 0.32 (Or, 0.16 + 0.16 = 0.32)

## 22.3 Venn diagrams

1. a)

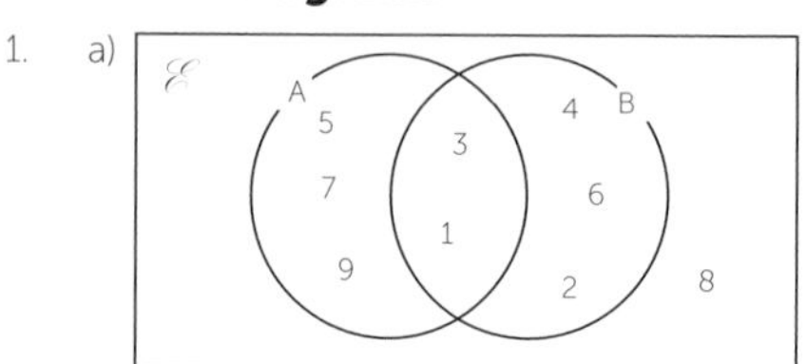


b) 1 and 3 are in the intersection

2. a) 11 b) 9 c) 3 d) 30

3. a) 8 + 10 + 6 + 4 = 28 b) 6 + 10 = 16 c) $\frac{4}{7}$

d) 10 e) they did both track and field events.

4. a) 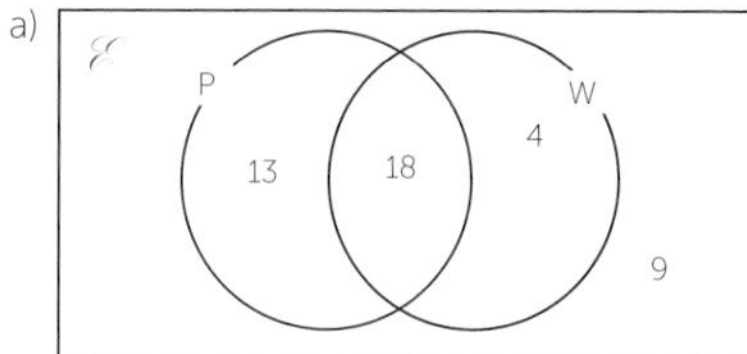


b) 13 + 18 + 4 = 35

c) these people either play or watch sport or do both.

5. a) 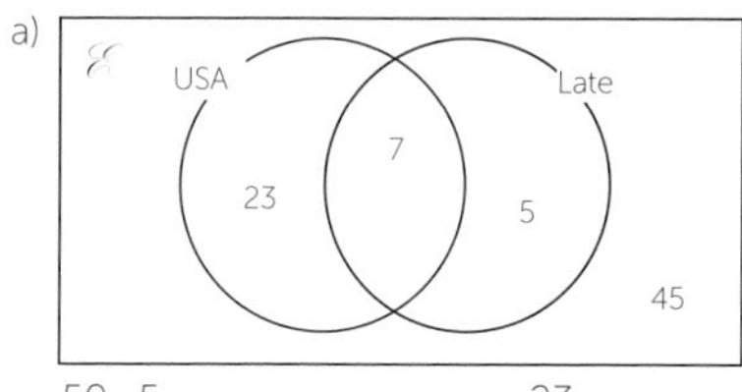


b) $\frac{50}{80}$, $\frac{5}{8}$ or 0.625 c) $\frac{23}{80}$

6. a) 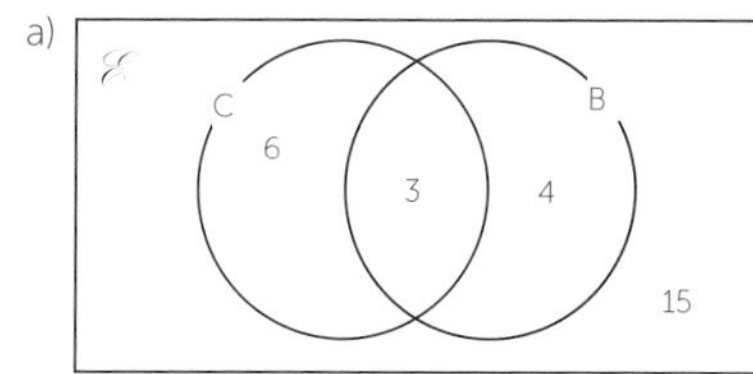


b) they are either in the band or in the choir or in both

c) $\frac{13}{28}$ or 0.46

7. a) 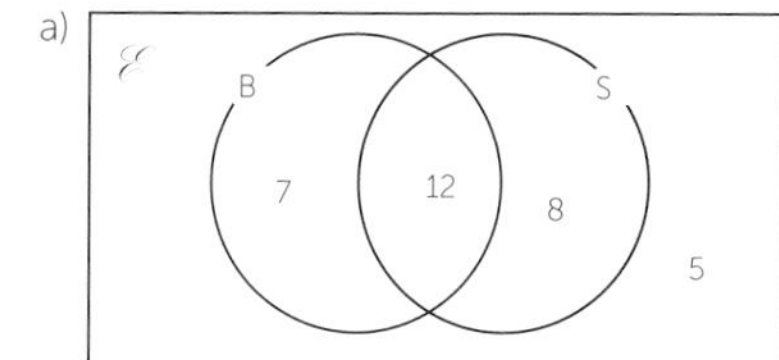


b) $\frac{5}{32}$ or 0.16

## 22.4 Dependent events

1. $\frac{3}{4}$ or 0.75

2. $\frac{13}{29}$ or 0.45

3. 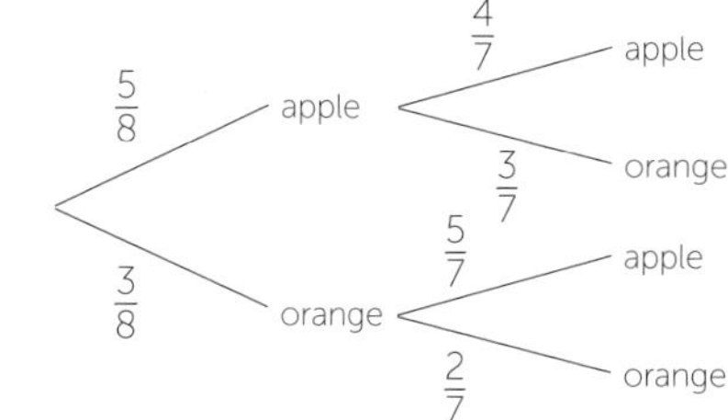


4. 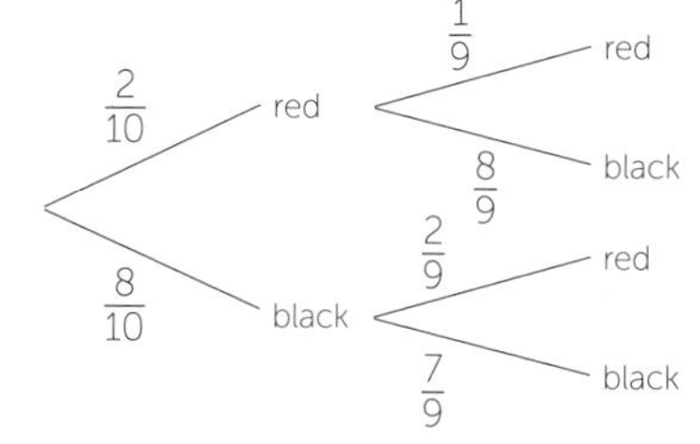


b) P(Black, black) = $\frac{8}{10} \times \frac{7}{9} = \frac{56}{90} = \frac{28}{45}$

5. P(same colour) = P(blue, blue) + P(black, black)

$= \frac{6}{12} \times \frac{5}{11} + \frac{6}{12} \times \frac{5}{11} = \frac{5}{11}$ so no, she is not correct.

P(different colour) = $\frac{6}{12} \times \frac{6}{11} + \frac{6}{12} \times \frac{6}{11} = \frac{6}{11}$

6. 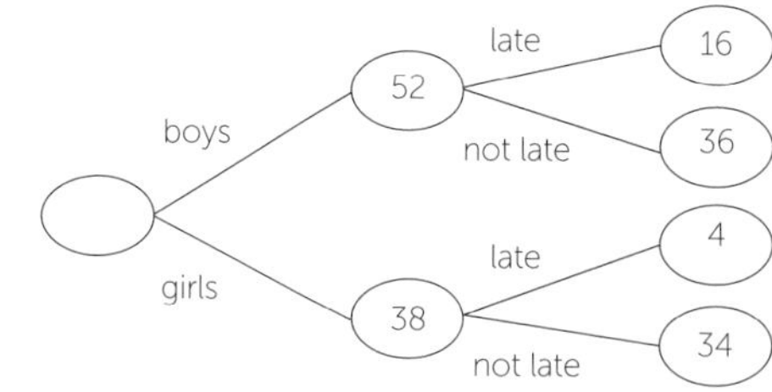


7. a) 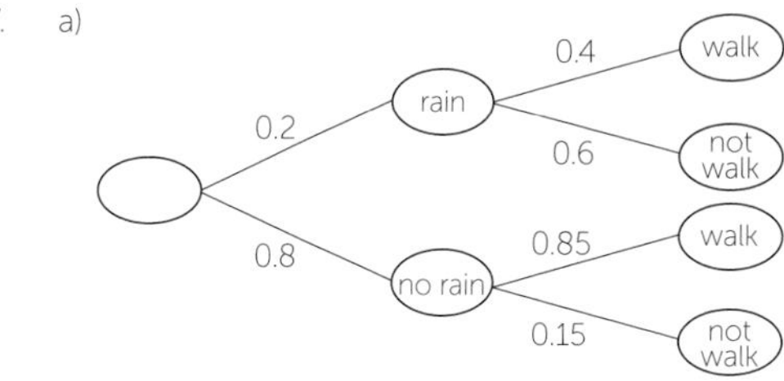


b) P(walk) =
P(rain, walk) + P(no rain, walk) = 0.2 × 0.4 + 0.8 × 0.85 = 0.76

8. a) 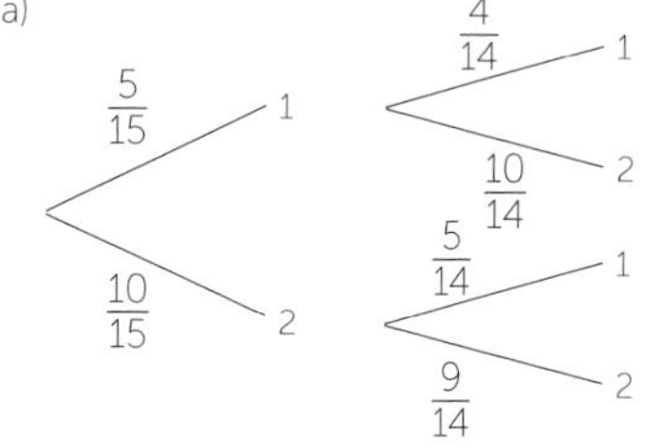


b) P(1, 1) + P(2,2) = $\frac{5}{15} \times \frac{4}{14} + \frac{10}{15} \times \frac{9}{14} = \frac{11}{21}$

c) P(score of 3) = P(1, 2) + P(2, 1) = $\frac{5}{15} \times \frac{10}{14} + \frac{10}{15} \times \frac{5}{14} = \frac{10}{21}$

## 22.5 Using tables and diagrams

1. a)

| | Biology | Chemistry | Physics | Total |
|---|---|---|---|---|
| **Female** | 17 | 18 | 11 | 46 |
| **Male** | 20 | 23 | 7 | 50 |
| **Total** | 37 | 41 | 18 | 96 |

b) $\frac{3}{16}$ c) $\frac{25}{48}$

2. a)

| | Yes | No | Total |
|---|---|---|---|
| **Men** | 149 | 143 | 292 |
| **Women** | 124 | 228 | 352 |
| **Total** | 273 | 371 | 644 |

b) $\frac{273}{644} = 0.424$ c) $\frac{292}{644} = 0.546$ d) $\frac{228}{371} = 0.615$

3. a)

| | regular | sugar-free | Total |
|---|---|---|---|
| **Girls** | 21 | 46 | 67 |
| **Boys** | 43 | 27 | 70 |
| **Total** | 64 | 73 | 137 |

b) $\frac{70}{137}$ c) $\frac{73}{137}$ d) $\frac{21}{67}$

4. a)

| | mobile | no mobile | total |
|---|---|---|---|
| **laptop** | 6 | 3 | 9 |
| **no laptop** | 12 | 4 | 16 |
| **total** | 18 | 7 | 25 |

b) 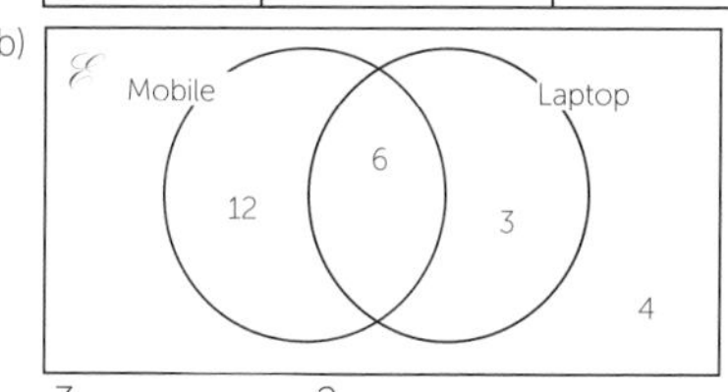


c) $\frac{3}{25}$ d) $\frac{2}{3}$

5. a) 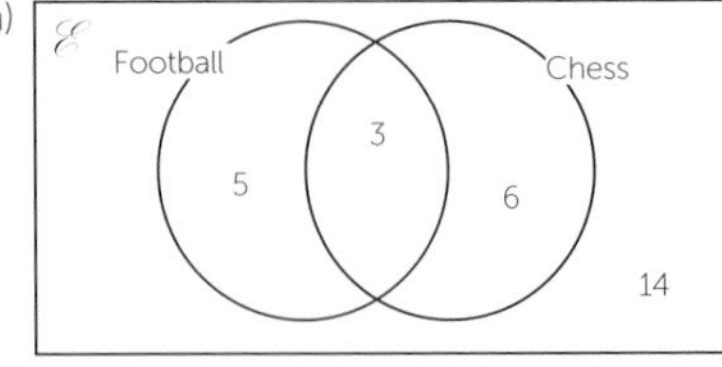


b) 14 c) $\frac{9}{28}$ d) $\frac{3}{8}$

6. a) 94

b)

| | Cat | No cat | total |
|---|---|---|---|
| **dog** | 12 | 33 | 45 |
| **no dog** | 24 | 25 | 49 |
| **total** | 36 | 58 | 94 |

c) $P(C) = \frac{36}{94}$ $P(C \cap D) = \frac{12}{94}$ $P(C \cup D) = \frac{69}{94}$

## 22.6 Solving probability problems

1. a) 0.02 b) 0.05 c) 0.93

2. a) 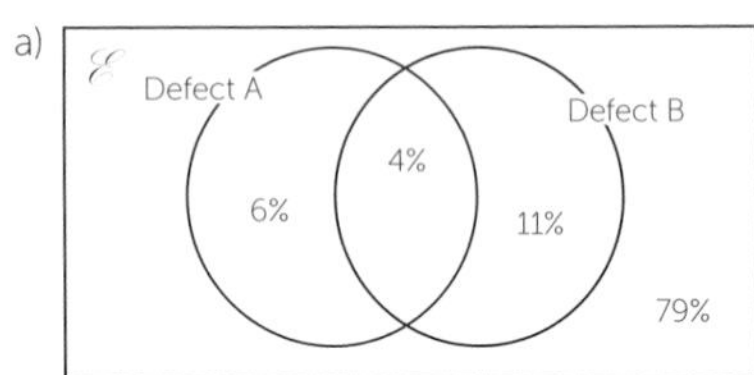


b) 0.06 c) 0.11 d) 0.79

3. No, he is not correct. For every 5 sticks, 1 is white so the probability of picking a white stick is $\frac{1}{5}$

4.  a) 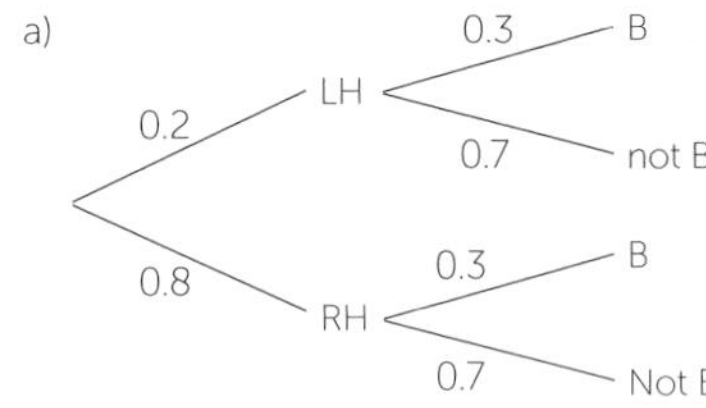


b) P(LH and B) = 0.2 × 0.3 = 0.06 c) P(either or both) = 0.44

5. No, she is not correct; she should have multiplied the probabilities. The answer is $\frac{1}{216}$

6.  a) 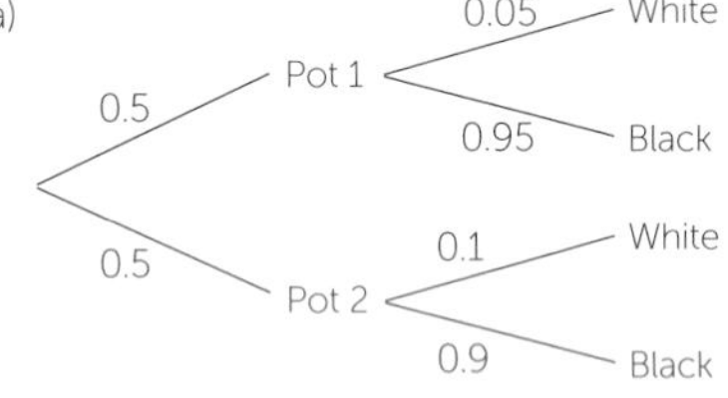


b) (0.5 × 0.05) + (0.5 × 0.1) = 0.075

7. a) 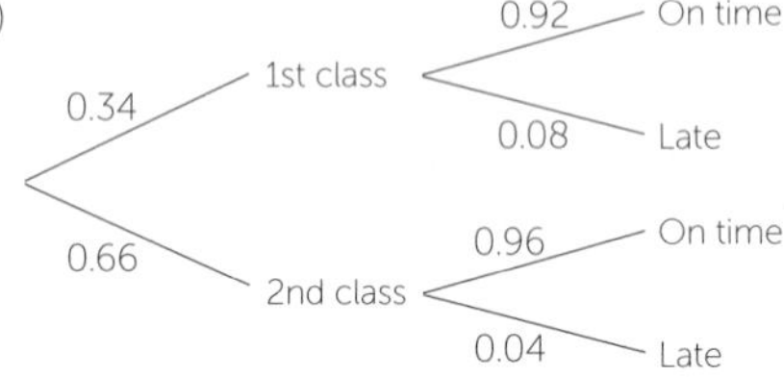


b) Yes, Lily is correct.
P(on time) = (0.34 × 0.92) + (0.66 × 0.96) = 0.946 4 = 94.64%.
This is less than 95%.

8. a) BBB BBG BGB BGG GGG GGB GBG GBB

b) $\frac{1}{8}$

c) $\frac{3}{8}$

9. a)

| | Red | Not red | Total |
|---|---|---|---|
| **2 door** | 0 | 5 | 5 |
| **4 door** | 6 | 7 | 13 |
| **5 door** | 6 | 11 | 17 |
| **Total** | 12 | 23 | 35 |

b) $\frac{13}{35}$ c) $\frac{6}{12}$ or $\frac{1}{2}$

# Section 23

## 23.1 Vectors

1. a) 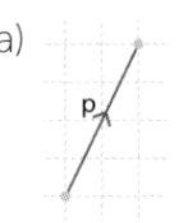


b) q

c) 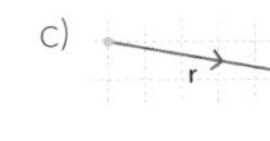


d) s

e) 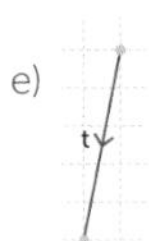


2. $\begin{pmatrix}1\\2\end{pmatrix}$ $\begin{pmatrix}-1\\2\end{pmatrix}$ $\begin{pmatrix}1\\-2\end{pmatrix}$ $\begin{pmatrix}-1\\-2\end{pmatrix}$ $\begin{pmatrix}2\\1\end{pmatrix}$ $\begin{pmatrix}-2\\1\end{pmatrix}$ $\begin{pmatrix}2\\-1\end{pmatrix}$ $\begin{pmatrix}-2\\-1\end{pmatrix}$

3. a) $\overrightarrow{MN} = \begin{pmatrix}2\\1\end{pmatrix}$, $\overrightarrow{NP} = \begin{pmatrix}3\\-6\end{pmatrix}$ and $\overrightarrow{PM} = \begin{pmatrix}-5\\5\end{pmatrix}$

b) $\overrightarrow{NM} = \begin{pmatrix}-2\\-1\end{pmatrix}$, $\overrightarrow{PN} = \begin{pmatrix}-3\\6\end{pmatrix}$ and $\overrightarrow{MP} = \begin{pmatrix}5\\-5\end{pmatrix}$

c) Each pair has equal coordinates and magnitude (length) but has opposite directions

4. a) 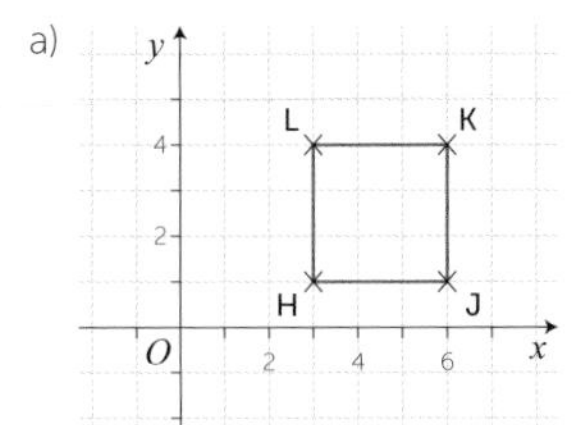


b) $\overrightarrow{HJ} = \begin{pmatrix}3\\0\end{pmatrix}$, $\overrightarrow{JK} = \begin{pmatrix}0\\3\end{pmatrix}$ c) $\overrightarrow{HL} = \begin{pmatrix}0\\3\end{pmatrix}$, $\overrightarrow{LK} = \begin{pmatrix}3\\0\end{pmatrix}$

d) $\overrightarrow{HJ} = \overrightarrow{LK}$ They are parallel e) $\overrightarrow{HL} = \overrightarrow{JK}$ They are parallel

5. a) 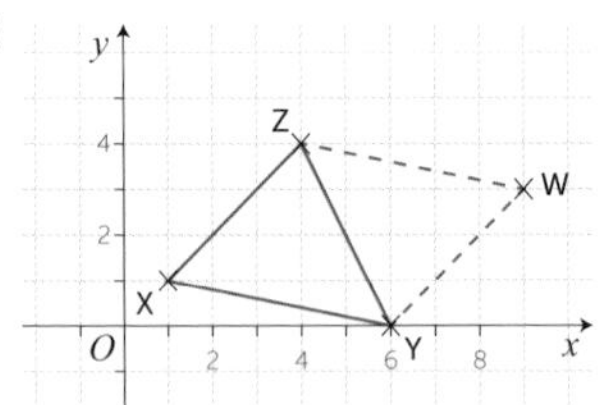


b) $\overrightarrow{XZ} = \begin{pmatrix}3\\3\end{pmatrix}$ c) W is (9, 3)

d) XYWZ is a parallelogram e) XY and ZW are equal and parallel

6. a) 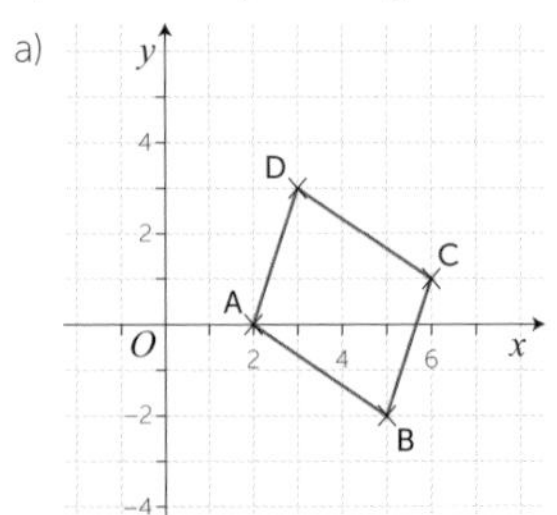


b) $BC = \begin{pmatrix}1\\3\end{pmatrix}$ c) C (6, 1)

7. $\overrightarrow{AC} = \begin{pmatrix}8\\6\end{pmatrix}$

8. a) $h + j = \begin{pmatrix}5\\4\end{pmatrix}$ and $j + h = \begin{pmatrix}5\\4\end{pmatrix}$ They are equal.

b) h + (j + k) = $\begin{pmatrix}1\\3\end{pmatrix}$ and (h + j) + k = $\begin{pmatrix}1\\3\end{pmatrix}$ They are equal.

c) h − j = $\begin{pmatrix}-1\\4\end{pmatrix}$ and j − h = $\begin{pmatrix}1\\-4\end{pmatrix}$ The vectors are equal in magnitude but opposite in direction.

9. 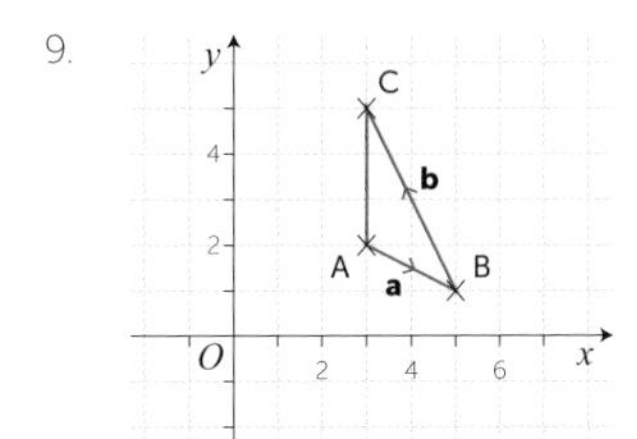


a) **a** = $\begin{pmatrix}2\\-1\end{pmatrix}$ b) C = (3, 4) c) $\overrightarrow{AC}$ = $\begin{pmatrix}0\\2\end{pmatrix}$ = **a** + **b**

10. a) C = (5, 5) b) **a** + **b** = $\begin{pmatrix}4\\3\end{pmatrix}$ It represents the vector $\overrightarrow{AC}$.

c) $\begin{pmatrix}-1\\-2\end{pmatrix}$ d) D = (3, 1) e) $\overrightarrow{AD}$ = $\begin{pmatrix}2\\-1\end{pmatrix}$

11. a) 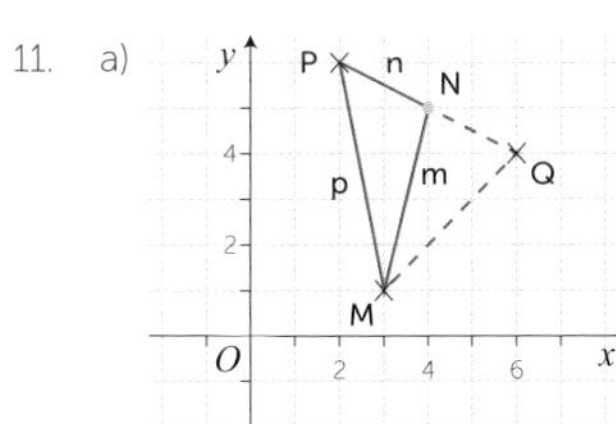


b) Q = (6, 4) c) R = (2, −3)

## 23.2 Parallel vectors

1. a) $\begin{pmatrix}12\\20\end{pmatrix}$ b) $\begin{pmatrix}1.5\\2.5\end{pmatrix}$ c) $\begin{pmatrix}-6\\-10\end{pmatrix}$ d) $\begin{pmatrix}-0.3\\-0.5\end{pmatrix}$

2. a) 2**b** b) −**b**

c) 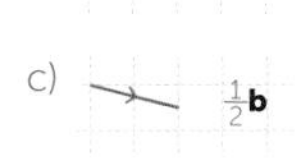


d) −3**b**

3. a) $\begin{pmatrix}-6\\9\end{pmatrix}$ b) $\begin{pmatrix}8\\-12\end{pmatrix}$ c) $\begin{pmatrix}30\\12\end{pmatrix}$ d) $\begin{pmatrix}-1\\1.5\end{pmatrix}$

e) $\begin{pmatrix}-5\\-2\end{pmatrix}$ f) $\begin{pmatrix}12.5\\5\end{pmatrix}$

4. a) $\overrightarrow{AB}$ = $\begin{pmatrix}1\\4\end{pmatrix}$, $\overrightarrow{CD}$ = $\begin{pmatrix}2\\8\end{pmatrix}$ b) $\overrightarrow{AB}$ = $\frac{1}{2}$**p**

c) $\overrightarrow{AB}$ and $\overrightarrow{CD}$ are parallel.

5. L has coordinates (6, 2)

6. a) **a** = 2**p** + 3**q** **b** = −**p** + 4**q** **c** = 5**p** − **q**
**d** = −3**p** + 12**q** **e** = 4**p** + 6**q** **f** = 10**p** − 2**q**
b) **a** and **e** are parallel, **b** and **d** are parallel, **c** and **f** are parallel.
c) **e** = 2**a**, **d** = 3**b**, **f** = 2**c**

## 23.3 Arcs and sectors

1. a) $5\pi$ cm b) $5\pi$ cm c) $\frac{3}{2}\pi$ m

2. a) $(5\pi + 20)$ cm b) $(5\pi + 10)$ cm c) $(\frac{3}{2}\pi + 18)$ m

3. a) $\frac{16}{9}\pi$ cm b) $7\pi$ m c) $2\pi$ cm

d) $\frac{96}{25}\pi$ m e) $\frac{416}{3}\pi$ mm f) $\frac{772}{25}\pi$ cm

4. $36\pi$ cm²

5. $\frac{375}{4}\pi$ m²

6. a) 72° b) $4\pi$ m

7. Radius = 5 cm, arc length = $\frac{5\pi}{6}$ cm

8. a) $\frac{1}{3}\pi p^2$ cm² b) $\frac{2}{3}\pi p$ cm

9. a) 280° b) $\frac{56\pi}{3}$ m

10. a) 2312 cm² b) 170 cm² c) 2143 cm²

## 23.4 Composite circle shapes

1. $(100 - 25\pi)$ m²

2. Perimeter = $10\pi$ cm, area = $\frac{75}{4}\pi$ cm²

3. $32\pi$ m²

4. $(4 + 12\pi)$ m

5. a) $128\pi$ m² b) $64\pi$ m² c) $32\pi$ m² d) $16\pi$ m²

6. Perimeter = $12 + 6\pi$ cm, area = $18\pi$ cm²

7. a) $(160 + 50\pi)$ cm b) $(4000 + 625\pi)$ cm²

8. Perimeter = $(9 + \frac{27}{2}\pi)$ m, area = $\frac{243}{8}\pi$ m²

9. a) $1250\pi$ cm b) 4 times

## 23.5 Surface area

1. 48 cm²

2. 571.176 cm²

3. a) $120\pi$ cm² b) $78\pi$ cm² c) $68\pi$ cm² d) $\frac{1575}{2}\pi$ cm²

4. A has surface area $130\pi$ cm², B has surface area $208\pi$ cm². B has the larger total surface area. They have the same curved surface area.

5. $900\pi$ cm²

6. 50 400 m²

7. a) Radius = 5 cm b) $90\pi$ cm² c) $115\pi$ cm²

8. a) $384\pi$ cm² b) $512\pi$ cm² c) $896\pi$ cm²

## 23.6 Volume of complex shapes

1. a) $72.9\pi$ m³ b) $45p^2q\pi$ cm³

2. No, she is not correct. The glass is wider at the top and therefore the volume in the top 3 cm is greater than $\frac{1}{4}$ of the total volume.

3. a) $105.966\pi$ cm³ b) £10.17

4. $450\pi$ cm³

5. $1.08 \times 10^{12}$ km³

6. $2.45 \times 10^{6}$ m³

7. a) $2700\pi$ cm³ b) $100\pi$ cm³ c) $2600\pi$ cm³

8. a) $(36\,000 + 9000\pi)$ cm³ b) $(9600 + 900\pi)$ cm³

# Section 24

## 24.1 Terminating and recurring decimals

1. a) $\frac{57}{100}$ b) $\frac{3}{20}$ c) $\frac{7}{10}$ d) $\frac{1}{8}$ e) $\frac{239}{1000}$

f) $\frac{29}{500}$ g) $\frac{17}{1000}$ h) $\frac{109}{1000}$ i) $1\frac{19}{50}$ j) $3\frac{4}{5}$

2. a) 0.43 b) 0.8 c) 0.7 d) 0.23 e) 0.06

f) 0.345 g) 0.053 h) 0.205 i) 1.23 j) 1.8

3. a) $\frac{3}{10}$ b) $\frac{1}{4}$ c) $\frac{9}{20}$ d) $\frac{29}{50}$ e) $\frac{3}{4}$

f) $\frac{13}{20}$ g) $\frac{18}{25}$ h) $\frac{1}{8}$ i) $\frac{3}{8}$ j) $1\frac{1}{5}$

4. a) $\frac{7}{50} = \frac{14}{100} = 14\%$ b) $\frac{9}{25} = \frac{36}{100} = 36\%$
   c) $\frac{7}{20} = \frac{35}{100} = 35\%$ d) $\frac{13}{40} = \frac{32.5}{100} = 32.5\%$

5. **$0.8 = \frac{4}{5}$** **$36\% = \frac{9}{25}$** **$0.375 = \frac{3}{8}$** **$5\% = \frac{1}{20}$**
   **$12.5\% = \frac{1}{8}$** **$0.12 = \frac{3}{25}$** **$0.225 = \frac{9}{40}$** **$0.96 = \frac{24}{25}$**

6. a) $\frac{3}{7} = 0.43 = 43\%$ b) $\frac{19}{30} = 0.63 = 63\%$ c) $\frac{13}{28} = 0.46 = 46\%$
   d) $\frac{14}{45} = 0.31 = 31\%$ e) $\frac{21}{32} = 0.66 = 66\%$

7. a) $\frac{7}{17} = 0.58\dot{3}$ b) $\frac{5}{8} = 0.625$ c) $\frac{1}{9} = 0.\dot{1}$ d) $\frac{1}{6} = 0.1\dot{6}$
   b) is the odd one out as it is not a recurring decimal.

8. a) $\frac{13}{16} > 0.76$ b) $0.65 > \frac{13}{21}$ c) $5.25 > 5.25\%$
   d) $18.75\% = \frac{3}{16}$ e) $\frac{3}{40} = 7.5\%$ f) $58.5\% > \frac{7}{12}$

9. $\frac{3}{8} = 37.5\ \%$
   37.5% < 38% of the audience are over 18.
   There are more 14 to 17 year olds than over 18 year olds.

10. 0.6875 litres

11. a) 0.090 909...; 0.181 818...; 0.272 727...; 0.363 636...; 0.454 545...
    The decimal part is the multiple of 9.
    b) 0.636 363...

## 24.2 Density and pressure

1. a) 3.8 km = 3800 m b) 4.2 tonne = 4200 kg
   c) 3400 ml = 340 *cl* d) 6700 g = 6.7 kg
   e) 1800 mm =180 cm f) 0.85 litres = 850 *ml*

2. a) False 790 m = 0.79 km b) True
   c) False 800 mm = 80 cm d) True
   e) True f) True

3. a) 5 litres = 5 000 *ml* b) 8900 m = 8.9 km
   c) 2500 g = 2.5 kg d) 4300 mm = 430 cm
   e) 600 cl = 6 litres f) 9.7 tonne = 9700 kg

4.

| substance | mass (g) | volume ($cm^3$) | density (g/$cm^3$) |
|---|---|---|---|
| **A** | 7600 | 200 | **38** |
| **B** | 6510 | 150 | **43.4** |
| **C** | 9800 | 140 | **70** |

5. a) 10.5 kg b) 8 litres

6.

| force (N) | area ($m^2$) | pressure ($N/m^2$) |
|---|---|---|
| 39 | 13 | **3** |
| 624 | **48** | 13 |
| **64** | 16 | 4 |
| 1449 | 80.5 | **18** |

7. a) 11.6 $N/cm^2$
   b) Fern is correct. The force and the area do not change so the pressure will remain the same. Area = 4 × 4 = 8 × 2

8. 1.6 $g/cm^3$.

9. a) 550 $N/cm^2$
   b) The area will be greater so the pressure will be lower.

10. The density is 19.32 $g/cm^3$ and does not change with volume. Hannah is incorrect.

11. Ben needs to find the mass of substance A and the mass of substance B, then add the masses together.
    Mass of Substance A = 30 × 18 = 540 g
    Mass of Substance B = 70 × 16 = 1120 g
    Mass of Substance C = 540 g + 1120 g = 1660 g

## 24.3 Compare lengths, areas and volumes

1. B) 9 cm or D) 16 cm

2. $x$ = 9 cm, $y$ = 5 cm

3. a) 5 $m^3$ = 5 000 000 $cm^3$ b) 8000 $cm^2$ = 800 000 $mm^2$
   c) 25 000 $cm^2$ = 2.5 $m^2$ d) 3500 ml = 3.5 litres
   e) 23 000 $mm^2$ = 230 $cm^2$ f) 520 $cm^3$ = 520 ml

4. 5: 1

5. 50 000 $cm^2$ = 5 $m^2$
   5000 $mm^2$ = 50 $cm^2$
   500 000 $cm^3$ = 0.5 $m^3$
   5000 $mm^3$ = 5 $cm^3$
   50 000 $cm^3$ = 50 litres

6. a) 2 : 5 b) 8 : 125

7. Volume of cube = 8 cm × 12 cm × 6 cm = 576 $cm^3$
   Half of the tank = 288 $cm^3$= 0.288 litres. Delphi is correct.

8. $x$ = 2 cm

9. One side is $\frac{5}{3}$ times bigger. The other side is $\frac{6}{4}$ times bigger. $\frac{5}{3} \neq \frac{6}{4}$

10. Area of 1st wall = 3 m × 3.5 m = 10.5 $m^2$.
    Area of 2nd wall = 2.8 m× 3.5 m = 9.8 $m^2$.
    Area of both walls = 10.5 $m^2$ + 9.8 $m^2$ = 20.3 $m^2$.
    Two coats of painting = 20.3 $m^2$ × 2 = 40.6 $m^2$
    Helen can paint 44$m^2$ of wall. She has enough paint.

11. Volume of tank = $(40\ cm)^3$= 64 000 $cm^3$= 64 000ml
    64 000 ÷ 200= 320 seconds = 5 minutes 20 seconds. Harry is correct.

## 24.4 Interpreting gradients

1. a) C b) B c) A

2. a) A took 55 minutes, B took 50 minutes, C took 40 minutes.
   b) Speed of A = 0.18 km/minute (2 d.p.) Speed of B = 0.2 km/minute
   Speed of C = 0.25 km/minute

3. a) = 0.5 $m/s^2$ b) = 1 $m/s^2$ c) = 1.3 $m/s^2$ d) = 0 $m/s^2$
   ascending order: d, a, b, c.

4. a) C b) −2 $m/s^2$

5. a) 24° in July b) 3 months (Jan, Feb, Dec)
   c) Yes, he is correct. Temperature rose by 8° in April and fell by 8° in September.

6. Colony A. The time to reach the same number of ants is less for Colony A.

7. a) 3.3 litres of fuel per hour (1 d.p.) b) 6 hours

8. Graph iii)

9. Both graphs show that the speed changes by 8 m/s in 4 seconds. The first graph shows a deceleration of 2 $m/s^2$. The second graph shows an acceleration of 2 $m/s^2$.

10. a) Selim travelled at a slower pace than Lira.
    b) Lira took less time than Selim to get to town.
    c) Lira caught up with Selim after 20 minutes.

## 24.5 Direct and inverse proportion

1. a) 27 g b) 243 g c) 297 g

2. a) 25p b) £2.25 c) £3

3. a) 800 g b) 480 g c) 300 g

4. B is the odd one out (direct proportion). All the others are in inverse proportion.

5. a) Yes. The graph is a straight line passing through the origin.
   b) 8 km c) 3.1 miles

6. Graph C

7. a) 21 days b) 21 days
   c) Everyone works at the same rate and for the same amount of time each week.
   They never have to wait while another part is completed.

8. a) 10 days b) 10 people
   c) Everyone is working at the same rate and for the same amount of time each day.
   They do not have to wait while the paint dries between coats.
   They can all be painting at the same time.
9. a) $D = 15$ b) $T = 14$
10. a) Only true if $P = Q$.
    b) Always true. $P = kQ$, so $2P = 2kQ$

## 24.6 Compound interest

1. a) £1157.63 b) £2339.72 c) £11 876.86
2. D £2122.42
3. a) £120 b) £371.64
4. a) £446.06
5. a) £50 000 b) 2.7% c) 3 years
6. a) Connor has used a simple interest formula.
   b) £1021.03 – £800 = £221.03
7. A and C are incorrect and should not be used.
8. C 8 years
9. £1622. 40 ÷ ($1.04^2$) = £1500
10. Account A = £1993.33, Account B = £2074.85, Account C = £2078.60
    Account C
11. a) Account A £80 638, Account B £80 714, Account B has more
    b) Account A £75 131, Account B £75 494
    No, Account B has more after 4 years.

## 24.7 Growth and decay problems

1. a) 1.03 b) 1.3 c) 0.93 d) 0.7
2. a) True b) True
   c) False. 0.88 × 1.09 d) False. 1.07 × 1.08
3. a) £143 000 b) £154 669 c) £203 534
4. a) 105 981 people b) 40 981 people
5. d) £5832
6. $\frac{5}{6} \div 20 \times 100 = 4.17\%$ (2 d.p.)
7. £3172.61
8. In the first year = 115% = 1.15
   In the second year = 1.15 × 0.96 = 1.104
   The value of the house has increased by 10.4%.
9. A 0.4096
10. % increase = 1500 ÷ 8500 × 100 = 17.65% (2 d.p.)
11. Saturday = $x$ time, Sunday = $1.3x$ time, Monday = $0.7 \times 1.3x = 0.91x$
    Jonathan uses less time on Monday compared to Saturday.

# Section 25

## 25.1 Linear equations

1. Trevor should have added 5 to the number ($x$) and then multiplied the result by 3, to get $3(x + 5) = 27$
2. a) $y = 4$ b) $a = 4.5$ c. $x = -3.5$
   d) $n = 5$ e) $p = 8$
3. $y = 38$
4. Iris changed the signs on the left hand side but not the right hand side, before adding 25 to both sides; answer should be:
   Subtract 25 from both sides: $-4q = -17$
   Divide both sides by –4: $q = 4.25$
5. $p = 2$
6. Kelly expanded the bracket incorrectly
   Ian missed a negative sign ($-3x$) after subtracting 23.
   The solution is:
   $23 - 3x = 2(x + 4)$
   Expand the bracket: (Kelly made an error here)
   $23 - 3x = 2x + 8$
   Subtract 23 from both sides: (Ian has written $3x$ instead of $-3x$)
   $-3x = 2x - 15$
   Subtract $2x$ from both sides:
   $-5x = -15$
   Divide both sides by –5:
   $x = 3$
7. $n = 14$
8. Angles are 69°, 69° and 42°
9. Angie is 8 years old.
10. $x = 4.25$

## 25.2 Simultaneous equations 1

1. What John has done is correct, but he hasn't found $y$.
   $x = 3, y = 2$
2. a) $x = 5, y = 3$ b) $a = -1, b = 3$
   c) $p = 2, q = -1$ d) $m = 0.5, n = 1.5$
3. a) $6x + 4y = 22$ b) $10x + 4y = 32$ c) $x = 2.5, y = 1.75$
4. coffee = £2.20 and tea = £1.40 so 1 coffee + 1 tea = £3.60
5. a) $x = 4, y = 2$ b) $m = -2, n = 4$
   c) $p = 1, q = -2$ d) $j = 7, k = -3$
6. $a = 11, b = 6$
7. Adult: £38 Child: £14
8. a) $a = 3, b = 1$ b) $x = 5, y = -1$
   c) $t = 2.5, u = 1.5$ d) $m = -4, n = -1$
9. $x = 3.5, y = 4.5$
10. $x = -1, y = 3$

## 25.3 Simultaneous equations 2

1. a) $x = 3, y = 3$ b) $p = 14, q = 7$
   c) $a = 6, b = 2$ d) $m = 1, n = 5$
2. Dylan has substituted incorrectly.
   It should be $3x + 2(2x) = 14$ so $x = 2$ and $y = 4$
3. a) $a = 6, b = 4$ b) $p = 4, q = 7$
   c) $x = 7, y = -2$ d) $t = -2, u = 1$
4. $a = 13, b = 4$
5. Red: 3.5 kg Blue: 10 kg
6. a) $x = -4, y = -13$ b) (–4,–13)

## 25.4 Solving quadratic equations

1. a) $y = -1$ or $-3$ b) $a = 1$ or 4 c) $n = -2$ or $-4$
   d) $d = -4$ or $-6$ e) $x = 3$ or 4
2. Harriet factorised incorrectly; the solution should be:
   $(x - 2)(x - 5) = 0$
   Solutions are $x = 2$ or $x = 5$
3. a) (3,0) or (5,0) b) $x = 3$ or 5
4. a) $q = -5$ or 2 b) $t = 0$ or $-5$
   c) $h = -2$ or 3 d) $y = -4$ or $-5$
5. Tyson's solution is correct but his workings are not; should be:
   $(n - 7)(n + 2) = 0$
   $n = 7$ and $-2$
6. a) $(x + 3)(x - 2) = 36$
   $x^2 + x - 6 = 36$
   $x^2 + x - 42 = 0$
   b) $x = -7$ or 6
   c) You cannot have a negative length so 6 is the only sensible answer
7. $y^2 - 2y - 8 = 7$
   $y^2 - 2y - 15 = 0$
   $(y - 5)(y + 3) = 0$
   $y = 5$ and $-3$
8. Intersects $x$-axis at (6,0) and (–5,0), $y$-axis at (0,–30)

## 25.5 Working with formulae

1. a) £11 b) £17.50 c) £28.30
   d) 40 kilometres e) 102 kilometres

2. 56 cm$^2$

3. a) 10 cakes b) 19 cakes c) 32 people
   d) 20 people require 10 cakes and 40 people require 15 cakes (the +5 is the issue)

4. Jasmine needs to divide the whole of $t + 3$ by 4.

   Answer should be $n = \frac{t+3}{4}$

5. a) $t = \frac{d}{7}$ b) $n = m - 9$ c) $h = \frac{c-10}{6}$
   d) $g = 10 - h$ e) $x = 2(y + 5)$ or $x = 2y + 10$

6. $h = \frac{c - xy}{12}$

7. a) $x = \frac{y+2}{4}$ b) $x = \frac{2y}{9}$ c) $x = \sqrt{y}$
   d) $x = \sqrt{\frac{7-y}{4}}$ e) $x = \sqrt{\frac{3y+1}{3}}$

8. $b = \frac{2A}{h} - a$

9. $r = \sqrt{\frac{2A}{\pi}}$

10. $x = \sqrt{4\pi - 7y}$

## 25.6 Limits of accuracy

1. a) 18 453 b) 18 450 c) 18 500
   d) 18 000 e) 20 000

2. Gerry has not found a number that is close to what there was initially, but has rounded the 4 up to a 5 correctly
   Paul has rounded to 2 decimal places not 2 significant figures
   Natalie should have rounded the 4 up to 5.
   Correct answer is 1500

3. 135

4. Any number from 399.5 up to but not including 400.5.
   e.g. 399.5, 400.4

5. 43 minutes 29 seconds

6. a) 75 s $\leq$ actual value < 85 s
   b) 12 250 m $\leq$ actual value < 12 350 m
   c) 33.5 cl $\leq$ actual value < 34.5 cl
   d) 12.25 cm $\leq$ actual value < 12.35 cm

7. Greatest bicycle length: 154.9 centimetres
   Least shed length: 1.5 m = 150 cm
   The bicycle is not guaranteed to fit in the shed

8. a) 20.689 655 172 b) 20.7 c) 20
   d) Truncated to 20 is the most sensible degree of accuracy because you can't get part of a coin.

9. 90 ÷ 7 = £12.85714286
   Clare's method would give everyone £13 (but £13 × 7 = £91 so someone will get £1 less than everyone else)
   Suggest truncating to 2 decimal places (nearest penny) as everyone will receive the same, with just 5p left over. Or, truncate to 1 decimal place, give everyone £12.80 and keep the extra 40p..

10. 74.75 cm$^2$ $\leq$ Area < 93.75 cm$^2$

# NOTES, DOODLES AND EXAM DATES

..........................................................................................

..........................................................................................

..........................................................................................

..........................................................................................

..........................................................................................

..........................................................................................

..........................................................................................

..........................................................................................

..........................................................................................

..........................................................................................

..........................................................................................

..........................................................................................

..........................................................................................

..........................................................................................

..........................................................................................

**Doodles**

**Exam dates**

Paper 1:

..................................

..................................

Paper 2:

..................................

..................................

Paper 3:

..................................

..................................

# CALCULATOR HACKS

## 1 Prime factorisation

SHIFT – FACT

Use this to find the prime factors of a number.

**Example**

To find the prime factors of 360:

**Key sequence**

360 = SHIFT °,,,

**Calculator screen shows**

## 2 Using table mode

**Mode 3**

For most calculations, the calculator will be in Mode 1 (COMP)

Mode 3 – TABLE can be used to fill in tables, for graphs.

Remember! Ensure that you go back to MODE 1 after filling in the table

**Example**

Complete the table of values for

$$y = x^2 + 2x + 4$$

| $x$ | -4 | -3.5 | -2 | -1 | 0 | 0.5 | 2 |
|---|---|---|---|---|---|---|---|
| $y$ | | | | | | | |

**Key sequence**

| 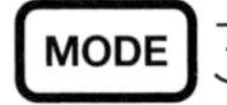 3 |  ) | $x^2$ +2 ALPHA ) +4 | = Start? | –4 = | End? 2 = | Step? 0.5 = |
|---|---|---|---|---|---|---|

The prompts Start? and End? require the inputs of the smallest and the largest value in the table (–4 and 2 in the example)

**Example**

The calculator gives you all y values for steps of 0.5 in x values.

The table on the exam paper shows unequal step sizes, so you must ensure that you read the correct y values from the calculator, to match the x values on the exam paper.

| $x$ | -4 | -3.5 | -2 | -1 | 0 | 0.5 | 2 |
|---|---|---|---|---|---|---|---|
| $y$ | 12 | 9.25 | 4 | 3 | 4 | 5.25 | 12 |

**Calculator screen shows**

| | X | F(X) |
|---|---|---|
| 1 | -4 | 12 |
| 2 | -3.5 | 9.25 |
| 3 | -3 | 7 |

-4.

## 3 Entering mixed numbers

### Example

When calculating

$$1\frac{1}{2} - \frac{3}{4}$$

a common error is to enter 1 then press and enter ...

 $\frac{1}{2}$

Instead, key the following sequence...

You should see three boxes on the screen

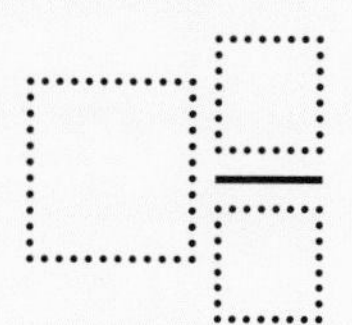

### Key sequence

1 ▶ 1 ▶ 2 ▶ - ⬚/⬚ 3 ▶ 4 =

The arrow in the key sequences shown is typically a large white button marked **REPLAY**

### Calculator screen shows

$1\frac{1}{2}-\frac{3}{4}$

$\frac{3}{4}$

## 4 Cancelling fractions

Express the following faction in its simplest form:

$$\frac{18}{27}$$

(Note: This may be the last part of a question on the calculator paper.)

### Key sequence

 18  27 = $\frac{2}{3}$

## 5 Are you in the correct TRIG mode for the exam?

Marks can be lost for not being in the correct trig. mode.

**Quick check**
When you key in the sequence

SIN (30)

### Do you see this?

sin(30)

$\frac{1}{2}$

**YES** - you are in the correct MODE
**NO** - you need to change the MODE

### Key sequence

To change the mode, key the following

 Option 3 

Go back and follow the instructions for checking your calculator is in the correct mode.

# CORE SKILLS AND COMMON MISTAKES

**There are six core skill areas in your Maths papers. Each of these areas are covered within this book and can be identified for each topic by the different tab colours of the specification references.**

**This is a summary of common errors made in past examinations. Avoid the same pitfalls and learn from the mistakes of others! Making these areas your strengths could give you an advantage.**

## Number N

1. Students' responses to some questions have many arithmetical errors, mainly in calculations requiring division.
2. Even on the calculator papers, some students use incorrect non-calculator methods, indicating they had no calculator (or were unable to use one correctly).
3. Rounding to a given number of significant figures poses problems for some students. Be confident with this.
4. Many students struggled with the concept of dividing fractions.
5. Money problems are tackled well, but questions involving other units, or a change of unit are poorly attempted.
6. Be prepared in your understanding of union or intersection, and their association with a Venn diagram.
7. Many misunderstandings relating to time were noticed, particularly when using a timetable or journey planning.
8. Students commonly do not recognise that an instruction to estimate an answer should trigger them to apply rounding. Any attempt to apply a complex calculation results in zero marks being awarded. Note that any attempt to round will gain some marks, not necessarily just to 1 significant figure.

## Algebra A

1. When negative values are involved, students' performance is generally weaker. For example, drawing a graph of $y = 1 - 4x$, or in calculating the values for a quadratic where the $x$ values are negative. Students should be more practised in using the symmetrical properties of a parabola to check their curve.
2. Rearranging formulae remains a weakness of many Foundation students and should be practised much more.
3. There seems to be little understanding of the relationships between equations and their graphs, for example, using the values of $m$ and $c$ on parallel graphs, when finding an equation of a straight line.
4. Methods of solving equations vary, but those who use the 'equation balancing' method tend to achieve more marks. Students should be reminded that it is rare to achieve full marks using trial and improvement methods and this method should be avoided. Students should also take care in using the correct order of operations.

## Ratio, Proportion and Rates of Change R

1. Many candidates attempt non-calculator methods for finding percentages on the calculator paper, leading to incorrect answers. Use the calculator for these questions or to check your answers.
2. On the non-calculator paper, percentages are mostly attempted by building up to the required percentage, and often candidates have difficulty in piecing together the parts.
3. Students often have difficulty when attempting questions using linked ratios (e.g. Given a : b and b : c, find a : c)
4. There are some instances of students failing to simplify ratios, even when asked to do so.
5. Scale diagrams are a weakness for many candidates.
6. Compound measures, such as speed, density, pressure and any context involving proportional units are frequently misunderstood.

## Geometry and Measures G

1. Recall of essential formulae remains a weakness, particularly those for areas of a triangle and trapezium and those related to a circle.
2. The use of correct mathematical language, for example, in geometrical reasoning and in transformation geometry is commonly seen. Students must be reminded that non-technical language will not gain any marks.
3. The handling and conversion of units is commonly misunderstood by Foundation candidates. Students, at both Foundation and Higher level, need to be reminded that there is usually one question in which they must state their units.
4. Mensuration work on problem solving continues to be challenging for students. They often mis-read the question and therefore miss out essential parts of the process for gaining a complete solution.
5. Finding the sum of the interior angles of a polygon requires more work, as many students assume it is 360° regardless of the number of sides of the polygon. It is calculated as the number of sides – 2 x 180°.

## Probability P Statistics S

1. When interpretation of a composite bar chart has been tested, students demonstrated significant weakness in interpreting it.
2. There is still evidence that protractors are being used inaccurately, or that students do not have a protractor, when drawing / interpreting pie charts.
3. While there is an improvement in writing criticism on statistical diagrams, students often write conflicting remarks, or comments that were too vague.

# EXAMINATION TIPS

**When you practise examination questions, work out your approximate grade using the following table. This table has been produced using a rounded average of past examination series for this GCSE. Be aware that boundaries vary by a few percentage points either side of those shown.**

**GCSE Maths: Foundation**

| Grade | 5 | 4 | 3 | 2 | 1 | U |
|---|---|---|---|---|---|---|
| Paper 1F (%) | 73 | 59 | 44 | 29 | 14 | 0 |
| Paper 1F (%) | 71 | 58 | 43 | 28 | 13 | 0 |
| Paper 1F (%) | 68 | 54 | 40 | 26 | 13 | 0 |

| Overall grade | 5 | 4 | 3 | 2 | 1 | U |
|---|---|---|---|---|---|---|
| F Tier (%) | 71 | 58 | 43 | 28 | 13 | 0 |

1. Read questions carefully. This includes any information such as tables, diagrams and graphs.
2. Remember to neatly and clearly cross out any work that you do not want to be marked. Do not scribble over it, rub it out or render it illegible in some other way.
3. Learn how to use your calculator, compasses and protractors correctly and take them to the exams. There is no calculator allowed in Paper 1F.
4. Show your workings. There are marks awarded for workings out on some questions, even if the answer is corrrect. Even the most basic of calculations or steps must be shown. This is particularly true of the calculator papers.

   These questions will typically state in the question:

   - "You must show all your working."
   - "Give reasons for your answer."
   - "Prove..."
5. Avoid using multiple methods to answer the same question. Examiners are instructed to award 0 marks for workings that are ambiguous, or where it is not clear which method leads to the answer given.

   If you change your mind on a method, you should cross out the previous working and show the intended method clearly.
6. Presentation matters. Good written communication helps the examiner to award you marks.

   Common issues include include:

   - Illegible handwriting can mean examiners don't award you marks if they can't confidently read your answers.
   - Some students write answers in a foreign language and therefore work cannot be marked - all answers are expected to be in English.
   - The numbers 4 and 9 are more commonly written ambiguously, also 1 and 7.
   - Over-writing to correct mistakes is becoming more common. Students are reminded again to cross out and re-write their answers.
7. Check through your answers if you have spare time. It is very easy to make a silly mistake that you could easily correct for a few extra marks. It might make the difference between two grades. If you go wrong somewhere, you may still be awarded some marks if the working out is there. It is also much easier to check your answers if you can see your working out. Remember to give units when asked to do so.

**Good luck!**

# INDEX